Adenosine and Adenine Nucleotides as Regulators of Cellular Function

Editor

John Whitfield Phillis, Ph.D. D.Sc., D.V.Sc.
Chairman and Professor of Physiology
Department of Physiology
Wayne State University Medical School
Detroit, Michigan

CRC Press
Boca Raton Ann Arbor Boston London

Library of Congress Cataloging-in-Publication Data

Adenosine and adenine nucleotides as regulators of cellular function/
   editor, John Whitfield Phillis.
         p.   cm.
   Includes bibliographical references and index.
   ISBN 0-8493-6928-2
   1. Adenosine—Physiological effect.   2. Purine nucleotides-
-Physiological effect.   3. Cellular signal transduction.
I. Phillis, J. W.
   [DNLM: 1. Adenine Nucleotides—physiology.   2. Adenosine-
-physiology.   QU 58 A228]
QP625.A27A32   1991
612.01579—dc20
DNLM/DLC
for Library of Congress   90-15182
                                                                         CIP

Direct all inquiries to CRC Press, Inc., 2000 Corporate Blvd., N.W., Boca Raton, Florida 33431.

© 1991 by CRC Press, Inc.

International Standard Book Number 0-8493-6928-2

Library of Congress Card Number 90-15182
Printed in the United States

PREFACE

The decision to prepare this volume was motivated by an awareness of the ever-expanding literature on the physiological roles of adenosine and the adenine nucleotides as regulators of cellular function in a variety of body systems. Research, especially that during the past two decades, has opened up new territories for the experimental study of cell regulation and has served to clarify the part that purines may play in controlling cell function. It is anticipated that this book, prepared with the assistance of a number of distinguished workers in the field, will provide a valuable source of information for investigators with a broad interest in the physiological functions of adenosine and the adenine nucleotides.

The subject matter has been organized to provide ready access to current research on purinergic regulation of cell function. The initial contributions embrace the historical development of research on adenosine and adenosine triphosphate. Following the historical introduction, various aspects of the formation, metabolism, release, transport, and receptors for endogenous adenosine and the adenine nucleotides are described. A comprehensive review of the functions of purines at the cellular level leads into a major section in which the role of adenosine in the regulation of flow in different vascular beds, and its role in angiogenesis is described. Sections on purinergic regulation of the cardiac, renal, gastrointestinal, and central nervous system follow, and the book concludes with a discussion of potential therapeutic avenues for purine use. In summary, the contents embrace almost every facet of this rapidly expanding field.

John W. Phillis

THE EDITOR

Dr. John Whitfield Phillis, B.V.Sc., Ph.D., D.Sc., D.V.Sc., is Department Chairman and Professor of Physiology at the Wayne State University Medical School in Detroit, Michigan.

Dr. Phillis obtained his training at Sydney University, receiving the B.V.Sc. (Hon.) degree in 1958 and obtained his Ph.D. degree in 1961 at the Australian National University, Canberra. In 1970 he was awarded a D.Sc. degree by Monash University, Melbourne and in 1976, a D.V.Sc. degree by Sydney University. He served as Lecturer/Senior Lecturer at Monash University from 1963 to 1969; as Visiting Professor in Biophysics at Indiana University in 1969; as Professor of Physiology at the Medical School of the University of Manitoba, 1970 to 1973; as Professor and Chairman of the Department of Physiology at the University of Saskatchewan, 1973 to 1981; and in 1981, he assumed his present position.

Dr. Phillis is a member of the Pharmacological and Physiological Societies of Great Britain, the American Physiological Society, the Society for Neuroscience, and the International Brain Research Organization. He was formerly the President of the Canadian Physiological Society. Dr. Phillis has been on the Medical Research Council of Canada, a Visiting Professor at several universities, and was a Wellcome Visiting Professor at Tulane University. He is a past Editor of the *Canadian Journal of Physiology and Pharmacology*, is presently Co-Editor of *Progress in Neurobiology,* and serves on the Editorial Boards of *General Pharmacology* and *International Journal of Purine and Pyrimidine Research.* He has been the recipient of research grants from the Canadian Medical Research Foundation, the National Institutes of Health, the American Heart Association, and the American Diabetes Association (Michigan Affiliate).

Dr. Phillis is the author of over 300 papers and book chapters and one book; he has edited two books. His current research interests relate to the pharmacology of chemical transmission in the brain and treatment of stroke.

CONTRIBUTORS

Robin A. Barraco, Ph.D.
Department of Physiology
Wayne State University Medical School
Detroit, Michigan

Christine Blazynski, Ph.D.
Biochemistry and Molecular Biophysics
Washington University School of
  Medicine
St. Louis, Missouri

Baigiang Cai, M.D.
Pulmonary-Critical Care Medicine
LSU Medical School
New Orleans, Louisiana

Martin K. Church, Ph.D., D.Sc.
Clinical Pharmacology
Southampton General Hospital
Southampton, England

Michael G. Collis, Ph.D.
Bioscience II
ICI Pharmaceuticals
Macclesfield, England

Michael A. Cook, Ph.D.
Pharmacology and Toxicology
University of Western Ontario
London, Ontario, Canada

Bruce N. Cronstein, M.D.
Department of Medicine
NYU Medical Center
New York, New York

Noel J. Cusack, Ph.D.
Department of Pharmacology
Whitby Research Inc.
Richmond, Virginia

Daniel J. Cushing, Ph.D.
Department of Cardiovascular
  Pharmacology
Lilly Research Laboratory
Eli Lilly and Company
Indianapolis, Indiana

Hugh H. Dalziel, Ph.D.
Department of Pharmacology
University of Nevada School of Medicine
Reno, Nevada

Michael Dragunow, Ph.D.
Department of Pharmacology
University of Auckland School of
  Medicine
Auckland, New Zealand

Chang Jian Feng, M.D.
Pulmonary-Critical Care Medicine
LSU Medical School
New Orleans, Louisiana

Karyn M. Forsyth, Ph.D.
Department of Pharmacology
University of Nevada School of Medicine
Reno, Nevada

Jonathan D. Geiger, Ph.D.
Department of Pharmacology and
  Therapeutics
University of Manitoba
Winnipeg, Manitoba, Canada

Richard D. Green, Ph.D.
Department of Pharmacology
University of Illinois
Chicago, Illinois

Harris J. Granger, M.D.
Department of Medical Physiology
College of Medicine
Texas A & M University
College Station, Texas

Mark W. Gorman, Ph.D.
Department of Physiology
Michigan State University
East Lansing, Michigan

Qingzhong Hao, M.D.
Pulmonary-Critical Care Medicine
LSU Medical School
New Orleans, Louisiana

Leif Hertz, M.D.
Department of Pharmacology
College of Medicine
University of Saskatchewan
Saskatoon, Saskatchewan, Canada

Katja Hoehn, M.D.
Department of Pharmacology
Dalhousie University
Halifax, Nova Scotia, Canada

Susanna M. O. Hourani, Ph.D.
Department of Biochemistry
University of Surrey
Guildford, England

Albert L. Hyman, M.D.
Department of Surgery
Tulane University Medical Center
New Orleans, Louisiana

Lana Kaiser, M.D.
Department of Physiology
Michigan State University
East Lansing, Michigan

Sharon S. Kelley, M.D., Ph.D.
Department of Physiology
Michigan State University
East Lansing, Michigan

Bobbi Langkamp-Henken, M.Sc.
Department of Physiology
University of Tennessee
Health Science Center
Memphis, Tennessee

W. Wayne Lautt, Ph.D.
Pharmacology and Therapeutics
University of Manitoba
Winnipeg, Manitoba, Canada

Howard L. Lippton, M.D.
Pulmonary-Critical Care Medicine
LSU Medical School
New Orleans, Louisiana

Paul J. Marangos, Ph.D.
Regentech Pharmaceuticals
San Diego, California

Cynthia J. Meininger, Ph.D.
Department of Medical Physiology
College of Medicine
Texas A & M University
College Station, Texas

Leonard P. Miller, Ph.D.
Research/CNS
Gensia Pharmaceutical Inc.
San Diego, California

S. Jamal Mustafa, Ph.D.
Department of Pharmacology
East Carolina University
Greenville, North Carolina

James I. Nagy, Ph.D.
Department of Physiology
University of Manitoba
Winnipeg, Manitoba, Canada

J. Arly Nelson, Ph.D.
Experimental Pediatrics
University of Texas
M. D. Anderson Cancer Center
Houston, Texas

W. H. Ng, M.Sc.
Clinical Pharmacology
Southampton General Hospital
Southampton, England

Rodolfo A. Padua, B.Sc.
Department of Physiology
University of Manitoba

Maria-Thereza R. Perez, Ph.D.
Department of Ophthalmology
University Hospital of Lund
Lund, Sweden

Teresa S. Priebe, Ph.D.
Experimental Pediatrics
University of Texas
M.D. Anderson Cancer Center
Houston, Texas

Kenneth G. Proctor, Ph.D.
Departments of Physiology and Surgery
University of Tennessee
Health Science Center
Memphis, Tennessee

Miodrag Radulovacki, Ph.D., M.D.
Department of Pharmacology
College of Medicine
University of Illinois
Chicago, Illinois

**Joaquim Alexandre Ribeiro, M.D.,
Ph.D.**
Laboratory of Pharmacology
Gulbenkian Institute of Science
Oeiras, Portugal

Karl A. Rudolphi, D.V.M.
Pharma Research
Hoechst AG Werk Kalle-Albert,
Wiesbaden, Germany

David Satchell, Ph.D.
Department of Zoology
University of Melbourne
Parkville, Australia

Jana Sawynok, Ph.D.
Department of Pharmacology
Dalhousie University
Halifax, Nova Scotia, Canada

Jürgen Schrader, Ph.D.
Department of Physiology
Heinrich-Heine-University
Düsseldorf, Germany

Ulrich Schwabe, M.D.
Pharmakologisches Institut
Universität Heidelberg
Heidelberg, Germany

Veronica M. Sciotti, Ph.D.
Department of Surgery
SUNY/Buffalo
Buffalo, New York

Eugene M. Silinsky, Ph.D.
Department of Pharmacology
Northwestern University Medical School
Chicago, Illinois

Alf Sollevi, M.D., Ph.D.
Department of Anesthesiology
Karolinska Hospital
Stockholm, Sweden

Harvey V. Sparks, Jr., M.D.
Department of Physiology
Michigan State University
East Lansing, Michigan

William S. Spielman, Ph.D.
Departments of Physiology and
  Biochemistry
Michigan State University
East Lansing, Michigan

Trevor W. Stone, Ph.D., D.Sc.
Department of Pharmacology
University of Glasgow
Glasgow, Scotland

David G. L. Van Wylen, Ph.D.
Department of Surgery
SUNY/Buffalo
Buffalo, New York

David P. Westfall, Ph.D.
Department of Pharmacology
University of Nevada School of Medicine
Reno, Nevada

Thomas D. White, Ph.D.
Department of Pharmacology
Dalhousie University
Halifax, Nova Scotia, Canada

H. Richard Winn, M.D.
Department of Neurological Surgery
Harborview Medical Center
Seattle, Washington

TABLE OF CONTENTS

I. Introduction

Chapter 1

HISTORICAL PERSPECTIVES — ADENOSINE

Daniel J. Cushing and S. Jamal Mustafa

TABLE OF CONTENTS

I. INTRODUCTION

The first documentation of the physiological effects of adenosine was in 1929 when Drury and Szent-Gyorgyi[1] reported that extracts from heart muscle, brain, kidney, and spleen had pronounced effects on cardiovascular function. The active substance in this extract was determined to be adenosine. This historic observation was the beginning of the scientific discipline known as adenosine physiology. Investigation of the effects of adenosine on the cardiovascular system and other organ systems, e.g., the central nervous system (CNS) and respiratory system, continued for the next 20 years.[2-4] Beginning in the mid-1950s and into the early 1960s Berne and co-workers[5,6] began to investigate the effects of adenosine on coronary blood flow. This work culminated in a working hypothesis contending that myocardial adenosine production was an important contributor in the metabolic regulation of coronary blood flow.[7] Berne and co-workers (and others) have broadened this hypothesis to include other organs including brain,[8,9] skeletal muscle,[10] and kidney.[11]

The next milestone in adenosine research was the identification of specific receptors for purine nucleosides (i.e., adenosine) and nucleotides (i.e., adenosine 5'-triphosphate [ATP]). Burnstock[12] suggested the terminology P_1 and P_2. The P_2 purinergic receptor was characterized by a higher affinity for ATP and adenosine 5'-diphosphate (ADP) than adenosine or adenosine 5'-monophosphate (AMP) and is discussed in more detail in the next chapter. The P_1 receptor by definition has a higher affinity for adenosine than ATP and ADP and is competitively blocked by the alkylxanthines. The P_1 receptor was subdivided based on adenylate cyclase response to adenosine.[13] Two adenosine-sensitive sites were defined: the "R"-site, which requires the integrity of the ribose moiety, and the "P"-site, which requires the integrity of the purine moiety. The "R"-site was later divided into "Ra" and "Ri" based on its ability to activate or inhibit adenylate cyclase, respectively.[14] Previously Van Calker et al.[15] proposed the terminology A_1 and A_2 for inhibition and activation of adenylate cyclase, respectively. The latter terminology has become the mainstay in the field, and extensions and subdivisions of that nomenclature have recently been made while keeping the logic of the system unchanged.[16-18]

Recently a large emphasis has been placed on the molecular identification of the receptors for adenosine. Isolation and purification of both A_1 and A_2 adenosine receptors has provided a solid framework for continued work in this area. In recent years it has become evident that the primary sequence of the adenosine receptors, and the development of complementary DNA for these receptors, will be available before the turn of the century.

The clinical usefulness of adenosine is gradually becoming a reality with the recent approval of adenosine for the treatment of supraventricular tachyarrhythmia[19-21] and clinical trials using adenosine for controlled hypotension during neurosurgery.[22,23] These milestones in adenosine pharmacology will hopefully pave the way for the development of more therapeutic applications for adenosine, its agonist analogs, and adenosine receptor antagonists.

II. CARDIOVASCULAR SYSTEM

Since the historic observation of Drury and Szent-Gyorgyi[1] that adenosine caused bradycardia, hypotension, coronary vasodilation, relaxation of the intestine, and sedation, a large body of information concerning the effects of adenosine on cardiovascular function has accumulated. The effects of adenosine on vascular resistance were again examined in the 1950s and early 1960s when Winbury et al.[24] and Wolf and Berne[5] reported that adenosine was a coronary vasodilator, and later Jacob and Berne[6] demonstrated that adenosine could rapidly penetrate the myocardial cell membrane. These observations were crucial support for the implication of adenosine as metabolic regulator of coronary blood flow during cardiac hypoxia.[7] This theory has been termed the "adenosine hypothesis" for local control of blood flow, and in this light, adenosine has been referred to as a "retaliatory metabolite".[25] Briefly, this concept contends that under hypoxic conditions ATP is broken down in cardiac myocytes and adenosine will be released, causing vasodilation. This dilation leads to an increase in the delivery of oxygen, which in turn decreases the utilization of ATP and thus returns adenosine release to normal levels. To date this hypothesis has been supported rather well; however, there are some concerns, (e.g., involvement of adenosine in reactive hyperemia), that are the subject of research in many laboratories around the world.

The effects of adenosine on cardiac function continued after the initial observations of Drury and Szent-Gyorgy.[1] In the 1930s Jezer et al.[26] and Honey et al.[27] demonstrated that adenosine impaired atrioventricular conduction, and studies on the direct chronotropic, inotropic, and dromotropic effects of adenosine continued for many years[3,28] until the outlook for its clinical usefulness waned. However, as a result of the work of Berne and co-workers, interest in the cardiovascular effects of adenosine has grown since the early 1970s[29-33] to culminate in the recent approval of adenosine for the treatment of supraventricular tachyarrhythmias[19-21] and clinical trials using adenosine for controlled hypotension in neurosurgery.[22,23]

Historically, the cardiac effects of adenosine were believed to be mediated by an A_1 adenosine receptor.[31] However, recent evidence suggests that the SA nodal bradycardia produced by adenosine is mediated by an "unusual" adenosine receptor.[34] Whether this represents a new subtype of the A_1 adenosine receptor or simply an A_1 receptor with unusual agonist-binding properties remain unknown and will undoubtedly be the topic of further investigation.

Pharmacologic characterization of the coronary adenosine receptor soon followed the physiological evidence in support of the role of adenosine in regulating coronary blood flow. The adenosine receptor mediating relaxation in coronary arteries from many species was determined to be extracellular in nature,[35,36] and further characterization indicated that it closely resembled an A_2 subtype receptor based on comparison of agonist potency profiles for relaxation with those for stimulation of adenylate cyclase in nonvascular tissue.[37-40] However, Baer and Vriend[41] suggested that such comparisons may not be valid and the use of such comparisons to classify smooth-muscle adenosine receptors must be viewed with caution.

Activation of adenylate cyclase in smooth muscle and the subsequent increase in cyclic AMP (cAMP) has been suggested to be the mechanism by which adenosine produces relaxation. Many workers have presented data which support the idea that adenosine increases cAMP production and subsequently activates a cAMP-dependent protein kinase.[42-46] However, other reports which do not support this idea also exist.[47-49] These discrepancies, along with the heterogeneity in agonist potency profiles for relaxation reported by Baer and Vriend,[41] suggest that the receptor-signaling mechanisms responsible for smooth-muscle relaxation are more complicated than originally suspected. Indeed, Mustafa[50,51] suggested the existence of more than one site for adenosine binding in dog coronary artery, and Ollinger and Kukovetz[52] reported a similar phenomenon in bovine coronary artery. The affinity states of these two binding sites, in both bovine and canine coronary artery, are similar to those reported by Bruns et al.[18] for the A_{2a} and A_{2b} receptors. Also, Daly et al.[53] indicated that A_2 adenosine receptors from different species have different activity profiles, depending upon the agonist tested, and further implicated the possible complication of intraspecies differences in A_2 adenosine receptors. Moreover, Leung et al.[54] suggested that coronary artery A_2 adenosine receptors differed from A_2 receptors in other tissues.

Recently Ramkumar et al.[46] indicated the existence of both A_1 and A_2 adenosine receptors in DDT_1-MF_2 clonal cells from Syrian hamster smooth muscle and demonstrated that both were coupled to adenylate cyclase, in an inhibitory and stimulatory manner, respectively. Along similar lines, McBean et al.[55] reported that adenosine, N^6-cyclohexyladenosine, 2-

chloroadenosine produced vasoconstriction in basilar vessels pretreated with 10^{-7} M 8-phenyltheophyline. It could be suggested that this constrictor response might be the result of an "A_1" receptor. Also, the concentration-response curve for 5'-N-ethylcarboxamido adenosine-induced relaxation reported by McBean et al.[55] was not in parallel with the other analogs examined. Taken together, the above-mentioned reports strongly suggest that adenosine receptor-mediated relaxation in smooth muscle is more complicated than agonist interaction with a single A_2 adenosine receptor.

Mechanisms other than adenylate cyclase activation could also be associated with adenosine-mediated relaxation. Kurtz[56] suggested that cyclic guanosine monophosphate production may be involved in adenosine-mediated relaxation, since adenosine increased guanylate cyclase activity. Inhibition of calcium influx by adenosine and its analogs has also been suggested to be a component in coronary relaxation.[57] However, calcium influx inhibition is not a requirement, since adenosine and its analogs are able to relax coronary artery in the absence of extracellular calcium.[58] Sabouni et al.[59] suggested that adenosine receptor-mediated hyperpolarization of coronary smooth muscle contributes to coronary relaxation, and Harder et al.[60] arrived at a similar conclusion.

The involvement of phospholipase C in adenosine-mediated relaxation has been proposed. Long and Stone[61] reported that adenosine attenuated inositol phosphate production in rat aorta and Docktrow and Lowenstein[62] demonstrated that adenosine was an inhibitor of phosphatidylinositol kinase. Inhibition of this enzyme will limit the production of phosphatidylinositol-4,5-biphosphate formation, the precursor for the purported contracting second messenger inositol triphosphate. Both of these groups concluded that a portion of the relaxation produced by adenosine could likely be the result of inhibition of the inositol phospholipid cascade.

Many second-messenger systems have been shown to be modified by adenosine and its analogs. It should be stressed that in our opinion no single messenger is completely responsible for adenosine-mediated relaxation. It is more reasonable to assume that a combination of messenger systems is involved, under varying physiological conditions, in the mechanism of adenosine action.

III. CENTRAL NERVOUS SYSTEM

Interest in the role of adenosine in CNS function stems from work that began in the mid-1950s. Holton and Holton[63] suggested a role for ATP in nerve transmission and Feldberg and Sherwood[4] reported that injection of adenosine and ATP produced muscular weakness, ataxia, and sleep. Holton[64] later demonstrated ATP-release during nerve stimulation.

The current school of thought with regard to the role of purines in the nervous system was outlined by Burnstock.[65] This hypothesis contends that ATP is the neurotransmitter of a third, nonadrenergic/noncholinergic, division of the sympathetic nervous system. This hypothesis is still the subject of some controversy with regard to whether ATP is an actual transmitter or a cotransmitter, however, the idea of purinergic modulation of neural function has been well accepted. The historical achievements that led to the formulation of this hypothesis are further detailed in the following chapter.

As mentioned earlier, adenosine has also been postulated by many investigators to be involved in neural transmission. Sattin and Rall[66] were the first to report the stimulation of adenylate cyclase by adenosine in brain slices and that the alkylxanthines were antagonists of this response. Adenosine and its agonist analogs produce sedation, analgesia, hypothermia, and muscular weakness, prevent seizure activity, and in most cases these effects are blocked by methylxanthines.[4,67-72] Certain CNS effects produced by methylxanthines are suspected to be the result of diminished adenosine activity, including self-mutilation, wakefulness, increased motor activity, and anxiety induction.[73-76]The mechanisms by which adenosine acts, involve alterations in neurotransmitter release and modulation of a variety of second-messenger systems including calcium and potassium fluxes adenylate cyclase, and phosphoinositide hydrolysis.[66,77-87] Adenosine has also been suggested to play a role in ethanol-induced motor incoordination.[88]

IV. RESPIRATORY SYSTEM

The first report describing the depressant effect of adenine nucleotides on respiration appeared in 1948.[89] This work was later confirmed by Green and Stoner.[3] This respiratory depressant effect of adenosine is believed to be the result of a direct effect of adenosine on brain centers controlling respiration.[90] Adenosine is produced in the brain in states of hypoxia,[9] and hypoxia has been demonstrated to depress respiration.[91] Caffeine and theophylline (two known adenosine receptor antagonists) have been known for many years to be respiratory stimulants,[92] but it was not until recently that the mechanism of their action was recognized as the antagonism of adenosine receptors.[93] These results have been applied to the treatment of the apnea associated with sudden infant death syndrome.[94,95]

Adenosine is also believed to be involved in asthma. Adenosine is produced in the lung due to hypoxia[96] and causes constriction of the airway smooth muscle.[97-99] Adenosine has been shown to cause bronchoconstriction in asthmatic, but not normal, subjects.[100] We have shown that in ragweed-sensitized rabbits, but not in normal rabbits, adenosine produces bronchoconstriction by its interaction with an A_1 subtype adenosine receptor.[101,102] These effects can be reversed by methylxanthine and nonmethylxanthine adenosine receptor antagonists.[100,102,103] Along with their adenosine receptor-blocking properties, methylxanthines inhibit phosphodiesterase activity, an action that itself can cause bronchial and tracheal relaxation.[104] Furthermore, enprofylline, a xanthine analog which inhibits phosphodiesterase, but has little adenosine receptor-blocking activity, is more potent than theophylline or caffeine as a bronchodilator or a tracheal relaxant.[104,105]

The release of neutrophil chemotactic factor in asthmatic patients challenged with adenosine has been suggested, and theophylline therapy attenuated this effect, suggesting the involvement of adenosine receptors.[106]

V. OTHER ORGANS AND TISSUES

Adenosine and its analogs have been shown to be potent inhibitors of the platelet aggregation produced by ADP and other agents.[107-109] This effect is thought to be the result of activation of an A_2 adenosine receptor and a resultant increase in cAMP production.[110-112] Furthermore, this antiaggretory effect of adenosine and its analogs is blocked by methylxanthines.[110,113]

Along with being an important physiological regulator of blood flow, respiration, and neural modulation, adenosine has been shown to play an important physiological role in lipolysis. The first report demonstrating the antilipolytic effect of adenosine appeared in 1961.[114] This work was subsequently supported and extended to suggest that the mechanism of action of adenosine involved is inhibition of adenylate cyclase activity.[115-123] This tissue has also been a useful model in which the molecular structure of the A_1 adenosine receptor has been studied.[124]

VI. FUTURE DIRECTIONS

Recently a large emphasis has been placed on the molecular identification of the receptors for adenosine. The A_1 adenosine receptor was the first to be characterized by photoaffinity labeling and is believed to be a 38-kDa protein.[124-126] This receptor was subsequently co-purified with a guanosine 5′-triphosphate (GTP) binding protein.[127] Isolation and purification of the A_2 adenosine receptor has been delayed by the lack of an appropriate ligand for both radioligand binding and photoaffinity cross-linking studies. However, recent evidence suggests that the tools may be available soon. Barrington et al.[128] identified the A_2 adenosine receptor binding subunit as a 45-kDa protein in striatum, and this same group suggested the existence of an A_2 adenosine receptor in DDT_1-MF_2 smooth-muscle cells with an apparent molecular weight of 42 kDa.[46] This work will certainly need independent validation, but it does establish that isolation and purification of the A_2 adenosine receptor will likely be accomplished in the near future. In the last few years it has become clear that the primary sequence of the adenosine receptors and the development of complementary DNA for these receptors will be available before the turn of the century. Undoubtedly the future reclassification of adenosine receptors, based on the information that will be gained through the use of molecular cloning techniques, will present investigators in this field with new insights and challenges.

VII. PERSPECTIVE

After the initial report of Drury and Szent-Gyorgyi[1] with regard to the involvement of purines in biological function, it was not until 1963 that adenosine received serious attention with the seminal work of Berne.[7] For the next decade workers devoted their time and resources to provide support for this hypothesis and established a firm physiological base on which this hypothesis stands. The 1970s was the era in adenosine research where the emphasis was on the metabolism, transport, and basic biochemistry of adenosine. The following decade (1980s), with the knowledge gained from previous years, witnessed the discovery and characterization of adenosine receptors. These years were also spent elucidating the second messengers by which adenosine may act to elicit its array of biological responses. The 1990s will lead this field into the realm of molecular pharmacology. We will bear witness to the deduction of the amino acid sequence for the adenosine receptors and ultimately the gene coding for these receptors. The significance of this work will depend entirely on how well we apply this knowledge to disease processes and use that information to develop therapeutic strategies to combat disease. These achievements will take this field into the early 21st century and hopefully the betterment of mankind by the centennial of the first report of the physiological actions of adenosine in 2029.

ACKNOWLEDGMENTS

We are grateful to the National Heart, Lung, and Blood Institute for their support over the years (Grant Number HL 27339-formerly HL 19202).

REFERENCES

1. **Drury, A. N. and Szent-Gyorgyi, A.,** The physiological activity of adenine compounds with especial reference to their action upon the mammalian heart, *J. Physiol.,* 68, 213, 1929.
2. **Drury, A. N.,** The physiological activity of nucleic acid and its derivatives, *Physiol. Rev.,* 16, 292, 1936.
3. **Green, H. N. and Stoner, H. B.,** *Biological Actions of the Adenosine Nucleotides,* Lewis, London, 1950.
4. **Feldberg, W. and Sherwood, S. L.,** Injections of drugs into lateral ventricle of the cat, *J. Physiol.,* 123, 148, 1954.
5. **Wolf, M. M. and Berne, R. M.,** Coronary vasodilator properties of purine and pyrimidine derivatives, *Circ. Res.,* 4, 343, 1956.
6. **Jacob, M. I. and Berne, R. M.,** Metabolism of purine derivatives by the isolated cat heart, *Am. J. Physiol.,* 198, 322, 1960.

7. **Berne, R. M.,** Cardiac nucleotides in hypoxia: a possible role in regulation of coronary blood flow, *Am. J. Physiol.,* 204, 317, 1963.

8. **Berne, R. M., Rubio, R., and Curnish, R.,** Release of adenosine from ischemic brain: effect on cerebral vascular resistance and incorporation into cerebral adenine nucleotides, *Circ. Res.,* 25, 262, 1974.

9. **Rubio, R., Berne, R. M., Bockman, E. L., and Curnish, R. R.,** Relationship between adenosine concentration and oxygen supply in rat brain, *Am. J. Physiol.,* 228, 1896, 1975.

10. **Dobson, J. G., Rubio, R., and Berne, R. M.,** Role of adenine nucleotides, adenosine and inorganic phosphate in the regulation of skeletal muscle blood flow, *Circ. Res.,* 29, 375, 1971.

11. **Osswald, H.,** Renal effects of adenosine and their inhibition by theophylline, *Naunyn-Schmiedeberg's Arch. Pharmacol.,* 288, 79, 1975.

12. **Burnstock, G.,** A basis for distinguishing two types of purinergic receptor, in *Cell Membrane Receptors for Drugs and Hormones: a Multidisciplinary Approach,* Bolis, L. and Straub, R., Eds., Raven Press, New York, 1978, 107.

13. **Londos, C. and Wolff, J.,** Two distinct adenosine-sensitive sites on adenylate cyclase, *Proc. Natl. Acad. Sci. U.S.A.,* 74, 5482, 1977.

14. **Londos, C., Cooper, D. M. F., and Wolff, J.,** Subclasses of adenosine receptors, *Proc. Natl. Acad. Sci. U.S.A.,* 77, 2551, 1980.

15. **Van Calker, D., Muller, M., and Hamprecht, B.,** Adenosine regulates via two different types of receptors, the accumulation of cyclic AMP in cultured brain cells, *J. Neurochem.,* 33, 999, 1979.

16. **Daly, J.W., Butts-Lamb, P., and Padgett, W.,** Subclasses of adenosine receptors in the central nervous system. Interaction with caffeine and related methylxanthines, *Cell. Mol. Neurobiol.,* 3, 69, 1983.

17. **Ribeiro, J. A. and Sebastiao, A. M.,** Adenosine receptors and calcium: basis for proposing a third (A$_3$) adenosine receptor, *Prog. Neurobiol.,* 26, 179, 1986.

18. **Bruns, R. F., Lu, G. H., and Pugsley, T. A.,** Characterization of the A$_2$ adenosine receptor labeled by [^{3}H]NECA in rat striatal membranes, *Mol. Pharmacol.,* 29, 331, 1986.

19. **DiMarco, J. P., Sellers, T. D., Lerman, B. B., Greenberg, M. L., Berne, R. M., and Belardinelli, L.,** Diagnostic and therapeutic use of adenosine in patients with supraventricular tachyarrhythmias, *J. Am. Coll. Cardiol.,* 6, 417, 1985.

20. **Overholt, E. D., Rheuban, K. S., Gutgesell, H. P., Lerman, B. B., and DiMarco, J. P.,** Usefulness of adenosine for arrhythmias in infants and children, *Am. J. Cardiol.,* 61, 336, 1988.

21. The Medical Letter, Adenosine, 32(821), 63, June 29, 1990.

22. **Sollevi, A., Lagerkranser, T., Irestedt, L., Gordon, E., and Lindquist, C.,** Controlled hypotension with adenosine in cerebral aneurysm surgery, *Anesthesiology,* 61, 400, 1984.

23. **Owall, A., Gordon, E., Lagerkranser, M., Lindquist, C., Rudehill, A., and Sollevi, A.,** Clinical experience with adenosine for controlled hypotension during cerebral aneurysm surgery, *Anesth. Analg., Cleveland,* 66, 229, 1987.

24. **Winbury, M. M., Papierski, D. H., Hemmer, M. L., and Hambourger, W. E.,** Coronary dilator action of the adenine-ATP series, *J. Pharmacol. Exp. Ther.,* 109, 255, 1953.

25. **Newby, A. C.,** Adenosine and the concept of retaliatory metabolites, *Trends Biochem. Sci.,* 9, 42, 1984.

26. **Jezer, A., Oppenheimer, B. S., and Schwartz, S. P.,** The effect of adenosine on cardiac irregularities in man, *Am. Heart J.,* 9, 252, 1933.

27. **Honey, R. M., Ritchie, W. T., and Thompson, W. A. R.,** The action of adenosine upon the human heart, *Q. J. Med.,* 23, 485, 1930.

28. **Wayne, E. J., Goodwin, J. F., and Stoner, H. B.,** The effect of adenosine triphosphate on the electro-cardiogram of man and animals, *Br. Heart J.,* 11, 55, 1949.

29. **Chiba, S. and Hashimoto, K.,** Differences in chronotropic and dromotropic responses of the SA and AV nodes to adenosine and acetylcholine, *Jpn. J. Pharmacol.,* 22, 273, 1972.

30. **Szentmiklosi, A. J., Nemeth, M., Szegi, J., Papp, J. G., and Szekeres, L.,** Effect of adenosine on sinoatrial and ventricular automaticity of the guinea pig, *Naunyn-Schmiedeberg's Arch. Pharmacol.,* 311, 147, 1980.

31. **Belardinelli, L., West, A., Crampton, R., and Berne, R. M.,** Chronotropic and dromotropic effects of adenosine, in *Regulatory Function of Adenosine,* Berne, R. M., Rall, T. W., and Rubio, R., Eds., Martinus Nijhoff, Boston, 1983, 377.

32. **Pelleg, A., Belhassen, B., Ilia, R., and Laniado, S.,** Comparative electrophysiologic effects of adenosine triphosphate and adenosine in the canine heart: influence of atropine, propranolol, vagotomy, dipyridamole and aminophylline, *Am. J. Cardiol.,* 55, 571, 1985.

33. **Fredholm, B. B. and Sollevi, A.,** Cardiovascular effects of adenosine, *Clin. Physiol.,* 6, 1, 1986.

34. **Belloni, F. L., Belardinelli, L., Halperin, C., and Hintze, T. H.,** An unusual receptor mediates adenosine-induced SA nodal bradycardia in dogs, *Am. J. Physiol.,* 256, H1553, 1989.

35. **Olsson, R. A., Davis, C. J., Khouri, E. M., and Patterson, R. E.,** Evidence for an adenosine receptor on the surface of dog coronary myocytes, *Circ. Res.,* 39, 93, 1976.

36. **Schrader, J., Nees, S., and Gerlach, E.,** Evidence for a cell surface adenosine receptor on coronary myocytes and atrial muscle cells, *Pfluegers Arch., 369,* 251, 1977.

37. **Kusachi, S., Thompson, R. D., and Olsson, R. A.,** Ligand selectivity of dog coronary adenosine receptor resembles that of adenylate cyclase stimulatory (Ra) receptors, *J. Pharmacol. Exp. Ther., 227,* 316, 1983.

38. **Mustafa, S. J. and Askar, A. O.,** Evidence suggesting an Ra-type adenosine receptor in bovine coronary arteries, *J. Pharmacol. Exp. Ther., 232,* 49, 1985.

39. **Ramagopal, M. V., Chitwood, R. W., and Mustafa, S. J.,** Evidence for an A_2 adenosine receptor in human coronary arteries, *Eur. J. Pharmacol., 151,* 483, 1988.

40. **Sabouni, M. H., Ramagopal, M. V., and Mustafa, S. J.,** Relaxation by adenosine and its analogs of potassium contracted human coronary arteries, *Naunyn-Schmiedeberg's Arch. Pharmacol., 341,* 388, 1990.

41. **Baer, H. P. and Vriend, R.,** Adenosine receptors in smooth muscle: structure-activity studies and the question of adenylate cyclase involvement in control of relaxation, *Can. J. Physiol. Pharmacol., 63,* 972, 1985.

42. **Kukovetz, W. R., Poch, G., Holzmann, S., Wurm, A., and Rinner, I.,** Role of cyclic nucleotides in adenosine-mediated regulation of coronary flow, *Adv. Cyclic Nucleotide Res., 9,* 397, 1978.

43. **Anand-Srivastava, M. B., Franks, D. J., Contin, M., and Genest, J.,** Presence of ''Ra'' and ''P''-site receptors for adenosine coupled to adenylate cyclase in cultured vascular smooth muscle cells, *Biochem. Biophys. Res. Commun., 108,* 213, 1982.

44. **Silver, P. J.,** Adenosine-mediated relaxation and activation of cyclic-AMP dependent protein kinase in coronary arterial smooth muscle, *J. Pharmacol. Exp. Ther., 228,* 342, 1984.

45. **Cassis, L. A., Loeb, A. L., and Peach, M. J.,** Mechanisms of adenosine- and ATP-induced relaxation in rabbit femoral artery: role of the endothelium and cyclic nucleotides, in *Topics and Perspectives in Adenosine Research,* Gerlach, E. and Becker, B. F., Eds., Springer-Verlag, Berlin, 1987, 486.

46. **Ramkumar, V., Barrington, W. W., Jacobson, K. A., and Stiles, G. L.,** Demonstration of both A_1 and A_2 adenosine receptors in DDT_1 MF_2 smooth muscle cells, *Mol. Pharmacol., 37,* 149, 1990.

47. **Herlihy, J. T., Bockman, E. L., Berne, R. M., and Rubio, R.,** Adenosine relaxation of isolated vascular smooth muscle, *Am. J. Physiol., 230,* 1239, 1976.

48. **Verhaeghe, R. H.,** Action of adenosine and adenine nucleotides on dogs' isolated veins, *Am. J. Physiol., 233,* H114, 1977.

49. **McKenzie S. G., Frew, R., and Baer, H-P.,** Characteristics of the relaxant response of adenosine and its analogs in intestinal smooth muscle, *Eur. J. Pharmacol., 41,* 183, 1977.

50. **Mustafa, S. J.,** Cellular and molecular mechanism(s) of coronary flow regulation by adenosine, *Mol. Cell. Biochem., 31,* 67, 1980.

51. **Mustafa, S. J.,** Adenosine receptors in the heart, in *Methods in Studying Cardiac Membranes,* Vol. 2, Dhalla, N. S., Ed., CRC Press, Boca Raton, FL, 1984, 182.

52. **Ollinger, P and Kukovetz, W. R.,** [^{3}H]Adenosine binding to bovine coronary arteries and myocardium, *Eur. J. Pharmacol., 93,* 35, 1983.

53. **Daly, J. W., Ukena, D., and Jacobson, K. A.,** Analogues of adenosine, theophylline and caffeine: selective interactions with A_1 and A_2 receptors, in *Topics and Perspectives in Adenosine Research,* Gerlach, E. and Becker, B. F., Eds., Springer-Verlag, Berlin, 1987, 23.

54. **Leung, E., Johnston, C., and Woodcock, E.,** An investigation of the receptors involved in the coronary vasodilatory effect of adenosine analogues, *Clin. Exp. Pharmacol. Physiol., 12,* 515, 1985.

55. **McBean, D. E., Harper, A. M., and Rudolphi, K. A.,** Effects of adenosine and its analogues on porcine basilar arteries: are only A_2 receptors involved?, *J. Cereb. Blood Flow Metab., 8,* 40, 1988.

56. **Kurtz, A.,** Adenosine stimulates guanylate cyclase activity in vascular smooth muscle cells, *J. Biol. Chem., 262,* 6296, 1987.

57. **Ramagopal, M. V. and Mustafa, S. J.,** Effects of adenosine and its analogs on calcium influx in coronary artery, *Am. J. Physiol. (Heart Circ. Physiol. 24) 255,* H1492, 1988.

58. **Ramagopal, M. V., Nakazawa, M., and Mustafa, S. J.,** Relaxing effects of adenosine in coronary artery in calcium-free medium, *Eur. J. Pharmacol., 159,* 33, 1989.

59. **Sabouni, M. H., Hargittai, P. T., Lieberman, E. M., and Mustafa, S. J.,** Evidence for adenosine receptor-mediated hyperpolarization in coronary smooth muscle, *Am. J. Physiol., 257,* H1750, 1989.

60. **Harder, D. R., Belardinelli, L., Sperelakis, N., Rubio, R., and Berne, R. M.,** Differential effects of adenosine and nitroglycerin on the action potentials of large and small coronary arteries, *Circ. Res., 44,* 176, 1979.

61. **Long, C. J. and Stone, T. W.,** Adenosine reduces agonist-induced production of inositol phosphates, *J. Pharm. Pharmacol., 39,* 1010, 1987.

62. **Docktrow, S. R. and Lowenstein, J. M.,** Inhibition of phosphatidylinositol kinase in vascular smooth muscle membranes by adenosine and related compounds, *Biochem. Pharmacol., 36,* 2255, 1987.

63. **Holton, F. A. and Holton, P.,** The capillary dilator substances in dry powders of spinal roots: possible role of adenosine triphosphate in chemical transmission from nerve endings, *J. Physiol. (London), 126,* 124, 1954.

64. **Holton, P.**, The liberation of adenosine triphosphate on antidromic stimulation of sensory nerves, *J. Physiol. (London)*, 145, 494, 1959.

65. **Burnstock, G.**, Purinergic nerves, *Pharm. Rev.*, 24, 509, 1972.

66. **Sattin, A. and Rall, T. W.**, The effect of adenosine and adenine nucleotides on the cyclic adenosine 3',5'-phosphate content of guinea pig cerebral cortex slices, *Mol. Pharmacol.*, 6, 12, 1970.

67. **Buday, P. V., Carr, C. J., and Miya, T. S.**, A pharmacologic study of some nucleosides and nucleotides, *J. Pharm. Pharmacol.*, 13, 290, 1961.

68. **Haulica, I., Ababei, L., Branisteanu, D., and Topoliceanu, F.**, Preliminary data on the possible hypnogenic role of adenosine, *J. Neurochem.*, 21, 1019, 1973.

69. **Maitre, M., Ciesielski, L., Lehmann, A., Kempf, E., and Mandel, P.**, Protective effect of adenosine and nicotinamide against audiogenic seizure, *Biochem. Pharmacol.*, 23, 2807, 1974.

70. **Dunwiddie, T. V. and Worth, T.**, Sedative and anticonvulsant effects of adenosine analogs in mouse and rat, *J. Pharmacol. Exp. Ther.*, 220, 70, 1982.

71. **Ahlijanian, M. K. and Takemori, A. E.**, Effects of (-)-N⁶(R-phenylisopropyladenosine (PIA) and caffeine on nociception and morphine-induced analgesia, tolerance and dependence in mice, *Eur. J. Pharmacol.*, 112, 171, 1985.

72. **Ahlijanian, M. K. and Takemori, A. E.**, Changes in adenosine receptor sensitivity in morphine-tolerant and -dependent mice, *J. Pharmacol. Exp. Ther.*, 236, 615, 1986.

73. **Barraco, R. A., Coffin, V. L., Altman, H. J., and Phillis, J. W.**, Central effects of adenosine analogues on locomotor activity in mice and antagonism of caffeine, *Brain Res.*, 272, 392, 1983.

74. **Minana, M. D., Portoles, M., Jorda, G., and Grisolia, S.**, Lesch-Nyhan syndrome, caffeine model: increase of purine and pyrimidine enzymes in rat brain, *J. Neurochem.*, 43, 1556, 1984.

75. **Snyder, S. H. and Sklar, P.**, Behavioral and molecular actions of caffeine: focus on adenosine, *J. Psychiatr. Res.*, 18, 91, 1984.

76. **Charney, D. S., Galloway, M. P., and Heninger, G. R.**, The effects of caffeine on plasma MPHG, subjective anxiety, autonomic symptoms and blood pressure in healthy humans, *Life Sci.*, 35, 135, 1984.

77. **Phillis, J. W., Kostopoulos, G. K., and Limacher, J. J.**, Depression of corticospinal cells by various purines and pyrimidines, *Can. J. Physiol. Pharmacol.*, 52, 1226, 1974.

78. **Phillis, J. W., Kostopoulos, G. K., and Limacher, J. J.**, A potent depressant action of adenosine derivatives on cerebral cortical neurons, *Eur. J. Pharm.*, 30, 125, 1975.

79. **Ribeiro, J. A., Sa-Almeida, A. M., and Namorado, J. M.**, Adenosine and adenosine triphosphate decrease ⁴⁵Ca uptake by synaptosomes stimulated by potassium, *Biochem. Pharmacol.*, 28, 1297, 1979.

80. **Hartzell, H. C.**, Adenosine receptors in frog sinus venosus: slow inhibitory potentials produced by adenine compounds and acetylcholine, *J. Physiol. (London)*, 293, 23, 1979.

81. **Henon, B. K., Turner, D. K., and McAfee, D. A.**, Adenosine receptors: electrophysiological actions at pre- and postsynaptic sites on mammalian neurons, *Soc. Neurosci. Abstr.*, 6, 257, 1980.

82. **Ribeiro, J. A.**, The decrease of neuromuscular transmission by adenosine depends on previous neuromuscular depression, *Arch. Int. Pharmacol.*, 255, 59, 1982.

83. **Hollingsworth, E. B., De La Cruz, R. A., and Daly, J. W.**, Accumulations of inositol phosphates and cyclic AMP in brain slices: synergistic interactions of histamine and 2-chloroadenosine, *Eur. J. Pharmacol.*, 122, 45, 1986.

84. **Petcoff, D. W. and Cooper, D. M. F.**, Adenosine receptor agonists inhibit inositol phosphate accumulation in rat striatal slices, *Eur. J. Pharmacol.*, 137, 269, 1987.

85. **Delahunty, T. M., Cronin, M. J., and Linden, J.**, Regulation of GH₃-cell function via adenosine A₁ receptors, *Biochem. J.*, 255, 69, 1988.

86. **Rubio, R., Bencherif, M., and Berne, R. M.**, Inositol phospholipid metabolism during and following synaptic activation: role of adenosine, *J. Neurochem.*, 52, 797, 1989.

87. **Kendall, D. A., and Firth, J. L.**, Inositol phospholipid hydrolysis in human brain; adenosine inhibition of the response to histamine, *Br. J. Pharmacol.*, 100, 37, 1990.

88. **Dar, M. S., Mustafa, S. J., and Wooles, W. R.**, Possible role of adenosine in the CNS effects of ethanol, *Life Sci.*, 33, 1363, 1983.

89. **Emmelin, N. and Feldberg W.**, Systemic effects of adenosine triphosphate, *Br. J. Pharmacol.*, 3, 273, 1948.

90. **Moss, I. R., Denavit-Saubie, M., Eldridge, F. L., Gillis, R. A., Herkenham, M., and Lahiri, S.**, Neuromodulators and transmitters in respiratory control, *Fed. Proc.*, 45, 2133, 1986.

91. **Watt, J. G., Dumke, P. R., and Comroe, J. H.**, Effects of inhalation of 100 percent and 14 percent oxygen upon respiration of unanesthetized dogs before and after chemoreceptor denervation, *Am. J. Physiol.*, 138, 610, 1943.

92. **Richmond, G. H.**, Action of caffeine and aminophylline as respiratory stimulants in man, *J. Appl. Physiol.*, 2, 16, 1949.

93. **Wessberg, P., Hedner, T., Persson, B., and Jonason, J.**, Adenosine mechanisms in the regulation of breathing in the rat, *Eur. J. Pharmacol.*, 106, 59, 1985.

94. **Aranda, J. V. and Turmen, T.,** Methylxanthines in apnea of prematurity, *Clin. Perinatol.,* 6, 87, 1979.
95. **Boutroy, M. J., Vert, P., Royer, R. J., Monin, P., and Royer, M. J.,** Caffeine, a metabolite of theophylline, during the treatment of apnea in the premature infant, *J. Pediatr.,* 94, 996, 1979.
96. **Mentzer, R. M., Rubio, R., and Berne, R. M.,** Release of adenosine from hypoxic canine lung tissue and its possible role in the pulmonary circulation, *Am. J. Physiol.,* 229, 1625, 1975.
97. **Fredholm, B. B., Brodin, K., and Strandberg, K.,** On the mechanism of relaxation of tracheal muscle by theophylline and other cyclic nucleotide phosphodiesterase inhibitors, *Acta Pharmacol. Toxicol.,* 45, 336, 1979.
98. **Karlsson, J. A., Kjellin, G., and Persson, C. G. A.,** Effect on tracheal smooth muscle of adenosine and methylxanthines, and their interaction, *J. Pharm. Pharmacol.,* 34, 788, 1982.
99. **Kroll, F., Karlsson, J. A., Persson, C. G. A., and Ryrfeldt, A.,** Interactions between xanthines, mepyramine and adenosine in the guinea pig lung, in *Antiasthma Xanthines and Adenosine,* Andersson, K. E. and Persson, C. G. A., Eds., Excerpta Medica, Amsterdam, 1985, 193.
100. **Cushley, M. J., Tattersfield, A. E., and Holgate, S. T.,** Inhaled adenosine and guanosine in normal and asthmatic subjects, *Br. J. Clin. Pharmacol.,* 15, 161, 1983.
101. **Ali, S., Mustafa, S. J., Bhaltia, S. C., Douglas, F. L., and Metzger, W. J.,** Effects of CGS-15943 on adenosine-induced broncho constriction in allergic rabbits, *FASEB J.,* 4, A613, 1990.
102. **Mustafa, S. J., Ali, S., and Metzger, W. J.,** Adenosine-induced bronchoconstriction in allergic rabbit: evidence for receptor involvement, *Jpn. J. Pharmacol.,* 52 (Suppl. II), 113P, 1990.
103. **Ali, S., Mustafa, S. J., Atkinson, L., Flanagan, R. C., Douglas, F. L., Kotake, A. N., and Metzger, W. J.,** Adenosine-induced bronchoconstriction in an allergic rabbit model and its antagonism by CGS-15943, a specific adenosine receptor antagonist, *FASEB J.,* 3, 1236, 1989.
104. **Brackett, L. E., Shamim, M. T., and Daly, J. W.,** The activity of caffeine, theophylline and enprofylline analogs as tracheal relaxants, *Biochem. Pharmacol.,* 39, 1897, 1990.
105. **Persson, C. G. A., Anderson, K. E., and Kjellin, G.,** Effects of enprofylline and theophylline may show the role of adenosine, *Life Sci.,* 38, 1057, 1986.
106. **Mustafa, S. J., Kukoly, C. A., Metzger, W. J., and Driver, A. G.,** Adenosine challenge causes the release of a high molecular weight neutrophil chemotactic factor in asthma, *Jpn. J. Pharmacol.,* 52 (Suppl. II), 113P, 1990.
107. **Born, G. V. R.,** Strong inhibition by 2-chloroadenosine of the aggregation of blood platelets by adenosine diphosphate, *Nature,* 202, 95, 1964.
108. **Kien, M., Belamarich, F. A., and Shepro, D.,** Effect of adenosine and related compounds on thrombocyte and platelet aggregation, *Am. J. Physiol.,* 220, 604, 1971.
109. **Haslam, R. J. and Rosson, G. M.,** Aggregation of human blood platelets by vasopressin, *Am. J. Physiol.,* 223, 958, 1972.
110. **Mills, D. C. B. and Smith, J. B.,** The influence of platelet aggregation of drugs that affect the accumulation of adenosine 3′,5′-cyclic monophosphate in platelets, *Biochem. J.,* 121, 185, 1971.
111. **Haslam, R. J. and Lynham, J. A.,** Activation and inhibition of blood platelet adenylate cyclase by adenosine or by 2-chloroadenosine, *Life Sci.,* 2 (Part II), 1143, 1972.
112. **Hutteman, E., Ukena, D., Lenschow, V., and Schwabe, U.,** Ra adenosine receptors in human platelets. Characterization of 5′-N-ethylcarboxamido-]^{3}H]adenosine binding in relation to adenylate cyclase activity, *Naunyn-Schmiedeberg's Arch. Pharmacol.,* 325, 226, 1984.
113. **Haslam, R. J. and Rossen, G. M.,** Effects of adenosine on levels of adenosine cyclic 3′,5′-monophosphate in human blood platelets in relation to adenosine incorporation and platelet aggregation, *Mol. Pharmacol.,* 11, 528, 1975.
114. **Dole, V. P.,** Effect of nucleic acid metabolites on lipolysis in adipose tissue, *J. Biol. Chem.,* 236, 3125, 1961.
115. **Kappeler, H.,** The pharmacology of Lipolysis. I. Mechanism of action of adenosine-containing nucleosides and nucleotides in adipose tissue lipolysis in vitro, *Diabetalogia,* 2, 52, 1966.
116. **Raben, M. S. and Matsuzaki, F.,** Effect of purines on epinephrine-induced lipolysis in adipose tissue, *J. Biol. Chem.,* 241, 4781, 1966.
117. **Davies, J. I.,** *In vitro* regulation of the lipolysis of adipose tissue, *Nature,* 218, 349, 1968.
118. **Pereira, J. N. and Holland, G. F.,** The effect of nicotinamide adenine dinucleotide on lipolysis in adipose tissue in vitro, *Experientia,* 22, 658, 1966.
119. **Blackard, W. G. and Cameron, T.,** Influence of nucleotide derivatives on lipolysis and lipid peroxide formation in vitro, *Metabolism,* 16, 91, 1967.
120. **Fain, J. N., Pointer, R. H., and Ward, W. F.,** Effects of adenosine nucleosides on adenylate cyclase, phosphodiesterase, cyclic adenosine monophosphate accumulation and lipolysis in fat cells, *J. Biol. Chem.,* 247, 6866, 1972.
121. **Schwabe, U., Ebert, R., and Erbler, H. C.,** Adenosine release from isolated fat cells and its significance for the effects of hormones on cyclic 3′,5′-AMP levels and lipolysis, *Naunyn-Schmiedeberg's Arch. Pharmacol.,* 276, 133, 1973.

13

122. **Londos, C., Cooper, D. M. F., Schlegel, W., and Rodbell, M.,** Adenosine analogs inhibit adipocyte adenylate cyclase by a GTP-dependent process: basis for actions of adenosine and methylxanthines on cyclic AMP production and lipolysis, *Proc. Natl. Acad. Sci. U.S.A.,* 75, 5362, 1978.
123. **Arch. J. R. S. and Newsholme, E. A.,** The control of metabolism and the hormonal role of adenosine, *Essays Biochem.,* 14, 82, 1978.
124. **Stiles, G. L., Daly, D. T., and Olsson, R. A.,** Photoaffinity labeling of A$_1$-adenosine receptors, *J. Biol. Chem.,* 260, 14569, 1985.
125. **Klotz, K., Cristallio, G., Grifantinio, M., and Vittorio, S.,** Photoaffinity labeling of A$_1$-adenosine receptors, *J. Biol. Chem.,* 260, 14659, 1985.
126. **Stiles, G. L.,** Photoaffinity cross-linked A$_1$ adenosine receptor-binding subunits, *J. Biol. Chem.,* 261, 10839, 1986.
127. **Munshi, R. and Linden, J.,** Co-purification of A$_1$ adenosine receptors and guanine nucleotide binding proteins from bovine brain, *J. Biol. Chem.,* 264, 14853, 1989.
128. **Barrington, W. W., Jacobson, K. A., Hutchison, A. J., Williams, M., and Stiles, G. L.,** Indentification of the A$_2$ adenosine receptor binding subunit by photoaffinity crosslinking, *Proc. Natl. Acad. Sci. U.S.A.,* 86, 6572, 1989.

SUGGESTED READINGS

Agarwal, K. C., Adenosine and platelet function, in *Role of Adenosine in Cerebral Metabolism and Blood Flow,* Stefanovich, V. and Okyayuz-Baklouti, I., Eds., VNU Science Press, Utrecht, The Netherlands, 1987, 107.

Arch, J. R. S. and Newsholme, E. A., The control of the metabolism and the hormonal role of adenosine, *Essays Biochem.,* 14, 82, 1978.

Baer, H. P. and Drummond, G. I., *Physiological and Regulatory Functions of Adenosine and Adenine Nucleotides,* Raven Press, New York, 1979.

Berne, R. M., Rall, T. W., and Rubio, R., *Regulatory Function of Adenosine,* Martinus Nijhoff, Boston, 1982.

Burnstock, G., Purinergic nerves, *Pharm. Rev.,* 24, 509, 1972.

Daly, J. W., Adenosine receptors, *Adv. Cyclic Nucleotide Prot. Phosphoryl. Res.,* 19, 29, 1985.

Dunwiddie, T. V., The physiological role of adenosine in the central nervous system, *Int. Rev. Neurobiol.,* 27, 63, 1985.

Fain, J. N. and Malbon, C. C., Regulation of adenylate cyclase by adenosine, *Mol. Cell. Biochem.,* 25, 143, 1979.

Gerlach, E. and Becker, B. F., *Topics and Perspectives in Adenosine Research,* Springer-Verlag, Berlin, 1987.

Green, H. N. and Stoner, H. B., *Biological Actions of the Adenine Nucleotides,* H. K. Lewis and Co., London, 1950.

Paton, D. M., *Adenosine and Adenine Nucleotides: Physiology and Pharmacology,* Taylor and Francis, London, 1988.

Phillis, J. W. and Wu, P. H., The role of adenosine and its nucleotides in central synaptic transmission, *Prog. Neurobiol.,* 16, 187, 1981.

Ribeiro, J. A., *Adenosine Receptors in the Nervous System,* Taylor and Francis, Basingstoke, England, 1989.

Ribeiro, J. A. and Sebastaio, A. M., Adenosine receptors and calcium: basis for proposing a third (A$_3$) adenosine receptor, *Prog. Neurobiol.,* 26, 179, 1986.

Stefanovich, V., Rudolphi, K., and Schubert, P., *Adenosine: Receptor and Modulation of Cell Function,* IRL Press, Oxford, 1985.

Williams, M., Purine receptors in mammalian tissues: pharmacology and functional significance, *Annu. Rev. Pharmacol. Toxicol.,* 27, 315, 1987.

Williams, M., *Adenosine and Adenosine Receptors,* Humana Press, Clifton, NJ, 1990.

Chapter 2

HISTORICAL PERSPECTIVES — ATP

David Satchell

TABLE OF CONTENTS

I. EARLY EXPERIMENTS WITH ATP AND RELATED NUCLEOTIDES

In 1914, Bass[1] detected adenine in blood and suggested that it was probably present as the nucleotide, adenylic acid or adenosine 5'-monophosphate (AMP). In 1920, Freund[2] obtained a similar substance from blood; it exhibited depressor properties and he described it as "fruhgift". Experiments on extracts of other tissues, especially muscle, also produced pharmacologically active substances. In 1929, Drury and Szent-Gyorgi[3] obtained a white powder from trichloroacetic acid extracts of cardiac muscle and identified it as adenylic acid. This substance was undoubtedly derived largely from the adenosine 5'-triphosphate (ATP) present in the original extract. They found in mammals that AMP produced a typical heart block, dilated the coronary arteries, lowered blood pressure, and diminished the amplitudes of contractions of isolated intestinal strips. This study, together with the work of Zipf[4] and Hoffman[5] in the following year, demonstrated that earlier biological activities of extracts were largely due to AMP and related compounds. At this time ATP had not been studied pharmacologically. However, the isolation of this compound in 1931 from skeletal muscle by Lohman[6] pointed to its potential as another active substance.

From the commencement of studies of the actions of ATP, it became apparent that the presence of the additional phosphates conferred differences in activity and often mode of action to adenosine and AMP. These differences were not to be resolved until separate receptors were determined more than 50 years later. In 1932, Deuticke[7] compared the effects of ATP, AMP, and adenosine on virgin guinea pig uterus and found that while all three caused contraction, the additional phosphates of ATP conferred the ability to cause a more rapid contraction. In 1933, Gaddum and Holtz[8] found that injection of ATP into perfused cat lungs caused vasodilatation at low concentration and a marked vasoconstriction together with a decrease in lung volume at high concentration. AMP and adenosine caused qualitatively similar effects, but were less potent.

A major problem in the retrospective interpretation of the early data on the actions of purine nucleotides and nucleosides is uncertainty as to the purity of compounds in use. This is due to the limitation of the techniques available at the time rather than a reflection on the skill and discernment of the workers. Gillespie[9], in 1934, was aware of this problem and took aliquots from one preparation of ATP and converted each to the different compounds he wished to study, namely, AMP, adenosine, ITP, and IMP. Gillespie, like Deuticke,[7] noted that ATP was more potent than AMP or adenosine in causing contraction of the isolated virgin guinea pig uterus. He observed that ATP caused a transitory relaxation followed by a more sustained contraction of the isolated rabbit small intestine. Other adenine nucleotides caused relaxations of this tissue, but only rarely caused a contractile phase. He noted that removal of phosphates as in AMP or adenosine decreased the properties of ATP described above, but enhanced activity in lowering blood pressure and dilating coronary vessels in cats. Deamination caused a marked decrease in activity; ITP and IMP were much less active than ATP and AMP, respectively, in experiments similiar to those above.

II. *IN VIVO* STUDIES

ATP has been studied widely *in vivo*. One reason for this is that even as early as the 1930s it had been observed that increasing the number of phosphate groupings enhanced nucleotide potency at several sites.[7-9] Subsequent to this the proposed role of ATP in muscle contraction and its crucial role in relation to the high-energy phosphate bond provided ATP with the aura of a potential panacea. This gave scientists and clinicians the impetus to inject ATP to see what it would do. The major findings are listed in Table 1.

TABLE 1
Effects of ATP *In Vivo*

Animal	Administration	Effect	Ref.
Cat and Rabbit	0.1 mg i.v.	Transient increase in blood pressure preceding a decrease[a]	9
Cat	4 μ mol i.v.	Fall in blood pressure[a]	10
Rat	i.v.	Fall in blood pressure ADP > ATP = AMP > ADO[ab]	11
	i.a.	Fall in blood pressure ATP > ADP > AMP > ADO[ab]	12
Dog	0.08—4 μmol/kg/min right atrium	Increase in heart rate, cardiac output, coronary flow, reduction in coronary resistance, peripheral resistance, and systemic blood pressure[a]	13
Rabbit	2.5 mg/kg i.v.	Caused dilatation of retinal arteries (AMP inactive)	14
Cat (chloralose AN)	0.2—0.4 mg i.a.	Steep fall in blood pressure, constriction of pulmonary vessels	18
Cat (decerebrate)	0.2—0.4 mg i.a.	As above plus probable central actions on skeletal muscle contraction, peristalsis, vomiting, micturition	
Cat	0.01—1 μg near carotid body	Hyperpnea and enhanced sensitivity of chemoreceptors to acetylcholine and $CaCl_2$	14
Human	10 mg i.a.	Increased blood flow in legs	17
	16—1000 μg/min i.v.	Large increase in forearm and hand blood flow equivalent to maximum achieved by histamine or acetylcholine	16
	5—40 mg i.v.	Increase in rate and depth of respiration and blood pressure; tachycardia at low dose; bradycardia and decrease in blood pressure at high dose	15
	2—4 mg bronchial artery	Tingling, bright patchy erythrema, raised temperature, and increased capillary permeability in arm	
	8—75 mg i.v., 25—40 mg i.a.	Frightening sensation in chest and hyperpnea, accompanied by probably increased pulmonary and splanchnic vascular resistance, fall in systolic and diastolic blood pressure, as well as renal blood flow and glomerular filtration rate	20
	Brachial artery	Vasodilatation of arm and hand with warmth and feeling of fullness	
Guinea pig	10—1000 μg/kg i.v.	Increased rate and depth of respiration, preceded by apnea, transitory fall in systolic blood pressure, bronchodilatation low dose, bronchoconstriction high dose	19
Dog	1—8 μg compounds intracoronary artery	Coronary dilatation; relative molar potency: ATP 100, ADP 95, AMP 28, ADO 25	21
Rat	17 μmol i.v.	Increased secretion of insulin	22
Human	i.v. not known	Used to combat flight fatigue in pilots	23

TABLE 1 (continued)
Effects of ATP *In Vivo*

Animal	Administration	Effect	Ref.
		Improvements in articulations of patients with polyarthritis	24
	ATP given with ACTH and vitamin C	Improvement in polyarthritis and bronchial asthma	14
	30 mg i.m. × 40 injections	Treatment of multiple sclerosis	25
	Prolonged treatment ATP and aneurin	Success claimed in treatment of multiple sclerosis	26
	20 mg i.m./d	Treatment of spastic and paralytic migraine	27
	10—50 mg i.v.	Treatment of myasthenia; improvement for a few minutes only	14
	20—100 mg i.v. or i.m.	Improved chronaxie of muscles paralyzed by poliomyelitis	28
	10 mg i.v. daily with strychnine	Treatment of Guillain-Barre syndrome in children	14
	5 mg i.v.	Enhanced weak contractions during childbirth	
	10 mg parenteral	Enhanced flow in perturbed coronary circulation	
	Not known	Therapeutic effect in nerve deafness	29
Mouse	Various salts of ATP 15—45 mg/100 g i.p.	Induced a state of shock preceded by fall in temperature	15
	50—250 mg/100 g i.p.	Dose ranges used to calculate ID_{50} death due to respiratory failure	
Rat	75 mg/100 g i.p.	Decrease in heart rate, blood pressure, temperature, urine flow	
Cat	1.0 mg i.v.	Rise in pulmonary blood pressure and systemic blood pressure before and after section of both vagi	
	1.0—2.0 mg i.v.	Sharp fall in intestinal volume followed by a later increase, apnea, negative chronotropic and inotropic effects[c]	
Rabbit	0.4—7.5 mg	Decrease in urine flow by 65 to 90%	
Dog	3.8—4.3 mg 1 kg/min i.v.	Reduced glomerular filtration rate with afferent arteriolar constriction, efferent arteriolar dilatation followed by hyperemia	31
Rat	10 mg i.p. before hemorrhage	Survival increased from 30 to 80%	32
Dog	6 mg/kg i.a. before hemorrhage	Survival increased from 20 to 80%	33
Mouse	200 mg/kg i.p.	Prolonged phenobarbital hypnosis, lowered body temperature	34
Cat	0.05—0.53 mg/kg i.a. 0.05 mg vertebral artery	Tetanic contraction tibial muscle[d] generalized skeletal muscle contraction[d]	35

[a]   A likely result from peripheral vasodilatation, pulmonary vasoconstriction, or both.
[b]   ATP unlike ADP largely inactivated by passage through the lungs following i.v. administration.
[c]   This may reflect vascular constriction, but occurred at the same time as maximum cardiac inhibition.
[d]   Actions on skeletal muscle are claimed to be due to central actions of ATP.

III. ATP AND SHOCK

At the beginning of World War II, Green and Stoner[30] and their British team were given the task of studying the role of ATP in the genesis of wound shock. It is presumed that the rationale underlying this approach was that damaged muscle could flood the individual with ATP. Indeed, the British Traumatic Shock Team 2 R.A.M.C. obtained conclusive evidence of nucleotide release from injured tissues of battle casualties. There was no detectable ATP or ADP in the serum of healthy individuals (Boettge et al.[14]). However, in 1950, Green and Stoner[30] reviewed the occurrence of adenine nucleotides in a wide range of tissues and noted, in agreement with the British traumatic shock team, that tissue injury caused the release of nucleotides into the extracellular space; from there it entered the bloodstream. It was known that so-called ''nucleotide shock'' could be generated by administration of high doses of ATP.[30] Furthermore, Green and his colleagues had established that the deep shock induced with ATP closely resembled traumatic and ischemic shock. They noted that raised plasma nucleotides had profound effects on many tissues and organs, the sum of which was reminiscent of the changes observed under conditions of shock. They also observed that ATP caused a fall in blood volume similar to that seen in all naturally occurring forms of shock. On heart, ATP at high concentration caused a striking depressant action; it dilated the systemic blood vessels and constricted pulmonary vessels. It relaxed intestine, splenic muscle, and bronchioles and inhibited renal function. They suggested that it depressed the respiratory center. ATP was the most powerful of the nucleotides tested and its local actions were mostly inhibitory. All of these actions were elicited by direct introduction of nucleotide into the bloodstream. Green and Stoner[30] reasoned therefore that it was the more potent ATP and not its breakdown products which was active in shock. They reported that ATP injection at a variety of sites could cause a shock-like state and, if injected in sufficient quantity, caused death due to respiratory inhibition.

It is a paradox that subsequent findings demonstrated that injection of ATP prior to hemorrhage conferred a marked increase in survival of the ensuing hemorrhagic shock in rat[22] and dog.[23] It may be relevent here that infusion of ATP causes a sustained release of corticosteroid hormones from the adrenal cortex (see Vogt.[36])

IV. DISTINCTION BETWEEN THE PHARMACOLOGICAL AND BIOCHEMICAL ACTIONS OF ATP

In 1941, Lippman[37] introduced the concept of the ''high-energy phosphate bond'' for ATP. This revolutionized scientific thought concerning the biological actions of this compound. ATP is unique in the metabolic processes of all organisms, since it has been selected by evolution as the key product of catabolism used in the many energy-requiring processes of cells. This undoubtedly influenced the interpretation of the pharmacology of ATP for the next 30 years. This was particularly important in experiments with muscle since it was not known with any certainty whether or not ATP could penetrate cell membranes and thus affect the contractile mechanisms.

In 1950, Green and Stoner[30] observed that it was vitally interesting that a substance with such high energy potential as ATP, when in an extracellular site, caused effects which produced lowering of the energy output of the body. Apart from the variable and potent effects of ATP on smooth muscle, it was known that close arterial injection of ATP caused contractions of skeletal muscle.[35] However, this was suggested to be due to the effects of a central action of ATP, since injection of ATP into the vertebral artery caused the same result.[35]

The physiological and pharmacological effects of ATP appear to be quite separate to its biochemical functions. It is a mistake to consider that the hydrolysis of the terminal phosphate of ATP after neurotransmission is a more wasteful process than occurs with the degradation of other neurohormones following transmission. It is also a mistake to assume that hydrolysis of the terminal phosphate of ATP is a requirement for pharmacological activity, although this was considered a strong possibility. In 1978, it was shown that the β,γ-methylene isostere of ATP (i.e., β,γ-methylene ATP), in which the terminal phosphate of ATP was stable to degradation, was just as potent a relaxant of intestinal smooth muscle as ATP.[38]

It is now well established that, at least in smooth muscle, ATP exerts its actions via receptors on the surface of the cells. It is unlikely that the intact ATP molecule can penetrate cell membranes under normal conditions.[39,40]

It seems, therefore, that the evolutionary process has harnessed the ubiquitous ATP molecule for physiological functions which are distinct from its biochemical functions.

V. SITE OF ACTION OF ATP

Since the pharmacological actions of ATP were distinct from its biochemical effects, and since it was unlikely that ATP could penetrate cell membranes yet was exceedingly active in tissues containing smooth muscle, the obvious conclusion was that it was acting via receptors present on the muscle cell surface (see also Axelsson et al.)[41,42] It was unlikely that the receptors were there by chance, but in many instances reflected a physiological regulatory function of the nucleotide.[43] It was well recognized that the actions of ATP differed quite markedly from those of AMP and adenosine. The concentration-effect curves to ATP and ADP on isolated guineapig taenia coli were similar, but were displaced to the left of the curves to AMP and adenosine (which were also similar) by almost two orders of magnitude.[44] Moreover, it had been demonstrated that repeated application of adenosine to isolated rabbit intestine left the intestine refractory to this substance, but did not affect the response to applied ATP.[43] This provided further evidence for the presence of different receptors for ATP and adenosine. In 1978, Burnstock[45] distinguished two types of purinergic receptor; the P_1 receptor had the order of potency: adenosine ⩾ AMP > ADP ⩾ ATP, while the P_2 receptor had the reverse order. Subsequent studies by Satchell and Maguire[46] on the actions of a congeneric series of ATP and adenosine analogs in intestinal smooth muscle demonstrated that the receptors for ATP and adenosine were separate with different structural requirements for agonist activity, e.g., epimerization or reduction of the 2-hydroxyl or substitution of bromine in position 8 of ATP and adenosine did not greatly modify action at the ATP or P_2 receptor, but substantially reduced or abolished activity at the adenosine or P_1 receptor. The acceptance of discrete receptor types for ATP and adenosine resolved a dilemma in the literature where ATP and adenosine caused opposite responses, e.g., in guinea pig bladder and chicken rectum.[43]

More recently, Burnstock and Kennedy[47] have distinguished two different receptor subtypes for ATP based on the order of potency of a series of ATP analogs on different tissues, P_{2x} receptors were largely excitatory with a potency order α,β-methylene ATP, β,γ-methylene ATP > ATP = 2-methylthio-ATP. P_{2y} receptors were largely inhibitory with a potency order 2-methylthio-ATP ⩾ ATP > α,β-methylene ATP, β,γ-methylene ATP.

VI. ONSET AND DURATION OF ACTION OF ATP

ATP in its actions on smooth muscle exhibits two properties which distinguish it from the vast majority of smooth muscle active substances. These properties are, firstly, the rapidity of onset of action and, secondly, the transient nature of the response. ATP took only 11 s to bring about maximum relaxation of strips of guinea pig taenia coli, in contrast to catecholamines, which took 30 s to elicit maximum relaxation.[43] In mammalian vas deferens preparations, ATP elicited rapid contractions[48] and is the only substance identified

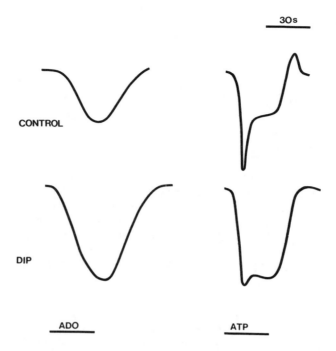

30s

CONTROL

DIP

ADO

ATP

FIGURE 1. Relaxations of guinea pig taenia coli strips. Responses to adenosine (ADO, 25 μM and ATP, 7 μM), before (upper panel) and after (lower panel) dipyridamole (DIP, 1 μM).

which is sufficiently rapid in onset of action to fulfill the requirements as a cotransmitter mediating the twitch response.

The response to ATP was transient in almost all visceral and vascular smooth muscle-containing preparations from the mammals examined.[48] Even in preparations of gut suspended in organ baths of large volume, the effect of ATP was transient. This finding was not readily explained; however, it was quite possible that rapidly formed metabolites were accumulating in the extracellular space and not exchanging with ATP in the bath medium.[48] The observations that stable analogs of ATP exhibit sustained responses[48] supports the idea that the transient nature of the response to ATP is associated with breakdown of the nucleotide.

VII. ARE THE RESPONSES TO APPLIED ATP DUE TO ATP ALONE OR DO METABOLITES CONTRIBUTE TO THE RESPONSE?

For many years research workers have been aware that *in vivo* administration of ATP was accompanied by rapid degradation; and anything apart from the most immediate results could be due to the effect of a metabolite. As discussed previously, ATP administered to tissues in organ baths is also broken down quickly. In strips of guinea pig taenia coli, adenosine (25 μM) caused a slowly developing relaxation, taking approximately 30 s to reach maximum. In contrast to this, ATP (7 μM) caused a relaxation, which was much more rapid in onset and took 11 s to reach maximum (Figure 1). The relaxation was not sustained, returning part of the way back to baseline, but not all the way back, leaving a reduced, delayed phase of the relaxation. The adenosine transport inhibitor dipyridamole (1 μM) potentiated relaxations to adenosine in the tissue; it also potentiated the delayed phase of the response to ATP (Figure 1), suggesting that adenosine formed from ATP during its contact time with the preparation was in fact contributing to the relaxation. Results of this type emphasize the effectiveness and rapidity with which ectoenzymes can convert ATP to adenosine and stress the care which must be taken in interpreting early data on ATP, even in *in vitro* studies.

VIII. CAN METABOLITES OF ATP ELICIT AN ATP RESPONSE?

In view of the discussion above, it would seem unlikely that administration or application of AMP or adenosine could elicit an ATP response, since in order to do so the administered substance would have to be converted to ATP, or at least to ADP, which mimics ATP at some sites. Albaum et al.[49] demonstrated a considerable increase in the level of blood ATP after rectal administration or intramuscular injection of AMP. Similar results were observed following injections of AMP in rabbits.[14] These results demonstrate the complicated nature of adenine nucleotide and nucleoside metabolism and emphasise the care which must be taken in interpreting the actions of the nucleotides and nucleosides, especially *in vivo*.

IX. PARTICIPATION OF ATP IN NEUROTRANSMISSION

In 1954, Holton[50] reported that ATP was released from rabbit ear during stimulation of the great auricular nerve. It was considered that the release was the result of sensory nerve stimulation. In 1969, Satchell et al.[51] provided evidence that a purine compound was a neurotransmitter in nonadrenergic inhibitory neurons in guinea pig intestine. AMP and adenosine were the compounds identified as coming from the preparations, although it was considered that these substances were metabolites of neuronally released ATP. This idea was consistent with further studies from the same laboratory which demonstrated that when ATP was added to perfused preparations, metabolites of the same composition appeared in the perfusate.[51]

In 1971, studies with labeled adenosine[52] provided evidence for specific release of label from nonadrenergic, noncholinergic nerves in gut in response to the ganglionic stimulant nicotine. It was also demonstrated in the same series of experiments that stimulation of perivascular sympathetic nerves to strips of gut released label as well. Thus the experiments not only provided evidence in favor of purinergic nonadrenergic, noncholinergic neurotransmission, but also provided early evidence in favor of ATP involvement in sympathetic nerve transmission, although this was not recognized at that time. In 1973, Nakanishi and Takeda[53] suggested that the transmitter released by hypogastric nerves to the guinea pig seminal vesicle could be a complex of ATP and noradrenaline.

In 1974, Su and Sum[54] and Su[55] determined the release of ATP following sympathetic nerve stimulation in a variety of blood vessels. They believed that it was stored in vesicles together with noradrenaline and released during the exocytotic process. While they recognized the possibility of a postsynaptic action of released ATP, they believed that its role was one of neuromodulation, whereby it was rapidly degraded to adenosine in the synapse prior to occupation of presynaptic receptors.

In 1977, Verhaeghe et al.[56] examined the inhibition of sympathetic neurotransmission in canine blood vessels by adenosine and ATP. Both substances effectively depressed the nerve-induced efflux of noradrenaline, but only the response to adenosine was antagonized by the P_1 receptor antagonist theophylline. The implication of this finding is that ATP was causing presynaptic inhibition via a P_2 receptor. It was also demonstrated that ATP could inhibit cholinergic neurotransmission at the skeletal neuromuscular junction[57] and in smooth muscle of the intestinal wall.[58]

In 1978, Westfall et al.[59] demonstrated that ^{3}H-purines were taken up and released together with noradrenaline from adrenergic nerves supplying the guinea pig vas deferens.

Studies from the same laboratory[60] 3 years later showed that the ATP P_2 receptor antagonist, arylazido aminopropionyl-ATP (ANAPP$_3$), blocked the initial twitch component of the response to sympathetic nerve stimulation. This was firm evidence supporting a role of ATP with noradrenaline in cotransmission in sympathetic nerves. Shortly afterwards, Sneddon and Burnstock[61] used α,β-methylene ATP to desensitize P_2 receptors in guinea pig vas deferens and specifically abolished the twitch component of the sympathetic nerve response as well the associated e.j.p. and the response to applied ATP.

X. RELEASE OF ATP FROM THE ADRENAL MEDULLA

In 1955, Hillarp et al.[62] reported the presence of high concentrations of ATP in cow adrenal medulla. It was stored together with catecholamines in chromaffin granules. In cat adrenal medulla, one molecule of ATP was stored with four molecules of catecholamines[63] and the combination was released, presumably in this ratio in response to splanchnic nerve stimulation or by application of acetylcholine or barium.[64,65] Collected perfusates contained largely AMP and some ATP, although addition to perfusion fluids of 1 to 2 mM EDTA reversed this ratio to one of largely ATP and relatively little AMP.[65]

Release of ATP by the adrenal medulla has been regarded as a curiosity rather than as a potential physiological regulatory process. Under these circumstances it may seem a coincidence that ATP and catecholamines are released by both sympathetic nerve terminals and the adrenal medulla under conditions of "fight and flight". Since ATP is likely to have a transmitter role in sympathetic nerve terminals, the question might well be asked: "How sure can we be that ATP or its metabolites are not released from the adrenal medulla in sufficient quantities to elicit responses elsewhere in the body?" The answer is we cannot be sure, and further experimental evidence is required. To provide some perspective as to a possible role of released nucleotide, peak release of ATP into the adrenal vein is estimated to be at a concentration of between 1 to 10 μM.[66] The blood passes into the posterior vena cava; however, from there it passes rapidly to the right side of the heart and to the lungs. Either of these organs could conceivably be affected, although an action at more distal sites having passed through the pulmonary capillary bed would be less likely. It will be recalled that ATP has a powerful constrictor effect on the lung vasculature (see Table 1). Problems in endeavoring to interpret this data arise since it is not known what changes would occur in the composition and amounts of nucleotide and nucleoside during transit. The transit time to lung could be as little as 10 s in most mammals. Moreover, dilution would occur at points of junction with the inferior and superior vena cavae. Should ATP be released during "fight and flight" in sufficient quantities to affect lung vasculature, then this would pose a dilemma since vascular constriction would be contrary to the needs of the animal at that time. Further experimental data are necessary in order to evaluate this interesting problem.

XI. ATP ANTAGONISTS

In 1950, Green and Stoner[30] examined the effects of antimalarial drugs on ATP-induced shock. They reported that quinine and mepacrine antagonized the negative chronotropic effects of ATP on guinea pig heart. Quinine also antagonized the depressor effect of ATP in cats. Administration of quinine at the same time as a shock-inducing dose of ATP in mice decreased the mortality and prevented the appearance of signs of ATP shock.[30] Quinidine, a stereoisomer of quinine, was reported to antagonize relaxations due to ATP in isolated strips of guineapig taenia coli.[43] However, it also abolished relaxations due to noradrenaline at concentrations less than were required to antagonize responses to ATP, indicating a lack of specificity of the drug.

In 1971, Rikimaru et al.[67] found that high concentrations of imidazole and phentolamine were required to abolish relaxations to ATP on guinea pig taenia coli. Further investigations[68] showed that imidazole also antagonized relaxations due to amyl nitrite, demonstrating a lack of specificity by imidazole. Phentolamine and other 2-substitued imidazolines, e.g., antazoline and tolazoline, antagonized responses to ATP in taenia more effectively than responses to amyl nitrite.[68] The imidazolines could be regarded as acting via P_{2y} receptors.

The methylxanthines are well-recognized antagonists of adenosine at P_1 receptors. In 1963, Nichols and Walaszek[69] found that caffeine antagonized the inhibitory effect of ATP on isolated rabbit jejunum, although it did not affect relaxations of the guinea pig taenia coli due to ATP.[43] The actions of methylxanthines as antagonists at P_1 receptors are consistent and predictable. At P_2 receptors they are inconsistent in causing blockade, a large part of which may be due to the rapid breakdown of ATP during its contact time with the preparation and consequent adenosine effects. An action of ATP per se at P_1 receptors is less likely. This information has been reviewed elsewhere.[46]

Other compounds have been found to antagonize the inhibitory effects of ATP on gut, in particular 2,2′-pyridylisatogen,[70] apamine,[71] and the dye reactive blue.2.[72] While all three compounds were effective antagonists of the effects of ATP, each lacked specificity in its action, although some specificity has been claimed for the latter compound.[72]

One of the most interesting compounds discovered which has specificity in antagonism at P_2 receptors is the photoaffinity analog of ATP, ANAPP$_3$.[73] This compound is most effective at P_{2x} receptors, and has been a useful tool in the identification of the role of ATP in cotransmission in vas deferens and in blood vessels.

Methylene-substituted analogs of ATP were found to be more active on intestinal smooth muscle than ATP.[74] The analog α,β-methylene ATP was used by Burnstock and co-workers.[61,75] to desensitize P_{2x} receptors in vas deferens and in blood vessels, and similar to the experiments with ANAPP$_3$ above, again provided valuable support for a cotransmitter role for ATP. This compound was not effective in causing desensitization at P_{2y} receptors.

XII. SIGNIFICANT ACTIONS OF ATP IN DIFFERENT TISSUES

A. THE VASCULATURE

In 1933, Gaddum and Holtz[8] examined the effects of perfused ATP, AMP, and adenosine on isolated lungs of cats and dogs. All three compounds had qualitatively the same effects in that they caused vasodilatation at low doses and vasoconstriction at high doses. They believed that the constriction observed at higher doses was due to an effect on the inflow. ATP was more potent in these actions than either AMP or adenosine. In 1948, Emmelin and Feldberg[18] also noted the marked susceptibility of the pulmonary vessels to injection of as little as 0.2 mg of ATP in cat. This effect was so strong that little blood could enter the left heart with a consequent fall in systemic blood pressure. A similar constriction was observed in response to addition of 50 mg to the arterial perfusate of isolated rabbit lungs (Lunde et al.[76]).

ATP was also found to cause vascular constriction at other sites. In 1951, Davies et al.[20] provided evidence suggesting that ATP caused a decrease in splanchnic blood flow in human; although in 1945 De Waele and Van de Velde[77] reported that ATP caused splanchnic vasodilation with constriction of the splenic capsule following injection of ATP into the jugular veins of dogs and rabbits. Recent studies on isolated preparations support the view that the usual effect of ATP on perfused mesenteric bed[78] and isolated mesenteric arterioles[79] is a constriction.

In contrast to the above findings, ATP is a potent dilator at most other sites. In general it causes dilatation of most peripheral blood vessels, including the hind limbs of dogs and cats[80] and the human hand and forearm.[16,80] Other studies in human showed that ATP caused a general decrease in peripheral vascular resistance,[30] reflecting the occurrence of peripheral vasodilatation. It also caused flushing of the skin[20,30] and at high concentration promoted an increase in capillary filtration.[16] However, exceptions were findings that ATP normally caused vasoconstriction in rabbit ear artery[81] and rat tail artery.[82] In 1956, Wolf and Berne[83] found that ATP and ADP were four times more potent than AMP and adenosine in causing an increase in coronary blood flow in dogs.

While historically the actions of ATP on the vasculature appear straightforward, more detailed studies on isolated vascular preparations have shown that in many preparations ATP has the potential to cause either vasoconstriction or vasodilatation, and this may depend upon whether the endothelium is present or absent,[78] whether or not the applied ATP can be broken down to adenosine, or whether or not ATP can act per se via P_1 receptors[84] and whether the preparation is at high or low tone.[78] Related to this also may be the occurrence or the respective size of populations of P_{2x} and P_{2y} receptors.

Students of the physiology and pharmacology of ATP may find it challenging to test a teleological question as to whether or not the effects of ATP on the vasculature are a logical and predictable result of a physiological role in the "fight and flight" process. This action would be reminiscent of the actions of catecholamines, and would occur either by release of nucleotide from sympathetic nerve terminals or from the adrenal medulla, or both. In this respect, it would appear generally that there was constriction of blood vessels to splanchnic areas and to the kidneys in response to ATP and to the vessels of tail and ears. This is consistent with the expectation of "fight and flight" where blood is shifted from areas where it is not immediately required to areas where it is urgently needed. Consistent also with this idea were the findings that ATP dilated somatic blood vessels. However, some of the actions of ATP do not fit in with this idea, the most noteworthy of which is the consistently observed constriction of the pulmonary vasculature. More information is needed to evaluate this interesting question further, especially in relation to the possibility that the proportion of ATP and noradrenaline released from sympathetic nerve terminals at different sites varies.

B. URINOGENITAL SYSTEM

In 1932, Deuticke[7] found that ATP, AMP, and adenosine all caused contraction of isolated preparations of guinea pig uterus. Later findings supported a contractile role for ATP and adenosine in mammalian uterine preparations (see Gillespie[9] and also References 85 to 89). More recently, Pennefather and Story[90] found that ATP caused contraction of the longitudinal myometrium from dioestrous guinea pig, but lacked this effect in the circular myometrium. In general, the sensitivity of this tissue varies markedly and depends upon whether tissues are taken from pregnant, nonpregnant, or virgin animals and on the species. The high degree of variability in sensitivity has discouraged many workers from studying this tissue in recent years. This is probably unfortunate because uterine function at childbirth is not well understood and adenine nucleotides and adenosine may well play a role.

ATP causes contractions in other smooth muscle containing tissues associated with the reproductive tract, in retractor penis of dog,[91] bull,[92] and monkey;[93] in guinea pig seminal vesicle;[94] and in the vas deferens[48,95,96] and bladder[97,98] of several mammalian species. ATP was the most potent of all the nucleotides and nucleosides studied.

C. ALIMENTARY SYSTEM

In 1934, Gillespie[9] reported that ATP caused a drop in tone in gut followed by an above-normal degree of contraction, then restoration of normal peristalsis. In the next decade, Lippman[37] proposed his high-energy phosphate bond concept and Szent Gyorgyi[99] demonstrated that ATP produced contractions when added to extruded threads of actomyosin. Buchthal and Kahlson[100] found that ATP, and not related nucleotides was potent in causing contraction in guinea pig small intestine. They proposed, in vogue with current thinking, that this action of ATP was a demonstration of the function of ATP as a humoral link in the release of contraction due to acetylcholine. It was not until a further decade had passed that Axelsson and colleagues[41,42] made a most significant finding, that the mechanical effects of ATP and related nucleotides appeared to be due to a cell-surface action, since they failed to affect glycerol-extracted smooth-muscle fibers.

Studies on the effects of ATP in intestinal segments found that it often caused biphasic responses as in the original experiments of Gillespie,[9] and even today it is not clear why this happens, although one possibility is that ATP causes differential effects in the circular and longitudinal muscle layers. For example, in guinea pig ileum, where ATP causes a biphasic response, it relaxed isolated circular muscle, yet contracted isolated longitudinal muscle.[48] However, the usual response in longitudinal layers of most intestine is a relaxation. The taenia coli of the guinea pig has been most widely studied. During the 1970s, quantitative studies were carried out on the effects of adenine nucleotides and adenosine on this tissue.[44] This was followed up by detailed studies on a series of more stable analogs of ATP and adenosine on isolated taenia coli preparations.[38,74]

Preparations of isolated segments of mammalian colon mostly relaxed in response to applied ATP, as did muscles to the coccyx.[43] In preparations of stomach and esophagus, the effects of ATP were variable, depending on the species and the site and disposition of the muscle layers.[43]

D. AIRWAYS

In 1931, Bennett and Drury[101] showed that adenine nucleotides and adenosine caused bronchodilatation in guinea pig *in vivo*. In 1963, Bianchi et al.[19] confirmed these *in vivo* findings and showed that ATP caused bronchodilatation in isolated perfused guinea pig lungs. *In vivo* bronchodilatation was observed in response to low doses of ATP although large doses caused bronchoconstriction. They also noted that low doses of ATP *in vivo* protected against the bronchoconstrictor effects of acetylcholine and histamine. Most recent studies have been carried out on isolated tracheal preparations where ATP and adenosine were found to be approximately equipotent in causing relaxation of the trachealis muscle.[102] However, these authors noted that the adenosine transport inhibitor dipyridamole potentiated relaxations due to ATP and adenosine equally, suggesting that ATP was rapidly broken down to adenosine and that adenosine was causing the response attributed to ATP. Moreover, these authors noted that the more stable methylene isosteres of ATP exhibited little activity on the trachealis muscle, suggesting that receptors for ATP were lacking in the tissue.

While ATP and adenosine relaxed transverse strips of trachea,[102,103] they contracted strips of trachea which were cut in a spiral.[103,104] An explanation for this apparent dilemma came with the finding that the trachealis muscle of the guinea pig has different muscle bands.[103] The traditionally recognized band passes transversely between the ends of adjacent cartilaginous hoops, and presumably this layer relaxes. The second layer is deeper and lies immediately below the submucosa; it passes from the C cartilages in a posterior direction towards the bifurcation of the bronchi. Presumably, this layer responded with a contraction.[103]

E. HEART

ATP has been found to consistently cause a powerful excitatory effect on frog heart followed by an inhibition.[14,105-108] Hoyle and Burnstock[108] suggested that the excitatory effects were mediated by P_2 receptors, while the inhibitory effects were mediated via P_1 receptors following degradation of ATP to adenosine (see also Reference 109). Hoyle and Burnstock[108] provided evidence that ATP was an excitatory cotransmitter released with adrenaline from nerve terminals supplying frog atria.

The effects of ATP in mammalia were much less consistent. Chronotropic and inotropic effects were predominantly negative on isolated mammalian atria of dog,[13] cat,[30] and rabbit.[13,30] On whole hearts, chronotropic effects were marked and largely obscured the nonetheless interesting inotropic effects. In Langendorf-perfused guinea pig and rat hearts, ATP had opposite effects,[14] suggesting a genuine species difference.

When administered *in vivo*, ATP caused a negative chronotropic effect in cat[13] and rat[30] and a positive chronotropic effect in dog,[13] human,[30] and rabbit.[9] However, it should be remembered that ATP has other powerful concomitant actions *in vivo* including systemic vasodilatation and pulmonary vasoconstriction, making it difficult to distinguish primary from secondary actions on heart rate. Relevant to this is the finding of Sydow and Ahlquist[110] that AMP and adenosine caused an increase in heart rate in anesthetized dogs. However, after atropine a decrease was observed, suggesting that vagal reflexes were involved. Of further relevance is a report[14] that ATP caused an increase in heart rate in conscious dogs, but a decrease in narcotized dogs; under the latter conditions its regulatory processes were probably inhibited.[14] A problem with interpretation of the data following systemic administration of ATP is the uncertainty as to the composition of nucleotide or nucleoside reaching the heart or other target organ.

An interesting and potentially very important function of ATP is its ability to restore normal beat in hearts in which the rhythm has been disturbed by a variety of experimental conditions.[14,111]

XIII. FUTURE DIRECTIONS

There are two important directions which should be taken in future studies on ATP. Firstly, the extent to which ATP is involved in physiological regulatory functions should be determined. An outstanding example of this is the role of ATP in sympathetic nerve co-transmission. This is particularly important in the vasculature, where the relative proportions of ATP to noradrenaline released by sympathetic nerve terminals at different sites are likely to vary, as are populations of receptor types. It will be interesting to determine whether or not such mechanisms have evolved to mediate the changes in regional blood flow that occur in "fight and flight". Such findings are likely to have clinical implications and will create wide interest. Secondly, potent and specific antagonists for ATP receptors need to be discovered. While a large number of compounds have been screened for this property, much of this research has not been undertaken in the most fruitful manner. It is essential that in order to study analogs of ATP for antagonist properties, they must first of all be highly stable to degradation, especially the phosphate chain, which requires a minimum of two phosphates for action at P_2 receptors. After all, it was this approach which provided the first purine receptor-desensitizing drugs. If chemists need encouragement to synthesize stable analogs of ATP, the reward will come in providing drugs as valuable tools to further examine the physiological actions of ATP, at different receptors, and especially in providing drugs with the potential of having clinically valuable actions in a number of disease states.

REFERENCES

1. **Bass, R.,** Uber die Purinkorper des Menschlichen Blutes und den Wirkungsmodus der 2-Phenyl-4-Cholincarbonsaure (Atophan), *Arch. Exp. Pathol. Pharmakol.*, 76, 40, 1914.
2. **Freund, H.,** Uber die pharmakologischen Wirkungen des defibrinierten Blutes, *Arch. Exp. Pathol. Pharmakol.*, 86, 267, 1920.
3. **Drury, A. N. and Szent-Gyorgyi, A.,** The physiological activity of adenine compounds with especial reference to their action upon mammalian heart, *J. Physiol. (London)*, 68, 213, 1929.
4. **Zipf, K.,** Uber die pharmakologische Wirkung des frisch defibrinierten Blutes, *Arch. Exp. Pathol. Pharmakol.*, 150, 91, 1930.
5. **Hoffman, W. S.,** The isolation of crystalline adenine nucleotide from blood, *J. Biol. Chem.*, 63, 675, 1925.

6. **Lohman, K.,** Darstellung der Adenylphosphosaure aus Muskulature, *Biochem. Z.*, 233, 466, 1931.
7. **Deuticke, H. J.,** Uber den Einfluss von Adenosin und adenosinphosphosauren auf isolierten Meerschweinuterus, *Pfluegers Arch.*, 230, 537, 1932.
8. **Gaddum, J. H. and Holtz, P.,** The localization of the action of drugs on the pulmonary vessels of dogs and cats, *J. Physiol. (London)*, 77, 139, 1933.
9. **Gillespie, J. H.,** The biological significance of the linkages in adenosine, *J. Physiol. (London)*, 80, 45, 1934.
10. **Flesher, J. W., Oester, Y., and Myers, T. C.,** Vasodepressor effects of adenosine phosphates, *Nature*, 185, 772, 1960.
11. **Gordon, D. B. and Hesse, D. H.,** Blood pressure lowering action of adenosine diphosphate and related compounds, *Am. J. Physiol.*, 201, 1123, 1961.
12. **Gordon, D. B.,** Basis of depressor action of adenosine phosphates, *Am. J. Physiol.*, 201, 1126, 1989.
13. **Rowe, G., Afonso, W. C., and Gartner, G.,** The systemic and coronary haemodynamic effects of ATP, *Am. Heart J.*, 64, 228, 1962.
14. **Boettge, K., Jaeger, K., and Mittenzwei, H.,** The system of adenylic acids: new results and problems, *Arzneim. Forsch.*, 7, 24, 1959.
15. **Stoner, H. B. and Green, H. N.,** Experimental limb ischaemia in man with special reference to the role of ATP, *Clin. Sci. (London)*, 5, 159, 1944.
16. **Duff, F., Patterson, G. C., and Shepard, J. T.,** A quantitative study of the response to adenosine triphosphate on the blood vessels of the human hand and forearm, *J. Physiol. (London)*, 64, 581, 1954.
17. **Hess, H.,** Die Wirkung von Adenylverbindungen auf die Durchblutung des ruhenden Skeletalmuskels des Menschen, *Klin. Wochenschr.*, 33, 525, 1955.
18. **Emmelin, N. and Feldberg, W.,** Systemic effects of ATP, *Br. J. Pharmacol.*, 3, 273, 1948.
19. **Bianchi, A., DeNatale, G., and Giaquinto, S.,** Effects of adenosine and its phosphorylated derivatives on the respiratory apparatus, *Arch. Int. Pharmacodyn. Ther.*, 145, 498, 1963.
20. **Davies, D., Gropper, A., and Schroeder, H.,** Circulatory and respiratory effects of adenosine triphosphate in man, *Circulation*, 3, 543, 1951.
21. **Winbury, M., Paplerski, D., Henmer, L., and Hambourger, W.,** Coronary dilator action of adenine-ATP series, *J. Pharmacol. Exp. Ther.*, 109, 225, 1953.
22. **Candela, L. R. and Garcia Fernandez, M.,** Stimulation of secretion of insulin by ATP, *Nature*, 197, 1210, 1963.
23. **Rotondo, G.,** On the use of adenosine triphosphoric acid in pilots with slight or initial flight fatigue, *Riv. Med. Aeronaut.*, 27, 176, 1964.
24. **Grenier, J., Bavay, J., and Lutier, F.,** L'ACTH peut-elle etre remplacee dans certain cas par l'acide adenosine triphosphorique (ou ATP), *La. Presse Med.*, 60, 619, 1952.
25. **Mankowski, B. and Slonimskaja, V.,** Die Therapie der progressiven Muskeldystrophie mit Adenosintriphosphosaure, *Zbl. Ges. Neurol.*, 132, 263, 1955.
26. **Bauer, H.,** Die Brenztraubensaurekonzentration im Blut von Multiplensklerosen und Kontrolfallen, *Biochem. Z.*, 60, 619, 1956.
27. **Muller, G. M.,** Adenosinverbindungen in der Migranetherapie, *Muench. Med. Woschenschr.*, 95, 273, 1953.
28. **Dobner, E.,** Die Bedeutung elektrischer Methoden in der Diagonostik und Therapie poliomyetischer Lahmungen, *Dtsch. Med. Wochenschr.*, 80, 153, 1955.
29. **Oksawa, R., Nakamura, N., and Takafugi, J.,** Therapeutic effect of ATP in the treatment of nerve deafness, *Otolaryngology (Tokyo)*, 33, 115, 1961.
30. **Green, H. N. and Stoner, H. B.,** *Biological Actions of Adenine Nucleotides*, Lewis, London, 1950.
31. **Houck, C., Bing, R., Craig, C., and Visscher, F.,** Renal hyperemia after intravenous infusion of AMP, adenosine or ATP in the dog, *Am. J. Physiol.*, 153, 159, 1948.
32. **Sharma, G. P. and Eisen, B.,** Protective effect of ATP in experimental haemorrhagic shock, *Surgery*, 59, 66, 1966.
33. **Tulat, S., Massion, W., and Schilling, J.,** The effect of ATP administration in irreversible shock, *Physiologist*, 6, 284, 1963.
34. **Lessin, A. and Parkes, M.,** The relation between sedation and body temperature in the mouse, *Br. J. Pharmacol.*, 12, 245, 1944.
35. **Buchthal, F., Engbach, L., Sten-knudsen, O., and Thomasen, E.,** Application of adenosine triphosphate and related compounds to the spinal cord of the cat, *J. Physiol. (London)*, 106, 3, 1947.
36. **Vogt, M.,** Cortical secretion of the isolated perfused adrenal, *J. Physiol. (London)*, 113, 129, 1951.
37. **Lippman, F.,** Metabolic generation and utilization of phosphate bond energy, *Enzymology*, 1, 99, 1941.
38. **Satchell, D. G. and Maguire, M. H.,** The contribution of adenosine to the inhibitory actions of adenine nucleotides on the guinea-pig taenia coli: studies with adenine nucleotide analogs and dipyridamole, *J. Pharmacol. Exp. Ther.*, 211, 626, 1879.
39. **Maguire, H. H.,** Personal communication.

40. **Patterson, A.,** Personal communication.
41. **Axelsson, J., Holmberg, B., and Hoberg, G.,** ATP and intestinal smooth muscle, *Acta Physiol. Scand.,* 68 (Suppl. 227), 1966.
42. **Axelsson, J. and Holmberg, B.,** The effects of extracellularly applied ATP and related compounds on the electrical and mechanical activity of the smooth muscle of the taenia coli of the guinea-pig, *Acta Physiol. Scand.,* 75, 149, 1969.
43. **Maguire, M. H. and Satchell, D. G.,** *Purinergic Receptors in Visceral Smooth Muscle,* Chapman Hall, London, 1981, 47.
44. **Satchell, D. G. and Burnstock, G.,** Quantitative studies of the release of purine compounds following stimulation of non-adrenergic inhibitory nerves in the stomach, *Biochem. Pharmacol.,* 20, 1694, 1971.
45. **Burnstock, G.,** A basis for distinguishing two types of purinergic receptor, in *Cell Membrane Receptors for Drugs and Hormones: A Multidisciplinary Approach,* Straub, R. W. and Bolis, L., Eds., Raven Press, New York, 1978, 107.
46. **Satchell, D. G. and Maguire, M. H.,** Evidence for separate receptors for ATP and adenosine in the guinea-pig taenia coli, *Eur. J. Pharmacol.,* 81, 669, 1982.
47. **Burnstock, G. and Kennedy, C.,** Is there a basis for distinguishing two types of P2-purinoceptor?, *Gen. Pharmacol.,* 16, 433, 1985.
48. **Satchell, D. G.,** The effects of ATP and related nucleotides on visceral smooth muscle, *Ann. N.Y. Acad. Sci.,* 603, 53, 1990.
49. **Albaum, H. G., Rottino, A., and Hoffman, G. T.,** Effect of adenylates on the level of adenosine triphosphate (ATP) in the blood of normal persons and patients with Hodgkin's disease, *J. Lab. Clin. Invest.,* 42, 255, 1953.
50. **Holton, P.,** The liberation of adenosine triphosphate on antidromic stimulation of sensory nerves, *J. Physiol (London),* 126, 124, 1954.
51. **Satchell, D. G., Burnstock, G., and Campbell, G.,** Evidence for a purine compound as the transmitter in non-adrenergic inhibitory nerves in gut, *Aust. J. Exp. Biol. Med. Sci.,* 470, 24, 1969.
52. **Su, C., Bevan, J., and Burnstock, G.,** Adenosine release during stimulation of enteric nerves, *Science,* 173, 337, 1971.
53. **Nakanishi, H. and Takeda, H.,** The possibility that adenosine triphosphate is an excitatory transmitter in the guinea-pig seminal vesicle, *Jpn. J. Pharmacol.,* 23, 479, 1973.
54. **Su, C. and Sum, C.,** The uptake and release of adenine derivatives in the rabbit portal vein, *Proc. West. Pharmacol. Soc.,* 17, 122, 1974.
55. **Su, C.,** Neurogenic release of purine compounds in blood vessels, *J. Pharmacol. Exp. Ther.,* 195, 159, 1975.
56. **Verhaeghe, R. H., Vanhoutte, P., and Shepard, J.,** Inhibition of sympathetic neurotransmission in canine blood vessels, *Circulation,* 40, 208, 1977.
57. **Ribiero, J. A. and Waljer, J.,** The effects of adenosine triphosphate and adenosine diphosphate on transmission at the rat and frog neuromuscular junctions, *Br. J. Pharmacol.,* 54, 213, 1975.
58. **Takagi, K. and Nakamura, T.,** Effect of dibutyryl cyclic adenosine monophosphate and adenosine triphosphate on acetylcholine output from cholinergic nerves in the guinea-pig ileum, *Jpn. J. Pharmacol.,* 22, 33, 1972.
59. **Westfall, D., Stitzel, R., and Rowe, J.,** The postjunctional effects and neural release of purine compounds in the guinea-pig vas deferens, *Eur. J. Pharmacol.,* 50, 27, 1978.
60. **Fedan, J. S., Hogaboom, G., O'Donnel, J., Colby, J., and Westfall, D. P.,** Contributions by purines to the neurogenic response of the vas deferens of the guinea-pig, *Eur. J. Pharmacol.,* 69, 41, 1981.
61. **Sneddon, P. and Burnstock, G.,** Inhibition of excitatory junction potentials in guinea-pig vas deferens by α,β-methylene ATP. Further evidence for ATP and noradrenaline as cotransmitters, *Eur. J. Pharmacol.,* 100, 85, 1984.
62. **Hillarp, N. A., Hogberg, B., and Nilson, B. F.,** Adenosine triphosphate in the adrenal medulla of the cat, *Nature (London),* 176, 1032, 1955.
63. **Douglas, W. W.,** Stimulus secretion coupling. The concept and clues from chromaffin and other cells, *Br. J. Pharmacol.,* 34, 451, 1968.
64. **Douglas, W. W., and Poisner, A.,** Evidence that the secretory adrenal chromaffin cell releases catecholamines directly from ATP-rich granules, *J. Physiol (London),* 183, 236, 1966.
65. **Douglas, W. W. and Poisner, A.,** On the relation between ATP splitting and secretion in the adrenal chromaffin cell, *J. Physiol. (London),* 183, 249, 1966.
66. **Marley, P.,** Personal communication.
67. **Rikimaru, A., Fukushi, Y., and Suzuki, T.,** Effects of imidazole and phentolamine on the relaxant responses of guinea-pig taenia coli to transmural stimulation and to adenosine triphosphate, *Tohoku. J. Exp. Med.,* 105, 199, 1971.

68. **Satchell, D. G., Burnstock, G., and Dann, P.,** Antagonism of the effects of purinergic nerve stimulation and exogenously applied ATP on the guinea-pig taenia coli by 2-substituted imidazolines and related compounds, *Eur. J. Pharmacol.,* 23, 264, 1973.

69. **Nichols, R. and Walaszek, E. J.,** Studies on the pharmacology of caffeine, *Fed. Proc., Fed. Am. Soc. Exp. Biol.,* 22, 308, 1963.

70. **Spedding, M., Sweetman, A. J., and Weetman, D. F.,** Antagonism of adenosine 5'-triphosphate-induced relaxation by 2-2'-pyrididylisatogen in the taenia of the guinea-pig caecum, *Br. J. Pharmacol.,* 53, 575, 1975.

71. **Vladimirova, I. A. and Shuba, M. F.,** The effect of strychnine, hydrostine and apamine on synaptic transmission in smooth muscle cells, *Neirofiziologiya,* 10, 295, 1978.

72. **Burnstock, G., Hopwood, A. M., Hoyle, C. H. V., et al.,** Reactive blue-2 selectively antagonizes the relaxant responses to ATP and its analogues which are mediated by the P2y-purinoceptors, *Br. J. Pharmacol.,* 89, 857P, 1986.

73. **Hogaboom, G. K., O'Donnell, J. P., and Fedan, J. S.,** Purinergic receptors: photoaffinity analog of adenosine triphosphate is a specific adenosine triphosphate antagonist, *Science,* 208, 1273, 1980.

74. **Satchell, D. G. and Maguire, M. H.,** Inhibitory effects of adenine nucleotide analogs on the isolated guinea-pig taenia coli, *J. Pharmacol. Exp. Ther.,* 195, 540, 1975.

75. **Burnstock, G. and Sneddon, P.,** Evidence for ATP and noradrenaline as cotransmitters in sympathetic nerves, *Clin. Sci.,* 68, 89S, 1985.

76. **Lunde, P. K. M., Waaler, B. A., and Walloe, L.,** The inhibitory effect of various phenols upon ATP-induced vasoconstriction in isolated perfused rabbit lungs, *Acta Physiol. Scand.,* 72, 331, 1968.

77. **de Waele, H. and Van de Velde, J.,** L'action vasculaire de l'acide adenosine triphosphorique, *Arch. Int. Pharmacodyn. Ther.,* 70, 228, 1945.

78. **Ralevic, V. and Burnstock, G.,** Actions mediated by P2-purinoceptor subtypes in the isolated perfused mesenteric bed of rats, *Br. J. Pharmacol.,* 95, 637, 1988.

79. **Burnstock, G. and Warland, J. J.,** P2-purinoceptors of two subtypes in the rabbit mesenteric artery: reactive blue 2 selectively inhibits responses mediated via the P2y- but not the P2x-purinoceptor, *Br. J. Pharmacol.,* 90, 383, 1987.

80. **Folkow, B.,** The vasodilator action of adenosine triphosphate, *Acta Physiol. Scand.,* 17, 311, 1949.

81. **Kennedy, C., Saville, V. L., and Burnstock, G.,** The contributions of noradrenaline and ATP to the responses of the rabbit central ear artery to sympathetic nerve stimulation depend on the parameters of stimulation, *Eur. J. Pharmacol.,* 122, 291, 1986.

82. **Sneddon, P. and Burnstock, G.,** ATP as a cotransmitter in rat tail artery, *Eur. J. Pharmacol.,* 106, 149, 1984.

83. **Wolf, M. M. and Berne, R. M.,** Coronary vasodilator properties of purine and pyrimidine derivatives, *Circ. Res.,* 4, 343, 1956.

84. **Kennedy, C. and Burnstock, G.,** ATP produces vasodilation via P1-purinoceptors and vasoconstriction via P2-purinoceptors in the isolated rabbit central ear artery, *Blood Vessels,* 22, 145, 1985.

85. **Mihich, E., Clarke, D. A., and Philips, F. S.,** Effect of adenosine analogs on isolated intestine and uterus, *J. Pharmacol. Exp. Ther.,* 111, 335, 1954.

86. **Watts, D. T.,** Stimulation of uterine muscle by adenosine triphosphate, *Am. J. Physiol.,* 173, 291, 1953.

87. **Bunday, P. V., Carr, C. J., and Miya, T. S.,** A pharmacological study of some nucleosides and nucleotides, *J. Pharm. Pharmacol.,* 13, 290, 1961.

88. **Daniel, E. E. and Irwin, J.,** On the mechanism whereby certain nucleotides produce contractions of smooth muscle, *Can. J. Physiol. Pharmacol.,* 43, 89, 1965.

89. **Stafford, A.,** Potentiation of adenosine and adenine nucleotides by dipyridamole, *Br. J. Pharmacol.,* 28, 218, 1966.

90. **Pennefather, J. N. and Story, M. E.,** The effect of ATP on isolated uterine preparations from the dioestrous guinea-pig, *Clin. Exp. Pharm. Physiol.,* Suppl. 11, 37, 1989.

91. **Luduena, F. P. and Grigas, E. O.,** Effect of some biological substances on dog retractor penis *in vitro,* *Arch. Int. Pharmacodyn. Ther.,* 196, 269, 1972.

92. **Klinge, E. and Sjostrand, N. O.,** Contraction and relaxation of the retractor penis muscle and the penile artery of the bull, *Acta Physiol. Scand.,* 420, 1, 1975.

93. **Klinge, E. and Sjostrand, N. O.,** Comparative study of some isolated mammalian smooth muscle effectors of penile erection, *Acta Physiol. Scand.,* 100, 354, 1977.

94. **Nakanishi, H. and Takeda, H.,** The possible role of adenosine triphosphate in chemical transmission between the hypogastric nerve ending and seminal vesicle in guinea-pig, *Jpn. J. Pharmacol.,* 23, 479, 1973.

95. **Hedqvist, P. and Fredholm, B. B.,** Effects of adenosine on adrenergic neurotransmission, prejunctional inhibition and post-junctional enhancement, *Naunyn-Schmiedebergs Arch. Pharmakol.,* 293, 217, 1976.

96. **Holck, M. I. and Marks, B. H.,** Purine nucleoside and nucleotide interactions on normal and subsensitive alpha adrenoceptor responsiveness in guinea-pig vas deferens, *J. Pharmacol. Exp. Ther.,* 205, 104, 1978.

97. **Ambache, N. and Zar, M. A.,** Non-cholinergic transmission by post ganglionic motor neurones in the mammalian bladder, *J. Physiol. (London),* 210, 761, 1970.
98. **Burnstock, G., Dumsday, B., and Smythe, A.,** Atropine resistant excitation of the urinary bladder: the possibility of transmission via nerves releasing a purine nucleotide, *Br. J. Pharmacol.,* 44, 451, 1972.
99. **Szent Gyorgyi, A.,** *The Chemistry of Muscular Contraction,* Academic Press, New York, 1947.
100. **Buchthal, F. and Kahlson, G.,** The motor effect of adenosine triphosphate and allied phosphorous compounds on smooth mammalian muscle, *Acta Physiol. Scand.,* 8, 325, 1944.
101. **Bennett, D. W. and Drury, A. N.,** Further observations relating to the physiological activity of adenosine compounds, *J. Physiol. (London),* 72, 288, 1931.
102. **Christie, J. and Satchell, D. G.,** Purine receptors in the trachea: is there a receptor for ATP?, *Br. J. Pharmacol.,* 70, 512, 1980.
103. **Smith, R. and Satchell, D. G.,** Adenosine causes contractions in spiral strips and relaxations in transverse strips of guinea-pig trachea, *Eur. J. Pharmacol.,* 101, 243, 1984.
104. **Fredholm, B., Brodkin, K., and Strandberg, K.,** On the mechanism of relaxation of tracheal muscle by theophylline and other cyclic nucleotide phosphodiesterase inhibitors, *Acta Pharmacol. Toxicol.,* 45, 336, 1979.
105. **Versprille, A.,** The influence of uridine nucleotides upon isolated frog hearts, *Pfluegers Arch.,* 177, 285, 1963.
106. **Flintney, F. W., Lamb, J. F., and Singh, J.,** The effects of ATP on the hypodynamic frog ventricle, *J. Physiol. (London),* 273, 50 P, 1977.
107. **Goto, M., Yatani, A., and Tsuda, Y.,** An analysis of the action of ATP and related compounds on membrane current and tension components of the bullfrog atrium, *Jpn. J. Physiol.,* 28, 611, 1977.
108. **Hoyle, C. V. and Burnstock, G.,** Evidence that ATP is a neurotransmitter in the frog heart, *Eur. J. Pharmacol.,* 124, 285, 1986.
109. **Burnstock, G. and Meghji, P.,** Distribution of P1 and P2 purinoceptors in the guinea-pig and frog heart, *Br. J. Pharmacol.,* 73, 879, 1981.
110. **Sydow, V. and Ahlquist, R. P.,** Cardiovascular actions of adenine compounds, *Fed. Proc., Fed. Am. Soc. Exp. Biol.,* 11, 395, 1953.
111. **Fischer, H. and Frohlicher, A.,** Isolated mammalian heart of cat and dog, *Experientia,* 4, 155, 1963.

II. Purine Receptors, Metabolism, Release, and Transport

Chapter 3

ADENOSINE RECEPTORS: LIGAND-BINDING STUDIES

Ulrich Schwabe

TABLE OF CONTENTS

I. INTRODUCTION

Adenosine receptors are integral membrane proteins in virtually all cell types and mediate the important physiological actions of adenosine. Much of the present information on adenosine receptors was originally achieved by adenylyl cyclase studies in which the relative potencies of agonists and antagonists were compared. Biochemical and pharmacological data suggest the existence of at least two subtypes of adenosine receptors which are coupled to different components of the adenylyl cyclase system.[1,2] The A_1 receptor inhibits adenylyl cyclase by interaction with a GTP-binding protein G_i, whereas the A_2 receptor acts via G_s to activate the enzyme. In addition to the cell-surface adenosine receptors, a so-called purine site (P-site) for adenosine is located on the cytoplasmic face of the catalytic subunit of the adenylyl cyclase, which mediates the inhibitory effects of high concentrations of adenosine. Recent data indicate that this secondary inhibitory site may be a physiological target for 3'-AMP and 2'-deoxy-3'-AMP.[3]

Over the last years evidence has been obtained that adenosine has effects that occur independently from cyclic AMP (cAMP). The best-characterized effector system is the potassium conductance in atrial cells and hippocampal neurons.[4-6] The receptor appears to be coupled to the channel via a pertussis toxin-sensitive G protein, termed G_K. Probably it is the same potassium channel which is activated by the muscarinic acetylcholine receptor, the so-called receptor-coupled potassium channel. Several other effector systems have been described which are modulated by A_1 adenosine receptors in a positive or negative manner. Adenosine inhibits voltage-sensitive calcium currents in hippocampus and other neuronal tissues.[7] The current may be carried via the N-type channel, which appears to be responsible for the calcium influx triggering transmitter release. Some authors believe that a special receptor subtype, a so-called A_3 receptor, is coupled to this calcium channel,[8] but the discussion about this matter is still controversial. Further effector systems appear to be coupled to the A_1 receptor, such as phospholipase C, which is either stimulated in rabbit cortical collecting tubule cells[9] or inhibited in GH_3 cells.[10] It has also been suggested that guanylyl cyclase activity in vascular smooth-muscle cells and a low K_m form of cAMP phosphodiesterase in fat cells may be modulated by A_1 adenosine receptors.[11,12]

An important factor contributing to the identification of adenosine receptors has been the observation that the methylxanthines theophylline and caffeine, in addition to the effects as phosphodiesterase inhibitors, are effective adenosine receptor antagonists.[13] However, they are virtually nonselective for A_1 and A_2 receptors[2,14,15] and thus have not been useful for categorizing adenosine receptor subtypes. More potent and highly selective antagonists were obtained by the synthesis of 8-substituted xanthine derivatives, such as the xanthine amine congener of 1,3-dipropyl-8-phenylxanthine (XAC) and the highly A_1-selective antagonist 8-cyclopentyl-1,3-dipropylxanthine (DPCPX).[16-18]

In addition, several structurally related classes of nonxanthine adenosine antagonists have been synthesized in recent years of which triazoloquinazolines, thiazoloquinolines, and triazoloquinoxalines appear to be the most promising compounds.[19-21] CGS 15943 is the most potent and the first A_2-selective adenosine receptor antagonist.[19] The 4-amino-[1,2,4]triazolo-[4,3-a]quinoxaline derivative CP-66713 had a high A_2 receptor affinity and a more than 450-fold A_2 selectivity.[21]

The development of selective agonists and antagonists enabled adenosine receptors to be studied by radioligand techniques. Labeling receptors directly with radioactive compounds with high specificity and specific activity has added a powerful tool to study drug affinity, receptor density, and kinetic constants. Perhaps most importantly, the receptor protein can be identified after solubilization and purification by affinity chromatography and other purification procedures. In this chapter, the recent progress in the use of radioligand binding studies to characterize adenosine receptors is summarized.

TABLE 1
Agonist Radioligands for A_1 Receptors

Radioligand	A_1 Affinity rat brain K_D (nM)	A_1 Selectivity A_2/A_1 (K_i/K_D)	Ref.
[³H]PIA	1.3	560	25, 29
[³H]CHA	1.3	390	24, 50
[³H]ClA	1.3	7	26, 50
[³H]CPA	0.5	2500	31, 29
[³H]CCPA	0.2	9750	32, 29
[¹²⁵I]HPIA	0.5		33
[¹²⁵I]ABA	1.9		34
[¹²⁵I]APNEA	2.0		35

II. A_1 ADENOSINE RECEPTORS

A. AGONIST RADIOLIGANDS

Radioligand techniques in brain membranes have first been used to characterize adenosine receptor sites. Early attempts using radioactively labeled adenosine were generally unsuccessful, since receptor binding could not always be separated from binding to metabolic enzymes and nonspecific binding of metabolic degradation products.[22,23] The first successful binding studies have been performed with the radiolabeled N^6-substituted adenosine derivatives [³H]CHA and [³H]PIA.[24,25] Less satisfactory results were obtained with [³H]2-chloroadenosine because of the probable instability of the ligand.[26,27] The binding of [³H]CHA and [³H]PIA was saturable, reversible, and displayed a high affinity for the membrane recognition sites in the nanomolar range. In addition, the binding was stereoselective, since R-PIA was 30- to 50-fold more potent than S-PIA in competing for the radioligand binding site. R-PIA was described as the first selective agonist for A_1 receptors[1] and is still considered a kind of standard agonist for this receptor subtype.

More recently, three additional compounds have been synthesized which are more potent and more selective than the original compounds. N^6-cyclopentyladenosine (CPA), 2-chloro-N^6-cyclopentyladenosine (CCPA), and S-N^6-(2-endo-norbornyl)adenosine (S-ENBA) have been reported to have higher potency and selectivity for A_1 adenosine receptors than R-PIA.[28-30] CCPA is 3- to 5-fold more potent at A_1 receptors than R-PIA and in addition has an almost 10,000-fold A_1 selectivity over the A_2 receptor.[29] S-ENBA has the same high affinity for A_1 receptors as CCPA, but is somewhat less A_1 selective.[30] [³H]CPA and [³H]CCPA have been introduced as highly selective radioligands for labeling A_1 receptors.[31,32] [³H]CCPA has a subnanomolar affinity and high selectivity for A_1 receptors, exhibiting virtually no affinity for A_2 receptors, and has proved to be a useful agonist radioligand to detect A_1 adenosine receptors also in tissues with very low receptor density.[32]

The next step was the introduction of radioiodinated compounds such as [¹²⁵I]R-N^6-p-hydroxyphenylisopropyladenosine ([¹²⁵I]HPIA), [¹²⁵I]N^6-p-aminobenzyladenosine ([¹²⁵I]ABA), and [¹²⁵I]N^6-2-(4-aminophenyl)ethyladenosine ([¹²⁵I]APNEA).[33-35] Iodinated ligands, which have a much higher specific radioactivity (2200 Ci/mmol) than tritiated compounds (20 to 100 Ci/mmol), permit the detection of A_1 receptors in smaller amounts of tissues or in tissues with very low receptor density. The agonist radioligands, which have been developed for labeling A_1 adenosine receptors, are listed in Table 1. In most cases A_1 affinity was determined by saturation experiments in rat brain membranes. A_1 selectivity vs. A_2 receptors was evaluated by the ratio A_2/A_1 (K_i/K_D), and the K_i values for A_2 receptors were determined by competition for radioligand binding of the unlabeled compounds to rat striatal membranes. At present, [³H]CCPA and [¹²⁵I]HPIA appear to be the most suitable radioligands for A_1

TABLE 2
Antagonist Radioligands for A_1 Receptors

Radioligand	A_1 Affinity rat brain K_D (nM)	Selectivity (A_2/A_1)	Nonspecific binding at K_D	Ref.
[³H]DPX	50	3	40%	24
[³H]XAC	1.2	7	20%	40
[³H]DPCPX	0.3	1100	1—3%	17, 18
[¹²⁵I]BW-A844U	0.4	1700	10%	41

receptors. Both can be used for the characterization of A_1 receptors in tissues with low receptor density and [³H]CCPA has the additional advantage of a very high A_1 selectivity.

The binding of agonist radioligands to A_1 receptors is influenced by interactions with the inhibitory G_i protein. Like other G protein-coupled receptors, A_1 adenosine receptors bind agonists with two distinct affinities. In the absence of guanine nucleotides (GTP and nonhydrolyzable analogs), A_1 receptors form agonist- and Mg^{2+}-stabilized receptor-G protein complexes which bind agonists with high affinity. GTP activates the G protein and converts the receptor to a low-affinity state for agonists, resulting in a 10- to 100-fold decrease of the potencies of agonists in competing for the receptor sites.[36,37] The high-affinity binding is also prevented by adenosine 5′-diphosphate ribosylation of the G protein alpha-subunit with pertussis toxin[38] or by alkylation with N-ethylmaleimide (NEM).[39] Agonist radioligands primarily label the high-affinity receptor sites since they are used in very low concentrations (0.5 to 1 nM) under usual experimental conditions.

B. ANTAGONIST RADIOLIGANDS

Several adenosine antagonists have also been radioactively labeled and used for the identification of A_1 receptors. 1,3-Diethyl-8-[³H]phenylxanthine ([³H]DPX) was the first antagonist radioligand used for the study of adenosine receptors.[24] However, it was not considered satisfactory as a radioligand because of unfavorable binding properties such as a relatively low affinity, high proportion of nonspecific binding, and low specific activity. Most of the limitations were overcome by the successful development of a [³H]xanthine amine congener of 1,3-dipropyl-8-phenylxanthine ([³H]XAC).[40] This compound had a 50-fold higher affinity for A_1 receptors than [³H]DPX. Further progress was obtained with the introduction of [³H]DPCPX and the first radioiodinated antagonist radioligand [¹²⁵I]3-(4-amino)phenethyl-1-propyl-8-cyclopentylxanthine ([¹²⁵I]BW-A844U).[17,18,41] As can be seen from Table 2, [³H]DPCPX and [¹²⁵I]BW-A844U are very potent A_1 antagonists with affinities in the subnanomolar range. Furthermore, both are specific antagonist ligands for A_1 receptors with an A_1 selectivity of 1100 and 1700 vs. A_2 receptors in rat striatum. [³H]DPCPX has a very low proportion of nonspecific binding (1 to 3%) and is now widely used as the preferred radioligand for the study of A_1 adenosine receptors. The high affinity and low nonspecific binding allowed the detection of A_1 receptors in tissues with low receptor densities such as the heart.[18,42] It should also be mentioned that [³H]DPCPX is the first radioligand suitable for the characterization of A_1 receptors in intact cells such as ventricular myocytes.[42]

Antagonist radioligands bind with similar affinity to both high- and low-affinity sites for agonists without differentiating between the two states. Therefore, the total receptor number is more easily estimated from antagonist binding, because it may be difficult to detect the low-affinity binding of agonist radioligands due to high levels of nonspecific binding which occur at the high agonist concentrations required to saturate these sites. On the other hand, the two affinity states for agonists are observed in competition experiments for the displacement of antagonist radioligands from A_1 receptors by agonists. The com-

TABLE 3
Photoaffinity and Affinity Ligands for A_1 Adenosine Receptors

Photoaffinity ligand	A_1 Affinity rat brain K_D (nM)	Incorporation	M_r value (kDa)	Ref.
Agonist Ligands				
[^{125}I]AHPIA	1.5	35%	35	46
[^{125}I]AZABA	1.1		36	47
[^{125}I]AZPNEA	2.0	25%	38	48
Antagonist Ligands				
[^{125}I]Azido-BW-A844U	0.4	4%	34	41
[^{3}H]m-DITC-XAC	27	90%	38	49

petition curves for agonists are shallow with apparent Hill coefficients substantially less than one (n_H 0.5 to 0.75), suggesting two states of different affinities. Guanine nucleotides shift the agonist competition curve to the right and steepen it with a conversion of all A_1 receptors to a low-affinity state.[18,37,38,43,44] Several studies have shown that guanine nucleotides increase antagonist radioligand binding.[39,43-45] The effects are more pronounced in solubilized than in membrane-bound receptors. It is suggested that uncoupling of A_1 receptors from G_i by guanine nucleotides or by inactivation of G_i with NEM results in an increased antagonist binding.

C. PHOTOAFFINITY LIGANDS

Another approach to study the properties of the membrane-bound receptors is photoaffinity labeling. Appropriate photoaffinity ligands can be radioiodinated and contain a photoactivatable group which, upon UV irradiation, leads to covalent incorporation of the drug into the ligand-binding unit of the receptor. Incorporation is detected by SDS-gel electrophoresis followed by autoradiography. Most of the photoaffinity probes for A_1 receptors have been developed by introduction of an azido moiety into adenosine derivatives with high affinity and selectivity for this receptor subtype. [^{125}I]R-2-azido-N^6-p-hydroxyphenylisopropyladenosine ([^{125}I]AHPIA) is a photolabile derivative of R-HPIA in which an azido group has been substituted at the 2 position of the purine ring.[46] A similar approach led to the development of [^{125}I]N^6-(4-azido-3-iodobenzyl)adenosine ([^{125}I]AZBA) and [^{125}I]N^6-2-(4-azido-3-iodophenyl)ethyladenosine ([^{125}I]AZPNEA), both of which have been synthesized from the parent compounds ABA and APNEA and possess a functional 4-azido group at the benzyl or phenyl ring, respectively.[47,48] In addition, the first antagonist photoaffinity label [^{125}I]azido-BW-A844U was synthesized from the corresponding 3-arylamine derivative of DPCPX.[41] All these compounds interacted with A_1 receptors with high affinity (K_D values 0.4 to 2 nM) and were specifically photoincorporated into the ligand-binding subunit of the A_1 receptor with M_r values of 34 to 38 kDa (Table 3). The highest efficiency of photolabeling was achieved with [^{125}I]AHPIA (35%), whereas the antagonist ligand [^{125}I]azido-BW-A844U resulted in a much lower efficiency (4%). Furthermore, the affinity label DITC-XAC has been developed,[49] which contains a highly reactive chemical group and was covalently incorporated into the A_1 receptor with high efficiency (90%).

III. A_2 ADENOSINE RECEPTORS

A. AGONIST RADIOLIGANDS

Until recently, only a few selective agonists for A_2 adenosine receptors have been identified. 5'-N-ethylcarboxamido adenosine (NECA) was originally proposed as a proto-

TABLE 4
Radioligands for A_2 Receptors

Radioligand	A_2 Affinity rat striatum K_D (nM)	A_2 Selectivity A_1/A_2 (K_i/K_D)	Ref.
Agonist Ligands			
[³H]NECA	10	0.6	50
[³H]CGS 21680	15.5	170	55
[¹²⁵I]PAPA-APEC	1.5	>500	57
Antagonist Ligands			
[³H]XAC	12	0.2	58
[³H]PD 115,199	2.6	5	59

typical agonist for A_2 receptors,[2] but it has a higher affinity for A_1 receptors in many tissues and therefore is not A_2 selective.[50] The first attempts to use [³H]NECA as a radioligand in liver and platelet membranes were unsuccessful because the major part of the [³H]NECA binding was to nonreceptor binding sites of high density which did not agree with the pharmacology of A_2 receptors.[51,52] A considerable improvement was achieved when [³H]NECA binding in rat striatum was studied in the presence of 50 nM CPA, which specifically eliminated the A_1 receptor component of the binding. Under this condition, [³H]NECA bound with high affinity (10 nM) to the striatal A_2 receptor and showed a structure-activity profile consistent with a labeling of the so-called high-affinity A_{2a} receptors.[50] [³H]NECA has also been employed for a successful characterization of A_2 receptors in the PC 12 pheochromocytoma cell line.[53] In human platelets, [³H]NECA could only be used as a radioligand when a low-affinity, nonreceptor, NECA-binding protein was separated from the A_2 receptor by gel chromatography.[54]

Recently, the highly A_2-selective agonist radioligand [³H]CGS 21680 has been introduced.[55] It is a NECA derivative with a large substituent at the 2 position of the purine moiety, 2-[p-(2-carboxyethyl)-phenethylamino]-5'-N-ethylcarboxamido adenosine. [³H]CGS 21680 bound with high affinity (K_D 15.5 nM) to striatal membranes with negligible specific binding obtained in rat cortical membranes. The high degree of A_2 selectivity (170-fold) and the high affinity make this newly developed NECA derivative the current ligand of choice for the characterization of high-affinity A_{2a} receptors (Table 4). CGS 21680 has no intrinsic activity to stimulate cAMP formation in rat hippocampus[56] and therefore is probably not a suitable radioligand for the characterization of the low-affinity A_{2b} receptor.

CGS 21680 has also been used to synthesize an iodinated radioligand for the A_2 receptor.[57] [¹²⁵I]PAPA-APEC was obtained by a functionalized congener approach and had a high affinity of 1.5 nM. Using this ligand and photoaffinity cross linking with the succinimidyl derivative SANPAH, the A_2 receptor binding subunit was identified as a peptide with an apparent molecular weight of 45 kDa.

B. ANTAGONIST RADIOLIGANDS

A_2 receptors have also been characterized using the antagonist radioligands [³H]XAC and [³H]N-[2-(dimethylamino)ethyl]-N-methyl-4-(2,3,6,7-tetrahydro-2,6-dioxo-1,3-dipropyl-1H-purine-8-yl)benzenesulfonamide ([³H]PD 115,199).[58,59] [³H]XAC is a nonselective radioligand and appears to bind with sufficient affinity (K_D 12 nM) to A_2 receptors of human platelet membranes. However, the binding is associated with a high proportion (75%) of nonspecific binding, which limits the practical use of this ligand.

[³H]PD 115,199 has been used for the characterization of high-affinity A_{2a} receptors in rat striatal membranes.[59] Specific labeling of the A_{2a} receptor can be observed when an appropriate concentration of the A_1-selective antagonist DPCPX is included in the binding reaction. The binding of this antagonist radioligand to rat brain membranes indicates a pharmacological profile and regional distribution similar to that obtained with [³H] CGS 21680 and is consistent with a specific labeling of the high-affinity A_{2a} receptor. Binding of [³H]PD 115,199 to the low-affinity A_{2b} receptors could not be detected.

IV. CONCLUSIONS

The techniques presented in this chapter indicate the usefulness of radioligand binding studies in investigation of the function and structure of adenosine receptors in the brain and peripheral systems. The data clearly support the existence of two different subtypes of adenosine receptors, the A_1 receptor with an apparent molecular weight of 34 kDa and the striatal A_2 receptor of 45 kDa.

The pharmacological profile of the A_1 receptor is best defined by the agonist potency order of CCPA >R-PIA >NECA >CGS 21680 and that of the A_2 receptor by NECA >CGS 21680 >R-PIA >CCPA. At present, the antagonist [³H]DPCPX appears to be the most suitable radioligand for labeling A_1 receptors. It has a high degree of A_1 selectivity, a low level of nonspecific binding, and a high specific radioactivity. [³H]DPCPX recognizes both affinity states of the receptor for agonists and was successfully used for binding studies in intact cells. The agonist radioligand [³H]CCPA has the highest affinity and selectivity for A_1 receptors and may be of value for the study of special problems which require agonist binding.

The characterization of A_2 adenosine receptors was considerably improved when [³H]CGS 21680 was introduced for the labeling of the high-affinity A_{2a} receptor in striatum. This radioligand is probably not useful for the characterization of the low-affinity A_{2b} receptor. [³H]NECA can only be used in A_2-selective cellular systems such as platelets after removal of the nonreceptor NECA-binding protein.

REFERENCES

1. **van Calker, D., Müller, M., and Hamprecht, B.,** Adenosine inhibits the accumulation of cyclic AMP in cultured brain cells, *Nature,* 276, 839, 1978.
2. **Londos, C., Cooper, D. M. F., and Wolff, J.,** Subclasses of external adenosine receptors, *Proc. Natl. Acad. Sci. U.S.A.,* 77, 2551, 1980.
3. **Johnson, R. A., Yeung, S.-M. H., Bushfield, M., Stübner, D., and Shoshani, I.,** Physiological and biochemical aspects of "P"-site-mediated inhibition of adenylyl cyclase, in *Purines in Cellular Signaling, Targets for New Drugs,* Jacobson, K. A., Daly, J. W., and Manganiello, V., Eds., Springer-Verlag, New York, 1990, 158.
4. **Belardinelli, L. and Isenberg, G.,** Isolated atrial myocytes: adenosine and acetylcholine increase potassium conductance, *Am. J. Physiol.,* 244, H734, 1983.
5. **Böhm, M., Brückner, R., Neumann, J., Schmitz, W., Scholz, H., and Starbatty, J.,** Role of guanine nucleotide-binding protein in the regulation by adenosine of cardiac potassium conductance and force of contraction. Evaluation with pertussis toxin, *Naunyn-Schmiedeberg's Arch. Pharmacol.,* 332, 403, 1986.
6. **Trussell, L. O. and Jackson, M. B.,** Adenosine-activated potassium conductance in cultured striatal neurons, *Proc. Natl. Acad. Sci. U.S.A.,* 82, 4857, 1985.
7. **Fredholm, B. B. and Dunwiddie, T. V.,** How does adenosine inhibit transmitter release?, *Trends Pharmacol. Sci.,* 9, 130, 1988.
8. **Ribeiro, J. A. and Sebastiao, A. M.,** Adenosine receptors and calcium: basis for proposing a third (A_3) adenosine receptor, *Proc. Neurobiol.,* 26, 179, 1986.

9. **Arend, L. J., Handler, J. S., Rhim, J. S., Gusovsky, F., and Spielman, W. S.**, Adenosine-sensitive phosphoinositide turnover in a newly established renal cell line, *Am. J. Physiol.*, F1067, 1989.

10. **Linden, J. and Delahunty, T. M.**, Receptors that inhibit phosphoinositide breakdown, *Trends Pharmacol. Sci.*, 10, 114, 1989.

11. **Kurtz, A.**, Adenosine stimulates guanylate cyclase activity in vascular smooth muscle cells, *J. Biol. Chem.*, 262, 6296, 1987.

12. **Elks, M. L., Jackson, M., Manganiello, V. C., and Vaughan, M.**, Effect of N^6-(L-2-phenylisopropyl)adenosine and insulin on cAMP metabolism in 3T3-L1 adipocytes, *Am. J. Physiol.*, 252, C342, 1987.

13. **Sattin, A. and Rall, T. W.**, The effect of adenosine and adenine nucleotides on the cyclic adenosine $3',5'$-phosphate content of guinea pig cerebral cortex slices, *Mol. Pharmacol.*, 6, 13, 1970.

14. **Bruns, R. F., Daly, J. W., and Snyder, S. H.**, Adenosine receptor binding: structure-activity analysis generates extremely potent xanthine antagonists, *Proc. Natl. Acad. Sci. U.S.A.*, 80, 2077, 1983.

15. **Schwabe, U., Ukena, D., and Lohse, M. J.**, Xanthine derivatives as antagonists at A_1 and A_2 adenosine receptors, *Naunyn-Schmiedeberg's Arch. Pharmacol.*, 330, 212, 1985.

16. **Jacobson, K. A., Kirk, K. L., Padgett, W. L., and Daly, J. W.**, A functionalized congener approach to adenosine receptor antagonists: amino acid conjugates of 1,3-dipropylxanthine, *Mol. Pharmacol.*, 29, 126, 1985.

17. **Bruns, R. F., Fergus, J. H., Badger, E. W., Bristol, J. A., Santay, L. A., Hartman, J. D., Hays, S. J., and Huang, C. C.**, Binding of the A_1-selective adenosine antagonist 8-cyclopentyl-1,3-dipropylxanthine to rat brain membranes, *Naunyn-Schmiedeberg's Arch. Pharmacol.*, 335, 59, 1987.

18. **Lohse, M. J., Klotz, K. N., Lindenborn-Fotinos, J., Reddington, M., Schwabe, U., and Olsson, R. A.**, 8-Cyclopentyl-1,3-dipropylxanthine (DPCPX)—a selective high affinity antagonist radioligand for A_1 adenosine receptors, *Naunyn-Schmiedeberg's Arch. Pharmacol.*, 336, 204, 1987.

19. **Williams, M., Francis, J., Ghai, G., Braunwalder, A., Psychoyos, S., Stone, G. A., and Cash, W. D.**, Biochemical characterization of the triazoloquinazoline, CGS 15943, a novel, nonxanthine adenosine antagonist, *J. Pharmacol. Exp. Ther.*, 241, 415, 1987.

20. **Bruns, R. F., Davis, R. E., Ninteman, F. W., Poschel, B. P. H., Wiley, J. N., and Heffner, T. G.**, Adenosine antagonists as pharmacological tools, in *Adenosine and Adenine Nucleotides: Physiology and Pharmacology*, Paton, D. M., Ed., Taylor & Francis, London, 1988, 39.

21. **Sarges, R., Howard, H. R., Browne, R. G., and Koe, B. K.**, 4-amino-[1,2,4]triazolo[4,3-a]quinoxalines—highly selective adenosine antagonists and potential antidepressants, in *Purines in Cellular Signaling, Targets for New Drugs*, Jacobson, K. A., Daly, J. W., and Manganiello, V., Eds., Springer-Verlag, New York, 1989, 417.

22. **Schwabe, U., Kiffe, H., Puchstein, C., and Trost, T.**, Specific binding of ^{3}H-adenosine to rat brain membranes, *Naunyn-Schmiedeberg's Arch Pharmacol.*, 310, 59, 1979.

23. **Schwabe, U.**, Direct binding studies of adenosine receptors, *Trends Pharmacol. Sci.*, 2, 299, 1981.

24. **Bruns, R. F., Daly, J. W., and Snyder, S. H.**, Adenosine receptors in brain membranes: binding of N^6-cyclohexyl[^{3}H]adenosine and 1,3-diethyl-8-[^{3}H]phenylxanthine, *Proc. Natl. Acad. Sci. U.S.A.*, 77, 5547, 1980.

25. **Schwabe, U. and Trost, T.**, Characterization of adenosine receptors in rat brain by (-) [^{3}H]N^6-phenylisopropyladenosine, *Naunyn-Schmiedeberg's Arch. Pharmacol.*, 313, 179, 1980.

26. **Williams, M. and Risley, E. A.**, Biochemical characterization of putative central purinergic receptors by using 2-chloro[^{3}H]-adenosine, a stable analog of adenosine, *Proc. Natl. Acad. Sci. U.S.A.*, 77, 6892, 1980.

27. **Wu, P. H., Phillis, J. W., Balls, K., and Rinaldi, B.**, Specific binding of 2-[^{3}H]chloroadenosine to rat brain cortical membranes, *Can. J. Physiol. Pharmacol.*, 58, 576, 1980.

28. **Moos, W. H., Szotek, D. S., and Bruns, R. F.**, N^6-Cycloadenosines. Potent A_1-selective adenosine agonists, *J. Med. Chem.*, 28, 1383, 1985.

29. **Lohse, M. J., Klotz, K. N., Schwabe, U., Cristalli, G., Vitorri, S., and Grifantini, M.**, 2-Chloro-N^6-cyclopentyladenosine: a highly selective agonist at A_1 adenosine receptors, *Naunyn-Schmiedeberg's Arch. Pharmacol.*, 337, 687, 1988.

30. **Trivedi, B. K., Bridges, A. J., Patt, W. C., Priebe, S. R., and Bruns, R. F.**, N^6-Bicycloalkyladenosines with unusually high potency and selectivity for the adenosine A_1 receptor, *J. Med. Chem.*, 32, 8, 1989.

31. **Williams, M., Braunwalder, A., and Erickson, T. J.**, Evaluation of the binding of the A-1 selective adenosine radioligand, cyclopentyladenosine (CPA), to rat brain tissue, *Naunyn-Schmiedeberg's Arch. Pharmacol.*, 332, 179, 1986.

32. **Klotz, K. N., Lohse, M. J., Schwabe, U., Cristalli, G., Vitorri, S., and Grifantini, M.**, 2-Chloro-N^6-[^{3}H]cyclopentyladenosine ([^{3}H]CCPA)—a high affinity agonist radioligand for A_1 adenosine receptors, *Naunyn-Schmiedeberg's Arch. Pharmacol.*, 340, 679, 1989.

33. **Schwabe, U., Lenschow, V., Ukena, D., Ferry, D. R., and Glossmann, H.**, [^{125}I]N^6-p-Hydroxyphenylisopropyladenosine, a new ligand for R_i adenosine receptors, *Naunyn-Schmiedeberg's Arch. Pharmacol.*, 321, 84, 1982.

34. **Linden, J., Patel, A., and Sadek, S.,** [^{125}I]Aminobenzyladenosine, a new radioligand with improved specific binding to adenosine receptors in heart, *Circ. Res.,* 56, 279, 1985.

35. **Stiles, G. L., Daly, D. T., and Olsson, R. A.,** The A_1 adenosine receptor, identification of the binding subunit by photoaffinity cross-linking, *J. Biol. Chem.,* 260, 10806, 1985.

36. **Goodman, R. R., Cooper, M. J., Gavish, M., and Snyder, S. H.,** Guanine nucleotide and cation regulation of the binding of [^{3}H]cyclohexyladenosine and [^{3}H]diethylphenylxanthine to adenosine A_1 receptors in brain membranes, *Mol. Pharmacol.,* 21, 329, 1982.

37. **Lohse, M. J., Lenschow, V., and Schwabe, U.,** Two affinity states of R_i adenosine receptors in brain membranes, analysis of guanine nucleotide and temperature effects on radioligand binding, *Mol. Pharmacol.,* 26, 1, 1984.

38. **Ramkumar, V. and Stiles, G. L.,** Reciprocal modulation of agonist and antagonist binding to A_1 adenosine receptors by guanine nucleotides is mediated via a pertussis toxin-sensitive G protein, *J. Pharmacol. Exp. Ther.,* 246, 1194, 1988.

39. **Yeung, S.-M. and Green, R. D.,** [^{3}H]5'-N-ethylcarboxamide adenosine binds to both R_a and R_i adenosine receptors in rat striatum, *Naunyn-Schmiedeberg's Arch. Pharmacol.,* 325, 218, 1984.

40. **Jacobson, K. A., Ukena, D., Kirk, K. L., and Daly, J. W.,** [^{3}H]Xanthine amine congener of 1,3-dipropyl-8-phenylxanthine: an antagonist radioligand for adenosine receptors, *Proc. Natl. Acad. Sci. U.S.A.,* 83, 4089, 1986.

41. **Patel, A., Craig, R. H., Daluge, S. M., and Linden, J.,** ^{125}I-BW-A844U, an antagonist radioligand with high affinity and selectivity for adenosine A_1 receptors, and ^{125}I-azido-BW-A844U, a photoaffinity label, *Mol. Pharmacol.,* 33, 585, 1988.

42. **Martens, D., Lohse, M. J., and Schwabe, U.,** [^{3}H]-8-Cyclopentyl-1,3-dipropylxanthine binding to A_1 adenosine receptors of intact rat ventricular myocytes, *Circ. Res.,* 63, 613, 1988.

43. **Stiles, G. L.,** A_1 Adenosine receptor-G protein coupling in bovine brain membranes: effects of guanine nucleotides, salt, and solubilization, *J. Neurochem.,* 51, 1592, 1988.

44. **Ströher, M., Nanoff, C., and Schütz, W.,** Differences in the GTP-regulation of membrane-bound and solubilized A_1 adenosine receptors, *Naunyn-Schmiedeberg's Arch. Pharmacol.,* 340, 87, 1989.

45. **Klotz, K. N., Keil, R., Zimmer, F. J., and Schwabe, U.,** Guanine nucleotide effects on 8-cyclopentyl-1,3-[^{3}H]dipropylxanthine binding to membrane-bound and solubilized A_1 adenosine receptors of rat brain, *J. Neurochem.,* 54, 1988, 1990.

46. **Klotz, K. N., Cristalli, G., Grifantini, M., Vitorri, S., and Lohse, M. J.,** Photoaffinity labeling of A_1 adenosine receptors, *J. Biol. Chem.,* 260, 14659, 1985.

47. **Choca, J. I., Kwatra, M. M., Hosey, M. M., and Green, R. D.,** Specific photoaffinity labeling of inhibitory adenosine receptors, *Biochem. Biophys. Res. Commun.,* 131, 115, 1985.

48. **Stiles, G. L., Daly, D. T., and Olsson, R. A.,** Characterization of the A_1 adenosine receptor-adenylate cyclase system of cerebral cortex using an agonist photoaffinity ligand, *J. Neurochem.,* 47, 1020, 1986.

49. **Stiles, G. L. and Jacobson, K. A.,** High affinity acylating antagonists for the A_1 adenosine receptor: identification of binding subunit, *Mol. Pharmacol.,* 34, 724, 1988.

50. **Bruns, R. F., Lu, G. H., and Pugsley, T. A.,** Characterization of the A_2 adenosine receptor labeled by [^{3}H]NECA in rat striatal membranes, *Mol. Pharmacol.,* 29, 331, 1986.

51. **Schütz, W., Tuisl, E., and Kraupp, O.,** Adenosine receptor agonists: binding and adenylate cyclase stimulation in rat liver plasma membranes, *Naunyn-Schmiedeberg's Arch. Pharmacol.,* 319, 34, 1982.

52. **Hüttemann, E., Ukena, D., Lenschow, V., and Schwabe, U.,** R_a Adenosine receptors in human platelets, *Naunyn-Schmiedeberg's Arch. Pharmacol.,* 325, 226, 1984.

53. **Williams, M., Abreu, M., Jarvis, M. F., and Noronha-Blob, L.,** Characterization of adenosine receptors in the PC 12 pheochromocytoma cell line using radioligand binding: evidence for A-2 selectivity, *J. Neurochem.,* 48, 498, 1987.

54. **Lohse, M. J., Elger, B., Lindenborn-Fotinos, J., Klotz, K. N., and Schwabe, U.,** Separation of solubilized A_2 adenosine receptors of human platelets from non-receptor [^{3}H]NECA binding sites by gel filtration, *Naunyn-Schmiedeberg's Arch. Pharmacol.,* 337, 64, 1988.

55. **Jarvis, M. F., Schulz, R., Hutchison, A. J., Do, U. H., Sills, M. A., and Williams, M.,** [^{3}H]CGS 21680, a selective A_2 adenosine receptor agonist directly labels A_2 receptors in rat brain, *J. Pharmacol. Exp. Ther.,* 251, 888, 1989.

56. **Lupica, C. R., Cass, W. A., Zahniser, N. R., and Dunwiddie, T. V.,** Effects of the selective adenosine A_2 receptor agonist CGS 21680 on in vitro electrophysiology, cAMP formation and dopamine release in rat hippocampus and striatum, *J. Pharmacol. Exp. Ther.,* 252, 1134, 1989.

57. **Barrington, W. W., Jacobson, K. A., Hutchison, A. J., Williams, M., and Stiles, G. L.,** Identification of the A_2 adenosine receptor binding subunit by photoaffinity crosslinking, *Proc. Natl. Acad. Sci. U.S.A.,* 86, 6572, 1989.

58. **Ukena, D., Jacobson, K. A., Kirk, K. L., and Daly, J. W.,** A [^{3}H]amine congener of 1,3-dipropyl-8-phenylxanthine, a new radioligand for A_2 adenosine receptors of human platelets, *FEBS Lett.,* 199, 269, 1986.

59. **Bruns, R. F., Fergus, J. H., Badger, E. W., Bristol, J. A., Santay, L. A., and Hays, S. J.,** PD 115,199: An antagonist ligand for adenosine A_2 receptors, *Naunyn-Schmiedeberg's Arch. Pharmacol.,* 335, 64, 1987.

Chapter 4

ADENOSINE RECEPTORS, ADENYLATE CYCLASE; RELATIONSHIPS TO PHARMACOLOGICAL ACTIONS OF ADENOSINE

Richard D. Green

TABLE OF CONTENTS

I. INTRODUCTION

Although numerous physiological/pharmacological effects of adenosine have been appreciated for several decades, the realization that these effects are receptor mediated occurred much more recently. It was perhaps because this realization occurred within the context of the effects of adenosine on cAMP levels that adenosine receptors (AdoRs) and the modulation of cAMP metabolism became closely associated. Within a historical context it is important to remember that the first report of a link between adenosine and cAMP was the report that adenosine elevates the cAMP content of slices of cerebral cortex, and that initially it was not clear if this elevation was due to the stimulation of adenylate cyclase or to the inhibition of a cyclic nucleotide phosphodiesterase (PDE).[1] Adenylate cyclase assays performed by Londos et al.[2] lead to the hypothesis of the existence of multiple types of extracellular AdoRs which mediate either the inhibition or the stimulation of adenylate cyclase. These receptors were termed R_i and R_a receptors, respectively.[2] An identical hypothesis was advanced by Van Calker et al.[3] based on experiments in which cAMP levels were measured. The latter workers coined the terms A_1 and A_2, where $A_1 \sim R_i$ and $A_2 \sim R_a$.

The first ligand binding studies on AdoRs postdated the aforementioned studies. Correlations of structure-activity relationships for the binding of different radioligands and for the antagonism of this binding are interpreted within the context of the receptor classifications developed in the experiments on cAMP levels/adenylate cyclase. Thus, it is concluded that the agonist radioligands such as R-phenylisopropyl adenosine ([³H]R-PIA), N^6-cyclohexyl adenosine ([³H]CHA), and [¹²⁵I]aminobenzyladenosine ([¹²⁵I]ABA), along with the antagonist radioligands [³H]8-cyclopentyl-1,3-dipropylxanthine ([³H]CPX), [³H]xanthine amine congener ([³H]XAC), and [¹²⁵I]3-(4 amino-3-iodo)phenylethyl-8-cyclopentyl-1-propyl xanthine ([¹²⁵I]BW-A844U) bind to A_1 AdoRs, while the agonists 5'-*n*-ethylcarboxamido adenosine ([³H]NECA, under restricted conditions) and [³H]CGS 21680 bind to A_2 AdoRs (see Chapter 3 in this volume).

It has been evident for some time that AdoRs do things other than couple to adenylate cyclase. It is because of this that the R_i, R_a nomenclature has fallen into disuse and most workers use the A_1, A_2 nomenclature. It is now clear that both the central nervous system (CNS)[4,5] and certain areas of the heart including the atria[6,7] and the sinoatrial (SA) node[8] contain AdoRs that couple to a K^+ channel. Kurtz[9] has presented convincing evidence that vascular smooth muscle contains AdoRs that couple to the stimulation of guanylate cyclase. There are several reports suggesting the presence of AdoRs that mediate the inhibition of PI turnover.[10] In addition, it now appears that AdoRs can lower cAMP levels by a second mechanism, stimulation of a membrane-bound PDE.[11,12] The AdoRs that mediate the effects of K^+ channels, guanylate cyclase, PDE, and PI turnover have all been classified as A_1 receptors.

We are therefore left with two basic possibilities: (1) there are multiple subtypes of "A_1" AdoRs or (2) A_1 AdoRs couple to more than one type of G protein and in doing so couple to different effectors. These possibilites are, of course, not mutually exclusive. The purpose of the present review is to try to draw together the state of our present knowledge of the relationships between the physiological effects of AdoR stimulation, cAMP metabolism, and AdoRs as defined by ligand binding. Because of space constraints the discussion will be limited to three tissues.

II. SPECIFIC TISSUES/ORGAN SYSTEMS

A. ADIPOSE TISSUE

The ability of adenosine to modulate cAMP levels and lipolysis in isolated fat cells has been long recognized.[13] Indeed, it is now known that large variations in the "basal" rate of lipolysis in isolated fat cells is largely due to variations in the amount of "contaminating" adenosine which is continually released from the fat cells and must be removed by the addition of adenosine deaminase in order to stabilize "ligand-free" rates of lipolysis.[14]

There has perhaps been more work characterizing AdoRs in fat cells than in any other cell type or tissue. Of obvious advantage is the fact that highly pure preparations of intact fat cells or fat-cell membranes are relatively easily obtainable. This contrasts to preparations of cerebral cortex or the heart which are usually composed of multiple cell types and membranes of various subcellular origin. The report of Trost from Schwabe's[15] laboratory in 1981 on the binding of [^{3}H]R-PIA to fat-cell membranes detailed one of the first ligand binding studies on AdoRs. Work of Londos et al.[2] reported about the same time established that the cAMP-lowering effect of adenosine analogs in isolated fat cells correlates with their ability to inhibit adenylate cyclase activity in fat-cell membranes and with their antilipolytic effect in intact isolated fat cells. More recent work suggests the adenosine analogs may also act at AdoRs in fat cells to stimulate a membrane-bound cyclic nucleotide phosphodiesterase.[11,12] Thus, the cAMP-lowering effect of AdoR agonists in fat cells may be due to both the inhibition of adenylate cyclase and the stimulation of a PDE.

Honnor et al.[14,16] measured A-kinase activity ratios in rat fat-cell extracts (activity of cAMP-dependent protein kinase $\pm$ cAMP) and interpreted these values as indicators of intracellular concentrations of cAMP. The correlation between A-kinase activity ratios and the rate of lipolysis when activity ratios varied between ~ 0.05 and 0.35 (in response to stimulators and inhibitors of lipolysis) was excellent; lipolysis was maximally stimulated at higher A-kinase activity ratios.[16]

While AdoR stimulation may affect lipolysis by other mechanisms, there can be little doubt of a causal relationship between the cAMP-lowering and antilipolytic effects of adenosine analogs in isolated adipocytes. However, this correlation does not directly address the physiological relevance of adenosine in the control of lipolysis. There have been several studies implicating a role of AdoR dysfunction in obesity. Perhaps the most exciting is the recent study of Vannucci et al.[17] These authors measured cAMP levels and lipolysis in adipocytes isolated from obese and lean Zucker rats. Both the rate of lipolysis and the cAMP level in the ligand-free state (+ adenosine deaminase) were depressed in the adipocytes from the obese rats. Interestingly, the addition of the AdoR antagonist 8-phenyltheophylline, or treatment with pertussis toxin, which blocks interactions between receptors and G proteins of the G_i/G_o family,[18] largely eliminated these differences. As suggested by the authors, these results are indicative of an alteration in A_1 AdoR-mediated inhibition of adenylate cyclase in the adipocytes of the obese rats. Specifically, there appeared to be an AdoR:G protein-mediated decrease in cAMP (inhibition of adenylate cyclase plus, perhaps, stimulation of a PDE) in the absence of an AdoR agonist. Such a possibility is in accord with the hypotheses that A_1 AdoRs are coupled to a G protein in the absence of an effector and that antagonists may bind to free receptors with high affinity and destabilize "precoupled" receptors with the concomitant expression of "negative efficacy".[19] Whether or not this is the case, this and related studies on adipocytes from aged rats[20] and obese mice[21] give support to the hypothesis that A_1 AdoRs on fat cells play a physiological role in the control of lipolysis. It is also interesting to note that Honnor et al.[14] observed seasonal variations in the A-kinase activity ratio and in the sensitivity of adipocytes to the antilipolytic effect of R-PIA. Inspection of these data and those reported for the adipocytes of lean and fat Zucker rats[17] suggest that adipocytes from "winter rats" behave similarly to those from fat Zucker rats, while adipocytes from "spring rats" behave similarly to adipocytes from lean Zucker rats. It would be interesting to determine the effects of 8-phenyltheophylline on A-kinase activity ratios in adipocytes isolated from winter and spring rats. One would predict that 8-phenyltheophylline might elevate the A-kinase activity ratios in the adipocytes from winter rats.

The correlation between A-kinase activity ratios and lipolysis in the presence of various lipolytic and antilipolytic agents previously discussed did not include data obtained in the presence of insulin. While the antilipolytic effect of insulin does correlate with A-kinase activity ratios at low (<0.2) activity ratios, at higher activity ratios the antilipolytic effect of insulin cannot be accounted for by concomitant decreases in the A-kinase activity ratio.[22] The finding that the potency of insulin for eliciting its "cAMP-independent" lipolytic effect varies with the occupation of the AdoRs suggests that these receptors may link to processes in adipocytes other than adenylate cyclase.[22]

That AdoRs in adipocytes couple to a process in a cAMP-independent fashion is now clearly established. Receptors that couple to G_s inhibit insulin-stimulated glucose transport, while receptors that couple to G_i (including A_1 receptors) increase insulin-stimulated glucose transport. While these agents affect A-kinase activity ratios under some conditions, under other conditions A_1 agonists (and other agents that act via G_s or G_i) have no effect on the A-kinase activity ratio while markedly affecting glucose transport.[23]

In summary, it appears that AdoRs in fat cells control cAMP levels via both the inhibition of adenylate cyclase and the stimulation of a membrane-bound PDE. In addition, fat-cell AdoRs appear to be able to modulate the effect of insulin on glucose transport in a cAMP-independent fashion. Studies on adipocytes from genetically obese rodents suggest that fat cell AdoRs play a physiological role (and perhaps a pathologic role) in the control of lipolysis. It is likely that the effect of AdoR activation on cAMP metabolism in fat cells is an important underlying mechanism in this regard.

B. CARDIAC TISSUE

It is well established that adenosine and adenosine analogs are depressant to most, if not all, aspects of cardiac function.[24] Most workers categorize the effects of adenosine on the myocardium as either direct or indirect. The former category is composed of effects of adenosine unrelated to cyclic nucleotide metabolism. The negative chronotropic effect on SA node cells[8] and the negative inotropic effect on atrial myocytes[6,7] are examples of direct effects. (The latter effect also involves a cAMP-dependent mechanism under some conditions, i.e., when cAMP is elevated.)[24] Both of these effects of AdoR activation are mediated by a G protein and involve the same K^+ channel that mediates the negative inotropic and chronotropic effects of ACh in the atria.[7] This pathway is not operative in ventricular tissue.

Indirect effects, sometimes called antiadrenergic effects, of adenosine are those involving cyclic nucleotide metabolism. These effects are not confined to atrial tissue, i.e., indirect effects are manifest in both the atria and the ventricles. A scheme summarizing the effects of beta receptor and AdoR stimulation in a hypothetical myocardial cell is given in Figure 1. Briefly, the beta adrenergic receptor and AdoR couple the stimulation and inhibition of adenylate cyclase to modulate the level of cAMP in the myocyte. A cAMP-dependent protein kinase (PKA) catalyzed phosphorylation of the sarcolemmal Ca^{++} channel controls channel function and thus the inotropic state of the tissue.[25] Similarly, phosphorylation of phospholamban (PL), a protein in the sarcoplasmic reticulum, is believed to modulate Ca^{++} pumping by this organelle.[26] Thus, the interactions between the two receptor types are at the level of cAMP.

In 1984, we reported studies on membranes prepared from newborn chick heart.[27] In these studies we reported that the cardiac AdoR has the characteristics of an A_1 receptor in that [³H]R-PIA binds with high affinity and R-PIA inhibits isoproterenol (ISO)-stimulated adenylate cyclase activity. We also showed that R-PIA inhibits isoproterenol-stimulated phosphorylation of PL. The latter observation was important in that our preparations were no doubt contaminated with membranes of nonmyocyte origin and PL is restricted to myocytes, i.e., this observation pinpointed an effect on membranes of myocyte origin.

The ability of AdoR agonists to lower ISO-elevated cAMP levels in both heart tissue

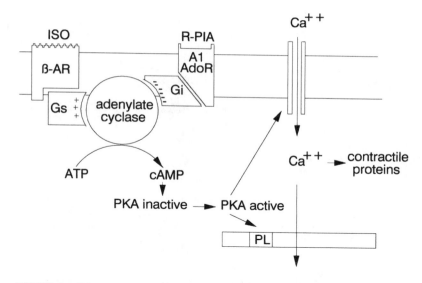

FIGURE 1. Scheme of relationships among β-adrenergic receptors (β-AR), A_1 AdoRs, and adenylate cyclase, and the subsequent effects of cAMP on $[Ca^{++}]_i$. PKA = cAMP-dependent protein kinase; PL = phospholamban (in sarcoplasmic reticulum).

TABLE 1
Cardiac Adenosine Receptors: Ligand Binding and Coupling to cAMP Metabolism

Preparation	Binding[a]	cAMP[b]	AC[b]	Ref.
Membranes/chick ventricle	[³H]R-PIA K_d 3—5 B_{max} 10	nd[c]	50	27
Membranes/chick atrial myocyte	[³H]CPX K_d 2 B_{max} 25	nd	27	29
Membranes/rat ventricle	[¹²⁵I]ABA K_d 11 B_{max} 15	nd	nd	33
Myocyte/chick ventricle	nd	30	nd	33
Membranes/rat ventricle	[¹²⁵I]HPIA K_d 1.1 B_{max} 18	nd	nd	32
Myocyte/rat ventricle	nd	30	nd	32
Myocyte/rat	nd	nd	40	30

[a] K_d in nM B_{max} in fmol/mg protein
[b]   Percent inhibition of ISO-elevated cAMP level or ISO-stimulated AC activity
[c]   nd = not determined

and isolated myocytes is clear. The mechanism(s) by which this cAMP-lowering effect is manifest is less so. While our observation that cardiac AdoRs mediate the inhibition of ISO-activated adenylate cyclase has been repeated in some laboratories,[28-30] other laboratories have not duplicated this finding.[31] This may be partially due to the study of impure membrane preparations,[31] but is also no doubt reflective of the ''uncoupling'' of receptors from adenylate cyclase which occurs to varying extents when membranes are prepared. Table 1 summarizes a limited number of reports on AdoRs in cardiac tissue in which ligand binding and cAMP levels, or adenylate cyclase were measured. In all cases the A_1-selective ligands bound with high affinity, suggesting that the cardiac AdoR is an A_1 receptor. This conclusion is further supported by the agonist profiles for the lowering of ISO-elevated cAMP levels (data not shown in Table 1). The reports shown are representative of those in which an AdoR-mediated inhibition of adenylate cyclase was demonstrable. We have been studying adenylate cyclase activity in detergent-permeabilized embryonic chick myocytes (H. Ma and R. D. Green, unpublished observations) in order to attempt to more clearly define receptor-mediated effects on adenylate cyclase activity. Both the stimulation of adenylate cyclase by ISO (Figure 2

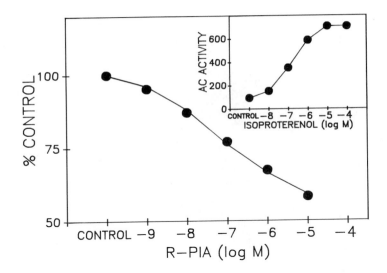

FIGURE 2.   Modulation of adenylate cyclase activity in permeabilized embryonic chick myocytes. Main panel: dose-response curve for R-PIA in the presence of 10 μM ISO. Inset: dose-response curve for ISO. Adenylate cyclase activity was measured using [^{32}P]deoxyATP as the substrate and rolipram + cyclic guanosine monophosphate (cGMP) to inhibit PDE activity.

inset) and the inhibition of this stimulated activity by R-PIA (Figure 2) are quantitatively larger in the permeabilized myocytes as compared to membrane preparations.[27] It is thus clear that cardiac AdoRs couple to both the inhibition of adenylate cyclase and the activation of a K^+ channel by a cAMP-independent mechanism. As adenosine is continually released in the myocardium and the rate of release is increased during exercise, it is likely that cardiac AdoRs play an important role in the control of cardiac function.[24]

C. CNS TISSUE

Any discussion of the physiological roles of adenosine and AdoRs cannot exclude a treatment of the CNS. Indeed, several chapters in this volume deal with this subject. Early in the development of the AdoR field, ligand binding studies established that A_1 AdoRs are widely distributed throughout the CNS, while A_2 receptors are largely confined to the striatum and mesolimbic area.[34,35] It is also established that adenylate cyclase activity in membrane preparations prepared from areas containing A_2 receptors as defined by ligand binding can be stimulated by adenosine analogs, while adenylate cyclase activity in membranes prepared from areas devoid of these binding sites cannot be stimulated.[35,36] Similarly, adenylate cyclase in membranes prepared from areas containing A_1 receptors as defined by ligand binding can be inhibited by adenosine analogs.[35,37] It is noteworthy that while adenosine analogs elevate the cAMP contents of slices of cerebral cortex and hippocampus, these areas have only A_1 receptors as defined by ligand binding and only inhibition of adenylate cyclase can be demonstrated in membranes prepared from these areas.[35,37] The existence of low-affinity A_2 receptors that uncouple from adenylate cyclase upon tissue disruption has been postulated to explain this paradox.[38]

It is now clear that AdoRs in the CNS couple to effectors other than adenylate cyclase. Notably, it has now been demonstrated that neurons prepared from striatum and hippocampus contain AdoRs that couple to the activation of a K^+ channel.[4,5] While this coupling involves a pertussis toxin-sensitive G protein, it does not appear to involve cAMP, i.e., the effect on channel activity is not indirect via an effect on adenylate cyclase.[4] Limited structure activity relationships suggest that the AdoR that mediates this response has the characteristics of an A_1 receptor.

Adenosine analogs have also been shown to inhibit calcium-dependent currents in dorsal root ganglia (DRG) cells isolated from embryonic mice[39] and 2-day-old rats.[40] Dolphin et al.[40] only studied one agonist, 2-chloroadenosine, but showed that this response was blocked by 8-phenyltheophylline. Low concentrations of 2-chloroadenosine inhibited forskolin-stimulated adenylate cyclase in membranes prepared from DRGs (higher concentrations stimulated), and these authors concluded that the effect of 2-chloroadenosine on the calcium conductance was via an A_1 receptor. While N^6-substituted adenosine analogs reduced the calcium conductance in the mouse DRGs and this effect was blocked by alkylxanthines, NECA was without effect.[39] These workers suggested that this response may be mediated by a novel AdoR, i.e., neither an A_1 nor an A_2 receptor.

A good deal of work has focused on pre- vs. postsynaptic effects of adenosine in the CNS. It is clear that depressant actions can involve both direct postsynaptic and presynaptic effects, i.e., inhibition of transmitter release. The relative importance of these two effects and the relationships between these effects and cAMP metabolism are obscure at present. As the effects of adenosine analogs on K^+ channels are not related to cAMP, and increases in K^+ channel activity could lead to both a direct depressant action on neuronal activity and to a decrease in transmitter release, cAMP may not be involved in either of these AdoR-mediated effects.

An argument, albeit weak, can be made for a cAMP-mediated effect of AdoR occupation in the striatum. Several years ago we reported that the injection of NECA into one caudate nucleus combined with the systemic administration of the dopamine receptor agonist apomorphine induces rotational behavior, the rotation being toward the NECA-injected side.[41] As NECA was more potent that R-PIA and the effect was antagonized by theophylline, we suggested that the response was mediated by A_1 receptors. We furthermore suggested that A_2 receptor stimulation is inhibitory to dopamine-receptor stimulation. Herrera-Marschitz et al.[42] have presented evidence to support this hypothesis. While the mechanisms for such proposed interactions remain to be determined, the possibility that an A_2 receptor modulation of intracellular levels of cAMP is involved remains tenable as, at present, the only known effect of A_2 receptor stimulation is the activation of adenylate cyclase. The importance of the striatal A_2 receptors in normal and pathologic striatal function remains to be further explored.

III. SUMMARY AND CONCLUSIONS

Although the above discussion is limited to three tissues, some general conclusions can be drawn. First, it would appear that some physiological responses to A_1-receptor stimulation are mediated by the inhibition of adenylate cyclase and the subsequent lowering in intracellular cAMP levels. The A_1 receptor-mediated antilipolytic effect in adipocytes and the negative inotropic effect in ventricular myocytes are two good examples of this mechanism. A good argument can also be made that the negative chronotropic effect of A_1-receptor stimulation at the SA node is mediated via a coupling to a K^+ channel. Current evidence suggests that the negative inotropic effect of A_1 receptor agonists in atrial tissue involves both pathways.

It is also clear that in spite of our extensive knowledge of the effects of adenosine and adenosine analogs on cAMP metabolism in neuronal (CNS) tissue, it is difficult to convincingly argue causal-effect relationships between these effects and the various physiological effects that are well documented.

If we cannot always form hard conclusions between the effects of AdoR agonists on cAMP metabolism and the physiological effects of these compounds, what can we say about the relationship between the sites detected by ligand binding and the sites which mediate the effects on cAMP metabolism? It could be argued that photoaffinity-labeling experiments have not revealed any marked heterogeneity in A_1 receptors in different tissues.[43-45] In addition, the finding that affinity chromatography on an agonist affinity gel results in the copurification of A_1 receptors and multiple types of G proteins could be used to argue that one type of A_1 receptor mediates different effects via different G proteins.[46] However, precedents from work on other receptors would mandate caution in this regard. For example, molecular cloning experiments on muscarinic receptors have revealed subtypes that preferentially couple to different effectors that were not predicted by ligand-binding and photoaffinity-labeling experiments.[47] Similar experiments with cloned AdoRs will be necessary in order to determine if subtypes of A_1 AdoRs exist, and, if so, to further probe the relationships between the receptors detected by ligand binding and those that couple to the various effector systems.

ACKNOWLEDGMENT

The unpublished results of experiments from the author's laboratory were supported by a grant from the National Science Foundation (BNS 8719594).

REFERENCES

1. **Sattin, A. and Rall, T. W.**, The effect of adenosine and adenine nucleotides on the cyclic 3′,5′-phosphate content of guinea pig cerebral cortex slices, *Mol. Pharmacol.*, 6, 13, 1970.
2. **Londos, C., Cooper, D. M. F., and Wolff, J.**, Subclasses of external adenosine receptors, *Proc. Natl. Acad. Sci. U.S.A.*, 77, 2551, 1980.
3. **Van Calker, D., Muller, M., and Hamprecht, B.**, Adenosine regulates via two different types of receptors, the accumulation of cyclic AMP in cultured brain cells, *J. Neurochem.*, 33, 999, 1979.
4. **Trussell, L. O. and Jackson, M. B.**, Dependence of an adenosine-activated potassium current on a GTP-binding protein in mammalian central neurons, *J. Neurosci.*, 7, 3306, 1987.
5. **Greene, R. W. and Haas, H. L.**, Adenosine action on CA1 pyramidal neurons in rat hippocampal slices, *J. Physiol. (London)*, 336, 119, 1985.
6. **Jochem, G. and Nawrath, H.**, Adenosine activates a potassium conductance in guinea-pig atrial muscle, *Experientia*, 39, 1347, 1983.
7. **Kurachi, Y., Nakajima, T., and Sugimoto, T.**, On the mechanism of activation of muscarinic K+ channels by adenosine in isolated atrial cells: involvement of GTP-binding proteins, *Pfluegers Arch.*, 407, 264, 1986.
8. **Belardinelli, L., Giles, W., and West, A.**, Ionic mechanism of adenosine actions in pacemaker cells from rabbit heart, *J. Physiol. (London)*, 405, 615, 1988.
9. **Kurtz, A.**, Adenosine stimulates guanylate cyclase activity in vascular smooth muscle cells, *J. Biol. Chem.*, 262, 6296, 1987.
10. **Linden, J. and Delahunty, T. M.**, Receptors that inhibit phosphoinositide breakdown, *TIPS*, 10, 114, 1989.
11. **de Mazancourt, P. and Giudicelli, Y.**, Guanine nucleotides and adenosine "R"-site analogs stimulate the membrane-bound low-K_m cyclic AMP phosphodiesterase of rat adipocytes, *FEBS Lett.*, 173, 385, 1984.
12. **Elks, M. L., Jackson, M., Manganiello, V. C., and Vaughn, M.**, Effect of N^6-(L-2-phenylisopropyl) adenosine and insulin on cAMP metabolism in 3T3-L1 adipocytes, *Am. J. Physiol.*, 252, C342, 1987.
13. **Schwabe, U., Ebert, R., and Erbler, H. C.**, Adenosine release from fat cells, effect on cyclic AMP levels and hormone actions, *Adv. Cyclic Nucleotide Res.*, 5, 569, 1975.
14. **Honnor, R. C., Dhillon, G. S., and Londos C.**, cAMP-dependent protein kinase and lipolysis in rat adipocytes. I. Cell preparation, manipulation, and predictability in behavior, *J. Biol. Chem.*, 260, 15122, 1985.
15. **Trost, T. and Schwabe, U.**, Adenosine receptors in fat cells, *Mol. Pharmacol.*, 19, 228, 1981.
16. **Honnor, R. C., Dhillon, G. S., and Londos, C.**, cAMP-dependent protein kinase and lipolysis in rat adipocytes. II. Definition of steady-state relationship with lipolytic and antilipolytic modulators, *J. Biol. Chem.*, 260, 15130, 1985.
17. **Vannucci, S. J., Klim, C. M., Martin, L. F., and LaNoue, K. F.**, A_1-adenosine receptor-mediated inhibition of adipocyte adenylate cyclase and lipolysis in Zucker rats, *Am. J. Physiol.*, 257, E871, 1989.

18. **Casey, P. J. and Gilman, A. G.,** G Protein involvement in receptor-effector coupling, *J. Biol. Chem.,* 263, 2577, 1988.
19. **Leung, E. and Green, R.,** Density gradient profiles of A$_1$ Adenosine receptors labelled by agonist and antagonist radioligands before and after detergent solubilization, *Mol. Pharmacol.,* 36, 412, 1989.
20. **Hoffman, B. B., Chang, H., Farahbakhsh, Z., and Reaven, G.,** Inhibition of lipolysis by adenosine is potentiated with age, *J. Clin. Invest.,* 74, 1750, 1984.
21. **Dehaye, J. P., Hebbelinck, M., Winand, J., and Christophe, J.,** Indirect evidence against a contribution of the guanine nucleotide-binding inhibitory component of adenylate cyclase to impaired lipolysis in the epidymal adipose tissue of congenitally obese (ob/ob) mice, *Horm. Metab. Res.,* 17, 333, 1985.
22. **Londos, C., Honnor, R. C., and Dhillon, G. S.,** cAMP-dependent protein kinase and lipolysis in rat adipocytes. III. Multiple modes of insulin regulation of lipolysis and regulation of insulin responses by adenylate cyclase regulators, *J. Biol. Chem.,* 260, 15139, 1985.
23. **Kuroda, M., Honnor, R. C., Cushman, S. W., Londos, C., and Simpson, I. A.,** Regulation of insulin-stimulated glucose transport in the isolated rat adipocyte, *J. Biol. Chem.,* 262, 245, 1987.
24. **Belardinelli, L., Linden, J., and Berne, R. M.,** The cardiac effects of adenosine, *Prog. Cardiovasc. Dis.,* 32, 73, 1989.
25. **Hosey, M. M. and Lazdunski, M.,** Calcium channels: molecular pharmacology, structure and regulation, *J. Membr. Biol.,* 104, 81, 1988.
26. **Tada, M. and Katz, A. M.,** Phosphorylation of the sarcoplasmic reticulum and sarcolemma, *Annu. Rev. Physiol.,* 44, 401, 1982.
27. **Hosey, M. M., McMahon, K. K., and Green, R. D.,** Inhibitory adenosine receptors in the heart: characterization by ligand binding studies and effects on β-adrenergic receptor stimulated adenylate cyclase and membrane protein phosphorylation, *J. Mol. Cell. Cardiol.,* 16, 931, 1984.
28. **Leung, E., Johnston, C. I., and Woodcock, E. A.,** Demonstration of adenylate cyclase coupled adenosine receptors in guinea pig ventricular membranes, *Biochem. Biophys. Res. Commun.,* 110, 208, 1983.
29. **Liang, B. T.,** Characterization of the adenosine receptor in cultured embryonic chick atrial myocytes: coupling to modulation of contractility and adenylate cyclase activity and identification by direct radioligand binding, *J. Pharmacol. Exp. Ther.,* 249, 775, 1989.
30. **Romano, F. D., MacDonald, S. G., and Dobson, J. G.,** Adenosine receptor coupling to adenylate cyclase of rat ventricular myocyte membranes, *Am. J. Physiol.,* 257, H1088, 1989.
31. **Schutz, W., Freissmuth, M., Hausleithner, V., and Tuisl, E.,** Cardiac sarcolemmal purity is essential for the verification of adenylate cyclase inhibition via A$_1$-adenosine receptors, *Naunyn-Schmiedeberg's Arch Pharmacol.,* 333, 156, 1986.
32. **Martens, D., Lohse, M. J., Rauch, B., and Schwabe, U.,** Pharmacological characterization of A$_1$ adenosine receptors in isolated rat ventricular myocytes, *Naunyn-Schmiedeberg's Arch. Pharmacol.,* 336, 342, 1987.
33. **Linden, J., Patel, A., and Sadek, S.,** [^{125}I]Aminobenzyladenosine, a new radioligand with improved specific binding to adenosine receptors in heart, *Circ. Res.,* 56, 279, 1985.
34. **Williams, M. and Risley, E. A.,** Biochemical characterization of putative central purinergic receptors by using 2-chloro[^{3}H]adenosine, a stable analog of adenosine, *Proc. Natl. Acad. Sci. U.S.A.,* 77, 6892, 1980.
35. **Yeung, S. M. H. and Green, R. D.,** [^{3}H]5'-N-Ethylcarboxamide adenosine binds to both R$_a$ and R$_i$ adenosine receptors in rat striatum, *Naunyn-Schmiedeberg's Arch. Pharmacol.,* 325, 218, 1984.
36. **Premont, J., Perez, M., Blanc, G., Tassin, J.-P., Thierry, A.-M., Herve, D., and Bockaert, J.,** Adenosine-sensitive adenylate cyclase in rat brain homogenates: kinetic characteristics, specificity, topological, subcellular and cellular distribution, *Mol. Pharmacol.,* 16, 790, 1979.
37. **Cooper, D. M. F., Londos, C., and Rodbell, M.,** Adenosine receptor-mediated inhibition of rat cerebral cortical adenylate cyclase by a GTP-dependent process, *Mol. Pharmacol.,* 18, 598, 1980.
38. **Daly, J. W., Butts-Lamb, P., and Padgett, W.,** Subclasses of adenosine receptors in the central nervous system. Interaction with caffeine and related methylxanthines, *Cell. Mol. Neurobiol.,* 1, 69, 1983.
39. **MacDonald, R. L., Skerritt, J. H., and Werz, M. A.,** Adenosine agonists reduce voltage-dependent calcium conductance of mouse sensory neurons in cell culture, *J. Physiol. (London),* 370, 75, 1986.
40. **Dolphin, A. C., Forda, S. R., and Scott, R. H.,** Calcium-dependent currents in cultured rat dorsal root ganglion neurons are inhibited by an adenosine analogue, *J. Physiol. (London),* 373, 47, 1986.
41. **Green, R. D., Proudfit, H. K., and Yeung S.-M. H.,** Modulation of striatal dopaminergic function by local injection of 5'-N-ethylcarboxamide adenosine, *Science,* 218, 58, 1982.
42. **Herrera-Marschitz, M., Casas, M., and Ungerstedt, U.,** Caffeine produces contralateral rotation in rats with unilateral dopamine denervation: comparisons with apomorphine-induced responses, *Psychopharmacology,* 94, 38, 1988.
43. **Leung, E., Kwatra, M. M., Hosey, M. M., and Green, R. D.,** Characterization of cardiac A$_1$ adenosine receptors by ligand binding and photoaffinity labeling, *J. Pharmacol. Exp. Ther.,* 244, 1150, 1988.
44. **Stiles, G. L.,** Photoaffinity cross-linked A$_1$ adenosine receptor-binding subunits. Homologous glycoprotein expression by different tissues, *J. Biol. Chem.,* 261, 10839, 1986.

45. **Klotz, K.-N. and Lohse, M. J.,** The glycoprotein nature of A_1 adenosine receptors, *Biochem. Biophys. Res. Commun.,* 140, 406, 1986.
46. **Munshi, R. and Linden, J.,** Co-purification of A_1 adenosine receptors and guanine nucleotide binding proteins from bovine brain, *J. Biol. Chem.,* 264, 14853, 1989.
47. **Peralta, E. G., Ashkenazi, A., Winslow, J. W., Ramachandran, J., and Capon, D. J.,** Differential regulation of PI hydrolysis and adenyl cyclase by muscarinic receptor subtypes, *Nature,* 34, 434, 1988.

Chapter 5

FORMATION AND METABOLISM OF ADENOSINE AND ADENINE NUCLEOTIDES IN CARDIAC TISSUE

Jürgen Schrader

TABLE OF CONTENTS

I. INTRODUCTION

In 1963, two groups independently made the observation that cardiac tissue contains measurable quantities of adenosine, the production of which is stimulated by hypoxia.[8,34] Biochemical studies on the metabolism of adenosine started only much later when sensitive methods for the quantitation of adenosine in biological samples had been developed. It was soon understood that the metabolism of adenosine is rather complex, involving at least four different enzymes (cytosolic 5'-nucleotidase, ecto-5'-nucleotidase, unspecific phosphatases, and S-adenosylhomocysteinehydrolase) which can form adenosine and another two enzymes (adenosine kinase and adenosine deaminase) which metabolize adenosine.[92] The extracellular adenosine concentration is therefore determined by six enzymes, including a specific transport step across the cell membrane.[78,92,93] Most of these reactions are highly regulated *in vivo* and the precise link between energy metabolism and adenosine formation is still a matter of debate.[74]

Most previous studies on the metabolism of adenosine were carried out in the heart (cardiomyocyte) and vascular cells (endothelium, smooth muscle). It is therefore not surprising that the cardiovascular system is presently the best understood in terms of the biochemistry and physiology of adenosine and adenine nucleotides. *In vitro* studies in the different cardiac cells revealed important metabolic differences, and details of the interplay between the different routes of adenosine metabolism at the level of the intact heart are now beginning to be elucidated. This chapter reviews the recent advances on the mechanisms involved in the intra- and extracellular formation of adenosine from ATP. Some aspects of the metabolism of adenosine and ATP have been previously reviewed.[36,74,92,93,99]

II. INTRACELLULAR PATHWAYS INVOLVED IN THE FORMATION OF ADENOSINE

Two principal enzymatic reactions are involved in the formation of adenosine: dephosphorylation of 5'-AMP by action of 5'-nucleotidase (EC 3.1.3.5) and alkaline phosphatase (EC 3.1.3.1), as well as hydrolysis of S-adenosyl-L-homocysteine (SAH) by SAH-hydrolase (3.3.1.1) to yield adenosine and L-homocysteine (Figure 1).

In most species and organs studied, 5'-nucleotidase is predominantly an ectoenzyme, serving in a cascade to extracellularly degrade ATP to adenosine.[83] A certain fraction of this enzyme is found to also occur in the cytosol and in lysosomes.[17,61,71,96,103] There are, however, important species differences.[63] Globally ischemic pigeon hearts produce adenosine rapidly despite the virtual absence of ecto-5'-nucleotidase.[70,71] The reported kinetic properties of the cytosolic 5'-nucleotidase[45,113] are adequate to fully explain its participation in the ischemia-induced adenosine formation in pigeon, but not rat, heart.[71]

The transmethylation pathway involves the transfer of the methyl group of adenosylmethionine (SAM) to a variety of methyl acceptors. About 80% of all cellular transmethylation reactions are coupled to SAM as a methyl donor and this reaction is not reversible.[102] In the metabolic steady state, the SAH formed by this pathway is continuously removed by SAH-hydrolase so that the formation of adenosine by this pathway is equivalent to the transmethylation rate from SAM to SAH.[58] SAH-hydrolase is an exclusively cytosolic enzyme and has been found to be ubiquitously distributed.[102] The reaction catalyzed by this enzyme is reversible, with the equilibrium lying far in the direction of SAH synthesis (equilibrium constant 10^{-6} M). Normally, however, the reaction proceeds in the direction of hydrolysis, because both reaction products (adenosine and L-homocysteine) are further metabolized by adenosine kinase or adenosine deaminase and methionine synthase or cystathionine β-synthase.

The relative contribution of the AMP and SAH pathways to the formation of adenosine

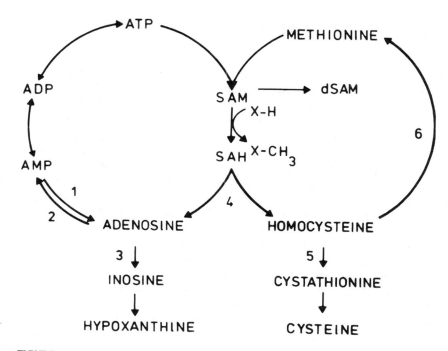

FIGURE 1. Pathway of adenosine and homocysteine production from the transmethylation pathway. SAM, S-adenosyl-L-methionine; dSAM, deoxySAM; SAH, S-adenosyl-L-homocysteine; X-H, methyl acceptor; and enzymes: (1) 5′-nucleotidase (EC 3.1.3.5); (2) adenosine kinase (EC 2.7.1.20); (3) adenosine deaminase (EC 3.5.4.4); (4) SAH-hydrolase (EC 3.3.1.1); (5) cystathionine β-synthase (EC 4.2.1.22); (6) 5′-methyltetrahydrofolate-homocysteine methyltransferase (EC 2.1.1.13).

is difficult to assess. The flux through the transmethylation pathway was determined to be about 1 nmol/min/g in the guinea pig heart,[58] 0.7 nmol/min/g in the isolated rat heart,[110] and 20 nmol/min/g in the liver *in situ*.[44] In the heart, the reported values by far exceed the rate of washout of adenosine by the coronary system so that most of the adenosine formed by this pathway is salvaged via adenosine kinase. Consistent with this view is the observation that inhibition of adenosine kinase greatly augments adenosine formation both in cardiac tissue[57] and in liver.[10] Conversely, inhibition of the SAH-hydrolase pathway with adenosine dialdehyde decreased global cardiac adenosine release by 34%.[23] Therefore, the generally accepted view that adenosine formation during normoxia arises only from 5′-AMP dephosphorylation appears to be no longer tenable. Flux data and inhibitor experiments suggest that cytosolic 5′-nucleotidase is normally substantially inhibited so that the transmethylation pathway and salvage via adenosine kinase dominate adenosine metabolism.

When adenosine formation is accelerated by hypoxia, the net flux of adenosine from the transmethylation pathway back into adenine nucleotides is reversed. The hypoxia-induced degradation of adenine nucleotides increases the free concentration of AMP.[16] In conjunction with disinhibition of 5′-nucleotidase,[57] catabolism to adenosine becomes quantitatively more important than the input by the transmethylation pathway and internal salvage of adenosine. Consistent with this view is the finding that the transmethylation pathway is essentially oxygen insensitive,[57] while degradation of adenine nucleotides to adenosine is well known to be critically dependent on tissue oxygenation.

III. COMPARTMENTATION OF ADENOSINE AND ADENINE NUCLEOTIDES

Radioactive prelabeling of a cellular metabolic pool has been a widely used technique to study changes in metabolic products in various cells and organs, thereby assessing precursor-product relationships. It is usually assumed that the precursor-product relationship is such that in the case of a homogeneous metabolic pool the specific radioactivity of the respective precursor and product is identical. In the early studies on the metabolism of adenosine, it was found that this is not true of ATP and adenosine in the isolated perfused heart, suggesting compartmentalization.[89] When adenine nucleotides of hearts were prelabeled by perfusion with radioactive adenosine or adenine, the specific radioactivity of adenosine liberated in the postlabeling phase exhibited a considerably higher specific activity than that of the precursor adenine nucleotides. This finding implied that released adenosine was derived from a small, but highly labeled pool of cardiac adenine nucleotides. It remained unclear, however, whether compartmentalisation occurred at the cellular or subcellular level. Later studies using autoradiography have shown that it is the vascular endothelium which is selectively prelabeled by perfusion with radioactive adenosine.[66,100] Cell fractionation studies[67] and multiple indicator dilution measurements[98,109] confirmed this conclusion. The endothelium therefore constitutes a highly active metabolic barrier for adenosine,[67] capable of trapping most of the infused adenosine both at the site of the capillaries and at the endothelium of large vessels.[54] The size of the endothelial adenine nucleotide pool was estimated to be 0.4 μmol/g heart,[53] which comprises about 8% of total cardiac adenine nucleotides. The endothelial contribution to total cardiac adenosine release was estimated to be 14% in the well-oxygenated heart.[53] This fraction considerably decreases when adenosine production is accelerated by hypoxia,[20] suggesting that adenosine is predominantly formed and released by the cardiomyocyte.

High-affinity adenosine transport sites have been characterized in vascular endothelial cells,[79] and this appears to be a common feature not only of microvascular endothelial cells, but also of macrovascular endothelium.[54] In the physiological concentration range of adenosine (0.012 to 1 μM), sequestration of adenosine is 90 to 92% in the perfused rat aorta.[54] In this model, 85% of the infused inosine (0.1 μM) becomes trapped by the endothelium, and this fraction is considerably higher than the uptake of inosine by cardiomycytes.[13] These findings suggest the presence of a two-tier salvage system in the endothelium. A high-affinity tier appears to be important for the salvage of adenosine by adenosine kinase at low substrate concentrations. A second low-affinity tier can salvage inosine and hypoxanthine, normally present in plasma in higher concentrations than adenosine.[6] Thus, endothelial adenosine deaminase and nucleoside phosphorylase extend the effective concentration range over which adenosine can be salvaged.[54]

One important functional implication of the endothelial barrier is that infused adenosine may not reach the underlying vascular smooth-muscle and parenchymal cells in biologically active concentrations. This would imply that the vasodilatory action of adenosine is endothelially mediated.[67] Alternatively, it was proposed[98] that the endothelial barrier for adenosine is not complete and that vascular smooth-muscle sensitivity to adenosine is higher than adenosine infusion studies have indicated. The magnitude of the endothelial concentration gradient for adenosine, however, not only depends on the permeability of the endothelium for adenosine, but also on the kinetic parameters of adenosine uptake from the subendothelial space by endothelial and smooth-muscle cells. Taking these parameters into consideration, the abluminal adenosine concentration was calculated to be 17 to 42% of the luminal adenosine concentration in the concentration range of 1 to 100 μM, respectively.[54] This implies that arteriolar smooth muscle in the physiological concentration range of adenosine is about five times more sensitive to adenosine than dose-response curves for infused adenosine suggest.

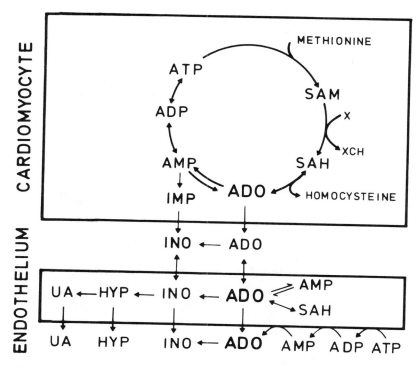

FIGURE 2. Routes of adenosine formation and degradation in cardiomyocytes and endothelium.

In Figure 2, our present knowledge on the various ATP and adenosine metabolizing reactions in cardiomyocytes and coronary endothelium is summarized. Cardiomyocytes lack adenosine deaminase, so that adenosine formed within this cellular compartment can only be released into the vascular space or salvaged via adenosine kinase. The endothelium is able to degrade adenosine down to uric acid.[6] Adenosine deaminase,[95] nucleoside phosphorylase,[86] and xanthine oxidase[35,47] have been shown to be exclusively endothelial enzymes. The reason for this cellular compartmentation of purine metabolism is presently not known.

IV. MECHANISMS THAT TRIGGER THE FORMATION OF ADENOSINE

It has been known for some time that hypoxia and ischemia are among the most potent stimuli which trigger the degradation of adenine nucleotides and thereby are linked to the formation of adenosine, a degradative product of high-energy phosphates.[8,34,107] Originally it was assumed that formation of adenosine is related to tissue PO_2, acting in a feedback-controlled system to adjust oxygen supply to the myocardial oxygen requirements.[8] Later it was found that adrenergic stimulation in the presence of adequate oxygen supply also enhanced cardiac adenosine production,[64,87] suggesting that adenosine formation may occur in direct proportion to the metabolic rate.[9]

The two principal mechanisms by which adenosine may be formed in order to maintain the ratio of oxygen supply to oxygen demand in equilibrium are schematically illustrated in Figure 3. According to model A, adenosine primarily signals tissue oxygen tension and acts in a feedback-controlled system. This form of the hypothesis in which tissue PO_2 is the controlled variable was termed the substrate form.[28] According to model B, the production of adenosine is coupled to the rate of energy metabolism, adenosine being formed in direct

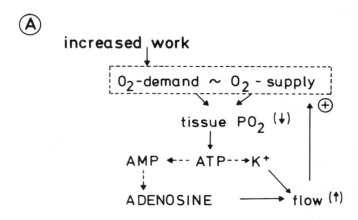

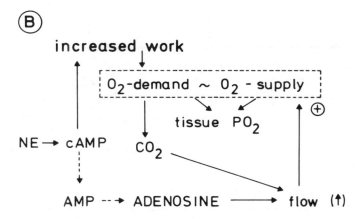

FIGURE 3.   Possible mechanisms by which adenosine may be formed by the heart in order to maintain the ratio of oxygen supply to oxygen demand in equilibrium when work of the heart (oxygen consumption) is increased (for details see text).

proportion to the amount of ATP utilized. In this model, the metabolite form of the hypothesis,[28] adenosine is not a primary feedback-controlled variable. Its rate of formation directly correlates with the oxygen consumption of the respective tissue.

A possible mechanism by which metabolic rate and adenosine formation are coupled could be through cAMP (Figure 3B). In the case of stimulation of the heart with catecholamines, this could involve the following sequence of events. The catecholamine-induced increase in tissue cAMP triggers the well-known increase in cardiac contractile force and thereby determines the energy requirements of the heart and at the same time increases flux through the cAMP $\rightarrow$ AMP pathway, finally leading to the formation of adenosine. Thus, the increase in a single metabolite (cAMP) could carry the information to signal changes in contractile force as well as elicit an increased oxygen supply to the heart by the vasodilatory action of adenosine. As a consequence of this control mechanism, tissue PO_2 would remain largely unchanged. Quantitative considerations based on experiments carried out in the isolated heart[5] suggest that model B may not be the dominant mechanism by which adenosine is formed. Similarly, β-adrenergic stimulation of human adipocytes was shown to enhance turnover of AMP,[50] but this effect on metabolic flux occurred with only little change in the tissue content of AMP and did not result in an augmented release of adenosine.[50]

An increasing line of evidence suggests that, at least in the heart, metabolic rate as such

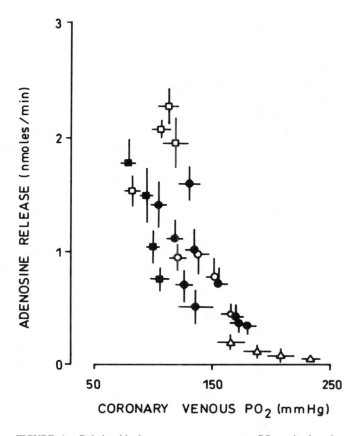

FIGURE 4. Relationship between coronary venous PO_2 and adenosine release in isolated perfused guinea pig heart. Oxygen consumption was stimulated with isoproterenol (ISO) 4 nM (●) at coronary perfusion pressure (CPP) 60 cm H_2O, ISO 2 nM (△) at CPP 90 cm H_2O, ISO 2 nM (○) at CPP 60 cm H_2O, ISO 2 nM (□) at CPP 35 cm H_2O, and norepinephrine 20 nM (■) at CPP 60 cm H_2O. (From Deussen, A. and Schrader, J., *J. Cell. Mol. Cardiol.*, in press.)

is not the major trigger leading to adenosine formation, but rather it is the imbalance between oxygen delivery via the coronary system and the oxygen demand of the heart. Increasing the work load of the heart with β-adrenergic stimulation and increasing cardiac afterload did not increase adenosine in the latter case because of a complete match of oxygen consumption with afterload-induced changes in coronary perfusion pressure.[4] When different inotropic stimuli were applied to enhance cardiac metabolic rate to the same extent, adenosine production was strictly dependent on the supply-to-demand ratio for oxygen.[5] Also, the phasic release of adenosine described after β-adrenergic stimulation[27] closely correlated with myocardial PO_2.[24] As is shown in Figure 4, a consistent relationship was found for β-adrenergic stimulation (isoproterenol), as well as for combined α- and β-adrenergic stimulation (norepinephrine), which holds true for the different degrees of coronary perfusion pressure tested. The dependence of adenosine on tissue oxygenation cannot be explained by the low oxygen-carrying capacity of the saline medium used in the above cited studies. Also, in the blood-perfused dog heart *in situ* the catecholamine-elicited release of adenosine was blunted by maintaining constant coronary perfusion pressure.[25] It thus appears that coronary venous adenosine only increases during β-adrenergic stimulation of the heart when other factors (low perfusion pressure, increased extravascular compression, α-adrenergic vasoconstriction) prevent an adequate rise of coronary flow and consequently lower myocardial oxygenation.

Relating cytosolic adenylates to the formation of adenosine,[16,42] it was suggested that adenosine production may be solely regulated by the concentration of free cytosolic AMP through the ATP potential. In this model, the activity of myokinase would translate changes in the ATP/ADP ratio into respective changes in free AMP. The increased availability of ADP would serve a dual purpose. Firstly, it would constitute the driving force for the formation of AMP and adenosine. Secondly, it would signal the mitochondria to increase oxidative phosphorylation.[41,46] Such a mechanism would help to adjust energy expenditure to ATP resynthesis at the cellular level and also, via adenosine, maintain oxygen supply balanced with oxygen requirements at the organ level. Recent NMR data, however, are not consistent with this view.[2,51] Despite a significant increase in cardiac workload and oxygen consumption, there were no detectable changes in the free-ADP concentration in the hemodynamic steady state.[2] Similarly, adenosine formation was recently shown to return to control values during continuous β-adrenergic stimulation of the heart.[25] Collectively, these data suggest that mechanisms other than ADP are responsible for the stimulation of cardiac respiration[62] and that adenosine cannot be responsible for the sustained increase in coronary flow.[24] Adenosine may become functionally important only when oxygen supply does not meet oxygen requirements.

Although hypoxia is the most potent stimulus for adenosine formation in most tissues, little is known as to the exact PO_2 below which adenosine formation is accelerated. Steady-state oxygen partial pressures in the medium above 2 mmHg have been reported to saturate oxidative phosphorylation in isolated cardiomyocytes.[112] Anoxic incubation enhanced cellular adenosine formation,[1] which most likely is formed intracellularly.[71,97] The critical PO_2 for adenosine was recently determined to be 3 mmHg in isolated, metabolically stable myocytes.[97] This value compares well with the frequency distribution of PO_2 (0 to 5 mmHg) measured in the blood-perfused dog heart with oxygen electrodes.[60] It must, however, be kept in mind that significant intramyocardial shear forces in combination with the caliber of the oxygen electrode may have influenced the measured PO_2 values in the beating heart.

V. TURNOVER OF ADENOSINE

A precise estimate of the cellular turnover of adenosine requires data on the pool size of adenosine as well as flux data of all input (SAH-hydrolase, 5'-nucleotidase) and output functions (adenosine kinase, adenosine deaminase), including washout of adenosine from the respective tissue. These data are only partially available in the literature. Nevertheless, an estimate is possible on the turnover of adenosine in the well-oxygenated heart. The pool size of free cytosolic adenosine was recently determined to be 0.061 nmol/g[21] for the normoxic guinea pig heart. Under metabolic steady-state conditions, the transmethylation pathway is the major determinant of intracellular adenosine formation and this was shown to proceed at 1.12 nmol/g in the same species.[58] From these data, the half-life of intracellular adenosine can be calculated to be 1.6 s. This value most likely underestimates the true half-life, since input by cytosolic 5'-nucleotidase and unspecific phosphatases was not considered. In functional terms, a short half-life would be a prerequisite for a regulatory metabolite such as adenosine, since changes in tissue and organ function ascribed to adenosine also occur in the range of seconds.

A direct estimate of the turnover of adenosine was recently obtained in plasma of human blood.[65] Using isotope-dilution techniques, the half-life of adenosine in the physiological concentration range of 80 nM to 1 μM was determined to be 0.6 to 1.5 s, respectively. There are, however, considerable species differences. The rate of removal of adenosine from plasma in dog blood was reported to be about 300 times slower (half-life = 3 min).[52,65]

The functional implication of these findings is that measurements of the release of adenosine in blood-perfused organs from arteriovenous differences may to some extent underestimate (dog) or even miss (human) the actual rate of adenosine formation by the respective organ.

VI. INDICES OF ADENOSINE FORMATION

Due to the rapid metabolism of adenosine and consequently its short half-life (1 to 2 s), special precautions must be taken when analyzing adenosine in tissue and body fluids. An additional problem arises from the fact that in the case of tissue adenosine measurements, most of this nucleoside is protein bound.[7,73,91] This fraction constitutes a biologically inert pool[7,73] that does not participate in the various metabolic reactions of adenosine (for review see Reference 102). Because of this high background of bound adenosine, changes in the concentration of free intracellular adenosine are difficult to measure by currently available extraction procedures. Small changes in free tissue adenosine may therefore go undetected.

Adenosine-binding proteins have been identified,[72] the most important being SAH-hydrolase. The mechanism of action of this enzyme and adenosine binding site has been identified.[43,77,84] Adenosine is tightly bound to SAH-hydrolase, is no longer available as a substrate for adenosine deaminase,[73] and dissociates with a half-life of several hours.[102]

Our laboratory has recently described a new method to circumvent the problem of protein-bound adenosine by using a kinetic approach.[21] Increases in endogenous free-adenosine levels, in the presence of homocysteine, result in a reversal of SAH-hydrolase activity from hydrolysis in the direction of synthesis. As a consequence, SAH accumulates in proportion to the free concentration of adenosine. Due to the fact that SAH increases at a constant rate[21] and is neither further metabolized[102] nor crosses cell membranes,[108] changes in the concentration of free adenosine are amplified by conversion to SAH. Most importantly, elevated plasma levels of L-homocysteine are hemodynamically ineffective. Using this technique, the free intracellular adenosine level in the normal heart was determined to be 0.08 μM, which increased to 2.0 μM during hypoxia.[21] The SAH technique permits us to detect changes in free tissue adenosine which are not detectable by conventional extraction procedures. Furthermore it permits the sensitive determination of the free-adenosine gradient from the endo- to epicardium under different functional conditions.[22] Using ^{11}C-labeled L-homocysteine, this technique recently permitted the local noninvasive assessment of adenosine in the heart by positron-emission tomography.[26] Two important requirements must be fulfilled to successfully apply the SAH technique. Firstly, SAH-hydrolase must be homogenously distributed within the tissue, a criterion fulfilled in the heart,[22] and most likely in other tissues as well.[102] Secondly, L-homocysteine must be applied in saturating concentration (above 200 μM) and therefore requires a certain residual blood flow to the tissue under study.

Another important technique to assess the formation of adenosine is by measuring its release into the surrounding medium in the case of isolated cells, or into the venous circulation (arteriovenous difference) in the case of a perfused organ. Cellular release of adenosine involves a transport step across the membrane along its concentration gradient.[48] A certain fraction of released adenosine is rapidly taken up again and/or is metabolized further extracellularly. In the case of a perfused organ, the barrier function of the endothelium,[98] significant levels of adenosine deaminase in plasma,[106] and a certain arterial adenosine plasma level[65] constitute further complicating factors. All these influences tend to underestimate the true rate of adenosine formation from a given tissue. Use of transport-blockers such as dipyridamole[69] cannot alleviate this problem, since these compounds block transmembrane transport of adenosine in both directions.

In order to gain access to the interstitial adenosine concentration in the heart, a pericardial-well technique was developed in which a saline medium is permitted to equilibrate with the epicardium.[55,64] The major problem with this technique is that it most likely reflects changes in the subepicardium, while most of the changes in adenosine occur in the mid- and sub-endocardium.[22] A modification of the pericardial-well technique is the direct collection of epicardial transudate in the isolated perfused heart.[29,49] Due to the increased interstitial flow rate with a saline medium, this method probably gives average values for any transmural metabolite gradient. For nonmuscle tissue an *in vivo* dialysis technique was developed.[105] This technique was recently applied to adipose tissue.[59]

VII. INTRACELLULAR AND EXTRACELLULAR FORMATION OF ADENOSINE

Adenosine may be formed intracellularly by the enzymatic reactions outlined above (Figures 1 and 2) or extracellularly from ATP by a cascade of highly active ectoenzymes.[82] Adenine nucleotides are released into the extracellular space during platelet aggregation,[33] catecholamine release from adrenal medulla,[111] neurotransmission,[14] and strenuous exercise.[30,31] In human platelets the concentration of ADP and ATP stored in dense granules is greater than 1 M,[104] and the concentration of ATP in blood leaking from punctured microvessels was determined to be as high as 20 μM.[11] Furthermore, selective release of up to 50% of the cellular content of ATP in cultured endothelial cells can be elicited by treating cells with thrombin[80] and granulocyte elastase.[56] The different effects, sources, and fate of extracellular ATP have been reviewed.[36]

In cardiac tissue, the absolute quantities of ATP released are rather small[32,76] and do not appear to be in the biologically active concentration range. It may be argued, however, that due to rapid extracellular degradation of adenine nucleotides, the concentration of ATP at its presumed intracardiac site of formation is considerably higher. Consistent with this view is the recent demonstration that effective inhibition of ecto-5'-nucleotidase with the ADP derivative, α,β-methylene-adenosine 5'-diphosphate (AOPCP), increased in isolated hearts the venous release of adenine nucleotides about tenfold without significantly changing the release of adenosine.[12] Most interestingly, the concentrations of released adenosine and adenine nucleotides in the presence of the inhibitor were in the same range. This finding appears to be functionally relevant since adenosine, AMP, ADP, and ATP are equipotent in the isolated heart preparation in dilating coronary resistance vessels.[15]

Cardiac ischemia augments the cardiac release of adenosine[88,90] and adenine nucleotides[12] to a similar extent. Peak concentrations of adenine nucleotides reached in the effluent perfusate after 1 min of cardiac ischemia are 0.1 μM, which is within the coronary vasodilatory range.[15] The effects of ischemia on adenine nucleotides cannot be mimicked by hypoxic perfusion, which is known to be associated with a substantial increase in the release of adenosine. Extracellular degradation of adenine nucleotides is therefore unlikely to be an important factor in the hypoxia-induced formation of adenosine. This confirms an earlier finding on the effects of AOPCP on cardiac adenosine formation.[96]

Plasma concentration of adenosine in human blood is in the nanomolar range[39,75] and increases threefold to fourfold after arterial occlusion.[65] Similarly, a careful study by Forrester[31] demonstrates that the basal level of ATP in human venous plasma was less than $2 \times 10^{-8} M$, but increased more than 50-fold during exercise and partial arterial occlusion. More recently, plasma concentrations of ATP greater than $10^{-6} M$ have been reported.[40] These values are likely to be far too high, since $10^{-6} M$ ATP would have caused massive peripheral dilatation. Hemolysis and nucleotide release by the anticoagulant used (EDTA) may have caused this result.

The pattern of extracellular dephosphorylation from ATP $\rightarrow$ ADP $\rightarrow$ AMP $\rightarrow$ adenosine was observed in vascular endothelial and smooth-muscle cells,[37,81] B lymphocytes,[3] and plasma of human blood.[17] Leukocyte ecto-ADPase activity contributes more than 80% to total enzyme activity in whole blood, but leukocytes have insignificant ability to degrade AMP to adenosine.[17] From the available kinetic data and the known ratio of surface area of blood cells to endothelial cells, it was suggested that the capacity to degrade adenine nucleotides within the large vessels resides equally at the luminal endothelial surface and in the flowing blood.[17] In microvessels, however, catabolism by endothelial cells predominates.[17] Interestingly, T lymphocytes, in contrast to B lymphocytes, are unable to degrade adenine nucleotides, and this metabolic difference was suggested to serve the communication between B and T cells through adenosine.[3]

VIII. CONCLUSIONS AND PERSPECTIVES

Studies on the mechanisms of adenosine formation in different tissues are important not only for the understanding of the various functions ascribed to this nucleoside, but also to pharmacologically interfere with its rate of formation. AICA-riboside pretreatment, probably through inhibition of AMP-deaminase, was recently shown to potentiate tissue accumulation of adenosine only in the ischemic myocardium.[38] This effect was paralleled by reduced neutrophil infiltration and increased local tissue perfusion, finally resulting in improved ventricular function. The concept of enhancing endogenous adenosine at sites where adenosine is most needed is an interesting concept which deserves further study.[94]

The recent finding that adenosine deaminase is present only in the endothelium, and not in myocytes[95] is most intriguing. Due to the cellular distribution of the above enzymes, the adenosine formed by myocytes can either be salvaged to AMP, or, after being released from the myocyte, degraded in the endothelial compartment to inosine, hypoxanthine, and finally uric acid.[6] Whether there is any metabolic or functional advantage associated with the selective catabolism of adenosine in the endothelium is presently not known. It also would be interesting to know whether or not this cellular compartmentation of adenosine is a general feature found also in other organs, such as the brain.

The formation of adenosine is linked directly to the net breakdown of cytosolic ATP. Conceptually, adenosine can be viewed as a locally acting metabolite that signals changes in the energy metabolism via changes in the supply-to-demand ratio for oxygen.[94] Because adenosine can prevent excessive ATP breakdown, it was termed a retaliatory metabolite,[68] which plays a homeostatic role in tissue energy metabolism.[94] Tissue oxygenation is the essential parameter that triggers the cardiac formation of adenosine under a variety of physiological conditions. While it may be true for the heart, this mechanism may not be equally important in neural tissue. An increase in nerve firing leads to a cosecretion of ATP, which is extracellularly degraded to adenosine[101] to finally induce autoinhibition after repeated stimulation.[85] This process appears to be essentially independent of oxygen. There is accumulating evidence that adenine nucleotides can be released from cells other than neural cells and platelets. As is discussed above, the intact endothelium is capable of liberating adenine nucleotides both under basal conditions and when stimulated. Little, however, is known at present as to the precise mechanism of release and whether or not released ATP carries any physiologically relevant message.

REFERENCES

1. **Altschuld, R. A., Gamelin, L. M., Kelley, R. E., Lambert, M. R., Apel, L. E., and Brierley, G. P.,** Degradation and resynthesis of adenine nucleotides in adult rat heart myocytes, *J. Biol. Chem.*, 262, 13527, 1987.
2. **Balaban, R. S., Kantor, H. L., Katz, L. A., and Briggs, R. W.,** Relation between work and phosphate metabolites in the in vivo paced mammalian heart, *Science,* 232, 1121, 1986.

3. **Barankiewicz, J., Dosch, H. M., and Cohen, A.,** Extracellular nucleotide catabolism in human B and T lymphocytes, *J. Biol. Chem.,* 263, 7094, 1988.
4. **Bardenheuer, H. and Schrader, J.,** Relationship between myocardial oxygen consumption, coronary flow and adenosine release in the improved isolated working heart preparation of guinea pigs, *Circ. Res.,* 51, 263, 1983.
5. **Bardenheuer, H. and Schrader, J.,** Supply-to-demand ratio for oxygen determines formation of adenosine by the heart, *Am. J. Physiol.,* 250, H173, 1986.
6. **Becker, B. F., and Gerlach, E.,** Uric acid, the major catabolite of cardiac adenine nucleotides and adenosine, originates in the coronary endothelium, in *Topics and Perspectives in Adenosine Research,* Gerlach, E. and Becker, B. F., Eds., Springer-Verlag, Berlin, 1987, 209.
7. **Belloni, F. L., Rubio, R., and Berne, M.,** Intracellular adenosine in isolated rat liver cells, *Pfluegers Arch.,* 400, 106, 1984.
8. **Berne, R. M.,** Cardiac nucleotides in hypoxia: possible role in regulation of coronary blood flow, *Am. J. Physiol.,* 204, 317, 1963.
9. **Berne, R. M.,** The role of adenosine in the regulation of coronary blood flow, *Circ. Res.,* 47, 807, 1980.
10. **Bontemps, F., Van der Berghe, G., and Hers, H. G.,** Evidence for a substrate cycle between AMP and adenosine in isolated hepatocytes, *Proc. Natl. Acad. Sci. U.S.A.,* 80, 2829, 1983.
11. **Born, G. V. R. and Kratzer, M. A. A.,** Source and concentration of extracellular adenosine triphosphate during haemostasis in rats, rabbits and man, *J. Physiol. (London),* 354, 419, 1984.
12. **Borst, M. and Schrader, J.,** Adenine nucleotide release from isolated perfused guinea pig hearts and extracellular formation of adenosine, *Circ. Res.,* in press.
13. **Bowditch, J., Brown, A. K., and Dow, J. W.,** Accumulation and salvage of adenosine and inosine by isolated mature cardiac myocytes, *Biochim. Biophys. Acta,* 844, 119, 1985.
14. **Burnstock, G.,** Neurotransmitters and trophic factors in the autonomic nervous system, *J. Physiol. (London),* 313, 1, 1981.
15. **Bünger, R., Haddy, F. J., and Gerlach, E.,** Coronary responses to dilating substances and competitive inhibition by theophylline in the isolated perfused guinea pig heart, *Pfluegers Arch.,* 358, 213, 1975.
16. **Bünger, R. and Soboll, S.,** Cytosolic adenylates and adenosine release in perfused working heart, *Eur. J. Biochem.,* 159, 203, 1986.
17. **Coade, St. B. and Pearson, J. D.,** Metabolism of adenine nucleotides in human blood, *Circ. Res.,* 65, 531, 1989.
18. **Collinson, A. R., Peuhkurinen, K. J., and Lowenstein, J. M.,** Regulation and function of 5'-nucleotidases, in *Topics and Perspectives in Adenosine Research,* Gerlach, E and Becker, B. F., Eds., Springer-Verlag, Berlin, 1987, 133.
19. **De la Haba, G. and Cantoni, G. L.,** The enzymatic synthesis of S-adenosyl-L-homocysteine from adenosine and homocysteine, *J. Biol. Chem.,* 234, 603, 1959.
20. **Deussen, A., Möser, G., and Schrader, J.,** Contribution of coronary endothelial cells to cardiac adenosine production, *Pfluegers Arch.,* 406, 608, 1986.
21. **Deussen, A., Borst, M., and Schrader, J.,** Formation of S-adenosylhomocysteine in the heart. I. An index of free intracellular adenosine, *Circ. Res.,* 63, 240, 1988.
22. **Deussen, A., Borst, M., Kroll, K., and Schrader, J.,** Formation of S-adenosylhomocysteine in the heart. II. A sensitive index for regional myocardial underperfusion, *Circ. Res.,* 63, 250, 1988.
23. **Deussen, A., Lloyd, H. G. E., and Schrader, J.,** Contribution of S-adenosylhomocysteine to cardiac adenosine formation, *J. Mol. Cell. Cardiol.,* 21, 773, 1989.
24. **Deussen, A. and Schrader, J.,** Cardiac adenosine production is linked to myocardial pO_2, *J. Cell. Mol. Cardiol.,* in press.
25. **Deussen, A., Walter, Ch., Borst, M., and Schrader, J.,** Transmural gradient of adenosine in canine heart during functional hyperemia, *Am. J. Physiol.,* in press.
26. **Deussen, A. and Schrader, J.,** Assessment of local tissue adenosine by measurement of S-adenosylhomocysteine (SAH), in *Adenosine and Adenine Nucleotides,* Imai, S. and Nakazawa, M., Eds., Elsevier, in press.
27. **De Witt, D. F., Wangler, R. D., Thompson, C. I., and Sparks, H. V., Jr.,** Phasic release of adenosine during steady state metabolic stimulation in the isolated guinea pig heart, *Circ. Res.,* 53, 636, 1983.
28. **Feigl, E. O.,** Coronary physiology, *Physiol. Rev.,* 66, 1, 1983.
29. **Fenton, R. A. and Dobson, J. G., Jr.,** Measurement by fluorescence of interstitial adenosine levels in normoxic, hypoxic, and ischemic perfused rat hearts, *Circ. Res.,* 60, 177, 1987.
30. **Forrester, T. and Lind, A.,** Adenosine triphosphate in the venous effluent and its relationship to exercise, *Fed. Proc., Fed. Am. Soc. Exp. Biol.,* 28, 1280, 1969.
31. **Forrester, T.,** An estimate of adenosine triphosphate release into the venous effluent from exercising human forearm muscle, *J. Physiol. (London),* 244, 611, 1972.
32. **Forrester, T. and Williams, C. A.,** Release of adenosine triphosphate from isolated adult heart cells in response to hypoxia, *J. Physiol.,* 268, 371, 1977.

33. **Gaader, A., Jonsen, J., Laland, S., Hellem, A., and Owren, P. A.,** Adenosine diphosphate in red cells as a factor in the adhesiveness of human blood platelets, *Nature,* 192, 531, 1961.

34. **Gerlach, E., Deuticke, F. J., and Dreisbach, R. H.,** Der Nucleotid-Abbau im Herzmuskel bei Sauerstoffmangel und seine mögliche Bedeutung für die Coronardurchblutung, *Naturwissenschaften,* 50, 228, 1963.

35. **Gerlach, E., Nees, S., and Becker, B. F.,** The vascular endothelium: a survey of some newly evolving biochemical and physiological features, *Basic Res. Cardiol.,* 80, 459, 1985.

36. **Gordon, J. L.,** Extracellular ATP: effects, sources and fate, *Biochem. J.,* 233, 309, 1986.

37. **Gordon, J. L., Pearson, J. D., and Slakey, L. L.,** The hydrolysis of extracellular adenine nucleotides by cultured endothelial cells from pig aorta, *J. Biol. Chem.,* 261, 15496, 1986.

38. **Gruber, H. E., Hoffer, M. E., McAllister, D. R.., Laikind, P. K., Lane, T. A., Schmid-Schoenbein, G. W., and Engler, R. L.,** Increased adenosine concentration in blood from ischemic myocardium by AICA riboside: effects on flow, granulocytes and injury, *Circulation,* 80, 1400, 1989.

39. **Hamm, Ch. W., Kupper, W., Brederhorst, R., Hilz, H., and Bleifeld, W.,** Quantitation of coronary venous adenosine in patients: limitations evaluated by radioimmunoassay, *Cardiovasc. Res.,* 22, 236, 1988.

40. **Harkness, R. A., Coade, S. B., and Webster, A. D. B.,** ATP, ADP and AMP in plasma from peripheral venous blood, *Clin. Chim. Acta,* 143, 91, 1984.

41. **Hassinen, I. E.,** Mitochondrial respiratory control in the myocardium, *Biochim. Biophys. Acta,* 853, 135, 1986.

42. **Headrick, J. P. and Willis, R. J.,** Adenosine formation and energy metabolism. A 31p-NMR study in isolated rat heart, *Am. J. Physiol.,* 258, H617, 1990.

43. **Hershfield, M. S. and Kredich, N. M.,** S-Adenosylhomocysteine hydrolase is an adenosine-binding protein: a target for adenosine toxicity, *Science,* 202, 757, 1978.

44. **Hoffman, J. L.,** The rate of transmethylation in mouse liver as measured by trapping S-adenosylhomocysteine, *Arch. Biochem. Biophys.,* 205, 132, 1980.

45. **Itoh, R., Oka, J., and Ozasa, H.,** Regulation of rat heart cytosol 5'-nucleotidase by adenylate energy charge, *Biochem. J.,* 235, 847, 1986.

46. **Jacobus, W. E.,** Respiratory control and the integration of heart high energy metabolism by mitochondrial creatine kinase, *Annu. Rev. Physiol.,* 47, 707, 1985.

47. **Jarasch, E. D., Grund, C., Bonder, G., Heid, H. W., Keenan, T. W., and Franke, W. W.,** Localisation of xanthine oxidase in mammary gland epithelium and capillary endothelium, *Cell,* 25, 67, 1981.

48. **Jarvis, S. M.,** Kinetic and molecular properties of nucleoside transporters in animal cells, in *Topics and Perspectives in Adenosine Research,* Gerlach, E. and Becker, B. F., Eds., Springer-Verlag, Berlin, 1987, 102.

49. **Kammermeier, H. and Wendtland, B.,** Interstitial fluid of isolated perfused rat hearts: glucose and lactate concentration, *J. Mol. Cell. Cardiol.,* 19, 167, 1987.

50. **Kather, H.,** Pathways of purine metabolism in human adipocytes, *J. Biol. Chem.,* 265, 96, 1990.

51. **Katz, L. A., Swain, J. A., Portman, M. A., and Balaban, R. S.,** Relation between phosphate metabolites and oxygen consumption of heart in vivo, *Am. J. Physiol.,* 256, H265, 1989.

52. **Klabunde, R. E. and Althouse, D. G.,** Adenosine metabolism in dog whole blood: effects of dipyridamole, *Life Sci.,* 28, 263, 1981.

53. **Kroll, K., Schrader, J., Piper, H. M., and Henrich, M.,** Release of adenosine and cyclic AMP from coronary endothelium in isolated guinea pig heart: relation to coronary flow, *Circ. Res.,* 60, 659, 1987.

54. **Kroll, K., Kelm, M. K. M., Bürrig, K.-F., and Schrader, J.,** Transendothelial transport and metabolism of adenosine and inosine in the intact rat aorta, *Circ. Res.,* 64, 1147, 1989.

55. **Kusachi, S. and Olsson, R. A.,** Pericardial superfusion to measure cardiac interstitial adenosine concentration, *Am. J. Physiol.,* 244, H458, 1983.

56. **Le Roy, E. C., Ager, A., and Gordon, J. L.,** Effects of neutrophil elastase and other proteases on porcine aortic endothelial prostaglandin production, adenine nucleotide release, and responses to vasoactive agents, *J. Clin. Invest.,* 74, 1003, 1984.

57. **Lloyd, H. G. E. and Schrader, J.,** The importance of the transmethylation pathway for adenosine metabolism in the heart, in *Topics and Perspectives in Adenosine Research,* Gerlach, E. and Becker, B. F., Eds., Springer-Verlag, Berlin, 1987, 199.

58. **Lloyd, H. G. E., Deussen, A., Wupperman, H., and Schrader, J.,** The transmethylation pathway as a source for adenosine in the isolated guinea pig heart, *Biochem. J.,* 252, 489, 1988.

59. **Lönnroth, P., Jansson, P. A., Fredholm, B. B., and Smith, U.,** Microdialysis of intercellular adenosine concentration in subcutaneous tissue in humans, *Am. J. Physiol.,* 256, E250, 1989.

60. **Lösse, B., Schuchardt, S., and Niederle, N.,** The oxygen pressure histogram in the left ventricular myocardium of the dog, *Pfluegers Arch.,* 356, 121, 1975.

61. **Lowenstein, J. M., Yu, M. K., and Naito, Y.,** Regulation of adenosine metabolism by 5'-nucleotidases, in *Regulatory Function of Adenosine,* Berne, R. M., Rall, T. W., and Rubio, R., Eds., Martinus Nijhoff, Boston, 1983, 117.

62. **McCormack, J. G. and Denton, R. M.**, Ca^{++} as a second messenger within mitochondria, *TIBS*, 11, 258, 1986.

63. **Meghji, P., Middleton, K. M., and Newby, A. C.**, Absolute rates of adenosine formation during ischemia in rat and pigeon hearts, *Biochem. J.*, 249, 695, 1988.

64. **Miller, W. L., Belardinelli, L., Bacchus, A., Foley, D. H., Rubio, R., and Berne, R. M.**, Canine myocardial adenosine and lactate production, oxygen consumption, and coronary blood flow during stellate ganglia stimulation, *Circ. Res.*, 45, 708, 1979.

65. **Möser, G. H., Schrader, J., and Deussen, A.**, Turnover of adenosine in plasma of human and dog blood, *Am. J. Physiol.*, 256, C799, 1989.

66. **Nees, S. and Gerlach, E.**, Adenine nucleotide and adenosine metabolism in cultured coronary endothelial cells: formation and release of adenine compounds and possible functional implications, in *Regulatory Function of Adenosine*, Berne, R. M., Rall, T. W., and Rubio, R., Eds., Martinus Nijhoff, Boston, 1983, 347.

67. **Nees, S., Herzog, V., Becker, B. F., Böck, B., Des Rosiers, C., and Gerlach, E.**, The coronary endothelium: a highly active metabolic barrier for adenosine, *Basic Res. Cardiol.*, 80, 515, 1985.

68. **Newby, A. C.**, Adenosine and the concept of "retaliatory metabolites", *Trends Biochem.*, 9, 42, 1984.

69. **Newby, A. C.**, How does dipyridamole elevate extracellular adenosine concentration? Predictions from a three-compartment model of adenosine formation and inactivation, *Biochem. J.*, 237, 845, 1986.

70. **Newby, A. C.**, The pigeon heart 5'-nucleotidase responsible for ischemia-induced adenosine formation, *Biochem. J.*, 253, 123, 1988.

71. **Newby, A. C., Worku, Y., and Meghji, P.**, Critical evaluation of the role of ecto- and cytosolic 5'-nucleotidase in adenosine formation, in *Topics and Perspectives in Adenosine Research*, Gerlach, E. and Becker, B. F., Eds., Springer-Verlag, Berlin, 1987, 155.

72. **Olsson, R. A., Vomacka, R. B., and Nixon, D. G.**, Adenosine-binding proteins in dog heart, in *Physiological and Regulatory Functions of Adenosine and Adenine Nucleotides*, Baer, H. P. and Drummond, G. I., Eds., Raven Press, New York, 1979, 297.

73. **Olsson, R. A., Saito, D., and Steinhart, C. R.**, Compartmentalization of the adenosine pool of dog and rat hearts, *Circ. Res.*, 50, 617, 1982.

74. **Olsson, R. A. and Bünger, R.**, Metabolic control of coronary blood flow, in *Progress in Cardiovascular Diseases*, Vol. 29, Sonnenblick, E. H. and Lesch, M., Eds., Grune & Stratton, New York, 1987, 369.

75. **Ontyd, J. and Schrader, J.**, Measurement of adenosine, inosine and hypoxanthine in human plasma, *J. Chromatogr.*, 307, 404, 1984.

76. **Paddle, B. M. and Burnstock, G.**, Release of ATP from perfused heart during coronary vasodilation, *Blood Vessels*, 11, 110, 1974.

77. **Palmer, J. L. and Abeles, R. H.**, The mechanism of action of S-adenosylhomocysteinase, *J. Biol. Chem.*, 254, 1217, 1979.

78. **Paterson, A. R. P., Jakobs, E. S., Ng, C. Y. C., and Odegard, R. D.**, Nucleoside transport inhibition in vitro and in vivo, in *Topics and Perspectives in Adenosine Research*, Gerlach, E. and Becker, B. F., Eds., Springer-Verlag, Berlin, 1987, 89.

79. **Pearson, J. D., Carleton, J. S., Hutchings, A., and Gordon, J. L.**, Uptake and metabolism of adenosine by pig aortic endothelial and smooth-muscle cells in culture, *Biochem. J.*, 170, 265, 1978.

80. **Pearson, J. D. and Gordon, J. L.**, Vascular endothelial and smooth muscle cells in culture selectively release adenine nucleotides, *Nature*, 281, 384, 1979.

81. **Pearson, J. D., Carleton, J. S., and Gordon, J. L.**, Metabolism of adenine nucleotides by ectoenzymes of vascular endothelial and smooth muscle cells in culture, *Biochem. J.*, 190, 421, 1980.

82. **Pearson, J. D.**, Ectonucleotidases: measurement of activities and use of inhibitors, in *Methods Used in Adenosine Research*, Paton, D. M., Ed., Plenum Press, New York, 1985, 83.

83. **Pearson, J. D. and Coade, S. B.**, Kinetics of endothelial cell ectonucleotidases, in *Topics and Perspectives in Adenosine Research*, Gerlach, E. and Becker, B. F., Eds., Springer-Verlag, Berlin, 1987, 145.

84. **Richards, H. H., Chiang, P. K., and Cantoni, G. L.**, Adenosylhomocysteine hydrolase, *J. Biol. Chem.*, 253, 4476, 1978.

85. **Richardson, P. J., Brown, S. J., Bailyes, E. M., and Luzio, J. P.**, Ectoenzymes control adenosine modulation of immunoisolated cholinergic synapses, *Nature (London)*, 327, 232, 1987.

86. **Rubio, R., Wiedmeir, T., and Berne, R. M.**, Nucleoside phosphorylase: localization and role in the myocardial distribution of purines, *Am. J. Physiol.*, 222, 550, 1972.

87. **Saito, D., Nixon, D. G., Vomacka, R. B., and Olsson, R. A.**, Relationship of cardiac oxygen usage, adenosine content, and coronary resistance in dogs, *Circ. Res.*, 47, 875, 1980.

88. **Saito, D., Steinhart, C. R., Nixon, D. G., and Olsson, R. A.**, Intracoronary adenosine deaminase reduces canine myocardial reactive hyperemia, *Circ. Res.*, 49, 1262, 1981.

89. **Schrader, J. and Gerlach, E.**, Compartmentation of cardiac adenine nucleotides and formation of adenosine, *Pfluegers Arch.*, 367, 129, 1976.

90. **Schrader, J., Haddy, F. J., and Gerlach, E.,** Release of adenosine, inosine and hypoxanthine from the isolated guinea pig heart during hypoxia, flow autoregulation and reactive hyperemia, *Pfluegers Arch.,* 396, 1, 1977.

91. **Schrader, J., Schütz, W., and Bardenheuer, H.,** Role of S-adenosylhomocysteine hydrolase in adenosine metabolism in mammalian heart, *Biochem. J.,* 196, 65, 1981.

92. **Schrader, J.,** Sites of action and production of adenosine in the heart, in *Purinergic Receptors,* Burnstock, G., Ed., Chapmann and Hall, 1981, 121.

93. **Schrader, J.,** Metabolism of adenosine and sites of production in the heart, in *Regulatory Function of Adenosine,* Berne, R. M., Rall, T. W., and Rubio, R., Eds., Martinus Nijhoff, Boston, 1983, 133.

94. **Schrader, J.,** Adenosine: a homeostatic metabolite in cardiac energy metabolism, *Circulation,* 81, 389, 1990.

95. **Schrader, W. P. and West, C. A.,** Localization of adenosine deaminase and adenosine deaminase complexing protein in rabbit heart. Implications for adenosine metabolism, *Circ. Res.,* 66, 754, 1990.

96. **Schütz, W., Schrader, J., and Gerlach, E.,** Different sites of adenosine formation in the heart, *Am. J. Physiol.,* 240, H963, 1981.

97. **Smolenski, R. T., Schrader, J., and DeGroot, H.,** Oxygen partial pressure and free intracellular adenosine of isolated cardiomyocytes, *Am. J. Physiol.,* in press.

98. **Sparks, H. V., Jr., De Witt, D. F., Wangler, R. D., Gorman, M. W., and Bassingthwaighte, J. B.,** Capillary transport of adenosine, *Fed. Proc., Fed. Am. Soc. Exp. Biol.,* 44, 2620, 1985.

99. **Sparks, H. V., Jr. and Bardenheuer, H.,** Regulation of adenosine formation by the heart, *Circ. Res.,* 58, 193, 1986.

100. **Stirling, C.,** Autoradiographic localization of H^3-adenosine (abstract), in *Regulatory Function of Adenosine,* Berne, R. M., Rall, T. W., and Rubio, R., eds., Martinus Nijhoff, Boston, 1983, 542.

101. **Stone, T. W. Newby, A. C., and Lloyd, H. G. E.,** Adenosine release, in The Adenosine Receptors, Vol. 6, Williams, M., Ed., Humana Press, Clifton, NJ, in press.

102. **Ueland, P. M.,** Pharmacological and biochemical aspects of S-adenosylhomocysteine and S-adenosylhomocysteine hydrolase, *Pharmacol. Rev.,* 34, 223, 1982.

103. **Truong, V. L., Collinson, A. R., and Lowenstein, J. M.,** 5'-Nucleosidases in rat heart, *Biochem. J.,* 253, 117, 1988.

104. **Ugurbil, K. and Holmsen, H.,** in *Platelets in Biology and Pathology,* Gordon, J. L., Ed., Elsevier Scientific, New York, 1981, 146.

105. **Ungerstedt, U.,** Microdialysis—a new bioanalytical sampling technique, *Curr. Separations,* 7, 43, 1986.

106. **van Belle, H.,** Uptake and deamination of adenosine by blood, *Biochim. Biophys. Acta,* 192, 124, 1969.

107. **van Bilsen, M., Van Der Vusse, G. J., Coumans, W. A., De Groot, M. J. M., Willemsen, P. H. M., and Reneman, R. S.,** Degradation of adenine nucleotides in ischemic and reperfused rat heart, *Am. J. Physiol.,* 257, H47, 1989.

108. **Walker, R. D. and Duerre, J. A.,** S-Adenosylhomocysteine metabolism in various species, *Can. J. Biochem.,* 53, 312, 1974.

109. **Wangler, R. D., Gorman, M. W., Wang, C. Y., De Witt, D. F., Chan, I. S., Bassingthwaighte, J. B., and Sparks, H. V.,** Transcapillary adenosine transport and interstitial adenosine concentration in guinea pig hearts, *Am. J. Physiol.,* 257, H89, 1989.

110. **Watkins, C. A. and Morgan, H. E.,** Relationship between rates of methylation and synthesis of heart protein, *J. Biol. Chem.,* 254, 693, 1979.

111. **Winkler, H.,** The composition of adrenal chromaffin granules: an assessment of controversial results, *Neuroscience,* 1, 65, 1976.

112. **Wittenberg, B. A. and Wittenberg, J. B.,** Oxygen pressure gradients in isolated cardiac myocytes, *J. Biol. Chem.,* 260, 6548, 1985.

113. **Worku, Y. and Newby, A. C.,** The mechanism of adenosine production of rat polymorphonuclear leucocytes, *Biochem. J.,* 214, 325, 1983.

Chapter 6

ADENOSINE DEAMINASE REGULATION OF PURINE ACTIONS

J. D. Geiger, R. A. Padua, and J. I. Nagy

TABLE OF CONTENTS

I. INTRODUCTION

The physiological actions that adenosine and its phosphorylated derivatives exert in various tissues are determined not only by purine receptor subtypes and the secondary signal transduction systems associated with these, but also by mechanisms governing purine availability *in vivo*.[1] Availability at receptors is determined to varying degrees by purine nucleoside and nucleotide release, reuptake, production, and degradation. Formulation of concepts concerning utilization of purine nucleosides as neuromodulators in the central nervous system (CNS) and regulators of cellular activity in peripheral tissues is dependent on a thorough understanding of purine metabolism, particularly that of adenosine. The major biochemical pathways that contribute to adenosine production and metabolism have been reviewed elsewhere.[1] These include recapture into the nucleotide pool by adenosine kinase (AK, EC 2.7.1.20) or hydrolysis by adenosine deaminase (ADA, EC 3.5.4.4), leading to the formation of inosine. Here we review some of the biochemical characteristics of ADA, draw attention to some of the pharmacological uses of this enzyme as well as those of its inhibitors, and discuss some of the proposed functional roles of ADA.

II. BIOCHEMICAL CONSIDERATIONS

The substrates adenosine and 2′-deoxyadenosine are hydrolyzed by ADA to yield the products inosine and 2′-deoxyinosine, respectively. Although some of the molecular mechanisms of this hydrolytic process and the biochemical properties of ADA isolated from tissues of several species are well established,[1] some unresolved issues relevant to considerations of the contribution of ADA to purine metabolism and, consequently, to the control of the cellular regulatory actions of purines are as follows:

1. ADA may be a member of a family of enzymes having purine nucleoside deaminase activity. For example, a nonspecific aminohydrolase capable of adenosine and 2′-deoxyadenosine deamination has been identified in splenic tissue and cultured human lymphoblasts from normal and ADA-deficient patients.[2,3] This enzyme is different from ADA in several respects; it has a 40-times greater K_m for adenosine, a different profile of relative substrate specificity, a lower pH optimum, greater heat stability, lower sensitivity to inactivation by heavy metals and ADA inhibitors, greater sensitivity to inhibition by adenine, and lacks cross reaction with antibodies against ADA. The possible existence of multiple forms of adenosine catabolic enzymes and the currently incomplete understanding of the biochemistry of these enzymes should be taken into account in attempts to dismiss or support any proposed involvement of ADA or adenosine deamination in restricting manifestations of adenosine-mediated cellular homeostasis.

2. ADA itself is thought to be predominantly a cytoplasmic enzyme and consists of a single 36-kDa polypeptide with which catalytic activity is associated. However, this low-molecular weight catalytic moiety can interact with what are termed ADA-binding proteins (ADA-BP) to form high-molecular weight aggregates up to 280 kDa in size.[4] These ADA-BP have been purified and localized immunohistochemically to cytoplasmic membranes and, in some cases, to external surfaces of cells.[5] It has been proposed that ADA-BP may anchor ADA at sites where the enzyme could exert control of adenosine levels near adenosine receptors.[6] Such a proposal is consistent with observations that a presumptive ecto-ADA is responsible for a significant fraction of total tissue extracellular adenosine deamination in brain,[7] vagus nerves,[8] lung,[9] coronary arterial endothelial cells,[10] and erythrocytes[11] (see Figure 1). Analyses of the functional role(s) of the proportion of ADA that is bound to cell-surface ADA-BP

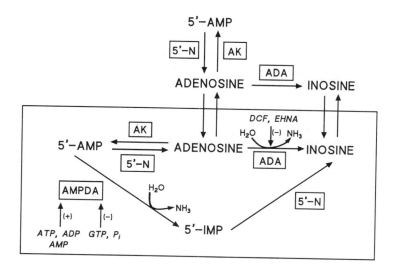

FIGURE 1. Schematic representation of extracellular and intracellular (inside box) metabolism of adenosine. Abbreviations for enzymes are as follows: 5'-nucleotidase (5'-N), adenosine kinase (AK), adenosine deaminase (ADA), and adenylate deaminase (AMPDA).

complexes and investigations of the influence of these complexes on adenosine levels and actions have been rendered less than straightforward by the enormous tissue variation in the percent occupancy of ADA-BP by ADA and the expression of ADA-BP in tissues of mouse, guinea pig, rabbit, and human, but seemingly not those of rat.

3. The gene for ADA in mouse is approximately 27 kb in length and encodes 1.5-, 1.7-, and 5.2-kb polyadenylated mRNAs, the 1.7 kb being the most abundant.[12,13] The promoter region of the mouse gene has been described as having novel characteristics in that it is GC rich, has multiple transcription-initiation sites, and lacks the TATA and CAAT boxes typical of many eukaryotic promoters. In contrast, the human ADA gene, which was found to be encoded by a single locus on the long arm of chromosome 20,[14] was approximately 32 kb in length and has a promoter with only one transcription-initiation site.[15] The now identified overall similarities between the promoter regions of ADA and those of "housekeeping" genes have added to generally held earlier views based on metabolic grounds that ADA is a constitutively expressed housekeeping enzyme.[16] However, it is difficult to reconcile such a ubiquitous housekeeping role for ADA with the findings that ADA activity varies by at least 50-fold among peripheral tissues, that steady-state ADA mRNA levels and transcription rates vary by as much as 10-fold among various cell types, that the enzyme increases in activity by 10- to 30-fold during cellular differentiation, and that ADA activity is high in rapidly dividing cells and, accordingly, has been implicated as an important regulator of cellular proliferation.[17] Moreover, the levels of ADA mRNA are posttranscriptionally regulated and closely parallel levels of ADA activity.[18] Therefore, ADA has either special functions in certain tissues and cells, or some cells have greater housekeeping needs for ADA. The former is further suggested by the pathological consequences of cellular ADA deficiency or excess. For example, severe combined immunodeficiency disease (SCID) is an autosomal recessive disease where 50 to 60% of sufferers completely lack ADA activity in erythrocytes and lymphoid cells. Curiously, however, ADA activity is severely reduced in other tissues, but only lymphocyte development and (primarily) T cell function are affected. Incidentally, despite early enthusiasm that

wasted mice were deficient in ADA, express immunological abnormalities, and thus might be an animal model for SCID, two reports failed to confirm the ADA deficiency.[1] Plasticity rather than constitutive rigidity in ADA expression is further suggested by increased ADA activity in patients suffering from a variety of seemingly unrelated illnesses, including congenital hypoplastic anemia, hereditary hemolytic anemia, leukemia, short-limbed dwarfism, reticular dysgenesis, infectious mononucleosis, hepatitis, and hepatocellular jaundice.[19-25]

III. ADA DISTRIBUTION AND EXPERIMENTAL USES

ADA activity is phylogenetically ubiquitous and widely distributed throughout mammalian tissues. In human, rat, and mouse, the highest levels of enzyme activity were found in thymus, spleen, placenta, and in organs comprising the gastrointestinal tract, whereas low activity was found in muscle, lung, and kidney (Table 1). In the rat CNS, immunohistochemical and biochemical studies have shown that ADA has a highly heterogeneous distribution.[1] ADA-positive neurons were found in dorsal root ganglia, spinal cord, retina, brainstem and spinal parasympathetic nuclei, and posterior hypothalamus.[26] The distribution of ADA-immunoreactive neural systems paralleled that of ADA activity as well as the distribution of adenosine-transport sites labeled autoradiographically with [³H]nitrobenzylthioinosine.[1] Taken together, these results provide information that may aid the study of the actions of adenosine in CNS neurons that express an abundance of ADA. However, we and others have reported substantial species differences in the distribution of ADA activity in the CNS and in the biochemical characteristics of the enzyme,[1,27] which has led to the suggestion that the properties of ADA may vary among animals. The significance of these differences is likely to remain hidden until their functional consequences to both metabolism and regulation of cellular function are determined.

A procedure that has been often used in many different *in vitro* and *in vivo* experimental paradigms is the addition or infusion of commercial preparations of ADA to promote the breakdown of adenosine. Although supplementation of tissue with exogenous ADA does not in itself provide evidence for or against the possible involvement of endogenous ADA in the regulation of purine levels or metabolism, it is quite possibly the currently most powerful and specific test for the mediation of biochemical or physiological processes by *endogenous* adenosine. Positive results of such tests are indicated when added ADA either suppresses a stimulatory response or alleviates an inhibitory action thought to be mediated by endogenous adenosine. Some examples of systems where the addition of exogenous ADA has been shown to influence biological responses are shown in Table 2. The application of this experimental approach requires consideration of several points. In *in vivo* studies, the amounts of ADA administered should be adequate to catabolize adenosine under basal conditions as well as conditions leading to increased adenosine concentrations and production rates. In addition, it should be emphasized that ADA may cross membranes and reach relevant sites of action slowly, if at all, and that dialysis or filtration of commercial ADA preparations might in some cases be necessary in order to remove contaminants which may cause prolonged hyperemic and shock responses.[28] In *in vitro* assays, the amount of ADA added may affect results if not carefully controlled.[29]

IV. ADA INHIBITION

Inhibition of enzymes such as those responsible for the synthesis and degradation of neurotransmitters and other physiologically active substances followed by assessment of the effects of enzyme eradication has been a staple approach in investigations of biological systems and one that has been extensively applied to studies of purine metabolism and

TABLE 1
Distribution of ADA Activity in Human, Rat, and Mouse Tissues

Tissue	Mouse	Rat	Human	Ref.
Alimentary				
Esophagus	770	—	—	59
Tongue	580	48	—	59
Stomach	190	(5110)	89	4
Forestomach	740	—	—	59
Glandular stomach	37	—	—	59
Intestine	20	—	—	59
Duodenum	(32,110)	(13,000)	127	4, 27
Jejunum	(12,200)	(13,240)	63	4, 27, 60
Pancreas	(1,100)	(3,090)	27	4, 27
Hematopoietic				
Thymus	362	62	283	4, 61
Spleen	97	54	128	4, 61
Erythrocytes	67	—	—	59
T cells	55	—	—	59
Bone marrow	21	—	—	59
Macrophages	19	—	—	59
B cells	12	—	—	59
Lymph node	—	—	23	4
Other				
Placenta	190	(28,840)	—	27, 59
Brain	10	6.8	—	61
Cerebral cortex	—	1.6	27	4
Cerebellum	—	2.2	9	4
Adrenal gland	—	15	21	4
Appendix	—	—	20	4
Testes	(1,300)	—	13	4, 27
Heart	(145)	(1430)	11	4, 27, 60
Liver	9	11.2	13	4, 60, 61
Skeletal muscle	(43)	(499)	8	4, 27, 60
Spinal cord	—	1.4	9	4
Lung	(662)	16.4	7	4, 27, 60
Kidney	(1,060)	150	6	4,27,60
Thyroid	10	6.8	4	4

Note: ADA activity was expressed as nanomoles per milligram of protein per minute except for those in parentheses which represent nanomoles per gram wet weight tissue per minute.

purinergic regulatory mechanisms. Given the involvement of adenosine in such mechanisms, the enzyme targeted for perturbation has frequently been ADA. Inhibition of ADA can serve three separate, though not necessarily mutually exclusive, purposes. Inactivation of the enzyme allows evaluation of the degree to which ADA contributes to adenosine metabolism, provides a means to assess the biological activities of endogenous adenosine in cases where the levels of this nucleoside and the expression of such activities are limited by catabolism via ADA, and facilitates analyses of the actions of exogenously added adenosine by preventing its unwanted breakdown in experimental preparations containing high levels of ADA. The compounds most commonly used to inactivate ADA are 2'-deoxycoformycin (DCF) and erythro-9-(2-hydroxy-3-nonyl)adenine (EHNA). These compounds are potent inhibitors of ADA, appear to be selective for the enzyme when used at appropriate concentrations, and are easily absorbed and distributed within tissues. Some examples of various *in vivo* and *in vitro* systems in which these inhibitors have been applied to study purine-mediated processes are given in Table 2.

TABLE 2
Effects of Exogenous ADA and ADA Inhibitors on Presumed Adenosine-Mediated Responses

Response	ADA amount	Effect(s)	ADA-I (conc)	Effect(s)	Ref.
Dopamine release	2.7 IU/ml	←	DCF (27 nM)	→	62
Acetylcholine release	10 µg/ml	←	DCF (10 µM)	→	63, 64
			EHNA (25 µM)	→	—
Adenosine release			EHNA (10 µM)	←	65
Serotonin release	10 µg/ml	←			66
Noradrenaline release	10 µg/ml	←			67
ACTH release	5 IU/ml	←			68
Inositol release	1 IU/ml	←			69
Noradrenaline synthesis	5 IU/ml	→			70
Field potentials	10—30 µg/ml	←			71
cAMP production					
Diazepam	200 IU/mg	→			72
Adenosine	0.5 IU/ml	→	DCF (3 µM)	→	67, 73
Glutamate	83 µg/ml	→			74
Noradrenaline	10 µg/ml	→			75
cAMP production					
Forskolin µg/ml	→			76	
Dopamine	0.1 IU/ml	→			77
cAMP inhibition					
CHA	1 IU/ml	←			78
Adenosine release hypoxia			EHNA (2 mg/g)	←	58, 79
			DCF (5 µg/kg)	←	
Adenosine production hypoxia	1.4 IU/ml	→	DCF (0.1 µg/kg)	←	80
Vasodilation hypoxia	5 IU/ml	→	EHNA (10 µg/kg)	←	81, 82
			DCF (0.1 µg/kg)	←	79
Vasodilation hypercapnia		←			79
Epileptiform activity	10 µg/ml		DCF (500 µg/kg)	→	83
Ischemic damage			DCF (1.5 µmol/min)	→	79, 84
			EDNA (3 µmol/min)	→	
Neutrophil-mediated injury	0.25 IU/ml	←	DCF (1 µM)	→	85
Uric acid formation	0.1 IU/min	←			86

Parameter	Concentration	Effect	Inhibitor	Effect	Ref.
Adenosine receptor binding			EHNA (61 nM)	→	87, 88
Hypothermia adenosine	5 IU/ml		DCF (5 nmol/h/week)	→	89
Rotation behavior apomorphine	1 µg/ml		DCF (10 mg/kg)	←	90
Relaxation taenia coli		→	EHNA (2 mg/kg)	→	43
Cellular respiration		←	EHNA (50 µM)	→	91
Adenosine cytotoxicity			EHNA (5 µM)	←	92
Gastric acid secretion	0.5 IU/ml	←	DCF	←	93
Isoproterenol-stimulated lipolysis	0.1 µg/ml	←			94
Noradrenaline-stimulated lipolysis	0.5 IU/ml	←			47
Insulin-stimulated glycolysis	5 µg/ml	←			95
Vascular resistance	6.5 IU/ml	←			96
Glucose uptake	6.5 IU/ml				96
Inhibition of gluconeogenesis			DCF (10 µM)	↑	97

Considerable attention has been devoted to DCF since it has been used clinically as an inhibitor to limit the metabolism of some immunosuppressive, antiviral, and antitumor purine analogs which are also substrates for ADA, and as a primary therapeutic agent for certain types of leukemia.[30] However, even at very low doses it has side effects which may arise from the aberrant actions of accumulated adenosine and 2'-deoxyadenosine.[1] DCF is a transition-state, noncompetitive inhibitor of ADA with an extraordinarily high potency for the enzyme ($K_i = 10^{-11}$ M). Given systemically, it accumulates quickly in areas with the highest levels of ADA activity, and at various times after administration produces various degrees of ADA inhibition in different tissues, which probably reflects differences in the turnover rates of the enzyme.[1,31] Although adenylate deaminase (AMPDA) is potently inhibited by DCF in some tissues, tests of *in vivo* and *in vitro* specificity of DCF for ADA vs. AMPDA in brain have shown that concentration thresholds for inhibition of AMPDA were many orders of magnitude higher than that for inhibition of ADA.[31] EHNA competitively inhibits ADA with a potency in the nanomolar range, but at concentrations around the 100-μM range, it has been shown to inhibit sperm axonemal dynein ATPase activity, actin-based cell motility, *de novo* nucleotide synthesis, and adenylate deaminase. The mechanisms underlying these higher concentration effects appear to be unrelated to ADA inhibition or to prevention of adenosine catabolism.[32,33]

While observations of an influence of ADA inhibition by, for example, DCF on physiological and biochemical processes listed in Table 2 may be considered as evidence for the involvement of adenosine in mediating these processes, a converse conclusion may be drawn by those more interested in the pharmacology of DCF. Thus, such observations in preparations where the involvement of adenosine is reasonably well established may be considered as at least indirect support for the notion that the pharmacological effects of DCF are mediated through adenosine. The strength of such circular arguments depends, of course, on the degree of certainty to which DCF biochemical specificity and adenosine physiological involvement are independently evaluated. It should be noted that although ADA inhibition has been shown to increase adenosine levels and release, it remains uncertain whether the physiological potentiations result directly from increased quantities of releasable stores of adenosine or indirectly from secondary effects on purine metabolism.

V. FUNCTIONAL ROLES

A. ADENOSINE SUPPRESSION SET POINT

It has been suggested in recent years that adenosine contributes to the establishment of cellular metabolic or functional tone. If this view is correct, then it may be speculated that the levels of ADA in certain tissues are co-regulated with adenosine receptors so as to optimize availability of adenosine according to desired needs for interaction with its receptors. In our studies, we found that DCF administered systemically to rats caused prolonged reductions in brain ADA activity, which did not appear to be due to a cellular toxic action. When a second dose of DCF was administered 14 d after an initial injection, the second treatment had virtually no inhibitory effect on ADA. These results suggested that the mechanism involved in the prolonged reduction of ADA activity was somehow saturated or inactivated. We speculated that ADA inhibition may lead to increased adenosine levels and decreased production of ADA by reducing neural activity and producing an aberrant elevation in what might be considered an "adenosine-suppression set point".[34] Rather than returning to some prevous balance, this new set point is ingrained as the *status quo* and requires decreased ADA activity for its continued maintenance. Alternatively, it may be speculated that in analogy with other transmitters, a DCF-induced increase in adenosine availability resulted in down regulation of adenosine receptors leading to a reduction in the presumptive "adenosine-suppression set point". In order to explain the DCF-induced prolonged reduction in ADA activity in this scenario, we postulate that such a reduction is undesirable and compensated by decreased catabolism of adenosine. If the long-term effects of DCF on ADA involve such mechanisms, then further studies of ADA may shed some light on processes governing the actions of adenosine.

B. INOSINE ACTIONS AND PRODUCTION

Although it is now clear that ADA is not evenly distributed in the CNS or peripheral tissues, the significance of this heterogeneity is not fully understood. We previously hypothesized that some central neurons which were found to express relatively greater quantities of ADA may have an added metabolic burden imposed by their possible utilization of purines in intercellular communication, and may require ADA to regulate adenosine levels and reduce possible cytotoxicity of adenosine and 2'-deoxyadenosine.[1,26] An additional possibility considered here is that the presence of at least some proportion of ADA reflects more a requirement for inosine synthesis than adenosine degradation. Based on its very low affinity for adenosine receptors, we and many others have referred to inosine as a physiologically inactive metabolite. However, there are indications in a variety of tissues that inosine has pharmacological actions distinct from, as well as in common with, adenosine. Actions of inosine that are similar to those of adenosine may be exerted indirectly by its competitive inhibition of adenosine uptake.

In CNS tissues, inosine has been shown to produce excitatory and inhibitory responses on cultured spinal neurons,[35] to suppress food intake by a mechanism apparently independent of adenosine,[36] to induce hypothermia, and to potentiate adenosine-induced vasodilatation of cerebral pial vessels probably through inhibition of adenosine transport.[37] That the levels of inosine reached in tissues as a result of hypoxic, anoxic, hemorrhagic, convulsant, and depolarization stimuli are physiologically active is supported by findings that the levels of inosine attained during, for example, seizures were sufficiently elevated to inhibit epileptiform activity, and administered inosine exhibited antiseizure potencies similar to those of adenosine.[38] At least part of this antiseizure activity may be mediated through benzodiazepine receptors for which inosine has been implicated as an endogenous ligand.[38] That inosine was the physiologically active species was supported by findings that agents such as phenytoin, which reportedly blocked seizure-induced rises in inosine levels, also blocked the elevated seizure thresholds subsequent to electroshock treatment. Since EHNA blocked the production of inosine and the presumed inosine-mediated rise in seizure threshold, it is likely that inosine was derived from degradation of adenosine via ADA, the activity of which was reported to increase, following seizures.[38]

In myocardium, findings with regard to inosine are particularly important given the high incidence of cardiac failure and ischemic disease. Inosine, which may be thought of as a local hormone, was found to increase blood flow, increase the chronotropic effects of norepinephrine, and produce positive inotropy without affecting heart rate, arterial pressure, or tissue energy charge.[39,40] Moreover, inosine release from myocardial tissue was found to increase perfusion rate, decrease infarct size after coronary occlusions, and generally increase overall cardiac contractility and function.[39] Based on results such as these, inosine has been referred to as an important new positive inotropic agent that has been used in human patients suffering from cardiac failure.[39] While it seems reasonably clear that inosine may produce pharmacologically and possibly physiologically important effects in heart, the mechanisms proposed by which these actions are mediated, i.e., β-adrenergic stimulation, increases in cAMP and/or decreased calcium require additional experimental challenge. Nevertheless, it appears clear that the protective effects of inosine were not being mediated through adenosine receptors.[39]

In other tissues, inosine has been shown to influence a variety of physiological responses. In the intestinal vasculature, inosine produced vasodilatation without affecting transcapillary fluid exchange by mechanism(s) both separate from and in common with adenosine.[41] In isolated rat hepatocytes, it activated ureagenesis with a potency about 60 times less than adenosine by a mechanism that was calcium dependent and blocked by glucagon, but not by adrenalin.[42] In addition, inosine with a potency only slightly less than that of adenosine relaxed tracheal smooth muscle[43] and increased the gluconeogenic effects of lactate (but not of glucagon or epinephrine) by mechanisms different from those for adenosine.[44] Although inosine was found to relax canine coronary and renal artery preparations by a mechanism distinct from similar effects produced by adenosine, in this case the potency of inosine was about three orders of magnitude less than that of adenosine.[45] Under certain conditions inosine, at concentrations insufficient to produce effects on its own, nevertheless potentiated the actions of adenosine in heart and intestine.[46] Particularly compelling evidence for the physiological effects of inosine as distinct from those of adenosine was obtained where it was shown that inosine, but not adenosine, inhibited both basal and stimulated rates of lipolysis in ADA-pretreated, isolated fat cells.[47]

Taken together, these findings suggest that inosine should *not* be considered a physiologically inert drug. The actions of inosine, some of which have been mentioned above, may be mediated by cell-surface receptors, but are more likely to be mediated, at least in the case of aortic smooth-muscle relaxation, intracellularly through nonadenosine "receptors", possibly through calcium.[48] Despite the fact that in many of these studies investigators failed to address the possibility that administered or generated inosine may have led to increased adenosine levels and actions through adenosine receptors following purine salvage, these studies strongly suggest that in addition to a possible role of ADA in adenosine inactivation, the enzyme may generate deaminated products having biochemically or physiologically relevant functions. As discussed below, however, it is important to point out that adenosine is not the sole, and in some cases not even the primary, source of inosine.

Adenosine may be produced either intracellularly or extracellularly through the dephosphorylation of AMP via AMP-specific cytosolic or ecto-5'-nucleotidases, respectively. Inosine, on the other hand, may be generated either intracellularly through deamination of AMP via cytosolic AMPDA to yield IMP followed by dephosphorylation of IMP via an IMP-specific 5'-nucleotidase, or both intracellularly and extracellularly through deamination of adenosine via cytosolic or ecto-ADA, respectively. The extent to which inosine arises from IMP dephosphorylation or adenosine deamination depends on a multitude of factors and varies with species, tissues, and tissue energy state. A significant proportion of inosine is produced via the AMPDA pathway under normal physiological conditions in brain, human erythrocytes, fat cells, and cultured endothelial cells,[49-52] as well as under altered physiological states such as in hypoxic endothelial cells and hepatocytes, in exercised skeletal muscle, and in ischemic brain.[51,53,54] Alternatively, ADA was the predominant pathway for inosine production in control and electrically stimulated brain, in control and metabolically stressed human T lymphoblastoid cells, in hypoxic hearts and myocytes, in human erythrocytes deprived of glucose, in stimulated vagus nerves, and in ischemic skeletal muscle.[8,49,50,54-56] With respect to species differences, inosine was generated principally through deamination of adenosine in ischemic heart of guinea pig, but through dephosphorylation in rat, rabbit, and frog heart.[57] In brain, inosine was produced following adenosine loading, ischemia, electroshock, and increased neuronal activity.[38,49,58] In some of these cases, it is likely that inosine production originated from the action of ADA, since the levels of adenosine were increased and those of inosine decreased by EHNA. In general, however, it appears that 5'-AMP degradation occurs by dephosphorylation under basal conditions, and by deamination leading to the production of IMP and ultimately inosine in tissues subjected to metabolically stressful conditions such as ischemia and hypoxia. This, of course, can be easily understood in terms of cellular attempts to maintain energy charge. Two points of note are as follows. First, a futile cycle where adenosine formed from 5'-AMP is immediately rephosphorylated by AK may lead to an underestimate of the importance of the AMP-dephosphorylation pathway. Second, although AMPDA activity in many tissues is often up to ten times higher than that of ADA, the ADA pathway may still prevail because the activity of the complex allosteric enzyme AMPDA is dependent on a variety of regulatory factors, including activation by ATP, ADP, AMP, alkali metals, and calcium, and on inhibition by GTP and Pi, all at physiologically attainable concentrations.[49]

ACKNOWLEDGMENTS

Our work cited here has been supported by grants from the Medical Research Council of Canada, Manitoba Health Research Council, University of Manitoba Faculty Fund, and Health Sciences Centre Research Foundation. J. D. Geiger, R. A. Padua, and J. I. Nagy are recipients of Scholarship, Studentship, and Scientist Awards from the Medical Research Council, respectively.

REFERENCES

1. **Geiger, J. D. and Nagy, J. I.**, Adenosine deaminase and [³H]nitrobenzylthioinosine as markers of adenosine metabolism and transport in central purinergic systems, in *Adenosine and Adenosine Receptors*, Williams, M., Ed., Humana Press, Clifton, NJ, 1990, 225.
2. **Schrader, W. P., Pollara, B., and Meuwissen, H. J.**, Characterization of the residual adenosine deaminating activity in the spleen of a patient with combined immunodeficiency disease and adenosine deaminase deficiency, *Proc. Natl. Acad. Sci. U.S.A.*, 75, 446, 1978.
3. **Daddona, P. E. and Kelley, W. M.**, Characteristics of an aminohydrolase distinct from adenosine deaminase in cultured human lymphoblasts, *Biochim. Biophys. Acta*, 658, 280, 1981.
4. **Van der Weyden, M. B. and Kelley, W. N.**, Human adenosine deaminase. Distribution and properties, *J. Biol. Chem.*, 251, 5448, 1976.
5. **Andy, R. J. and Kornfeld, R.**, The adenosine deaminase binding protein of human skin fibroblasts is located on the cell surface, *J. Biol. Chem.*, 257, 7922, 1982.
6. **Schrader, W. P. and West, C. A.**, Adenosine deaminase complexing proteins are localized in exocrine glands of the rabbit, *J. Histochem. Cytochem.*, 33, 508, 1985.
7. **Franco, R., Canela, E. I., and Bozal, J.**, Heterogeneous localization of some purine enzymes in subcellular fractions of rat brain and cerebellum, *Neurochem. Res.*, 11, 423, 1986.
8. **Maire, J. C., Medilanski, J., and Straub, R. W.**, Release of adenosine, inosine and hypoxanthine from rabbit non-myelinated nerve fibers at rest and during activity, *J. Physiol.*, 357, 67, 1984.
9. **Hellewell, P. G. and Pearson, J. D.**, Metabolism of circulating adenosine by the porcine isolated perfused lung, *Circ. Res.*, 53, 1, 1983.
10. **Meghji, P., Middleton, K., Hassall, C. J. S., Phillips, M. I., and Newby, A. C.**, Evidence for extracellular deamination of adenosine in the rat heart, *Int. J. Biochem.*, 20, 1335, 1988.
11. **Bielat, K. and Tritsch, G. L.**, Ecto-enzyme activity of human erythrocyte adenosine deaminase, *Mol. Cell. Biochem.*, 86, 135, 1989.
12. **Ingolia, D. E., Al-Ubaidi, M. R., Yeung, C. Y., Bigo, H. A., Wright, D. A., and Kellems, R. E.**, Molecular cloning of the murine adenosine deaminase gene from a genetically enriched source: identification and characterization of the promotion region, *Mol. Cell. Biol.*, 6, 4458, 1986.
13. **Yeung, . Y., Ingolia, D. E., Roth, D. B., Shoemaker, C., Al-Ubaidi, M. R., Yen, J. Y., Ching, C., Bobonis, C., Kaufman, R. J., and Kellems, R. E.**, Identification of functional murine adenosine deaminase cDNA clones by complementation in *Escherichia coli*, *J. Biol. Chem.*, 260, 10299, 1985.
14. **Tischfield, J. A., Creagan, R. P., Nichols, E. A, and Ruddle, F. H.**, Assignment of a gene for adenosine deaminase to human chromosome 20, *Hum. Hered.*, 24, 1, 1974.
15. **Wiginton, D. A., Kaplan, D. J., States, C. J., Akeson, A. L., Perme, L. M., Bilyk, I. J., Vaughn, A. J., Lattier, D. L., and Hutton, J. J.**, Complete sequence and structure of the gene for human adenosine deaminase, *Biochemistry*, 25, 8234, 1986.
16. **Valerio, D., Duyvesteyn, M. G. C., Dekker, B. M. M., Weeda, G., Berkvens, Th.M., van der Boorn, L., van Ormondt, H., and van der Eb, A. J.**, Adenosine deaminase: characterization and expression of a gene with a remarkable promoter, *EMBO J.*, 4, 437, 1985.

17. **Hoang, T. and Bergeron, M.**, Adenosine deaminase in renal ontogeny and compensatory hypertrophy in the rat, *Cell Tissue Kinet.*, 16, 59, 1983.

18. **Berkvens, T. M., Schoute, F., van Ormondt, H., Merra Khan, P., and Van der Eb, A. J.**, Adenosine deaminase gene expression is regulated posttranscriptionally in the nucleus, *Nucleic Acids Res.*, 16, 3255, 1988.

19. **Koehler, L. H. and Benz, E. J.**, Serum adenosine deaminase: methodology and clinical applications, *Clin. Chem.*, 8, 133, 1962.

20. **Goldberg, D. M.**, Serum adenosine deaminase in the differential diagnosis of jaundice, *Br. Med. J.*, 1, 353, 1965.

21. **Glader, B. E., Backer, K., and Diamond, L. K.**, Elevated erythrocyte adenosine deaminase activity in congenital hypoplastic anemia, *N. Engl. J. Med.*, 309, 1486, 1983.

22. **Valentine, W. M., Pablia, D. E., Tartaglia, A. P., and Gilsanz, F.**, Hereditary hemolytic anemia with increased red cell adenosine deaminase (45 to 70-fold) and decreased adenosine triphosphate, *Science*, 195, 783, 1977.

23. **Morisaki, T., Fujii, H., and Miwa, S.**, Adenosine deaminase (ADA) in leukemia. Clinical value of plasma ADA activity and characterization of leukemic cell ADA, *Am. J. Hematol.*, 19, 37, 1985.

24. **Cowan, M. J. and Ammann, A. J.**, Immunodeficiency syndromes associated with inherited metabolic disorders, *Clin. Hematol.*, 10, 139, 1981.

25. **Ownby, D. R., Pizzo, S., Blackmon, L., Gall, S. A., and Buckley, R. H.**, Severe combined immunodeficiency with leukopenia (reticular dysgenesis) in siblings: immunologic and histopathologic findings, *J. Pediatr.*, 89, 382, 1976.

26. **Nagy, J. I., Geiger, J. D., and Staines, W. A.**, Adenosine deaminase and purinergic neuroregulation, *Neurochem. Int.*, 211, 1990.

27. **Brady, T. G. and O'Donovan, C. I.**, A study of the tissue distribution of adenosine deaminase in six mammal species, *Comp. Biochem. Physiol.*, 14, 101, 1965.

28. **Olsson, R. A.**, Intra-arterial adenosine deaminase. A tool for assessing physiologic functions of adenosine, in *Topics and Perspectives in Adenosine Research*, Gerlach, E. and Becker, B. F., Eds., Springer-Verlag, Berlin, 1987, 438.

29. **Linden, J.**, Adenosine deaminase for removing adenosine. How much is enough?, *Trends Pharmacol. Sci.*, 10, 260, 1989.

30. **Cummings, F. J., Crabtree, G. W., Wiemann, M. C., Spremulli, E. N., Parks, R. E., and Calabresi, P.**, Clinical, pharmacologic and immunologic effects of 2'-deoxycoformycin, *Clin. Pharmacol. Ther.*, 44, 501, 1988.

31. **Padua, R., Geiger, J. D., Dambock, S., and Nagy, J. I.**, 2'-Deoxycoformycin inhibition of adenosine deaminase in rat brain: *in vivo* and *in vitro* analysis of specificity, potency and enzyme recovery, *J. Neurochem.*, 54, 1169, 1990.

32. **Schliwa, M., Ezzell, R. M., and Euteneuer, U.**, Erythro-9-[3-(2-hydroxynonyl)]adenine is an effective inhibitor of cell motility and actin assembly, *Proc. Natl. Acad. Sci. U.S.A.*, 81, 6044, 1984.

33. **Carabaza, A., Ricart, M. D., Mor, A., Buinovart, J. J., and Ciudad, C. J.**, Role of AMP on the activation of glycogen synthase and phosphorylase by adenosine, fructose, and glutamine in rat hepatocytes, *J. Biol. Chem.*, 265, 2724, 1990.

34. **Geiger, J. D., Padua, R. A., and Nagy, J. I.**, Adenosine deaminase and adenosine transport in the CNS, in *Purines in Cellular Signalling. Targets for New Drugs*, Jacobson, K., Daly, J., and Manganiello, V., Eds., Springer-Verlag, New York, 1990, 20.

35. **MacDonald, J. F., Barker, J. L., Paul, S. M., Marangos, P. J., and Skolnick, P.**, Inosine may be an endogenous ligand for benzodiazepine receptors on cultured spinal neurons, *Science*, 205, 715, 1979.

36. **Levine, A. S. and Morley, J. E.**, Purinergic regulation of food intake, *Science*, 217, 77, 1982.

37. **Ngai, A. C., Monsen, M. R., Ibayashi, S., Ko, K. R., and Winn, H. R.**, Effect of inosine on pial arterioles: potentiation of adenosine-induced vasodilation, *Am. J. Physiol.*, 256, H603, 1989.

38. **Dragunow, M.**, Purinergic mechanisms in epilepsy, *Progr. Neurobiol.*, 31, 85, 1988.

39. **Czarnecki, W. and Czarnecki, A.**, Haemodynamic effects of inosine. A new drug for failing human heart, *Pharmacol. Res.*, 21, 587, 1989.

40. **Samet, M. K. and Rutledge, C. O.**, Antagonism of the positive chronotropic effect of norepinephrine by purine nucleosides in rat atria, *J. Pharmacol. Exp. Ther.*, 232, 106, 1985.

41. **Granger, D. N., Valleau, J. D., Parker, R. E., Lane, R. S., and Taylor, A. E.**, Effects of adenosine on intestinal hemodynamics, oxygen delivery, and capillary fluid exchange, *Am. J. Physiol.*, 235, H707, 1978.

42. **Guinzberg, P. R., Laguna, I., Zentella, A., Guzman, R., and Pina, E.**, Effect of adenosine and inosine on ureagenesis in hepatocytes, *Biochem. J.*, 245, 371, 1987.

43. **Satchell, D.**, Use of purine nucleotide and nucleoside metabolising enzymes as tools to determine the presence of purinergic nerve transmission in smooth muscle, *Adv. Exp. Med. Biol.*, 253B, 435, 1989.

44. **De Pina, M. Z., Diaz-Cruz, A., Guinzberg, R., and Pina, E.,** "Hormone-like" effect of adenosine and inosine on gluconeogenesis from lactate in isolated hepatocytes, *Life Sci.*, 45, 2269, 1989.

45. **Sinclair, R. J., Randall, J. R., Wise, G. E., and Jones, C. E.,** Response of isolated renal artery rings to adenosine and inosine, *Drug Dev. Res.*, 6, 391, 1985.

46. **Rossi, F., Lampa, E., Giordano, L., Marfella, A., Ariello, B., Matera, M. G., DeCarlo, R., and Marmo, E.,** Interactions between inosine and adenosine: experimental researches, *Res. Commun. Chem. Pathol. Pharmacol.*, 35, 397, 1982.

47. **Gaion, R. M., Dorigo, P., and Gambarotto, L.,** A reexamination of the effects induced by adenosine and its degradation products on rat fat cell lipolysis, *Biochem. Pharmacol.*, 37, 3215, 1988.

48. **Collis, M. G., Palmer, D. B., and Baxter, G. S.,** Evidence that the intracellular effects of adenosine in the guinea-pig aorta are mediated by inosine, *Eur. J. Pharmacol.*, 121, 141, 1986.

49. **Schultz, V. and Lowenstein, J. M.,** The purine nucleotide cycle. Studies of ammonia production and interconversions of adenine and hypoxanthine nucleotides and nucleosides by rat brain *in situ*, *J. Biol. Chem.*, 253, 1938, 1978.

50. **Bontemps, F., Van den Berghe, G., and Hers, H. G.,** Pathways of adenine nucleotide catabolism in erythrocytes, *J. Clin. Invest.*, 77, 824, 1986.

51. **Shryock, J. C., Rubio, R., and Berne, R. M.,** Release of adenosine from pig aortic endothelial cells during hypoxia and metabolic inhibition, *Am. J. Physiol.*, 254, H223, 1988.

52. **Kather, H.,** Purine accumulation in human fat cell suspensions. Evidence that human adipocytes release inosine and hypoxanthine rather than adenosine, *J. Biol. Chem.*, 263, 8803, 1988.

53. **Vincent, M. F., Van den Berge, G., and Hers, H. G.,** The pathway of adenine nucleotide catabolism and its control in isolated hepatocytes subjected to anoxia, *Biochem. J.*, 202, 117, 1982.

54. **Matthias, R. F. and Busch, E. W.,** Abbau der Purinnukleotide is ischamischen Gehirn und Muskelgeweden von Kaninchen, *Hoppe-Seyler's Z. Physiol. Chem.*, 350, 1410, 1969.

55. **Van Belle, H.,** Myocardial purines during ischemia, reperfusion and pharmacological protection, *Mol. Physiol.*, 8, 615, 1985.

56. **Newby, A. C., Holmquist, C. A., Illingsworth, J., and Pearson, J. D.,** The control of adenosine concentration in polymorphonuclear leucocytes, cultured heart cells and isolated perfused heart from the rat, *Biochem. J.*, 214, 317 1983.

57. **Van Belle, H., Wynants, J., and Goossens, F.,** Formation and release of nucleosides in the ischemic myocardium. Is the guinea-pig the exception?, *Basic Res. Cardiol.*, 80, 653, 1985.

58. **Zetterstrom, T., Vernet, L., Ungerstedt, U., Tossman, U., Jonzon, B., and Fredholm, B. B.,** Purine levels in the intact rat brain. Studies with an implanted perfused hollow fibre, *Neurosci. Lett.*, 29, 111, 1982.

59. **Kellems, R. E.,** Tissue Distribution of ADA Activity in Mice, brochure of the Department of Biochemistry Graduate Program, Baylor University, Waco, TX, 1989, 37.

60. **Arch, J. R. S. and Newsholme, E. A.,** Activities and some properties of 5'-nucleotidase, adenosine kinase and adenosine deaminase in tissues from vertebrates and invertebrates in relation to the control of the concentration and the physiological role of adenosine, *Biochem. J.*, 174, 965, 1978.

61. **Willis, E. H., Carson, D. A., and Shultz, L. D.,** Adenosine deaminase activity in recipients of bone marrow from immunodeficient mice homozygous for the wasted mutation, *Biochem. Biophys. Res. Commun.*, 145, 581, 1987.

62. **Michaelis, M. L., Michaelis, E. K., and Myers, S. L.,** Adenosine modulation of synaptosomal dopamine release, *Life Sci.*, 24, 2083, 1979.

63. **Jackisch, R., Fehr, R., and Hertting, G.,** Adenosine: an endogenous modulator of hippocampal noradrenaline release, *Neuropharmacology*, 24, 499, 1985.

64. **Sebastiao, A. M. and Ribeiro, J. A.,** On the adenosine receptor and adenosine inactivation at the rat diaphragm neuromuscular junction, *Br. J. Pharmacol.*, 94, 109, 1988.

65. **Green, R. D.,** Release of adenosine by C1300 neuroblastoma cells in tissue culture, *J. Supramol. Struct.*, 13, 175, 1980.

66. **Feuerstein, T., Hertting, G., and Jackisch, R.,** Modulation of hippocampal serotonin (5-HT) release by endogenous adenosine, *Eur. J. Pharmacol.*, 107, 233, 1985.

67. **Nimit, Y., Skolnick, P., and Daly, J.,** Adenosine and cyclic AMP in rat cerebral cortical slices: effects of adenosine uptake inhibitor and adenosine deaminase inhibitors, *J. Neurochem.*, 36, 908, 1981.

68. **Anand-Serivastava, M. B., Cantin, M., and Gutkowska, J.,** Adenosine regulates the release of adrenocorticotrophic hormone (ACTH) from cultured anterior pituitary cells, *Mol. Cell. Biochem.*, 89, 21, 1989.

69. **Rubio, R., Bencherif, M., and Berne, R. M.,** Inositol phospholipid metabolism during and following synaptic activation; role of adenosine, *J. Neurochem.*, 52, 797, 1989.

70. **Birch, P. J. and Fillenz, M.,** Adenosine receptor and beta-adrenoceptor stimulation increases noradrenaline synthesis in hippocampal synaptosomes, *Neurochem. Int.*, 8, 165, 1986.

71. **Dunwiddie, T. V. and Hoffer, B. J.,** Adenine nucleotides and synaptic transmission in the *in vitro* rat hippocampus, *Br. J. Pharmacol.*, 69, 59, 1980.

72. **York, M. J. and Davies, L. P.**, The effect of diazepam on adenosine uptake and adenosine-stimulated adenylate cyclase in guinea-pig brain, *Can. J. Physiol. Pharmacol.*, 60, 302, 1982.

73. **Kobayashi, K., Kuroda, Y., and Yoshioka, M.**, Change of cyclic AMP level in synaptosomes from cerebral cortex; increase by adenosine derivatives, *J. Neurochem.*, 36, 86, 1981.

74. **Rall, T. W. and Lehne, R. A.**, Ontogeny of adenosine 3',5'-monophosphate metabolism in guinea pig cerebral cortex. II. Development of responses to L-glutamate in the presence of adenosine or histamine, *Mol. Cell. Biochem.*, 73, 157, 1987.

75. **Schwabe, U., Ohga, Y., and Daly, J. W.**, The role of calcium in the regulation of cyclic nucleotide levels in brain slices of rat and guinea pig, *Naunyn-Schmiedeberg's Arch. Pharmacol.*, 302, 141, 1978.

76. **Murphy, M. G. and Byczko, Z.**, Effects of adenosine analogues on basal, prostaglandin E_1- and forskolin-stimulated cyclic AMP formation in intact neuroblastoma cells, *Biochem. Pharmacol.*, 38, 3289, 1989.

77. **Paes de Carvalho, R. and de Mello, F. G.**, Expression of A_1 adenosine receptors modulating dopamine-dependent cyclic AMP accumulation in the chick embryo retina, *J. Neurochem.*, 44, 845, 1985.

78. **Blazynski, C.**, Adenosine A_1 receptor-mediated inhibition of adenylate cyclase in rabbit retina, *J. Neurosci.*, 7, 2522, 1987.

79. **Phillis, J. W.**, Adenosine in the control of the cerebral circulation, *Cerebrovasc. Brain Metab. Rev.*, 1, 26, 1989.

80. **Belloni, F. L., Rubio, R., and Berne, R. M.**, Intracellular adenosine in isolated rat liver cells, *Pfluegers Arch.*, 400, 106, 1984.

81. **Kontos, H. A. and Wei, E. P.**, Role of adenosine in cerebral arteriolar dilation from arterial hypoxia, *Fed. Proc., Fed. Am. Soc. Exp. Biol.*, 40, 454, 1981.

82. **Dora, E.**, Effect of theophylline treatment on the functional hyperaemic and hypoxic responses of cerebrocortical microcirculation, *Acta Physiol. Hung.*, 68, 183, 1986.

83. **Ault, B. and Wang, C. M.**, Adenosine inhibits epileptiform activity arising in hippocampal area CA3, *Br. J. Pharmacol.*, 87, 695, 1986.

84. **Dhasmana, J. P., Digerness, S. F., Geckle, J. M., Ng, T. C., Glickson, J. D., and Blackstone, E. H.**, Effect of adenosine deaminase inhibitors on the heart's functional and biochemical recovery from ischemia: a study utilizing the isolated rat heart adapted to ^{31}P nuclear magnetic resonance, *J. Cardiovasc. Pharmacol.*, 5, 1040, 1983.

85. **Cronstein, B. N., Levin, R. I., Belanoff, J., Weissman, G., and Hirschhorn, R.**, Adenosine: an endogenous inhibitor of neutrophil-mediated injury to endothelial cells, *J. Clin. Invest.*, 78, 760, 1986.

86. **O'Neill, R. D.**, Adenosine modulation of striatal neurotransmitter release monitored *in vivo* using voltammetry, *Neurosci. Lett.*, 63, 11, 1986.

87. **Williams, M. and Risley, E. A.**, Biochemical characterization of putative central purinergic receptors by using 2-chloro[^{3}H]adenosine, a stable analog of adenosine, *Proc. Nat. Acad. Sci. U.S.A.*, 77, 6892, 1980.

88. **Porter, N. M., Radulovacki, M., and Green, R. D.**, Desensitization of adenosine and dopamine receptors in rat brain after treatment with adenosine analogs, *J. Pharmacol. Exp. Ther.*, 244, 218, 1988.

89. **Davies, L. P., Baird-Lambert, J., and Jamieson, D. D.**, Potentiation of pharmacological responses to adenosine, in vitro and in vivo, *Gen. Pharmacol.*, 13, 27, 1982.

90. **Fredholm, B. B., Herrera-Marschitz, M., Jonzon, B., Lindstrom, K., and Ungerstedt, U.**, On the mechanism by which methylxanthines enhance apomorphine-induced rotation behavior in the rat, *Pharmacol. Biochem. Behav.*, 19, 535, 1983.

91. **Schimmel, R. J., Elliott, M. E., and Dehmel, V. C.**, Interactions between adenosine and alpha-1 adrenergic agonists in regulation of respiration in hamster brown adipocytes, *Mol. Pharmacol.*, 32, 26, 1987.

92. **Snyder, F. F., Hershfield, M. S., and Seegmiller, J. E.**, Cytotoxic and metabolic effects of adenosine and adenine on human lymphoblasts, *Cancer Res.*, 38, 2357, 1978.

93. **Gerber, J. G. and Payne, N. A.**, Endogenous adenosine modulates gastric acid secretion to histamine in canine parietal cells, *J. Pharmacol. Exp. Ther.*, 244, 190, 1988.

94. **Szillat, D. and Bukowiecki, L. J.**, Control of brown adipose tissue lipolysis and respiration by adenosine, *Am. J. Physiol.*, 245, E555, 1983.

95. **Leighton, B. Lozeman, F. J., Vlachonikolis, G., Challiss, R. A. J., Pitcher, J. A., and Newsholme, E. A.**, Effects of adenosine deaminase on the sensitivity of glucose transport, glycolysis and glycogen synthesis to insulin in muscles of the rat, *Int. J. Biochem.*, 20, 23, 1988.

96. **Martin, S. E. and Bockman, E. L.**, Adenosine regulates blood flow and glucose uptake in adipose tissue of dogs, *Am. J. Physiol.*, 250, H1127, 1986.

97. **Lavoinne, A., Buc, H. A., Claeyssens, S., Pinosa, M., and Matray, F.**, The mechanism by which adenosine decreases gluconeogenesis from lactate in isolated rat hepatocytes, *Biochem. J.*, 246, 449, 1987.

Chapter 7

NUCLEOSIDE TRANSPORT IN CELLS: KINETICS AND INHIBITOR EFFECTS

Leif Hertz

TABLE OF CONTENTS

I. INTRODUCTION: WHY TRANSPORT NUCLEOSIDES

It is well established that the animal body as a whole does not require exogenous supplies of purines or pyrimidines (which, together with the hexose ribose, are the constituents of the nucleosides adenosine, uridine, guanosine, cytosine, and thymidine), but can synthesize these compounds from the products of protein and carbohydrate metabolism.[1]

In spite of the ability of mammalian cells to synthesize nucleosides, the presence of transport mechanisms for these compounds across cell membranes is well established. Their presence is of importance for several reasons: (1) some cells, e.g., erythrocytes, leukocytes, bone marrow cells, cells in the gastrointestinal tract, and perhaps some brain cells,[2,3] are deficient in *de novo* synthesis of purines or pyrimidines and thus need to accumulate purines or pyrimidines which have either been ingested or produced in other cells; (2) the physiological role of certain cells, e.g., in the kidney or in the choroid plexus, is to transport different compounds (which may include nucleosides) across a barrier between different compartments; and (3) some cells, predominantly in the central nervous (CNS) and cardiovascular systems, express receptors for nucleosides on their cell surface, and cellular uptake is an important mechanism to terminate the action of these transmitters.

Elucidation of transport mechanisms for nucleosides have to a large extent been obtained using erythrocytes and other blood cells as well as transformed cells, often in culture. Most of these cell types have been found to behave in a similar manner. This work has led to a considerable amount of solid information and has established some basic concepts, as will be discussed in more detail in Section III. More recently, some authors have attempted to apply these findings and concepts in studies of cells from the CNS, whereas others have studied transport in neural cells in an analogous way to transport of the amino acid transmitters, glutamate, and γ-aminobutyric acid (GABA). These two different approaches have led to very different concepts and conclusions and will be discussed in detail in Section IV. However, an understanding of the methodologies and terminologies used in transport studies is important for this discussion; they will be briefly described in the following section (for more detailed discussion, see, e.g., References 4 to 6), since some confusion exists in this area.

II. TRANSPORT: PRINCIPLES AND DEFINITION

A. GENERAL PRINCIPLES

Virtually all detailed investigations of accumulation across cell membranes are carried out by adding a radioactively labeled compound, e.g., [³H]- or [¹⁴C]adenosine, to the medium (extracellular phase) and measuring the content of radioactivity in the cells as a function of (1) the length of the incubation time and (2) the extracellular concentration. This is especially easy in monolayers of cultured cells adherent to a culture dish where the medium is easily removed by washing and no problems result from extracellular diffusion. However, identical principles can be applied to nonadherent isolated cells and synaptosomes.[7] Brain slices or intact muscle can be studied in a similar manner, but extracellular diffusion within the tissue may create a concentration gradient[8] and, therefore, difficulties in the interpretation. Determination of accumulated radioactivity in the cells (DPM per milligram protein) and of the specific activity (DPM per nanomole) in the medium allows a simple calculation of the accumulated amount of the compound into the cells (DPM $\times$ mg^{-1}/DPM $\times$ nmol^{-1}). The ratio between the volume of the extracellular phase (e.g., tissue culture medium in the dish) and the amount of tissue is often, but not always, so high that accumulation of a small amount of label from the medium into the tissue and release of initially unlabeled intracellular compound will not alter to any significant extent the specific activity of the labeled compound or its concentration in the medium.

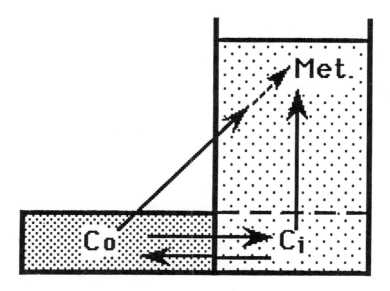

FIGURE 1. Cartoon describing transport of a soluble compound from the extracellular (:::::) to the intracellular (· · · ·) water space. Diffusion (facilitated or simple) can increase the intracellular concentration of the compound (C_i) up to, but not beyond, the extracellular (C_o) level (stippled line), and is generally bidirectional. A continuing inwardly directed diffusional net transport can be achieved by intracellular metabolism, reducing the intracellular level of the *unmetabolized* compound and thus maintaining a concentration gradient between C_i and C_o. A higher intracellular (C_i) than extracellular (C_o) concentration of the *compound itself* can be achieved by active transport. Whether or not an active-transport compound is secondarily metabolized (stippled arrow) has no influence on the rate of the transmembrane transport.

Release from cells can be studied in a manner which in principle is similar, i.e., by loading the cells with the isotope in question and following the subsequent accumulation of the labeled compound (or its metabolites) into a nonradioactive medium. Alternatively, the release of the compound as such can be measured. This can be done *in vivo* by measuring contents of the compound being released from the brain surface and accumulated in a cortical cup.[9]

B. TRANSPORT VS. UPTAKE

The definition of "transport" used in this review will be "passage of permeant molecules across the cell membrane". This is a wider definition than that used in most studies, e.g., in erythrocytes in which the definition has been limited to "transporter-mediated passage"; the present definition thus does not exclude a possible nontransporter-mediated transport, i.e., simple diffusion. This difference in definition probably reflects a different focus of research interest, i.e., a primary interest in membrane transport function in the research on erythrocytes vs. focus on transport as a means to remove components from the extracellular space in the present review. According to the present definition, transport may occur by three different mechanisms: (1) simple diffusion, (2) facilitated diffusion, and (3) active transport. Some characteristics of these processes are illustrated in Figure 1. Diffusion, whether facilitated or simple, is nonconcentrative, i.e., *net increase* in the intracellular content of the compound can occur only until extracellular (C_o) and intracellular (C_i) concentrations are identical (stippled line in Figure 1). After this has been established, bidirectional fluxes will still take place, but lead to no net changes. Simple unidirectional diffusion is nonsaturable when the concentration is increased, is not inhibitable by transport inhibitors, is unaffected by transport of the same, or a chemically closely related compound in the opposite direction, and is neither sodium dependent nor stereospecific. Both facilitated diffusion and active transport are saturable, inhibited by specific drugs, and occasionally stereospecific.[5,10] Active transport is energy requiring and very often occurs against a concentration gradient (heavy arrow between compartments o and i in Figure 1), whereas this is not so in the case of facilitated diffusion. Facilitated diffusion may occur as a "homoexchange", i.e., a coupled one-to-one exchange between intracellular and extracellular molecules. Experimentally, this is most easily indicated by a change of transport in one direction by procedures altering unidirectional transport in the opposite direction. This mechanism can, in some cases (i.e., if there initially is a higher intracellular than extracellular concentration of the compound under investigation), lead to a concentrative uptake of *label*, but not of the compound as such. Correspondingly, a compound may be exchanged in this manner with a chemically closely related compound (hetero-exchange). Some active transport processes, e.g., that catalyzed by the Na^+,K^+-ATPase, are also coupled with transport in the opposite direction, but in this case of another species, e.g., an inward transport of K^+ coupled with an outward transport of Na^+. The ion gradient created by this exchange can provide the energy for active uptake of e.g., transmitter amino acids, which thus becomes dependent upon the presence of Na^+ and K^+.[6] A similar system might be used for active transport of nucleosides in some cell types.[11]

Transport of permeant metabolizable compounds like glutamate and adenosine into intact cells will often be followed by intracellular conversion of the transported compound to metabolites retaining the label. After active transport, the total amount of transported compound in the cell will be independent of the extent of the subsequent metabolic conversion (stippled intracellular arrow in Figure 1), as also mentioned by Le Hir and Dubach.[11] However, since uptake by facilitated diffusion only occurs until the concentration of the transported compound in the intracellular water phase is similar to that in the incubation medium (extracellular fluid), net uptake across the cell membrane by facilitated diffusion can be greatly enhanced by metabolic conversion of the compound (solid intracellular arrow in Figure 1), which will maintain an inwardly directed gradient of the nonmetabolized compound. The process resulting from the sum of permeant transport by facilitated diffusion across the cell membrane and subsequent intracellular metabolism in the cells is often defined as "uptake". In erythrocytes and cells behaving in a similar manner, the transport by facilitated diffusion itself is in most cases faster than subsequent metabolic conversion. Thus, as illustrated in Figure 2, transport studied after ultrashort periods of incubation, e.g., a few seconds will be faster (≈ 20 pmol/μl cell water in the period 0 to 10 s or 120 pmol/min [Figure 2A] than uptake, studied over minutes or in some cases even hours (≈ 400 pmol/μl cell water in the period 0 to 6 min or 70 pmol/min [Figure 2B]). In the latter case, the actual time during which the transport occurs at the rate determined by the faster facilitated diffusion (Figure 2B) is short compared to the longer period during which the limiting factor is the rate of the metabolic conversion; therefore, the uptake rate after longer incubation times will, for all practical purposes, be determined by the metabolic rate during the uptake.

C. FACILITATED DIFFUSION

A strict distinction between transport by facilitated diffusion, which has to be measured during very short periods, and metabolism-driven uptake is of theoretical importance. The distinction between transport and uptake is also of practical importance since the transport by facilitated diffusion can be regulated and may be dramatically reduced, e.g., by the inhibitors dipyridamole or nitrobenzylthioinosine (NBTI). Under these conditions it may become rate limiting, leading to a slower accumulation and metabolism of extracellular nucleoside. This can be of therapeutic value, e.g., in cancer chemotherapy.[5,12] Facilitated diffusion is in many cases studied not only during an ultrashort incubation period (seconds

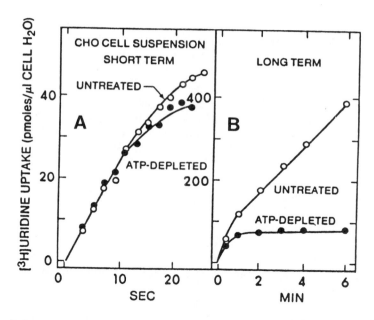

FIGURE 2. Accumulation of labeled uridine by cultured cells. The cells were either used without further treatment or were depleted of ATP by incubation at 37° for 10 min in medium containing 5 mM cyanide and 5 mM iodoacetate. The uptake of [5-³H]uridine is shown as a function of the length of the exposure to the radioisotope. (From Paterson, A. R. P., Kolassa, N., and Cass, C. E., *Pharmacol. Ther.*, 12, 515, 1981. With permission.)

or fractions of seconds), but also in nonmetabolizing cells, obtained either by selection of mutants, inhibition of energy metabolism by administration of metabolic poisons (e.g., cyanide, iodoacetate), or choice of a substrate that is not metabolized. It should obviously be ascertained that the abolishment of energy metabolism has no deleterious effect on transport mechanisms (facilitated diffusion or active transport) themselves. That the abolishment of adenosine 5'-triphosphate (ATP) formation by metabolic inhibitors has no such effect on the uptake of uridine into cultured Chinese hamster ovary cells, which is known to occur by facilitated diffusion, is illustrated in Figure 2A, showing similar transport rates in normal and in poisoned cells.[12] However, metabolism (at least into the nucleotides) and thus continued net uptake of label is prevented in the cells exposed to cyanide and iodoacetate. Thus, the curve describing long-term accumulation into poisoned cells (Figure 2B) indicates uptake by one mechanism only, i.e., diffusion.

The steady-state level of labeled uridine shown in Figure 2B for the ATP-depleted cells (after similar intracellular and extracellular concentrations have been established) does not indicate that no more *unidirectional* diffusion of uridine occurs into the cells, only that no more *net* accumulation of radioactivity takes place. At the maintained level of labeled uridine in the cells, influx (unidirectional inward transport) of labeled compound will continue, but it will equal efflux (unidirectional rate of release) of labeled compound. With unaltered [³H]uridine concentration and specific activity in the medium, influx of radioactivity will remain constant throughout the experiment (unless the cells are reacting to the experimental conditions, e.g., deteriorating). At zero time no "back-flux" of label will occur since no labeled uridine is found intracellularly at this stage, either because uridine is absent in cells which have not been exposed to extracellular uridine ("zero-trans" flux in the terminology of Plagemann and Wohlhueter)[13] or because the intracellular uridine pool is nonlabeled until the start of the experiment ("exchange diffusion" in the terminology of Plagemann and

Wohlhueter).[13] However, with continuing accumulation of [3H]uridine an increasing amount of labeled uridine will be found intracellularly and efflux of label will increase. This will result in a less rapid *net* uptake of radioactivity. Finally, after so much [3H]uridine has been accumulated that the concentrations of labeled uridine in the intracellular and extracellular water phases are identical, no further increase in radioactivity will occur, i.e., the intracellular radioactivity has reached a constant level. For this reason, only the initial, rectilinear part of the curve describing uptake into ATP-depleted cells in Figure 2B indicates the rate of unidirectional transport into the cells at a time when little or no "back-flux" of isotope occurs, whereas the steady-state level after longer incubations indicates the total content of labeled intracellular compound.

Several inhibitors of transport by facilitated diffusion exist;[5,12,13] they will, in general, affect both influx and efflux.

D. ACTIVE UPTAKE

Uptake by active transport is also often preceded by diffusion which leads to a more rapid initial accumulation. Since this process, in contrast to the active uptake itself, is not concentrative (Figure 1), the importance of the diffusional transport to the total transport will be negligible unless the cells are suddenly exposed to a high extracellular concentration. This is illustrated in Figure 3, showing adenosine accumulation into mouse neurons and astrocytes in primary cultures at extracellular adenosine concentrations of 1 and 250 μM. An initial, almost immediate diffusional uptake up to a level corresponding to the extracellular concentration is prominent at the high — but not at the low — concentration and is followed by a maintained active uptake, as will be discussed in Section IV. Again, a constant, continued influx of the labeled compound will generally not be mirrored by a continuous straight line of increase in intracellular radioactivity since — with time — the "back-flux" of labeled compound increases until it finally becomes equal to influx. Due to a much larger intracellular content as a result of the active accumulation, this takes place later and at a higher level than that corresponding to similar concentrations in the intracellular and extra-cellular water phase (Figure 3). It is only if the accumulated compound, or all metabolites of this compound which remain labeled, are quantitatively retained in the tissue, that the initial slope will continue without any tendency to level out. However, if efflux of the metabolites also occurs and efflux of radioactivity finally equals influx, a steady-state level will eventually be reached. As in the case of diffusion, influx should therefore be determined from the early rectilinear part of the curves describing the active uptake. In some cases (e.g., 250 μM adenosine in Figure 3) this means *after diffusional net uptake has ceased.* In other cases (1 μM adenosine in Figure 3) the small diffusional uptake is of no significance. As already stated, subsequent metabolism will have no effect on the rate of the active transport. The active uptake is often regulable and may thus become a target for pharmacological intervention. Studies of drug effects on active uptake must be performed during the time period when active uptake prevails (minutes to hours). Transport studies over seconds or fractions of a second measure diffusion and will be useless. It is thus essential to know whether transport into a specific cell type is regulated by facilitated diffusion or by active transport in order to design appropriate experimental conditions. In this context it should be emphasized that the time course for the accumulation of a transported compound does not in itself allow any distinction between "metabolism-driven" uptake and active uptake. This can be seen from the fact that the two first components of the uptake curves in Figures 2 (for non-ATP-depleted cells) and 3 are, in principle, identical. Poisoning the cells with inhibitors of energy metabolism to deplete stores of energy-rich compounds like ATP, required for metabolic conversion, is obviously not permitted when active transport is studied. This is in sharp contrast to the legitimacy and usefulness of this procedure to distinguish between facilitated diffusion and metabolism-driven uptake. It should be emphasized that

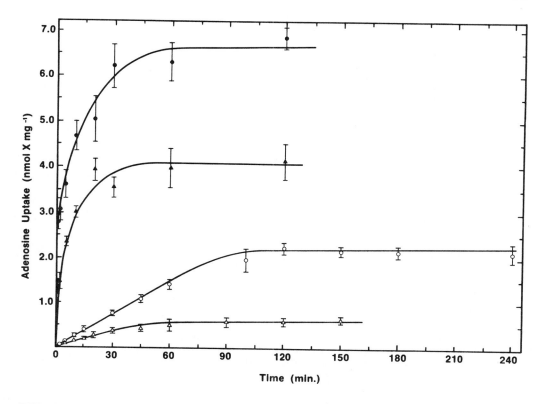

FIGURE 3. Accumulation of labeled adenosine into cultured neurons (triangles) and astrocytes (circles) as a function of the length of exposure to the radioisotope in a medium containing either 1.0 (open symbols) or 250 μM (closed symbols) adenosine. No "zero-time blanks" were subtracted, leading to an initial, almost immediate diffusional uptake into the intracellular water phase. This part of the uptake is only obvious at the high (250-μM) extracellular concentration and amounts to ≈2.5 nmol/mg protein in astrocytes (water space ≈10 μl/mg protein) and somewhat less in neurons. The subsequent, relatively rectilinear uptake represents the rate of active transport. Data are from Reference 14, but in the corresponding Figure in this reference "zero-time blanks" had been subtracted. (From Bender, A. S. and Hertz, L., *Neurochem. Res.*, 11, 1507, 1986. With permission.)

the ability of metabolic inhibitors to abolish a certain component of the accumulation (e.g., in Figure 2B) per se allows no conclusion whether a concentrative uptake is by active transport or is "metabolism-driven", since both nucleoside metabolism and active uptake are ATP dependent. Use of ATP-depleted cells for the study of transport of compounds which are accumulated by an active concentrative transport mechanism (e.g., adenosine uptake into neurons) is at best meaningless and at worst provides misleading information. The only exception to this is the case where it is specifically indicated that the aim of the study is to obtain information of the diffusional process under conditions when active transport is abolished.

E. SATURABILITY

In studies of saturability, transport rates are examined as a function of concentration. This means rates of accumulation of radioactivity determined at the initial rectilinear phase of the accumulation during a time period when the process in question dominates the accumulation, i.e., influx during seconds or fractions of a second for studies of facilitated diffusion and much longer time periods for active uptake. For studies of the latter it may be attempted to exclude the initial diffusional part of the uptake by subtraction of "zero-time blanks". Since the diffusional uptake for many compounds is insignificant, at least at

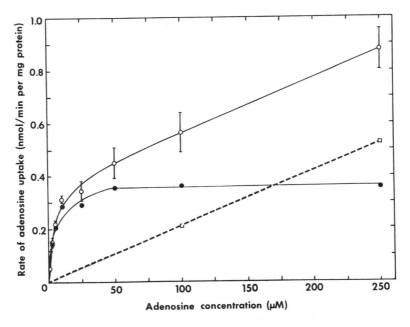

FIGURE 4.   Rate of accumulation of labeled adenosine (measured during a 5-min period) into cultured astrocytes. The open circles indicate *total* uptake rate, the open squares the uptake rate at saturating concentrations, displaced in parallel to originate at zero, i.e., the nonsaturable uptake rate, and the closed circles the curve obtained by subtraction of the nonsaturable uptake rate from the total uptake rate, yielding uptake rates for the *saturable* component. (Modified from Hertz, L., *J. Neurochem.*, 31, 55, 1978. With permission.)

low concentrations (e.g., 1 μM adenosine in Figure 3), such a procedure is not always important, especially if the active uptake rates are high. Moreover, the rate of simple unidirectional diffusional transport is directly proportional to the extracellular concentration. This is illustrated by the correlation between transport rate and adenosine concentration in Figure 4 at concentrations above ≈ 50 μM. In contrast, the increase in transport rate occurs more steeply with an increase in adenosine concentration below ≈ 50 μM. This is on account of an intense, high-affinity uptake mechanism which is virtually saturated, in Figure 4 around 50 μM. The continued rise in adenosine uptake at concentrations above ≈ 50 μM represent either diffusion into the water space of the cells (which could have been excluded by a subtraction of suitable "zero-time blanks") or a low-affinity uptake.[6] Another line, parallel to the curve above 50 μM adenosine, but displaced to pass through the zero time-zero uptake point (the stippled line), represents the nonsaturable uptake separately and was subtracted from the actually measured values in Figure 4 in order to determine saturation kinetics. This subtraction yielded the lower solid curve in Figure 4, which conforms to the equation $V = V_{max}[1/(1 + K_m/S)]$, i.e. the Michaelis-Menten equation. The two values describing the course of the saturable uptake is the V_{max}, i.e., the uptake rate at saturating concentrations (≈ 0.3 nmol/min/mg protein), and the K_m, i.e., the concentration yielding an uptake rate equal to one half of V_{max} (≈ 4 μM). The resolution of data similar to those shown in Figure 4 is now generally done by computer analysis, yielding not only V_{max} values and K_m values (occasionally for more than one component), but also a constant, k, describing the nonsaturable uptake (k × concentration), if relevant. In such programs, each of these three parameters can be determined on the basis of all data and obtained with standard errors.

As a result of the relatively low affinity of most nucleosides for the facilitated diffusion carrier (Km generally >100 μM), compared to the much higher affinity both for nucleotide formation and for active uptake (generally 1 to 10 μM), it cannot be concluded with certainty whether the stippled line in Figure 4 represents nonsaturable simple diffusion or facilitated diffusion with a K_m well above 250 μM. However, it is possible to evaluate effects of an inhibitor on diffusion (facilitated or simple) in actively metabolizing or actively transporting cells by measuring total accumulation at high substrate concentrations in the presence or absence of the inhibitor. If the inhibitor decreases accumulation of the compound *specifically* within this concentration range, the uptake must be by facilitated diffusion.

F. SUMMARY

Studies of the time course for the accumulation of a radioactive compound allow determination of transmembrane transport. Diffusion (be it simple or facilitated) can occur only until the intracellular concentration equals the extracellular concentration. Intracellular metabolism of the accumulated compound may greatly enhance the accumulated amount, but the rate with which this increase occurs is often less than the rate at which transmembrane transport as such can occur. The same goal of increasing uptake capacity is achieved by active transport. Intracellular metabolism does not affect either amount or rate of active uptake. Determination of transport rates by aid of radioactively labeled isotopes must be performed under such conditions that the "back-flux" of labeled compound is negligible (initial uptake rates for the process in question). The correlation between total transport rates and concentration is determined by all processes contributing to the uptake, both saturable and nonsaturable. Constants describing the correlation between concentration and uptake rates (V_{max} and K_m values for one or more saturable components; k for a diffusional component) can be obtained by graphical analysis or, more accurately and simply, by computer analysis.

III. NUCLEOSIDE TRANSPORT IN ERYTHROCYTES AND CELLS BEHAVING IN A SIMILAR MANNER

A very substantial amount of information in this area has been obtained over 2 decades, especially by the groups of Paterson et al.[12] and Plagemann et al.[5,13] The findings and conclusions of these groups have repeatedly been authoritatively reviewed. This section will only summarize some of the most important findings and conclusions of this work and the resulting concepts.

A key conclusion is that the overwhelmingly major part of transmembrane transport in these groups of cells, which often are deficient in their ability to synthesize purines[2,3] or may be involved in interorgan transport of nucleosides,[3] occurs as a saturable, nonconcentrative transport mechanism which is inhibitable by specific antagonists, i.e., by facilitated diffusion.[5,12,13,16] Simple diffusion appears to play a quantitatively much more minor role.[5,13] With the exception of certain mutant cells (which have been of considerable experimental interest) these cells normally metabolize nucleosides, especially adenosine, by an ATP-dependent nucleotide synthesis, but also by a non energy-requiring deamination. This metabolism is of great importance for transmembrane transport by keeping the intracellular concentration of, e.g., adenosine itself low and thus maintain an inwardly directed concentration gradient. As already discussed, the metabolism can be inhibited by ATP depletion, which has no immediate effect on the facilitated diffusion (Figure 2A), occasionally together with inhibition of deamination by a specific inhibitor. Thus, either mutant cells or ATP-depleted cells can be used to "dissect" facilitated diffusion out from the tandem of facilitated diffusion and subsequent metabolism.

Once metabolism has been "dissected" out (e.g., in ATP-depleted cells), facilitated diffusion cannot increase the total concentration of accumulated radioactivity in the intracellular water phase fluid (and thus in the tissue, provided no swelling occurs) above that in the extracellular level (Figure 2B). Thus, no further net accumulation of nucleoside occurs in the *ATP-depleted cells* after equilibration with the medium, whereas accumulation continues, generally at a somewhat decreased rate (dictated by rate-limiting metabolism), in *the untreated cells*. Whether or not (or when) the slope of the curve describing the continuing metabolism-driven accumulation of radioactivity will decrease to approach a steady-state content of radioactive nucleoside plus its metabolites (mainly the latter) depends upon the further fate of the metabolites as well as upon their pool sizes.

In the example illustrated in Figure 2, metabolism is clearly slower than facilitated diffusion. This is not necessarily always so, but the maximum rate at which total accumulation can occur in these cells is the V_{max} value for facilitated diffusion. Experimental modifications, e.g., a decrease in nucleoside concentration (which affects facilitated diffusion [high K_m value] and metabolism [low K_m value, at least for formation of nucleotides] differently), might convert a "metabolism-limited" uptake to a diffusion-limited uptake.

It should be kept in mind that the part of the uptake of labeled adenosine into erythrocytes and related cells which is abolished by ATP depletion might conceptually represent either continuous, metabolism-driven uptake by facilitated diffusion (as described above) or uptake by active transport. Convincing evidence that the ATP-dependent uptake in these cells represents metabolism-driven facilitated diffusion was first obtained by Lum et al.,[17] who demonstrated that the measured uptake can be quantitatively accounted for by the measured metabolite formation. The characteristics of such a nucleoside uptake are well suited for the purpose of supplying substrates according to the need of their utilization, since increased metabolism automatically increases uptake. This mechanism is analogous to that existing for glucose uptake into mammalian cells. It is difficult to entirely rule out the presence of a high-affinity uptake of, e.g., adenosine, but if it exists it plays a minor role in the total uptake of adenosine by human erythrocytes and various types of transformed cultured cells that have been examined in detail.[5] The only report of a concentrative nucleoside uptake into any of these types of cells (spleen cells) is that by Darnowski,[18] but this finding could not be confirmed by Plagemann and Woffendin,[19] although they did confirm that part of the uptake is sodium dependent.

The facilitated diffusion into erythrocytes and similarly behaving cells is nonconcentrative, unaffected by ATP depletion (Figure 2), saturable ($K_m > 50$ to $100~\mu M$), has high V_{max} values and broad substrate specificity,[5,12,13,20] and in general appears to be independent of the presence of sodium (see, however, above). It is very potently inhibited (IC_{50} in the low nanomolar or subnanomolar range) by the nucleoside analogs nitrobenzylthioguanosine (NBTG) and nitrobenzylthioinosine (NBTI), although with some species differences (see Table 4 in Reference 5). Other compounds, e.g., dipyridamole, papaverine, and several benzodiazepines, also inhibit nucleoside uptake by facilitated diffusion, in some cases competitively and in others noncompetitively, with IC_{50} values in the low micromolar range (see Table 5 in Reference 5).

Inhibition by benzodiazepines of facilitated diffusion of uridine into human erythrocytes has been studied in considerable detail by Hammond et al.[21] The advantage of using uridine is that this nucleoside is not metabolized in human erythrocytes. Both influx and efflux are decreased by benzodiazepines. The inhibition is not potent with K_i values ranging between 8 and 83 μM for RO5-4864 (a peripheral-type benzodiazepine), diazepam, clonazepam, and lorazepam, indicated according to decreasing rank order of potency. This rank order corresponds to that with which they inhibit binding of NBTI to red blood cells and of diazepam binding to guineapig heart tissue[21] or intact mouse astrocytes;[22] however, the effect on benzodiazepine binding is much more potent. The rank order is distinctly different from that with which benzodiazepines displace diazepam from neuronal membranes and intact mouse neurons in primary cultures.[22] Table 1 shows that the affinity of the ($-$) isomer RO11-6893 for the NBTI binding (and thus probably also for adenosine uptake) is significantly higher than that of the ($+$) isomer RO11-6896.

TABLE 1
Inhibition by Each of the Two Stereoisomers (+) RO11-6896 and (−) RO11-6893 of Nitrobenzylthioinosine (NBTI) Binding to Erythrocytes, Adenosine Uptake Into Neurons, Astrocytes, and Synaptosomes, and of Diazepam Binding to Neurons and Astrocytes

Parameter	(+) RO11-6896	(−) RO11-6893	Ref.
IC_{50} NBTI binding to erythrocytes (μM)	277	39[a]	21
IC_{20} Adenosine uptake in synaptosomes (μM)	0.06[a]	20	46
IC_{50} Diazepam binding to neurons (nM)	20[a]	80	23
IC_{50} Adenosine uptake in neurons	48	4[a]	14
IC_{50} Adenosine uptake in astrocytes (μM)	20	7 (NS)[b]	14
IC_{50} Diazepam binding to astrocytes (nM)	170	172 (NS)[b]	23

Note: Note that all values are expressed as IC_{50} values, except adenosine uptake into synaptosomes, which is an IC_{20} value, and that all concentrations are μM except for diazepam binding, which is nM. Hammond et al.[21] used a NBTI concentration of 0.35 nM, Bender and Hertz[14,23] a diazepam concentration of 1.8 nM, and Phillis et al.[46] an adenosine concentration of 0.2 nM.

[a] Preferred isomer.
[b] NS: Difference between the two stereoisomers not statistically significant.

Great caution should be taken before concluding that ATP-dependent nucleoside uptake in other cell types, including cells in the central nervous system (CNS), similarly occurs by facilitated diffusion and subsequent metabolism. This is especially so since concentrative, sodium-dependent, energy-requiring nucleoside transport has been convincingly demonstrated, e.g. in rat kidney cell brushborder vesicles, rabbit choroid plexus, and rabbit and guineapig intestinal cells, during the last 10 years.[5,11,23,24] Le Hir and Dubach[11] clearly stated that (1) this nucleoside transport carrier is different from that found in most other cell types; (2) its main function is to transport a nucleoside from one side of the membrane to the other (in this case to reabsorb nucleosides from the tubular fluid), rather than fueling intracellular metabolism; (3) this transport system may be more widespread than it presently (1985) appears. The concentrative, sodium-dependent nucleoside transport is not or only weakly inhibitable by NBTI and NBTG, which may even in some cases increase the cell content of the nucleoside by inhibiting efflux. The latter finding suggests the joint presence of nucleoside transporters of two types in these cells, of which one is concentrative, sodium dependent, and not significantly affected by these inhibitors and the other a bidirectional, facilitated diffusion, potently affected by the inhibitor.[24] In the choroid plexus both systems are inhibited by diazepam, but the effect is most potent on the diffusional transport.[23]

IV. NUCLEOSIDE TRANSPORT IN CELLS IN THE CENTRAL NERVOUS SYSTEM

A. TIME COURSE

The most important difference between nucleoside transport in blood cells and in cells from the CNS is that there is an intense, easily demonstrable active uptake, at least of the nucleoside adenosine, into cells from the CNS. This applies both to neurons and astrocytes.[25,26] Experimental evidence for such an active uptake is presented below. In addition to the active uptake there is a diffusional mechanism which, at least at low extracellular concentrations, will be of negligible importance.

Evidence for the existence of an active transport mechanism for adenosine into neurons is presented in Table 2 and Figure 5. Table 2 shows *cell contents* (not labeling) of adenosine

TABLE 2
Cellular Contents (Pool Sizes) of Adenosine and Its Metabolites (Nanomoles per Milligram Protein)

Compound	5 s		15 min		30 min		60 min	
	Control	Inhibitor	Control	Inhibitor	Control	Inhibitor	Control	Inhibitor
ADO	0.10 ± 0.03	5.05 ± 0.02	0.10 ± 0.05	7.07 ± 0.85	0.23 ± 0.10	6.94 ± 1.09	0.25 ± 0.11	7.27 ± 1.76
ATP + ADP	20.8 ± 1.2	28.5 ± 4.1	23.7 ± 2.5	29.6 ± 6.3	24.5 ± 2.8	31.0 ± 5.5	20.6 ± 1.1	33.0 ± 4.1
AMP	7.35 ± 1.14	6.91 ± 1.62	6.39 ± 0.73	6.12 ± 0.81	7.79 ± 1.26	5.37 ± 0.20	5.20 ± 0.14	6.16 ± 0.84
INO	4.40 ± 0.95	2.66 ± 0.49	4.03 ± 0.48	0.53 ± 0.18	3.65 ± 0.39	0.55 ± 0.23	3.10 ± 0.68	0.52 ± 0.29
HYP	2.97 ± 1.07	2.73 ± 0.68	4.42 ± 1.46	4.14 ± 1.93	2.58 ± 0.63	2.55 ± 0.60	2.48 ± 0.64	2.62 ± 0.62

Note: ADO: Adenosine; ATP: adenosine triphosphate; ADP: adenosine diphosphate; AMP: adenosine monophosphate; INO: inosine; HYP: hypoxanthine.

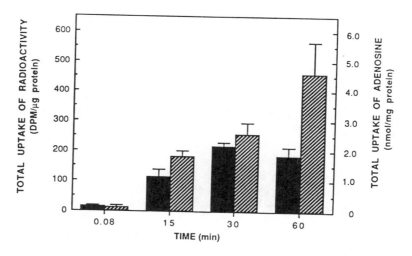

FIGURE 5. Total accumulation of labeled adenosine and its metabolites (nanomoles per milligram protein) into cultured neurons during incubation with (hachured bars) and without (solid bars) the adenosine deaminase inhibitor, 2'-deoxycoformycin. The adenosine concentration was 10 μM. Note that the time periods indicated are only relatively equidistant from 0.08 to 30 min. (From Hertz, L. and Matz, H., *Neurochem. Res.*, 14, 755, 1989. With permission.)

and its major metabolites in primary cultures of cortical neurons during incubation in the presence of 10 μM adenosine with or without the adenosine deaminase inhibitor deoxycoformycin. At the concentration used (10 μM), this inhibitor has a relatively specific inhibitory effect on the adenosine deaminase. It can be seen that the intracellular content of adenosine itself during prolonged (60-min) incubation remains relatively constant at 0.1 to 0.2 nmol/mg protein in the control cultures and that the variability in adenosine content is quite pronounced (reflecting the uncertainty of determination of such low quantities). In the presence of deoxycoformycin the adenosine content is very considerably increased, whereas the inosine content is decreased. The increase in adenosine content is obvious already 5 s after exposure to the inhibitor and it reaches a steady state before or around 15 min. From then onwards, the concentration of adenosine remains steady at about 7 nmol/mg protein. The intracellular water content of these cells is probably around 7 μl/mg protein (White et al.[26a] unpublished results) and certainly not above 10 μl/mg protein. Calculation of an intracellular concentration using the latter value, which will result in the minimum estimated concentration of adenosine in the intracellular water phase, shows that the content of adenosine during the period 15 to 60 min has increased to a level of at least 7 nmol adenosine per 10 μl intracellular water, which equals 700 nmol/ml water or an intracellular adenosine concentration of 700 μM. This is 70 times more than the extracellular adenosine concentration and provides unequivocal proof of an active, concentrative uptake of adenosine into the neurons. In the absence of the deaminase inhibitor the adenosine concentration does not rise significantly above 0.1 nmol/mg protein, yielding an intracellular adenosine concentration close to the extracellular concentration of 10 μM. Thus, the reason for the high adenosine concentration in the presence of the inhibitor is not a faulty technique, causing a degradation of adenosine nucleotides before the measurements.

In neurons at least 50% of the accumulated adenosine is metabolized to the deaminated products inosine and hypoxanthine.[27] Interference with deaminase activity therefore abolishes a major part of adenosine metabolism. If this metabolism was of importance in "driving" the transmembrane accumulation of adenosine, it would be expected that the total uptake of radioactivity in adenosine plus all its metabolites would decrease in the presence of the deaminase inhibitor. That this is not the case (but that the total uptake of radioactivity rather is increased in the presence of deoxycoformycin) can be seen from Figure 5, showing the total uptake of radioactivity after varying times of incubation with 10 μM adenosine.[25] This finding lends further support to the conclusion that adenosine is accumulated into neurons by a continuous, active, concentrative uptake, not by metabolism-driven, facilitated diffusion; therefore, long-term studies are not only permissible, but required to determine the true continued transport rate of adenosine into neurons. However, due to the increasing efflux of radioactivity("back-flux") when the intracellular specific activity increases (see Section II), the transport should be determined only during the period of a rectilinear increase (0 to 30 min in Figure 5). It can be seen that this uptake amounts to 0.10 nmol/min/mg protein at an extracellular adenosine concentration of 10 μM (Figure 5). The initial uptake within the first 5 s uptake is even faster (≈0.15 nmol/mg protein after 5 s, or ≈1.0 nmol/min/mg protein). This uptake represents the sum of diffusional uptake and active uptake; moreover, it cannot be ignored that remnants of adherent medium may contribute to the measured radioactivity. This initial uptake rate would dominate in the total accumulation, *if it were able to continue for a longer period*; however, after 5 s the diffusional uptake has ceased and, if this value is used to calculate transmembrane permeation of adenosine during normal *in vivo* conditions, the transport rate will be overestimated by a factor of ≈10. Clearance of extracellular adenosine into neurons by diffusion will therefore only be of importance at extracellular concentrations of adenosine high enough to saturate the active uptake system (see below) and under these conditions, especially if the extracellular concentrations of adenosine are abruptly increased to a high value.

Adenosine is accumulated not only into neurons, but also into astrocytes in primary cultures.[14,15,26] Studies on the effect of deoxycoformycin on adenosine uptake into primary cultures of brain astrocytes, where deamination plays a relatively smaller role than in neurons,[27] have yielded results similar to those observed in neurons. The total concentration of adenosine in the water phase of the cells, calculated as in the case of the neurons and under similar experimental conditions, becomes at least 100 μM, i.e., ten times higher than the extracellular concentration, when they are exposed to the deaminase inhibitor; again, the total radioactivity of adenosine plus all its metabolites is increased rather than decreased in the presence of deoxycoformycin and, again, the initial diffusional uptake during the first few seconds is faster than the continued active uptake, which amounts to ≈0.1 nmol/min/mg protein at 10 μM adenosine extracellularly.[26]

Essentially similar time courses for accumulation of labeled adenosine into primary cultures of chick astrocytes exposed to 25 μM [³H] adenosine or chick neurons exposed to 4 μM [³H] adenosine were observed by Thampy and Barnes.[28,29] The continued accumulation in nonpoisoned cells was ≈0.05 to 0.1 nmol/min/mg protein. Again, there was a short-lasting, much faster accumulation of adenosine which was over as soon as the adenosine concentration in the intracellular water phase, calculated as above, was similar to that in the incubation medium. A ≈tenfold lower rate of continued accumulation was found by Geiger et al.[30] and Johnston and Geiger[31] in "isolated brain cells", probably partly as a result of cell damage during the isolation and partly as a result of the use of a lower adenosine concentration (1 μM). Again, there was a much faster initial (first 5 s) uptake (Figure 6) until the intracellular concentration of labeled adenosine was approximately similar to that in the incubation medium.[31] Unfortunately, these authors used an incubation time of 15 s (i.e., a 5 s *mainly* — but not exclusively — diffusional uptake followed by a 10-s, mainly active, partly sodium-dependent uptake — see below) for subsequent investigation of transport kinetics and drug inhibition. At the concentration used (1 μM adenosine), one half of the total uptake after 15 s must have occurred by rapid diffusional transport dominating during the first 5 s, and the remainder by a less rapid active-transport system accounting for the major part of the uptake after the first 5 s (Figure 6). Accordingly, complex kinetics

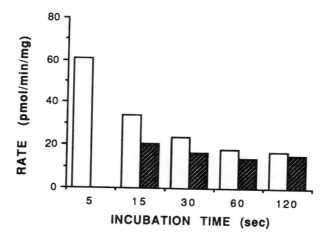

FIGURE 6. Rates of accumulation of labeled adenosine into "dissociated brain cells" as a function of the length of exposure to the radioisotope. Rates are indicated both as average rates for the total period from the beginning of the incubation (open bars) and as average rates from the termination of the first 5 s (hatched bars). Note that the total 15-s accumulation rate is the combined result of a very high immediate uptake rate (0 to 5 s) and a subsequent uptake rate (after the first 5 s), which is almost unaltered up to 120 s. (Modified from data in Reference 31)

for both saturation of adenosine uptake and drug inhibition were obtained, as will be discussed below. Most of the studies by Thampy and Barnes[28,29] were also done using incubation periods well below 1 min. In most of these experiments, the authors used cells poisoned with inhibitors of energy metabolism to prevent subsequent metabolism. This procedure will also abolish active transport (which, as discussed above, is the dominant means of adenosine uptake into neurons and astrocytes in primary cultures), but in these experiments at least some useful information is provided because *one* process only (diffusion, which probably is facilitated, as will be discussed below) is measured.

It has repeatedly been observed that the long-term adenosine uptake into different brain preparations or into synaptosomes from the electric organ of Torpedo is decreased in the absence of sodium or in the presence of ouabain.[32-34] These findings provide additional support for the concept that the nucleoside is accumulated by active transport, although evidence has been found for a sodium-dependent, nonconcentrative transport in spleen cells (see above). Johnston and Geiger[31] similarly observed that sodium depletion causes a small, but significant, decrease in adenosine accumulation into "isolated brain cells". This decrease was distinct after long-term incubation (1 to 3 min), but not after 5 s of incubation. This supports the previously suggested conclusion that the 5-s uptake mainly represents diffusional uptake, whereas active uptake makes a considerable contribution ($\approx$50%) to the uptake after 15 s and is dominant after still longer incubation periods.

B. SATURABILITY

Long-term (5-min) active transport rates for adenosine at different concentrations into primary cultures of astrocytes were shown in Figure 4.[15] From the curve representing the saturable part of the uptake it can be seen that the V_{max} for the uptake is $\approx$0.3 nmol/min/mg protein and the $K_m \approx 4 \mu M$. Relatively similar values for long-term uptake were reported by Bender and Hertz[14] in mouse astrocytes, by Thampy and Barnes[28] for nonpoisoned chick astrocytes, and by Lewin and Bleck[35] for glioma cells. The K_m value (6 μM) for the long-term uptake into mouse neurons in primary cultures is similar to that in astrocytes, but the V_{max} value appears to be slightly lower in the neurons.[14] The K_m value in the neurons is also comparable to that of 1 to 5 μM observed by Bender et al.[33] in synaptosomes, by Miras-Portugal et al.[36] in freshly isolated or cultured bovine chromaffin cells (7 μM), and by Shimizu et al.[37] and Davies and Hambley[38] in brain slices (10 to 20 μM), but lower than that (140 μM) reported by Banay-Schwartz et al.[34] in the latter preparation. The V_{max} values for adenosine uptake into chromaffin cells[36] and brain slices[37] are relatively similar to those in cultured neurons and astrocytes,[14,15] whereas the V_{max} (2 to 7 pmol/min/mg protein) is considerably lower in synaptosomes.[33] These values were obtained by measurement of long-term uptake. Provided the uptakes in all these preparations, as in cultured neurons and astrocytes, is by active uptake, not by metabolism-driven facilitated diffusion, this is the only appropriate procedure.

The use of a 15-s accumulation period by Geiger et al.,[30] which as previously mentioned, yields an uptake, of which approximately one half is by diffusion and the other half apparently by active, sodium-dependent, energy-requiring uptake, led to kinetic evidence for a two-component system of which one had very low and the other very high V_{max} and K_m values. The high-affinity system (low K_m; low V_{max}) has approximately the same affinity (Km ≈ 1 μM), but a much lower V_{max} (25 pmol/min/mg protein) than the corresponding systems in cultured neurons or astrocytes. The low-affinity system (high V_{max} and K_m) is very similar to that described by Thampy and Barnes[28] for ATP-depleted astrocytes. Both of these low-affinity uptake systems had maximal uptake rates (V_{max}) of ≈ 10 nmol/min/mg protein and K_m values of ≈ 0.5 mM. These values are comparable to those usually observed for facilitated diffusion into erythrocytes and related cells. Little difference was found in short-term uptake between normal and ATP-depleted astrocytes (exposed to iodoacetate and cyanide),[28] which is consistent with the finding that ATP depletion has little or no effect on facilitated diffusion (Figure 2) and obviously none on simple diffusion.

The diffusional uptake rates determined by Thampy and Barnes[28] should also be compared to the nonsaturable adenosine accumulation in astrocytes at high (>50 to 100 μM) adenosine concentrations shown in Figure 4. From this figure it can be seen that the total (i.e., saturable plus nonsaturable) uptake into these cells is increased from ≈ 0.55 nmol/min/mg protein at 100 μM adenosine to ≈ 0.90 nmol/min/mg protein at 250 μM. Since the saturable uptake ($K_m \approx 4$ μM) is virtually unaffected by an increase in the adenosine concentration within this range, this corresponds to an increase in nonsaturable uptake of 350 pmol/min/mg protein when the concentration in the medium is increased by 150 μM or 20 to 25 pmol/min/mg protein each time the medium concentration is increased by 10 μM. This value, which may represent simple diffusion or facilitated diffusion with a high K_m value, is five to six times lower than the initial uptake rates observed by Thampy and Barnes[28] (an uptake rate of ≈ 130 pmol/min/mg protein at 10 μM calculated from a K_m value of 370 μM and a V_{max} value of 10,000 pmol/min/mg protein). This relatively good concordance suggests that both types of measurement (initial uptake rate in posioned cells and alterations in long-term total uptake into normal cells at concentrations high enough to saturate the saturable uptake) may provide reliable information about diffusional uptake.

In ATP-depleted neurons, and using an incubation period of 5 to 25 s and an adenosine concentration of 4 μM, Thampy and Barnes[29] observed much lower V_{max} values (150 to 160 pmol/min/mg protein) and much higher affinity (K_m 6 to 13 μM) than in astrocytes. Approximately similar values were reported for the saturable part of a short-term (<30-s) adenosine uptake into synaptosomes by Barberis et al.[39] These values are unusually low for facilitated diffusion of nucleosides,[5] but, due to the low K_m value, the calculated uptake at 10 μM adenosine (≈ 80 pmol/min/mg protein) is comparable to that in astrocytes. A lower value (≈ 10 pmol/min/mg protein) for diffusional uptake at 10 μM adenosine can be calculated from the saturation curve for adenosine uptake in neurons observed by Bender and Hertz,[14] but in these experiments there was no tendency towards any saturation of the uptake rate within the concentration studied (up to 250 μM). Although the actual value may be considerably underestimated since "zero-time" uptakes were subtracted, the lack of saturability does not support the concept by Thampy and Barnes[29] of a low K_m value. This conclusion is further supported by the finding of a nonsaturable (up to 100 μM) uptake into synaptosomes, transporting approximately 40 pmol/min/mg protein.[38]

C. EFFECTS OF ADENOSINE TRANSPORT INHIBITORS

Like the sodium-dependent nucleoside uptake in other cells, the active uptake of adenosine into most CNS preparations has a low sensitivity to NTBI and NTBG. Thus, the long-term uptake in both astrocytes and neurons in primary cultures is inhibited with IC_{50} values in the low micromolar range.[23] The potency of these compounds is similar in synaptosomes.[40] Since the short (15-s) incubation time employed by Geiger et al.[30] leads to about equal amounts of adenosine uptake via active, sodium-dependent[31] transport and by diffusional uptake, it is no surprise that two distinctively different K_i values were observed for NTBI, one of which is about 100 μM.

The short-term diffusional uptake of adenosine is much more effectively inhibited by NBTI,[28,29] i.e., with K_i values between 10 nM (chick neurons in primary cultures) and 100 nM (chick astrocytes in primary cultures). This uptake may also constitute the part of the uptake in the studies by Geiger et al.[30] which was inhibited by NTBI with a K_i value of ≈400 nM. However, the K_i value for dipyridamole was even lower. The finding that the diffusional uptake is inhibitable by classical inhibitors of adenosine transport strongly suggests that this part of the uptake of adenosine is by facilitated diffusion, not by simple diffusion. It is an indication in the same direction that papaverine blocks the nonsaturable (up to 250 μM) uptake of adenosine into primary cultures of astrocytes, which becomes apparent at extracellular adenosine concentrations high enough to saturate the high-affinity, active uptake.[15]

An inhibition of adenosine uptake into neurons and astrocytes by benzodiazepines is of special interest, since the electrophysiological effects of adenosine on brain are distinctly enhanced in the presence of benzodiazepines; this effect is antagonized by the adenosine antagonist, theophylline,[41] leading to the suggestion that some of the actions on CNS activity exerted by benzodiazepines (maybe especially the anticonflict effect)[42] are due to inhibition of adenosine uptake.[43,44] Such an inhibition of adenosine uptake by benzodiazepines has been shown in many preparations, including brain slices,[45] synaptosomes,[46] and astrocytes and neurons in primary cultures.[14] The rank order with which different benzodiazepines inhibit adenosine uptake is identical in neurons and astrocytes. It is, however, distinctly different from that with which they displace benzodiazepines from both neuronal and astrocytic binding sites,[47] indicating that the benzodiazepine receptor is not identical to the adenosine uptake site in either neurons or astrocytes. Moreover, both flumazenil (a neuronal-type benzodiazepine antagonist) and PK 11195 (a peripheral-type [astrocytic] benzodiazepine antagonist) inhibit, on their own, adenosine uptake into cultured neurons and astrocytes and do not decrease the inhibition of the uptake by diazepam when added together with this drug.[47] As is shown in Table 1, the (+) stereoisomer of the benzodiazepines (RO 11-6896 (+) and RO 11-6893 (−) is more potent in inhibition of adenosine uptake in synaptosomes[46] and in inhibition of benzodiazepine binding to neuronal membranes and to primary cultures of neurons[47] or to brain tissue in vivo.[48] This stereoisomer is also more effective therapeutically.[48] However, (−) RO 11-6896 is the more potent inhibitor of adenosine uptake into cultured neurons,[14] a stereoisomer preference similar to that found for NBTI binding to erythrocytes.[21] The two stereoisomers are equipotent in inhibition of both adenosine uptake (IC_{50} 7 to 20 μM) and of diazepam binding (IC_{50} 170 to 172 nM) in astrocytes (Table 1).[14,22]

In addition to the inhibition of adenosine uptake, benzodiazepines also inhibit adenosine

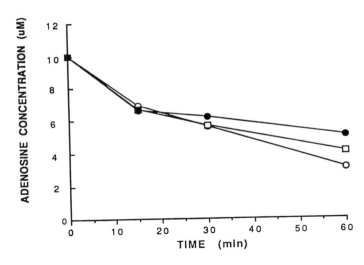

FIGURE 7. Adenosine concentration in the incubation medium of cultured astrocytes as a function of the length of the incubation under control conditions (open circles) and in the presence of 10 μM (closed circles) or 1 μM (open squares) diazepam. (From Huang, R. and Hertz, L., unpublished experiments.)

metabolism in astrocytes.[49] This is not the reason for the inhibited uptake, since the accumulation of label retained in the intracellular pool of adenosine itself is greatly decreased in the presence of 10 μM diazepam after long-term incubation. This effect could contribute to the enhancement of adenosine action *in vivo*, since it helps to maintain the extracellular concentration of adenosine (Huang and Hertz,[49a] unpublished experiments). This is illustrated in Figure 7, showing that the adenosine concentration in the incubation medium rapidly decreases during incubation of primary cultures of astrocytes in a medium which initially contained 10 μM of adenosine and that this decrease is partly (and significantly) inhibited by diazepam (at both 1 and 10 μM). At the same time, a normally occurring concomitant increase in hypoxanthine, one of the deaminated adenosine metabolites, is decreased by the benzodiazepine.[49a] This effect of benzodiazepines might conceptually be exerted either intracellularly or on an ectodeaminase. Since diazepam has no effect on the adenosine concentration in the medium in the presence of the adenosine uptake inhibitor, dipyridamole, it is likely that the inhibition is intracellular. It can be expected that the deaminase inhibitor deoxycoformycin will exert a similar action. This, together with the protective action of extracellular adenosine against ischemia, might explain the ability of deoxycoformycin to protect neurons against ischemic brain damage.[50]

V. NUCLEOSIDE TRANSPORT IN CELLS IN THE CARDIOVASCULAR SYSTEM

A. ADENOSINE EFFECTS ON THE CARDIOVASCULAR SYSTEM

In addition to its effect on the CNS, adenosine also exerts direct effects on cells in the cardiovascular system. The best established effect is probably a pronounced vasodilation, and it cannot be excluded that several adenosine effects on CNS function described in the literature could be secondary to the hypotension evoked by the vasodilation.

Several vascular beds are dilated by adenosine, e.g., cerebral vessels, coronary vessels, and intestinal vessels (Chapter 14). As in the brain, the action of adenosine is probably to a large extent terminated by uptake of adenosine into adjacent cells. Such an uptake has been demonstrated in brain microvessels.[51,52]

Atrioventricular conduction is affected by exogenous adenosine (Chapter 13) and it has recently been shown that release of endogenous adenosine can be observed electrocardiographically in humans administered dipyridamole.[53] Adenosine also exerts an inhibitory effect on traumatically induced ventricular automaticity in the rat, which is enhanced by diazepam; this effect can be inhibited by an adenosine antagonist (theophylline) and by the peripheral-type benzodiazepine blocker PK 11195, but not by the central-type benzodiazepine antagonist RO 15-1788.[54,55] Other studies have shown that benzodiazepines potentiate adenosine effects in isolated cardiac and smooth muscle.[56]

B. TRANSPORT KINETICS

As in brain cells, studies of adenosine transport in cardiac muscle have been performed following two different approaches. The short-term uptake (measured during a 30-s period) into cultured rat myocytes was found by Ford and Rovetto[57] to be saturable with a K_m value of 6 μM and a V_{max} of 575 pmol/min/mg protein (at 24°C), Both these values are far below corresponding data in erythrocytes, but quite similar to the values observed in neurons and astrocytes. In addition, there is a nonsaturable uptake which, at a concentration of ≈ 200 μM adenosine, accounts for approximately one half of the total uptake. At the extracellular adenosine levels under normal conditions (≈ 0.1 μM) and even during hypoxia (low micromolar range), the nonsaturable uptake will contribute very little to the total uptake.[57] Already after 30 s the adenosine metabolites constituted at least two thirds of the total accumulated radioactivity, except at very high extracellular adenosine concentrations.

Approximately similar kinetic constants (K_m 11.1 μM; V_{max} 1.7 nmol/min/10^6 cells) were obtained by Bowditch et al.[58] using 5-min uptake periods for adenosine uptake into cultured rat cardiac myocytes. In contrast to Ford and Rovetto,[57] who found that the uptake was rectilinear for only 1 min (and claimed that the use of longer uptake periods cause an underestimation of the uptake rate), Bowditch et al.[58] reported a rectilinear uptake during the entire 5-min period. From the kinetic constants it can be calculated that the uptake at 10 μM must have amounted to about 4 nmol/10^6 cells after 5 min. On the assumption that 10^6 cells equal 6 mg protein[59] and that the cells contain a maximum of 10 μl water per milligram protein, the total accumulation of adenosine plus its metabolites must, at this time amount to ≈ 80 μM, of which about one third may be found in adenosine itself, i.e., possibly a slightly higher concentration than in the medium. The 30-s uptake of adenosine determined by Ford and Rovetto[57] is almost abolished in the presence of 10 μM NBTI, but no information was given about the effect of lower concentrations. In contrast, the 5-min uptake was reported to have a K_i of 2 nM.[58]

Incorporation of 2 μM labeled adenosine, measured over a 5-min period, into cultured rat aortic smooth muscle, is low (5 to 10 pmol/min/10^6 cells)[60,61] and appears to be nonsaturable. However, endothelial cells from rat aorta show a ten times faster uptake at 2 μM; this uptake occurs partly by a high-affinity uptake system with a K_m of ≈ 3 μM and a V_{max} of ≈ 90 pmol/min/10^6 cells and partly by a low-affinity uptake (K_m 250 μM; V_{max} 300 pmol/min/10^6 cells), which quantitatively plays no role at adenosine concentrations below 10 μM. Dipyridamole inhibits the high-affinity uptake with an IC_{50} of <100 nM, but has little or no effect on the low-affinity uptake, even at 100 μM.[60] Recent *in vivo* experiments have confirmed that endothelial cells, in this case in the arteries of dog hindlimbs, show an avid uptake of adenosine.[62]

C. DIFFUSION VS. ACTIVE UPTAKE

The nonsaturable uptake rate into smooth muscle suggests that this uptake is not by active transport. Its potent inhibition by NBTI indicates that it occurs by facilitated diffusion rather than by simple diffusion. The uptake into endothelial cells is considerably faster. Whether its high-affinity component represents active uptake or facilitated diffusion cannot be established from the data given, but the low K_m tends to support the concept of an active uptake.

The uptake into heart muscle is also saturable. It is followed by an intense and rapid metabolism, and it is unknown whether or not there is any concentrative uptake of adenosine itself under the experimental conditions used. Unfortunately, the data on inhibition by NBTI were inconclusive and it is unknown whether or not the uptake requires the presence of sodium. It is therefore an open question whether it occurs by facilitated diffusion or by active transport, although the low K_m value, again, is more suggestive of an active uptake than of facilitated diffusion. Most importantly, the characteristics of the adenosine uptake, including the rapid deamination of accumulated adenosine, suggest that it should be relatively simple to obtain an unequivocal answer to this question using experimental strategies similar to those which have been applied to cultured neurons and astrocytes.

VI. CONCLUDING REMARKS

Historically, determination of adenosine transport has been fraught with difficulties, methodological problems, and misinterpretations. However, about a decade ago, the necessity became obvious of strictly distinguishing between transport and uptake, with uptake indicating a tandem of intracellular metabolism and facilitated diffusion along a concentration gradient, maintained as a result of the metabolic conversion of adenosine. It has been clearly stated in the literature that this applies to the situation in erythrocytes and certain other cell types, several of them transformed cell lines, whereas it might not apply to other cell types, which may display active uptake mechanisms. Such an active uptake could be energetically driven by a transcellular sodium gradient, and thus be sodium dependent. A sodium-dependent adenosine uptake has been described during the last decade in several brain preparations. Nevertheless, it has been categorically advocated that adenosine transport in brain preparations must be determined in a similar manner as in erythrocytes, i.e., using ultra-short incubation times and/or cells poisoned with metabolic inhibitors. The absence of any major metabolism of adenosine was taken as the criterion of suitable experimental conditions.[30,31] The recent demonstration that adenosine appears to be accumulated by an active, energy-dependent adenosine transport into both neurons and astrocytes makes this criterion and the experimental approach used invalid. However, it should be strongly emphasized that whereas this applies to brain cells and certain other cells, e.g., kidney cells and gastrointestinal cells, it may *not* apply to those cell types where the mechanisms of adenosine have not yet been identified.

One reason that the mechanism of adenosine transport appears to be so fundamentally different in different cell types is probably the many functional roles of this compound. Facilitated diffusion, regulated by metabolism, is well suited for cells needing adenosine as a metabolic substrate. However, a similar mechanism would not be well suited for regulation of extracellular concentrations in a tissue like brain, where cells express receptors for adenosine, and where an active uptake of adenosine may constitute an important mechanism for regulation of the extracellular concentration. The same argument could be made for heart cells, but the means of adenosine transport in these cells still have to be established. The low rate of adenosine uptake in vascular smooth muscle and its apparent lack of regulation may seem peculiar. However, the role of these cells might be to respond to demands from adjacent parenchymal cells, signaled by adenosine release, without interfering with the signal itself.

Inhibitors of adenosine transport have been of crucial importance for the elucidation of mechanisms for adenosine transport. They may also become of therapeutic value, although the situation is complicated by the fact that facilitated diffusion generally is affected bidirectionally. Moreover, active transport and facilitated diffusion appear to be affected by similar drugs, occasionally (e.g., in the case of NBTI) with a difference in potency. This may be of special pharmacological importance for adenosine transport across biological membranes between different compartments, e.g., choroid plexus (and maybe endothelial cells), where uptake from one side occurs by a different mechanism from release towards the other side. However, also in cells which do not have such specific barrier function, efflux is probably of much greater importance for regulation of extracellular adenosine concentrations than it appears from this review. The virtual exclusion of this topic is mainly a result of less information being available about efflux of adenosine itself and its permeable metabolites. Such data are hard to establish for these rapidly metabolized compounds. The observation that one of the ways by which benzodiazepines enhance the physiological action of adenosine is by inhibiting intracellular degradation and thus maintaining a higher extracellular adenosine concentration suggests that adenosine efflux contributes to the regulation of extracellular adenosine concentration in the brain. Further characterization of efflux processes for adenosine and its metabolites may be one of the major challenges in adenosine research during the 1990s. It is to be expected that substantial efforts will be devoted to such research because of an increasing amount of evidence that adenosine, its analogs, and inhibitors of its transport may be of major pharmacological relevance not only in connection with further understanding of the mechanisms of benzodiazepine actions, but also for cytoprotection against ischemic events, especially in the heart and in the brain.

REFERENCES

1. **Hartman, S. C. and Buchanan, J. M.,** Nucleic acids, purines, pyridimines (nucleotide synthesis), *Annu. Rev. Biochem.,* 28, 365, 1959.
2. **Murray, A. W.,** The biological significance of purine salvage in rabbit bone marrow cells, human leucocytes, blood platelets and erythrocytes, *Annu. Rev. Biochem.,* 40, 811, 1971.
3. **Pritchard, J. B., O'Connor, N. O., Oliver, J. M., and Berlin, R. D.,** Uptake and supply of purine compounds to the liver, *Am. J. Physiol.,* 229, 967, 1975.
4. **Stein, W. D.,** *Transport and Diffusion Across Cell Membrane,* Academic Press, Orlando, FL, 1986.
5. **Plagemann, P. G. W., Wohlhueter, R. M., and Woffendin, C.,** Nucleoside and nucleobase transport in animal cells, *Biochim. Biophys. Acta,* 947, 405, 1988.
6. **Christensen, H. N.,** Distinguishing amino acid transport systems of a given cell or tissue, *Meth. Enzymol.,* 173, 576, 1989.
7. **Wohlhueter, R. M. and Plagemann, P. G. W.,** Measurement of transport versus metabolism in cultured cells, *Meth. Enzymol.,* 173, 714, 1989.
8. **Mohrman, D. E.,** Adenosine handling in interstitia of cremaster muscle studied by bioassay, *Am. J. Physiol.,* 254, 369, 1988.
9. **Phillis, J. W., O'Regan, M. H., and Walter, G. A.,** Effects of deoxycoformycin on adenosine, inosine, hypoxanthine, xanthine, and uric acid release from the hypoxemic rat cerebral cortex, *J. Cereb. Blood Flow Metab.,* 8, 733, 1988.
10. **Gati, W. P., Dagnino, L., and Paterson, A. R. P.,** Enantiomeric selectivity of adenosine transport systems in mouse erythrocytes and L1210 cells, *Biochem. J.,* 263, 957, 1989.
11. **Le Hir, M. and Dubach, U. C.,** Uphill transport of pyrimidine nucleosides in renal brush border vesicles, *Pfluegers Arch.,* 404, 238, 1985.
12. **Paterson, A. R. P., Kolassa, N., and Cass, C. E.,** Transport of nucleoside drugs in animal cells, *Pharmacol. Ther.,* 12, 515, 1981.
13. **Plagemann, P. G. W. and Wohlhueter, R. M.,** Permeation of nucleosides, nucleic acid bases, and nucleotides in animal cells, *Curr. Top. Membr. Transp.,* 14, 226, 1980.
14. **Bender, A. S. and Hertz, L.,** Similarities of adenosine uptake systems in astrocytes and neurons in primary cultures, *Neurochem. Res.,* 11, 1507, 1986.
15. **Hertz, L.,** Kinetics of adenosine uptake into astrocytes, *J. Neurochem.,* 31, 55, 1978.
16. **Oliver, J. M. and Paterson, A. R. P.,** Nucleoside transport I. A mediated process in human erythrocytes, *Can. J. Biochem.,* 49, 262, 1971.

17. **Lum, C. T., Marz, R., Plagemann, P. G., and Wohlhueter, R. M.,** Adenosine transport and metabolism in mouse leukemia cells and in canine thymocytes and peripheral blood leukocytes, *J. Cell. Physiol.*, 101, 173, 1979.

18. **Darnowski, J. W.,** Concentrative uridine transport by murine splenocytes: kinetics, substrate specificity, and sodium dependency, *Cancer Res.*, 47, 2614, 1987.

19. **Plagemann, P. G. W. and Woffendin, C.,** Na$^+$-dependent and -independent transport of uridine and its phosphorylation in mouse spleen cells, *Biochim. Biophys. Acta*, 981, 315, 1989.

20. **Plagemann, P. G. W. and Wohlhueter, R. M.,** Nucleoside transport in cultured mammalian cells. Multiple forms with different sensitivity to inhibition by nitrobenzylthioninosine or hypoxanthine, *Biochim. Biophys. Acta*, 773, 39, 1984.

21. **Hammond, J. R., Jarvis, S. M., Paterson, A. R. P., and Clanachan, A. S.,** Benzodiazepine inhibition of nucleoside transport in human erythrocytes, *Biochem. Pharmacol.*, 32, 1229, 1983.

22. **Bender, A. S. and Hertz, L.,** Pharmacological characterization of diazepam receptors in neurons and astrocytes in primary cultures, *J. Neurosci. Res.*, 18, 366, 1987.

23. **Spector, J.,** Thymidine transport and metabolism in choroid plexus: effect of diazepam and thiopental, *J. Pharm. Exp. Ther.*, 235, 16, 1985.

24. **Jakobs, E. S. and Paterson, A. R. P.,** Sodium-dependent concentrative nucleoside transport in cultured intestinal epithelial cells, *Biochem. Biophys. Res. Commun.*, 140, 1028, 1986.

25. **Hertz, L. and Matz, H.,** Inhibition of adenosine deaminase activity reveals an intense active transport of adenosine into neurons in primary cultures, *Neurochem. Res.*, 14, 755, 1989.

26. **Matz, H. and Hertz, L.,** Effects of an adenosine deaminase inhibitor on active uptake and metabolism of adenosine in astrocytes in primary cultures, *Brain Res.*, 515, 168, 1990.

26a. **White, H. S., Hertz, L., and Woodbury, D. M.,** unpublished results.

27. **Matz, H. and Hertz, L.,** Adenosine metabolism in neurons and astrocytes in primary cultures, *J. Neurosci. Res.*, 24, 260, 1989.

28. **Thampy, K. G. and Barnes, E. M., Jr.,** Adenosine transport by cultured glial cells from chick embryo brain, *Arch. Biochem. Biophys.*, 220, 340, 1983.

29. **Thampy, K. G. and Barnes, E. M., Jr.,** Adenosine transport by primary cultures of neurons from chick embryo brain, *J. Neurochem.*, 40, 874, 1983.

30. **Geiger, J. D., Johnston, M. E., and Yago, V.,** Pharmacological characterization of rapidly accumulated adenosine by dissociated brain cells from adult rat, *J. Neurochem.*, 51, 283, 1988.

31. **Johnston, M. E. and Geiger, J. D.,** Sodium-dependent uptake of nucleosides by dissociated brain cells from the rat, *J. Neurochem.*, 52, 75, 1989.

32. **Meunier, F. M. and Morel, N.,** Adenosine uptake by cholinergic synaptosomes from Torpedo electric organ, *J. Neurochem.*, 31, 845, 1978.

33. **Bender, A. S., Wu, P. H., and Phillis, J. W.,** The characterization of [3H]adenosine uptake into cerebral cortical synaptosomes, *J. Neurochem.*, 35, 629, 1980.

34. **Banay-Schwartz, M., de Guzman, T., and Lajtha, A.,** Nucleotide uptake by slices of mouse brain, *J. Neurochem.*, 35, 544, 1980.

35. **Lewin, B. and Bleck, V.,** Uptake and release of adenosine by cultured astrocytoma cells, *J. Neurochem.*, 33, 365, 1979.

36. **Miras-Portugal, M. T., Torres, M., Rotlan, P., and Aunis, D.,** Adenosine transport in bovine chromaffin cells in culture, *J. Biochem.*, 261, 1712, 1986.

37. **Shimizu, H., Tanaka, S., and Kodama, T.,** Adenosine kinase of mammalian brain: partial purification and its role for the uptake of adenosine, *J. Neurochem.*, 19, 687, 1972.

38. **Davies, L. P. and Hambley, J. W.,** Regional distribution of adenosine uptake into guinea pig brain slices and the effect of some inhibitors: evidence for nitrobenzylthioinosine-sensitive and insensitive sites?, *Neurochem. Int.*, 8, 103, 1986.

39. **Barberis, C., Minn, A., and Gayet, J.,** Adenosine transport into guinea-pig synaptosomes, *J. Neurochem.*, 36, 347, 1981.

40. **Phillis, J. W. and Wu, P. H.,** Nitrobenzylthinosine inhibition of adenosine uptake in guinea pig brain, *J. Pharm. Pharmacol.*, 35, 540, 1983.

41. **Phillis, J. W., Edstrom, J. P., Ellis, S. W., and Kirkpatrick, J. R.,** Theophylline antagonizes flurazepam-induced depression of cerebral cortical neurons, *Can. J. Physiol. Pharmacol.*, 57, 917, 1979.

42. **Polc, P., Bonetti, E. P., Pieri, L., Cumin, R., Angioi, R. M., Mohler, H., and Haefely, W. E.,** Caffeine antagonizes several central effects of diazepam, *Life Sci.*, 28, 2265, 1981.

43. **Phillis, J. W. and O'Regan, M. H.,** The role of adenosine in the central actions of the benzodiazepines, *Prog. Neuropsychopharmacol. Biol. Psych.*, 12, 389, 1988.

44. **Phillis, J. W. and O'Regan, M. H.,** Benzodiazepine interaction with adenosine systems explains some anomalies in GABA hypothesis, *Trends Pharmacol. Sci.*, 9, 153, 1988.

45. **Mah, H. D. and Daly, J. W.,** Adenosine-dependent formation of cyclic AMP in brain slices, *Pharmacol. Res. Commun.*, 8, 65, 1976.

46. **Phillis, J. W., Bender, A. S., and Wu, P. H.,** Benzodiazepines inhibit adenosine uptake in rat brain synaptosomes, *Brain Res.,* 195, 494, 1980.
47. **Bender, A. S. and Hertz, L.,** Dissimilarities between benzodiazepine binding sites and adenosine uptake sites in astrocytes and neurons in primary cultures, *J. Neurosci. Res.,* 17, 154, 1987.
48. **Duka, T. Hollt, V. and Herz, A.,** In vivo receptor occupation by benzodiazepines and correlation with the pharmacological effects, *Brain Res.,* 179, 147, 1979.
49. **Matz, H. and Hertz, L.,** Effects of diazepam on adenosine metabolism of astrocytes, *Trans. Am. Soc. Neurochem.,* 20, 229, 1989.
49a. **Huang, R. and Hertz, L.,** unpublished experiments.
50. **Phillis, J. W. and O'Regan, M. H.,** Deoxycoformycin antagonizes ischemia-induced neuronal degeneration, *Brain Res. Bull.,* 22, 537, 1989.
51. **Wu, P. H. and Phillis, J. W.,** Uptake of adenosine by isolated rat brain capillaries, *J. Neurochem.,* 697, 1982.
52. **Stefanovich, V.,** Uptake of adenosine by isolated bovine cortex microvessels, *Neurochem. Res.,* 8, 1459, 1983.
53. **Lerman, B. B., Wesley, R. C., and Belardinelli, L.,** Electrophysiologic effects of dipyridamole on atrioventricular nodal conduction and supraventricular tachycardia. Role of endogenous adenosine, *Circulation,* 80, 1536, 1989.
54. **Ruiz, F., Hernandez, J., and Ribeiro, J. A.,** Theophylline antagonizes the effect of diazepam on ventricular automaticity, *Eur. J. Pharmacol.,* 155, 205, 1988.
55. **Ruiz, F., Hernandez, J., and Perez, D.,** The effect of diazepam on ventricular automaticity induced by local injury. Evidence of involvement of "peripheral type": benzodiazepine receptors, *J. Pharm. Pharmacol.,* 41, 306, 1989.
56. **Clanachan, A. S. and Marshall, R. J.,** Potentiation of the effects of adenosine on isolated cardiac and smooth muscle by diazepam, *Br. J. Pharmacol.,* 71, 459, 1980.
57. **Ford, D. A. and Rovetto, M. J.,** Rat cardiac myocyte adenosine transport and metabolism, *Am. J. Physiol.,* 252, H54, 1987.
58. **Bowditch, J., Brown, A. K., and Down, J. W.,** Accumulation and salvage of adenosine and inosine by isolated mature cardiac myocytes, *Biochim. Biophys. Acta,* 844, 119, 1985.
59. **Walker, E. J., Burns, J. H., and Dow, J. W.,** Amino acid transport and protein synthesis in energetically-stable calcium-tolerant isolated cardiac myocytes, *Biochim. Biophys. Acta,* 721, 280, 1982.
60. **Pearson, J. D., Carleton, J. S., Hutchings, A., and Gordon, J. L.,** Uptake and metabolism of adenosine by pig aortic endothelial and smooth-muscle cells in culture, *J. Biochem.,* 170, 265, 1978.
61. **Belloni, F. L., Bruttig, S. P., Rubio, R., and Berne, R. M.,** Uptake and release of adenosine by cultured rat aortic smooth muscle, *Microvasc. Res.,* 32, 200, 1986.
62. **Gorman, M. W., Bassingthwaighte, J. B., Olsson, R. A., and Sparks, H. V.,** Endothelial cell uptake of adenosine in canine skeletal muscle, *Am. J. Physiol.,* 250, H482, 1986.

Chapter 8

ADENOSINE AND ADENINE NUCLEOTIDES IN TISSUES AND PERFUSATES

Thomas D. White and Katja Hoehn

TABLE OF CONTENTS

I. INTRODUCTION

There is abundant evidence that both adenosine 5′-triphosphate (ATP) and adenosine, acting at specific extracellular receptors, elicit important actions throughout the body.[1-3] In order to exert these effects, the purines must first be released from cells. The idea that these substances, particularly ATP which is generally conserved by cells for its essential intracellular functions, are released has met considerable resistance over the years. Nevertheless, it is now recognized that ATP can exit certain cells to act at extracellular P_2 purinoceptors or, following extracellular metabolism by a series of ectonucleotidases, provide a source of adenosine to act at P_1 purinoceptors.

ATP is stored with acetylcholine within cholinergic vesicles[4] and with noradrenaline in sympathetic nerves,[5,6] and there is evidence that it can be coreleased with these substances to function as a cotransmitter. It is also cosecreted in large amounts with 5-HT from platelets[7] and with catecholamines from adrenal chromaffin cells.[8] ATP is released from certain nonsecretory cells such as the Torpedo electric organ,[9] skeletal muscle,[10] cardiac myocytes,[11-13] and vascular endothelial cells.[14] The mechanism by which this release occurs remains unknown. Carrier-mediated transport systems for nucleotides have been described in the cell membranes of some tissues;[15-18] if such systems can be clearly established, they could provide a means whereby nucleotides such as ATP might exit nonsecretory cells.

Unlike ATP, there is no evidence that adenosine is stored in synaptic vesicles and secreted as such from nerves. In this sense, neuronally released adenosine is more likely to function as a neuromodulator than as a transmitter. A bidirectional, facilitated-diffusional transporter for nucleosides has been described in various tissue including nerves.[19-21] Uptake of adenosine via this transporter plays an important role in terminating the extracellular actions of adenosine when the extracellular concentration exceeds the intracellular concentration. However, it also appears to facilitate the efflux of adenosine from cells when the intracellular concentration becomes elevated.[22-24] Release of adenosine via the nucleoside transporter could provide an important source of extracellular adenosine to exert protective effects during ischemia, hypoxia, and increased work load where intracellular adenosine levels would increase. This could be particularly important in the brain and cardiovascular system. Extracellular adenosine can also be derived from released ATP, adenosine 5′-diphosphate (ADP) or adenosine 5′-monophosphate (5′-AMP) by ectonucleotidases as discussed above. There is some evidence that cyclic AMP (cAMP) may be released from certain tissues[25,26] and ectophosphodiesterases have been described which could convert released cAMP to adenosine extracellularly.[25-27]

In the remainder of this chapter, some potential sources and evidence for the release of ATP and adenosine in the nervous and cardiovascular systems will be considered. Where possible, the reader will be referred to appropriate reviews.

II. RELEASE OF PURINES IN THE CENTRAL NERVOUS SYSTEM

A. ADENOSINE RELEASE

Adenosine is an important inhibitory neuromodulator of neuronal activity in the brain[3,28] and neuronally derived adenosine might also affect cerebral blood flow. The first studies of depolarization-evoked, radiolabeled adenosine release from cortical slices and synaptosomal beds were conducted in the early 1970s.[29-31] Subsequently, depolarization-evoked release of radiolabeled purines was demonstrated *in vivo*[32-34] and in response to various other stimuli such as ischemia[35] and exposure to morphine[36,37] and excitatory amino acid (EAA) agonists.[34,38,39]

More recently, both UV and fluorescence high-pressure liquid chromatography methods for detecting adenosine have been developed.[40,41] Depolarization-evoked release of endogenous adenosine has been measured from rat striatal slices[42] and from synaptosomes prepared from whole brain.[43] In synaptosomes, basal adenosine release appears to arise from the extracellular metabolism of a released nucleotide, whereas much of the veratridine- and most of the K^+-evoked release occurs as adenosine per se.[43] Evoked release of adenosine from brain preparations appears to occur at least in part on the nucleoside transporter, insofar as it is diminished by the nucleoside transport inhibitor dipyridamole.[24] In addition to the facilitated-diffusional nucleoside transporter, a high-affinity, Na^+-dependent, active transport system for adenosine has recently been described in brain cells.[44] This system might also be capable of releasing adenosine when the normal inward-directed electrochemical gradient for Na^+ is reversed (e.g., during depolarization, hypoxia, ischemia, etc.).

Brain interstitial adenosine is elevated during various pathological conditions such as ischemia, hypoxia, and systemic hypotension,[45-50] where it has been proposed to mediate cerebral vasodilation and provide neuroprotection against excessive energy demands. Acute administration of pharmacologically relevant concentrations of ethanol potentiates K^+-evoked adenosine release from rat cerebellar synaptosomes,[51] a finding consistent with the possibility that cerebellar adenosine may be involved in the mediation of some ethanol-induced motor disturbances.

The EAA, glutamate, releases adenosine from slices of rat parietal cortex.[52,53] Because exogenously administered adenosine and its analogs protect against EAA-induced neuronal cell death,[54,55] it has been suggested that released adenosine may protect against excitotoxicity.[56] Glutamate-evoked release of adenosine involves both N-methyl-D-aspartate (NMDA) and non-NMDA receptors,[53] and the specific EAA agonists, NMDA, kainate, and quisqualate all release adenosine.[57] Although the total amount of adenosine released by these agonists is not diminished in the absence of Ca^{2+}, kainate appears to release adenosine from separate Ca^{2+}-dependent and -independent pools.[57] NMDA is 33 times more potent in releasing adenosine than [^{3}H]noradrenaline from cortical slices, and block of adenosine release (but not [^{3}H]noradrenaline release) by uncompetitive NMDA antagonists such as Mg^{2+} and MK-801 is overcome at high concentrations of NMDA.[58] Maximal NMDA-evoked release of adenosine from cortical slices apparently occurs when relatively few receptors are occupied and before significant generation of action potentials occurs in the slices.[58] Consequently, it appears unlikely that adenosine, released during NMDA receptor activation, would be an effective antiexcitotoxic agent against excessive NMDA receptor stimulation. Rather, it seems more likely that released adenosine might provide an inhibitory threshold that must be overcome before NMDA-mediated transmission can proceed maximally; adenosine may help to maintain the selectivity of NMDA-mediated processes such as learning, memory, and synaptic plasticity in the cortex.

L-Glutamate also releases adenosine from rat cortical synaptosomes, but, unlike the situation in slices, this release is mediated by the Na^+-dependent, high-affinity transport of L-glutamate into the synaptosomes and not by EAA receptors.[59] The adenosine detected in the medium following exposure of synaptosomes to L-glutamate is derived from a released nucleotide which is neither ATP nor cAMP. This nonreceptor-mediated process is not observed in intact cortical slices,[53] possibly because it is overwhelmed by receptor-mediated release. Nevertheless, this release arises directly from glutamatergic nerve terminals, so that the adenosine would be in an appropriate location to act at presynaptic receptors and inhibit further release of glutamate.[28]

Adenosine exerts antinociceptive effects in the spinal cord, and the antinociceptive actions of morphine appear to be mediated indirectly by adenosine, insofar as they are diminished by adenosine antagonists.[60,61] In further support of this, morphine releases adenosine from spinal synaptosomes and from intrathecally perfused rat spinal cords *in vivo*. The spinal antinociceptive actions of 5-HT also appear to be mediated by adenosine, but in this case a nucleotide is released which is subsequently converted to adenosine by ectonucleotidases. Both morphine and 5-HT appear to release their purines from capsaicin-sensitive primary afferents. Pharmacological manipulation of spinal adenosine release may provide new methods for controlling pain.

Although there is clear evidence that adenosine is released from nerves in the central nervous system (CNS), it is possible that glia may also release adenosine. Adenosine is apparently released when glial cell cultures are exposed to electrical field stimulation[62] or to metabolic poisons.[63] In the latter case, adenosine appears to be released on the nucleoside transporter.

B. ATP RELEASE

Application of either adenosine or ATP decreases neuronal firing in the CNS,[64] presumably by acting at P_1 purinoceptors. However, ATP, unlike adenosine, can also produce excitatory effects,[64] raising the possibility that ATP might function as an excitatory transmitter in the brain. Moreover, extracellular ATP could act as a substrate for ectoprotein kinase[65] and it appears to activate phospholipase C in glial[66,67] and anterior pituitary cultures.[68] Synaptosomes prepared from whole rat brain release ATP when they are depolarized by K^+ or veratridine[69,70] and direct electrical stimulation of rat sensory motor cortex releases ATP into cortical cups *in vivo*.[71] Most of the ATP released from whole-brain synaptosomes does not appear to originate primarily from catecholaminergic nerves because release is unaffected when these nerves are destroyed by pretreatment with intraventricular 6-hydroxydopamine.[72] Although experiments with botulinum toxin A suggested that ATP release from whole-brain synaptosomes did not occur primarily from cholinergic synaptosomes,[73] Richardson and Brown[74] have shown that ATP is released from affinity-purified cholinergic synaptosomes during exposure to K^+ or veratridine.

ATP is also released from primary cultures of CNS neurons when they are depolarized with K^+ or veratridine.[75] An ectoprotein kinase may utilize the released ATP to phosphorylate specific proteins on the neuronal membranes. ATP is released presynaptically during electrical stimulation of the Schaffer collaterals in rat hippocampal slices.[76] Although Wieraszko and Seyfried[77] have presented evidence suggesting that application of ATP potentiates synaptic transmission in pyramidal CA1 cells in the hippocampus, Stone and Cusack,[78] using stable analogs of ATP, were unable to demonstrate the presence of P_2 purinoceptors in CA1 hippocampal neurons. Further investigations should be conducted to determine possible functions for released ATP in the hippocampus.

Finally, there is evidence that ATP might function as an excitatory transmitter in low-threshold, primary afferent inputs to the spinal dorsal horn.[79-80] In this regard, antidromic stimulation of the great auricular sensory nerve releases ATP into venous perfusates[81] and depolarization releases much more ATP from dorsal than from ventral spinal cord synaptosomes.[82]

III. RELEASE OF PURINES IN THE PERIPHERAL NERVOUS SYSTEM

The actions and release of purines in the autonomic nervous system have been reviewed extensively elsewhere.[1,2]

A. ATP RELEASE FROM SYMPATHETIC NERVES

There is convincing evidence that ATP is a cotransmitter with noradrenaline in the sympathetic nerves innervating the vas deferens.[1,2] Release of ATP has been detected when the hypogastric nerve and a small portion of the prostatic smooth muscle are stimulated with a suction electrode.[83] Release is not due to contraction of the vas, nor is it due to postsynaptic receptor stimulation because it persists following block of postsynaptic responses with prazosin and α,β-methylene ATP. Release of ATP is abolished following chemical sympathectomy with 6-hydroxydopamine, but is unaffected by reserpinization which depletes sympathetic vesicles of catecholamines, but not ATP.[84]

ATP may also function as an excitatory cotransmitter with noradrenaline in the sympathetic nerves innervating blood vessels,[1,2] although convincing evidence that ATP is released from these nerves has been difficult to obtain. Sedaa et al.[85] have recently reported that endogenous purines are released during transmural electrical stimulation of rabbit aorta. Although much of the released purine was detected as adenosine, it was likely derived from released ATP. They found that 95% of the purine release appeared to arise from the endothelium as a consequence of activation of α-adrenoceptors, about 7% as a result of α-receptor activation on smooth muscle, and the remaining 3% from the nerves themselves. However, the amount of ATP released from the nerves seemed consistent with a possible cotransmitter function for ATP in these sympathetic nerves.

Although ATP does not appear to be the nonadrenergic, noncholinergic (NANC) transmitter at ileal smooth muscle,[1,2] ATP depolarizes and hyperpolarizes nerves intrinsic to the ileal myenteric plexus.[86] ATP is released when myenteric synaptosomes are depolarized or exposed to nicotinic agonists and 5-HT and much of this appears to originate from noradrenergic varicosities, insofar as release is diminished substantially following chemical sympathectomy with 6-hydroxydopamine.[1,2] Surprisingly, release of [³H]noradrenaline, but not ATP, is modulated by presynaptic α_2 receptors.[87] This and other evidence reviewed elsewhere[2,24] raises the possibility that ATP and noradrenaline are not costored within the same synaptic vesicles, but may coexist within separate releasable pools in noradrenergic sympathetic nerves.

B. ATP RELEASE FROM PARASYMPATHETIC NERVES

ATP may be a noncholinergic excitatory transmitter in the pelvic nerves innervating the bladder,[1,2] and neurogenic release of ATP into perfusates has been detected when guinea pig,[88] but not rabbit,[89] detrusor strips are electrically stimulated transmurally.

C. ATP RELEASE FROM INTRINSIC ENTERIC NERVES

There is evidence that ATP may be a NANC transmitter in the guinea pig taenia coli, and in the duodenum, colon, and rectum of other species.[1,2] However, ATP does not appear to be the NANC transmitter that mediates neurogenic relaxations of the stomach, ileum, or canine ileocolonic junction. In the latter case, recent evidence suggests that NO may mediate the NANC responses.[90]

In the guinea pig taenia coli, release of ATP during transmural electrical stimulation has been reported,[88] although a more recent study failed to detect ATP release except under conditions which apparently directly depolarized the nerve and muscle.[91]

D. ATP RELEASE FROM MOTOR NERVES

ATP may function as a transmitter in motor neurons. Ewald[92] showed that ATP potentiates the action of acetylcholine in the rat hemidiaphragm. Later studies have shown that ATP depolarizes cultured chick myotubes.[93-95] Although Hume and Honig[95] were unable to detect single channel openings when ATP was applied to chick muscle in the absence of voltage-gated channel blockers, Hume and Thomas[96] later repeated these studies in the presence of appropriate current blockers and showed that ATP activates a size-selective conductance to small ions such as Na^+, K^+, and Cl^-. ATP appears to act at a receptor distinct from the nicotinic receptor in chick muscle,[96] but it may interact directly with nicotinic receptors at frog skeletal endplates[97] and Xenopus myotomal muscle cells.[98] It has been suggested that released ATP may itself act presynaptically to inhibit the release of acetylcholine from motor nerves,[99,100] but other studies suggest that activation of presynaptic P_1 purinoceptors on the motor terminals by adenosine plays a dominant role in purinergic modulation of motor nerve function.[101,102]

There is evidence that ATP is released from the rat phrenic nerve[103] and from cholinergic motor nerve synaptosomes isolated from the Torpedo electric organ.[104] Surprisingly, botulinum toxin A inhibits the release of acetylcholine from Torpedo synaptosomes without affecting ATP release,[105] suggesting that ATP may not be stored and/or cosecreted from the same vesicles as acetylcholine. There is also evidence that ATP may be released postsynaptically from target organs such as striated muscle[10] and the Torpedo electric organ.[9] Once released, ATP could exert excitatory effects postsynaptically and/or, following conversion to adenosine, act presynaptically to inhibit transmitter release.

IV. RELEASE OF PURINES IN THE CARDIOVASCULAR SYSTEM

The sources and functions of purines in the cardiovascular system have been reviewed recently[1,2] and will only be summarized here. The reader is referred to these reviews for specific references.

A. RELEASE OF PURINES IN BLOOD VESSELS

ATP is released as a cotransmitter with noradrenaline from sympathetic nerves innervating blood vessels (see Section III.A, above). In addition, there is evidence that ATP can be released from vascular endothelial cells, platelets, and, in the case of coronary vessels, from heart cells. Diadenosine tetraphosphate is also released from platelets and may provide a stable source of intraluminal ATP to produce endothelium-mediated relaxations of blood vessels. Intraluminal ATP and ADP act at P_2 purinoceptors on endothelial cells to release NO and PGI_2, both of which promote relaxation of blood vessels.

Adenosine, derived from released ATP or released as such from vascular constituents, relaxes blood vessels by actions at A_2 receptors located on the smooth muscle of most vessels; it may also produce endothelium-dependent dilations in some vessels. Adenosine is released from the heart during periods of ischemia and hypoxia, where it produces coronary vasodilation as well as inotropic and chronotropic effects as discussed below. At least some of this adenosine may be derived from released nucleotide. Facilitated diffusion into erythrocytes probably provides an important means of removing circulating adenosine. Endogenous extracellular adenosine autoregulates hepatic and cerebral blood flow during periods of ischemia and hypoxia.

B. RELEASE OF PURINES IN THE HEART

Although it is possible that ATP is released from vagal cholinergic or sympathetic adrenergic neurons in the heart, this has not yet been demonstrated. On the other hand, ATP is released from perfused hearts and isolated myocytes during periods of ischemia and hypoxia. ATP increases intracellular Ca^{2+} in ventricular myocytes, which could produce positive inotropic effects. Released ATP is rapidly degraded by ectonucleotidases to adenosine, which produces negative inotropic effects by acting at A_1 receptors in the ventricle, and negative chronotropic and dromotropic effects through A_1 receptors in the sinoatrial and atrioventricular nodes, respectively.

V. CONCLUSIONS

Purines are also released from other tissues not reviewed here.[1,2] The clinical use of ATP and adenosine agonists and antagonists has been hampered by the ubiquitous actions that these compounds exert on various organ systems, thereby producing numerous undesirable side effects. The development of techniques to manipulate the extracellular levels of endogenous adenosine and ATP in specific tissues could provide important new therapeutic approaches to various diseases.

REFERENCES

1. **White, T. D.,** Role of adenine compounds in autonomic neurotransmission, *Pharmacol. Ther.,* 38, 129, 1988.
2. **White, T. D.,** Role of ATP and adenosine in the autonomic nervous system, in *Novel Peripheral Neurotransmitters,* Bell, C., Ed., Pergamon Journals, Oxford, in press.
3. **Stone, T. W.,** Purine receptors and their pharmacological roles, *Adv. Drug Res.,* 18, 291, 1989.
4. **Dowdall, M. J., Boyne, A. F., and Whittaker, V. P.,** Adenosine triphosphate—a constituent of cholinergic synaptic vesicles, *Biochem. J.,* 140, 1, 1974.
5. **Lägercrantz, H. and Stjärne, L.,** Evidence that most noradrenaline is stored without ATP in sympathetic large dense core vesicles, *Nature,* 249, 843, 1974.
6. **Fried, G., Lägercrantz, H., and Hökfelt, T.,** Improved isolation of small noradrenergic vesicles from rat seminal ducts following castration. A density gradient centrifugation and morphological study, *Neuroscience,* 8, 1271, 1978.
7. **Detwiler, T. C. and Feinman, R. D.,** Kinetics of the thrombin-induced release of adenosine triphosphate by platelets. Comparison with release of calcium, *Biochemistry,* 12, 2462, 1973.
8. **White, T. D., Bourke, J. E., and Livett, B. G.,** Direct and continuous detection of ATP secretion from primary cultures of bovine adrenal chromaffin cells, *J. Neurochem.,* 49, 1266, 1987.
9. **Israel, M., Lesbats, B., Meunier, F. M., and Stinnakre, J.,** Postsynaptic release of adenosine triphosphate induced by single impulse transmitter action, *Proc. R. Soc. (Ser. B,)* 193, 461, 1976.
10. **Abood, L. G., Koketsu, K., and Miyamoto, S.,** Outflux of various phosphates during membrane depolarization of excitable tissues, *Am. J. Physiol,* 202, 469, 1962.
11. **Forrester, T. and Williams, C. A.,** Release of adenosine triphosphate from isolated adult heart cells in response to hypoxia, *J. Physiol. (London),* 268, 371, 1977.
12. **Forrester, T.,** Adenosine or adenosine triphosphate?, in *Vasodilatation,* Vanhoutte, P. M. and Leusen, I., Eds., Raven Press, New York, 1981, 250.
13. **Darius, H., Stahl, G. L., and Lefer, A. M.,** Pharmacologic modulation of ATP release from isolated rat hearts in response to vasoconstrictor stimuli using a continuous flow technique, *J. Pharmacol. Exp. Ther.,* 240, 542, 1987.
14. **Pearson, J. D. and Gordon, J. L.,** Vascular endothelial and smooth muscle cells in culture selectively release adenine nucleotides, *Nature,* 281, 284, 1979.
15. **Chaudry, I. H., Clemens, M. G., and Baue, A. E.,** Uptake of ATP by tissues, in *Purines: Pharmacology and Physiological Roles,* Stone, T. W., Ed., Macmillan, London, 1985, 115.
16. **Elgavish, A. and Elgavish, G. A.,** Evidence for the presence of an ATP transport system in brush-border membrane vesicles isolated from the kidney cortex, *Biochim. Biophys. Acta,* 812, 595, 1985.
17. **Sun, A. Y. and Lee, D. Z.,** Synaptosomal ADP uptake, *J. Neurochem. Suppl.,* 44, S90, 1985.
18. **Lindgren, C. A. and Smith, D. O.,** Increased presynaptic ATP levels coupled to synaptic activity at the crayfish neuromuscular junction, *J. Neurosci.,* 6, 2644, 1986.
19. **Paterson, A. R. P., Harley, E. R., and Cass, C.,** Measurement and inhibition of membrane transport of adenosine, in *Methods in Pharmacology. Vol. 6, Methods Used in Adenosine Research,* Paton, D. M., Ed., Plenum Press, New York, 1985, 165.
20. **Wu, P. H. and Phillis, J. W.,** Uptake by central nervous tissues as a mechanism for the regulation of extracellular adenosine concentrations, *Neurochem. Int.,* 6, 613, 1984.
21. **Deckert, J., Morgan, P. F., and Marangos, P. J.,** Adenosine uptake site heterogeneity in the mammalian CNS? Uptake inhibitors as probes and potential neuropharmaceuticals, *Life Sci.,* 42, 1331, 1988.
22. **Jonzon, B. and Fredholm, B. B.,** Release of purines, noradrenaline, and GABA from rat hippocampal slices by field stimulation, *J. Neurochem.,* 44, 217, 1985.
23. **Wohlheuter, R. M. and Plagemann, P. G. W.,** On the functional symmetry of nucleoside transport in mammalian cells, *Biochim. Biophys. Acta,* 689, 249, 1982.
24. **White, T. D. and MacDonald, W. F.,** Neural release of ATP and adenosine, *Ann. N.Y. Acad. Sci.,* 603, 287, 1990.
25. **Rosberg, S., Selstam, G., and Isaksson, O.,** Characterization of the metabolism of exogenous cyclic AMP by perfused rat heart and incubated prepubertal rat ovary, *Acta Physiol. Scand.,* 94, 522, 1975.

26. **Rosenberg, P. A. and Dichter, M. A.**, Extracellular cAMP accumulation and degradation in rat cerebral cortex in dissociated cell culture, *J. Neurosci.*, 9, 2654, 1989.

27. **Selstam, G. and Rosberg, S.**, Stimulatory effect of FSH in vitro on the extracellularly active cyclic AMP phosphodiesterase in the prepubertal rat ovary, *Acta Endocrinol.*, 81, 563, 1976.

28. **Dunwiddie, T. V.**, The physiological role of adenosine in the central nervous system, *Int. Rev. Neurobiol.*, 27, 63, 1985.

29. **Shimizu, H., Creveling, C. R., and Daly, J.**, Stimulated formation of adenosine $3',5'$-cyclic phosphate in cerebral cortex: synergism between electrical activity and biogenic amines, *Proc. Natl. Acad. Sci. U.S.A.*, 65, 1033, 1970.

30. **Pull, I,. and McIlwain, H.**, Adenine derivatives as neurohumoral agents in the brain: the quantities liberated on excitation of superfused cerebral tissues, *Biochem. J.*, 130, 975, 1972.

31. **Kuroda, Y. and McIlwain, H.**, Uptake and release of ^{14}C-adenine derivatives at beds of mammalian cortical synaptosomes in a superfusion system, *J. Neurochem.*, 22, 691, 1974.

32. **Sulakhe, P. V. and Phillis, J. W.**, The release of [^{3}H]adenosine and its derivatives from cat sensory motor cortex, *Life Sci.*, 17, 551, 1975.

33. **Schubert, P., Lee, K., West, M., Deadwyler, S., and Lynch, G.**, Stimulation-dependent release of ^{3}H-adenosine derivatives from central axon terminals to target neurons, *Nature*, 260, 541, 1976.

34. **Jhamandas, K. and Dumbrille, A.**, Regional release of [^{3}H]adenine derivatives from rat brain in vivo: effect of excitatory amino acids, opiate agonists, and benzodiazepines, *Can. J. Physiol. Pharmacol.*, 58, 1262, 1980.

35. **Berne, R. M., Rubio, R., and Curnish, R. R.**, Release of adenosine from ischemic brain; effect on cerebral vascular resistance and incorporation into cerebral adenine nucleotides, *Circ. Res.*, 35, 262, 1974.

36. **Phillis, J. W., Jiang Zhigen, G., Chelak, B. J., and Wu, P. H.**, Morphine enhances adenosine release from the in vivo rat cerebral cortex, *Eur. J. Pharmacol.*, 65, 97, 1979.

37. **Stone, T. W.**, The effects of morphine and methionine-enkephalin on the release of purines from cerebral cortex slices of rats and mice, *Br. J. Phamacol.*, 74, 171, 1981.

38. **Pull, I. and McIlwain, H.**, Actions of neurohumoral agents and cerebral metabolites on the output of adenine derivatives from superfused tissues of the brain, *J. Neurochem.*, 24, 695, 1975.

39. **Perkins, M. N. and Stone, T. V.**, Quinolinic acid: regional variations in neuronal sensitivity, *Brain Res.*, 259, 172, 1983.

40. **Fredholm, B. B. and Sollevi, A.**, The release of adenosine and inosine from canine subcutaneous adipose tissue by nerve stimulation and noradrenaline, *J. Physiol. (London)*, 313, 351, 1981.

41. **Wojcik, W. J. and Neff, N. H.**, Adenosine measurement by a rapid HPLC-fluorometric method: induced changes of adenosine content in regions of rat brain, *J. Neurochem.*, 41, 759, 1982.

42. **Wojcik, W. J. and Neff, N. H.**, Location of adenosine release and adenosine A$_2$ receptors to rat striatal neurons, *Life Sci.*, 33, 755, 1983.

43. **MacDonald, W. F. and White, T. D.**, Nature of extrasynaptosomal accumulation of endogenous adenosine evoked by K$^+$ and veratridine, *J. Neurochem.*, 45 791, 1985.

44. **Johnston, M. E. and Geiger, J. D.**, Sodium-dependent uptake of nucleosides by dissociated brain cells from the rat, *J. Neurochem.*, 52, 75, 1989.

45. **Zetterström, T., Vernet, L., Ungerstedt, U., Tossman, U., Jonzon, B., and Fredholm, B. B.**, Purine levels in the intact rat brain; studies with an implanted perfused hollow fibre, *Neurosci. Lett.*, 29, 111, 1982.

46. **Van Wylen, D. G. L., Park, T. S., Rubio, R., and Berne, R. M.**, Increases in cerebral interstitial fluid adenosine concentration during hypoxia, local potassium infusion, and ischemia, *J. Cereb. Blood Flow Metab.*, 6, 222, 1986.

47. **Hagberg, H., Andersson, P., Lacarewicz, J., Jacobson, I., Butcher, S., and Sandberg, M.**, Extracellular adenosine, inosine, hypoxanthine, and xanthine in relation to tissue nucleotides and purines in rat striatum during transient ischemia, *J. Neurochem.*, 49, 227, 1987.

48. **Park, T. S., Van Wylen, D. G. L., Rubio, R., and Berne, R. M.**, Brain interstitial adenosine and sagittal sinus blood flow during systemic hypotension in piglet, *J. Cereb. Blood Flow Metab.*, 8, 822, 1988.

49. **Phillis, J. W., O'Regan, M. H., and Walter, G. A.**, Effects of deoxycoformycin on adenosine, inosine, hypoxanthine, xanthine, and uric acid release from the hypoxemic rat cerebral cortex, *J. Cereb. Blood Flow Metab.*, 8, 733, 1988.

50. **Phillis, J. W., O'Regan, M. H., and Walter, G. A.**, Effects of two nucleoside transport inhibitors, dipyridamole and soluflazine, on purine release from the rat cerebral cortex, *Brain Res.*, 481, 309, 1989.

51. **Clark, M. and Dar, M. S.**, Effect of acute ethanol on release of endogenous adenosine from rat cerebellar synaptosomes, *J. Neurochem.*, 52, 1859, 1989.

52. **Hoehn, K. and White, T. D.**, Evoked release of endogenous adenosine from rat cortical slices by K$^+$ and glutamate, *Brain Res.*, 478, 149, 1989.

53. **Hoehn, K. and White, T. D.,** Role of excitatory amino acid receptors in K$^+$-and glutamate-evoked release of endogenous adenosine from rat cortical slices, *J. Neurochem.*, 54, 256, 1990.

54. **Evans, M. C., Swan, J. H., and Meldrum, B. S.,** An adenosine analogue, 2-chloroadenosine, protects against long term development of ischaemic cell loss in the rat hippocampus, *Neurosci. Lett.*, 83, 287, 1987.

55. **Connick, J. H. and Stone, T. W.,** Quinolinic acid neurotoxicity: protection by intracerebral phenyliso-propyladenosine (PIA) and potentiation by hypotension, *Neurosci. Lett.*, 101, 191, 1989.

56. **Dragunow, M. and Faull, R. L. M.,** Neuroprotective effects of adenosine, *Trends Neurosci.*, 9, 193, 1988.

57. **Hoehn, K. and White, T. D.,** NMDA, kainate, and quisqualate release endogenous adenosine from rat cortical slices, *Neuroscience*, 39, 441, 1990.

58. **Hoehn, K., Craig, C. G., and White, T. D.,** A comparison of NMDA-evoked release of adenosine and [^{3}H]norepinephrine from rat cortical slices, *J. Pharmacol. Exp. Ther.*, 255, 174, 1990.

59. **Hoehn, K. and White, T. D.,** Glutamate-evoked release of endogenous adenosine from rat cortical synaptosomes is mediated by glutamate uptake and not by receptors, *J. Neurochem.*, 54, 1716, 1990.

60. **Sawynok, J., Sweeney, M. I., and White, T. D.,** Adenosine release may mediate spinal analgesia by morphine, *Trends Pharmacol. Sci.*, 10, 186, 1989.

61. **Sawynok, J. and Sweeney, M. I.,** The role of purines in nociception, *Neuroscience*, 32, 557, 1989.

62. **Caciagli, F., Ciccarelli, R., Di Iori, P., Ballerini, P., and Tacconelli, L.,** Cultures of glial cells release purines under field electrical stimulation: the possible ionic mechanisms, *Pharmacol. Res. Commun.*, 20, 935, 1988.

63. **Meghi, P., Tuttle, J. B., and Rubio, R.,** Adenosine formation and release by embryonic chick neurons and glia in cell culture, *J. Neurochem.*, 53, 1852, 1989.

64. **Phillis, J. W. and Wu, P. H.,** Role of adenosine and its nucleotides in central synaptic transmission, *Progr. Neurobiol.*, 16, 187, 1981.

65. **Ehrlich, Y. H., Davis, T. B., Bock, E., Kornecki, E., and Lenox, R. H.,** Ecto-protein kinase activity on the external surface of neural cells, *Nature*, 320, 67, 1986.

66. **Gebicke-Haerter, P. J., Wurster, S., Schobert, A., and Hertting, G.,** P$_2$-purinoceptor induced pros-taglandin synthesis in primary rat astrocyte cultures, *Naunyn-Schmiedeberg's Arch. Pharmacol.*, 338, 704, 1988.

67. **Pearce, B., Murphy, S., Jeremy, J., Morrow, C., and Dandona, P.,** ATP-evoked calcium mobilisation and prostanoid release from astrocytes: P$_2$-purinergic receptors linked to phosphoinositide hydrolysis, *J. Neurochem.*, 50, 936, 1989.

68. **Van der Merve, R. A., Wakefield, I. K., Fine, J., Millar, R. P., and Davidson, J. S.,** Extracellular adenosine triphosphate activates phospholipase C and mobilizes intracellular calcium in primary cultures of sheep anterior pituitary cells, *FEBS Lett.*, 243, 333, 1989.

69. **White, T. D.,** Direct detection of depolarisation-induced release of ATP from a synaptosomal preparation, *Nature*, 267, 67, 1977.

70. **White, T. D.,** Release of ATP from a synaptosomal preparation by elevated extracellular K$^+$ and by veratridine, *J. Neurochem.*, 30, 329, 1978.

71. **Wu, P. H. and Phillis, J. W.,** Distribution and release of adenosine triphosphate in rat brain, *Neurochem. Res.*, 3, 563, 1978.

72. **Potter, P. E. and White, T. D.,** Lack of effect of 6-hydroxydopamine pretreatment on depolarization-induced release of ATP from rat brain synaptosomes, *Eur. J. Pharmacol.*, 80, 143, 1982.

73. **White, T. D., Potter, P., and Wonnacott, S.,** Depolarisation-induced release of ATP from cortical synaptosomes is not associated with acetylcholine release, *J. Neurochem.*, 34, 1129, 1980.

74. **Richardson, P. J. and Brown, S. J.,** ATP release from affinity-purified rat cholinergic nerve terminals, *J. Neurochem.*, 48, 622, 1987.

75. **Zhang, J., Kornecki, E., Jackman, J., and Ehrlich, Y. H.,** ATP secretion and extracellular protein phosphorylation by CNS neurons in primary culture, *Brain Res. Bull.*, 21, 459, 1988.

76. **Wieraszko, A., Goldsmith, G., and Seyfried, T. N.,** Stimulation-dependent release of adenosine tri-phosphate from hippocampal slices, *Brain Res.*, 485, 244, 1989.

77. **Wieraszko, A. and Seyfried, T. N.,** ATP-induced synaptic potentiation in hippocampal slices, *Brain Res.*, 491, 356, 1989.

78. **Stone, T. W. and Cusack, N. J.,** Absence of P$_2$-purinoceptors in hippocampal pathways, *Br. J. Pharmacol.*, 97, 631, 1989.

79. **Fyffe, R. E. W. and Perl, E. R.,** Is ATP a central synaptic mediator for certain primary afferent fibres from mammalian skin?, *Proc. Natl. Acad. Sci. U.S.A.*, 81, 6890, 1984.

80. **Salter, M. W. and Henry, J. L.,** Effects of adenosine 5'-monophosphate and adenosine 5'-triphosphate on functionally identified units in the cat spinal dorsal horn. Evidence for a differential effect of adenosine 5'-triphosphate on nociceptive vs non-nociceptive units, *Neuroscience*, 15, 815, 1985.

81. **Holton, P.,** The liberation of adenosine triphosphate on antidromic stimulation of sensory nerves, *J. Physiol. (London),* 145, 494, 1959.

82. **White, T. D., Downie, J. W., and Leslie, R. A.,** Characteristics of K^+- and veratridine-induced release of ATP from synaptosomes prepared from dorsal and ventral spinal cord, *Brain Res.,* 334, 372, 1985.

83. **Lew, M. J. and White, T. D.,** Release of endogenous ATP during sympathetic nerve stimulation, *Br. J. Pharmacol.,* 92, 349, 1987.

84. **Kirkpatrick, K. and Burnstock, G.,** Sympathetic nerve-mediated release of ATP from the guinea-pig vas deferens is unaffected by reserpine, *Eur. J. Pharmacol.,* 138, 207, 1987.

85. **Sedaa, K. O., Bjur, R. A., Shinozuka, K., and Westfall, D. P.,** Nerve and drug-induced release of adenine nucleosides and nucleotides from rabbit aorta, *J. Pharmacol. Exp. Ther.,* 252, 1060, 1990.

86. **Katayama, Y. and Morita, K.,** Adenosine 5'-triphosphate modulates membrane potassium conductance in guinea-pig myenteric neurones, *J. Physiol. (London),* 498, 373, 1989.

87. **Hammond, J. R., MacDonald, W. F., and White, T. D.,** Evoked secretion of [^{3}H]noradrenaline and ATP from nerve varicosities isolated from the myenteric plexus of the guinea pig ileum, *Can. J. Physiol. Pharmacol.,* 66, 369, 1988.

88. **Burnstock, G., Cocks, T., Kasakov, L., and Wong, H.,** Direct evidence for ATP release from non-adrenergic, non-cholinergic ('purinergic') nerves in the guinea-pig taenia coli and bladder, *Eur. J. Pharmacol.,* 49, 145, 1978.

89. **Chaudhry, A., Downie, J. W., and White, T. D.,** Tetrodotoxin-resistant release of ATP from superfused rabbit detrusor muscle during electrical field stimulation in the presence of luciferin-luciferase, *Can. J. Physiol. Pharmacol.,* 62, 153, 1984.

90. **Bult, H., Boeckxstaens, G. E., Pelckmans, P. A., Jordaens, F. H., Van Maercke, Y. M., and Herman, A. G.,** Nitric oxide as an inhibitory non-adrenergic non-cholinergic neurotransmitter, *Nature,* 345, 346, 1990.

91. **White, T. D., Potter, P., Moody, C., and Burnstock, G.,** Tetrodoxin-resistant release of ATP from guinea-pig taenia coli and vas deferens during electrical field stimulation in the presence of luciferin-luciferase, *Can. J. Physiol. Pharmacol.,* 59, 1094, 1981.

92. **Ewald, D. A.,** Potentiation of postjunctional cholinergic sensitivity of rat diaphragm muscle by high-energy-phosphate adenine nuceotides, *J. Membr. Biol.,* 29, 47, 1976.

93. **Kolb, H. and Wakelam, M. J. O.,** Transmitter-like action of ATP on patched membranes of cultured myoblasts and myotubes, *Nature,* 303, 621, 1983.

94. **Häggbladd, J., Eriksson, H., and Heilbronn, E.,** ATP-induced cation influx in myotubes is additive to cholinergic agonist action, *Acta. Physiol. Scand.,* 125, 389, 1985.

95. **Hume, R. I. and Honig, M. G.,** Excitatory action of ATP on embryonic chick muscle, *J. Neurosci.,* 6, 681, 1986.

96. **Hume, R. I. and Thomas, S. A.,** Multiple actions of adenosine 5'-triphosphate on chick skeletal muscle, *J. Physiol. (London),* 406, 503, 1988.

97. **Akasu, T., Hirai, K., and Koketsu, K.,** Increase of acetylcholine-receptor sensitivity by adenosine triphosphate: a novel action of ATP on ACh-sensitivity, *Br. J. Pharmacol.,* 74, 505, 1981.

98. **Igusa, Y.,** Adenosine 5'-triphosphate activates acetylcholine receptor channels in cultured Xenopus myotomal muscle cells, *J. Physiol. (London),* 405, 169, 1988.

99. **Silinsky, E. M. and Ginsborg, B. L.,** Inhibition of acetylcholine release from preganglionic frog nerves by ATP but not adenosine, *Nature,* 305, 327, 1983.

100. **Lindgren, C. A. and Smith, D. O.,** Extracellular ATP modulates calcium uptake and transmitter release at the neuromuscular junction, *J. Neurosci.,* 7, 1567, 1987.

101. **Ribeiro, J. A. and Sebastiao, A. M.,** On the role, inactivation and origin of endogenous adenosine at the frog neuromuscular junction, *J. Physiol. (London),* 384, 571, 1987.

102. **Branisteanu, D. D., Branisteanu, D. D. D., Covic, A., Brailoiu, E., Serban, D. N., and Haulica, I. D.,** Adenosine effects upon the spontaneous quantal transmitter release at the frog neuromuscular junction in the presence of protein kinase C-blocking and -activating agents, *Neurosci. Lett.,* 98, 96, 1989.

103. **Silinsky, E. M.,** On the association between transmitter secretion and the release of adenine dinucleotides from mammalian motor nerve terminals, *J. Physiol. (London),* 247, 145, 1975.

104. **Morel, N. and Meunier, F. M.,** Simultaneous release of acetylcholine and ATP from stimulated cholinergic synaptosomes, *J. Neurochem.,* 36, 1766, 1981.

105. **Marsal, J., Egea, G., Solsona, C., Rabasseda, X., and Blasi, J.,** Botulinum toxin type A blocks the morphological changes induced by chemical stimulation on the presynaptic membrane of Torpedo synaptosomes, *Proc. Natl. Acad. Sci.,* 86, 372, 1989.

III. Functions of Purines at the Cellular Level

Chapter 9

ADENOSINE, ADENINE NUCLEOTIDES, AND PLATELET FUNCTION

Noel J. Cusack and Susanna M. O. Hourani

TABLE OF CONTENTS

I. INTRODUCTION

Purinergic regulation of human platelet function is mediated by the opposing actions of adenosine and adenine nucleotides at two separate receptors, an adenine nucleotide receptor which when stimulated transforms platelets to an activated state, and an adenosine receptor which on stimulation inhibits platelet activation. The platelet adenine nucleotide receptor has been subsumed into the P_2 purinoceptor nomenclature as a P_{2T} subtype, and is unique since, of the endogenous nucleotides, only adenosine 5′-diphosphate (ADP) is an agonist, while adenosine 5′-triphosphate (ATP) is a competitive and specific antagonist. Adenosine, on the other hand, is a nonspecific, noncompetitive inhibitor of platelet function and the adenosine receptor is of the A_2 subtype.

II. STRUCTURE, FUNCTION, AND RESPONSES OF PLATELETS

In vitro, stirred suspensions of platelets challenged with aggregating agents change shape from discoid to spherical morphology, become self-adhesive (aggregate), and release the contents of their granules, in that order. Platelets possess three types of storage granules: the dense granules, which contain high concentrations of adenine nucleotides as well as 5-hydroxytryptamine and calcium; the alpha granules, which contain proteins involved in the coagulation cascade and in blood vessel function, including fibrinogen, which is essential for platelet aggregation and fibrin formation; and lysosomes containing a variety of hydrolytic enzymes. Low concentrations of strong aggregating agents cause reversible "primary" aggregation, but higher concentrations cause irreversible "secondary" aggregation which is associated with induction of the prostaglandin cascade and release of the contents of storage granules. In the case of ADP, the act of aggregation itself seems to be the main impetus for prostaglandin synthesis and the release reaction, released ADP and thromboxane A_2 enhancing secondary aggregation by ADP and by other platelet activators. Fibrinogen and calcium are absolute requirements for aggregation, and platelet activation by ADP is associated with a change in the conformation of the glycoprotein IIb-IIIa complex, resulting in an exposure of fibrinogen binding sites which require the presence of calcium before binding can take place. Interplatelet binding of fibrinogen mediates reversible aggregation, which becomes irreversible if the fibrinogen links are reinforced by adhesive proteins such as thrombospondin released from alpha granules.

III. STIMULUS-RESPONSE COUPLING

A. ADENINE NUCLEOTIDE RECEPTOR

Agents that act at receptors to cause platelet activation generally act via the phospholipase C pathway to generate from membrane-bound phosphatidyl inositol diphosphate the second messengers m-inositol, 1,4,5-tris phosphate (IP_3), which releases Ca^{2+} from intracellular stores, and diacylglycerol (DAG), while agents acting at receptors to inhibit platelet function act via stimulation of adenylate cyclase to generate adenosine 3′,5′-cyclic monophosphate (cAMP).[1] These effects are coupled to receptor occupation via guanine nucleotide binding proteins (G proteins), and are probably mediated by protein phosphorylation. The role of Ca^{2+} in platelet activation is suggested by the finding that the calcium ionophore, A23187, mimics the effects of stimulatory agonists such as ADP,[2] and studies in platelets permeabilized by high-voltage discharges show that intracellular Ca^{2+} activates platelets.[3] However, ADP is only a poor or ineffective stimulator of inositol phospholipid turnover,[1] and although ADP does cause significant increases in cytosolic Ca^{2+}, there is evidence that this is due mainly to Ca^{2+} influx rather than from mobilization from internal stores.[4,5] The opening of receptor-operated Ca^{2+} channels closely coupled to ADP receptors has recently been confirmed by studies using platelets loaded with the fluorescent Ca^{2+} indicator Quin-2, which showed that extracellular Mn^{2+} could enter ADP-activated platelets and quench the fluorescence,[6] as well as by a recent successful and technically demanding patch-clamp study on platelets.[7] Channel opening by ADP in Quin-2-loaded platelets was in accord with the pharmacology of the ADP receptor, as ATP, AMP, and adenosine did not cause increases in fluorescence, and ATP antagonized fluorescence induced by ADP.[4]

ADP has other effects which may or may not be involved in platelet activation. ADP stimulates Na^+/H^+ exchange, but overall acidification of the platelet cytoplasm was observed, and since amiloride at concentrations that block Na^+/H^+ exchange does not inhibit ADP-induced platelet aggregation, this exchange seems unrelated to aggregation.[8] ADP is also a potent noncompetitive inhibitor of adenylate cyclase, a process that is regulated by a Gi protein and is GTP dependent, and can inhibit by up to 90% the formation of cAMP induced by the adenosine or prostaglandin E_1 stimulated enzyme.[9] Intracellular inhibitors of adenylate cyclase do not themselves induce aggregation, nor do they enhance aggregation induced by ADP, suggesting that inhibition by ADP of adenylate cyclase is not a major contributor to platelet activation, although its role may be to facilitate aggregation when stimulators of the enzyme are present.[10] ADP-induced aggregation is associated with an increase in intracellular concentration of guanosine $3',5'$-cyclic monophosphate (cGMP), but this appears to be a consequence rather than a cause of aggregation and it may be a feedback inhibitor of platelet activation.[1]

B. ADENOSINE

Although the mechanism by which ADP causes human platelet aggregation is still not fully understood, it is clear that inhibition of platelet function by adenosine is via stimulation of adenylate cyclase. The effect of elevated concentrations of cAMP is to nonspecifically and noncompetitively inhibit platelet activation induced by ADP (and by all other aggregating agents);[1,10] increased concentrations of cAMP inhibit and reverse ADP-induced increases in intracellular Ca^{2+} concentration and increase the activity of the dense tubule Ca^{2+}-ATPase responsible for calcium sequestration.[1] Platelet adenylate cyclase is coupled to a Gs regulatory protein following receptor occupancy by adenosine, but adenylate cyclase can also be inhibited by adenosine acting at an intracellular P site.[9,11] As expected, inhibition of cAMP phosphodiesterase by papaverine or by methylxanthines leads to inhibition of ADP-induced aggregation,[9] and increases in intracellular cAMP concentration stimulate a cAMP-dependent protein kinase, which phosphorylates several platelet proteins, possibly including myosin light-chain kinase as well as proteins presumably involved in Ca^{2+} mobilization and sequestration.[1]

IV. STRUCTURE-ACTIVITY RELATIONSHIPS

A. ADENINE NUCLEOTIDE RECEPTOR
1. Agonists

The adenine nucleotide receptor is very sensitive to alterations to the structure of ADP.[12] There is an absolute requirement for the adenine base, so that guanosine $5'$-diphosphate, cytidine $5'$-diphosphate, and uridine $5'$-diphosphate are inactive. Substituents on the C^8 or N^6 position lead to greatly reduced potency,[13] but a variety of substituents are tolerated at the C^2 position,[13,14] which is a useful point of attachment for side chains to generate various affinity ligands for the P_{2T} purinoceptor.[15] 2-Chloro-ADP and 2-methylthio-ADP, as well as that of the photoaffinity analog 2-azido-ADP,[16] are up to 10-fold more potent than ADP at aggregating human platelets, and are, interestingly, up to 30-fold more potent as inhibitors of stimulated adenylate cyclase in intact platelets than as aggregating agents.[16] This could reflect either differing efficacies at the same receptor mediating both actions of ADP, or that two different receptors exist, but comparison of K_i values of a variety of competitive antagonists for each action of ADP suggest that both actions, inhibition of stimulated adenylate cyclase and induction of aggregation, are mediated by one adenine nucleotide receptor.[17]

Modifications to the D-ribose sugar, including removal or inversion of hydroxyl groups, leads to loss of activity. The adenine nucleotide receptor is stereospecific because the L-enantiomers, L-ADP, 2-chloro-L-ADP, and 2-azido-L-ADP, in which the D-ribose sugar has been replaced by L-ribose, are devoid of agonist activity.[14]

The 5′-diphosphate chain is required for full agonist potency, addition (ATP and its analogs) or subtraction (AMP and its analogs) generating antagonists. Replacement of one bridging oxygen by methylene as in homo-ADP and α,β-methylene-ADP, by imido as in α,β-imido-ADP, or of a terminal ionized oxygen by fluorine as in ADP-β-F, leaves analogs with weak or no agonist potency.[18] Replacement of an ionized oxygen by ionized sulfur generates ADP-α-S and ADP-β-S, both of which are partial agonists for platelet aggregation with intrinsic activities of 0.75.[19,20] Stereoselectivity is exhibited towards the diphosphate chain, since the Sp diastereoisomer of ADP-α-S is fivefold more potent than its Rp diastereoisomer.[20] ADP-β-S, like ADP, inhibits stimulated adenylate cyclase in intact platelets, but here too is only a partial agonist with an intrinsic activity of 0.5.[19] ADP-α-S, in contrast, does not inhibit stimulated adenylate cyclase, but does antagonize competitively this action of ADP, and again stereoselectivity is exhibited as Sp ADP-α-S is fivefold more potent than Rp ADP-α-S.[20] Adenosine inhibits ADP-α-S-induced aggregation more than it inhibits ADP-β-S-induced aggregation, presumably because ADP-α-S is unable to oppose stimulation of adenylate cyclase by adenosine.[21]

2. Antagonists

All convincing antagonists at the adenine nucleotide receptor are nucleotides, which complicates their assessment since the possibility exists of their dephosphorylation to either ADP or to ADP analogs, which can either enhance aggregation or, in unstirred platelets, lead to receptor desensitization, or to adenosine or its analogs which can inhibit nonspecifically ADP-induced aggregation.[12] ATP is a competitive antagonist of ADP-induced aggregation of human platelets with a K_i of about 20 μM, and AMP is a very weak antagonist.[22] ATP also inhibits competitively all of the active ADP analogs mentioned in the previous section, i.e., 2-azido-ADP, 2-chloro-ADP, 2-methylthio-ADP, ADP-α-S, and ADP-β-S, with a K_i value of about 20 μM confirming their action at the same adenine nucleotide receptor.[16,19,20]

The adenine nucleotide receptor allows more latitude for preservation of antagonist activity during alterations to the structure of nucleotides. There is still a requirement for the adenine base, but substitutions may be made at the C^8 position as in 8-bromo-ADP,[23] and at the C^2 position as in 2-chloro-ATP,[17] An absolute requirement still exists for the D-ribose sugar, as L-ATP and L-AMP are inactive.[14] However, various substituents or replacements on the phosphate moiety of AMP, ADP, or ATP have provided antagonists, with the most potent of them being the Sp diastereoisomer of adenosine 5′-(1-thiotriphosphate) (ATP-α-S), having a K_i of 4 μM.[17] Those that have been shown by Schild analysis to be competitive antagonists of ADP-induced aggregation and of ADP-induced inhibition of stimulated adenylate cyclase are β,γ-methylene-ATP, P_1P_5-diadenosine pentaphosphate, 2-chloroadenosine 5′-monophosphorothioate, γ-fluoro-ATP, α,β-dichloromethylene-ADP, Rp ATP-α-S, and Sp ATP-α-s.[17,24]

The behavior of C^2 alkylthio analogs of AMP and ATP is anomalous, in that they antagonize specifically, but noncompetitively, ADP-induced aggregation, achieving only 50% inhibition.[25] Further studies of this activity, using 2-methylthio-β,γ-methylene ATP, an analog that cannot be dephosphorylated to 2-methylthio-ADP, confirmed the specific, but noncompetitive antagonism of ADP-induced aggregation, yet demonstrated a complete antagonism of ADP-induced inhibition of stimulated adenylate cyclase.[26] In this light, it is interesting to note that 2-methylthio-β,γ-methylene-ATP antagonizes platelet aggregation induced by ADP-β-S, which itself also inhibits stimulated adenylate cyclase, but not aggregation induced by ADP-α-S, which does not inhibit the enzyme, and may indicate a role after all for adenylate cyclase inhibition during platelet activation by ADP.[26]

B. ADENOSINE RECEPTOR
1. Agonists

Many analogs of adenosine have been tested as inhibitors of platelet aggregation, which has usually been induced by ADP.[12] The adenine base framework is required for maximal agonist activity, guanosine, uridine, and cytosine being inactive.[12] Of the nitrogen atoms, those at the 3, 6, and 9 positions are essential, but the N^1 nitrogen may be replaced by methylene (1-deazaadenosine),[27] the N^7 nitrogen replaced by a methylene bearing a nitrile substituent providing the C^8 position has an amino group (6-amino-toyocamycin),[28] or by a methylene bearing a carboxamido group (sangivamycin).[28] Substituents on the N^1 position diminish potency, while some adenosine analogs with substituents at the C^2 position are potent inhibitors of platelet aggregation, notably 2-fluoroadenosine, 2-chloroadenosine, 2-bromoadenosine,[29] the photolyzable 2-azidoadenosine,[14] and a host of 2-thioadenosine derivatives, among the most potent of which is 2-cycloheptylthioadenosine.[30] The N^6-position tolerates monosubstitution by a variety of aryl, cycloalkyl, or unsaturated alkyl groups, generating potent analogs such as N^6-phenyladenosine, N^6-cyclohexyladenosine, and N^6-allyladenosine.[31] There is an absolute requirement for the D-ribose sugar, as replacement by L-ribose generates L-adenosine, which is inactive,[14] and the β-configuration of the linkage of D-ribose to adenine is also essential, as α-adenosine is inactive.[28] The ring oxygen in ribose may be replaced by methylene as in carbocyclic adenosine,[28] but the 2′,3′-cis hydroxyls are essential, as replacement of ribose by arabinose, xylose, or lyxose leads to total loss of activity.[28,29] The 4′-hydroxymethylene group may be replaced by certain carboxamide moieties, as in 5′-N-methylcarboxamidoadenosine (MECA), 5′-N-cyclopropylcarboxamidoadenosine (CPCA), 5′-N-ethylcarboxamidoadenosine (NECA), and carboxamidoadenosine itself (NCA).[32] NECA is the most potent adenosine analog, being about fivefold more potent than adenosine at inhibiting ADP-induced platelet aggregation, with an IC_{50} of about 0.14 μM, while its unnatural L-enantiomer, L-NECA, is inactive.[33] Increasing inhibition of ADP-induced aggregation by NECA was matched by a parallel increase in cAMP content of intact platelets, with an EC_{50} of 0.95 μM, and which was competitively inhibited by theophylline with a K_i value of about 8 μM.[32] Similar correlations of inhibition of aggregation with cAMP content have been demonstrated for 2-chloroadenosine and 2-azidoadenosine.[14]

In membranes prepared from homogenates of human platelets, NECA stimulates adenylate cyclase about fourfold over basal values, with an EC_{50} value of 0.5 μM, while CPCA is twofold more potent, MECA one third as potent, and NCA one tenth as potent at stimulating the enzyme.[34] Substitution of NECA at the N^6 position leads to a reduction in potency in most cases, to values intermediate between those of NECA and N^6-substituted adenosine analogs. 2-Chloroadenosine has an EC_{50} of about 1.7 μM and is therefore twofold more potent than adenosine itself, which has an EC_{50} value of about 0.82 μM.[34] 2-Aminoadenosine, 2-azidoadenosine, 2-hydrazinoadenosine, 2-methyladenosine, and N^6-cyclohexyladenosine also stimulate adenylate cyclase activity in membranes, as does N^6-phenyladenosine, which has an EC_{50} value of about 6 μM.[10,11]

2. Antagonists

The great majority of adenosine receptor antagonists are derivatives of methylxanthines, and so demonstration of antagonism of adenosine receptor-mediated inhibition of platelet aggregation is complicated by the unknown extent of blockade of cAMP phosphodiesterase activity, which in itself would cause inhibition of aggregation. However, antagonism of adenosine receptor-mediated accumulation of cAMP in the presence of total blockade of phosphodiesterase activity (e.g., by papaverine) is a convenient method.[9] Theophylline and caffeine inhibit adenosine-stimulated accumulation of cAMP with IC_{50} values of 25 μM and 72 μM, respectively;[9] and theophylline inhibits NECA-stimulated cAMP accumulation with an IC_{50} value of 8 μM.[33] In purified membranes from platelet homogenates, theophylline and caffeine inhibit NECA-stimulated adenylate cyclase, with K_i values of about 14 and 30 μM, respectively.[35] Enlargement of the N^1 or N^3 substituent enhances potency, so that N^1-propargyl-3,7-dimethylxanthine (DMPX) and 7-methyl-1,3-dipropylxanthine are seven- to tenfold more potent than caffeine.[35] Substitution at the C^8 position by a phenyl group also improves antagonist potency, and exploitation of the possibilities of substitution of the 8-phenyl group lead to the development of xanthine amine congener (XAC), which is over 1700-fold more potent than caffeine, having a K_i of 0.024 μM derived by Schild analysis.[36] Some nonxanthine adenosine receptor antagonists have been reported, and NECA-stimulated adenylate cyclase in platelet membranes is antagonized by a series of benzodipyrazoles, the most potent of which is 1,7-dihydro-3,5,8-trimethylbenzo[1,2-c:5,4-c']dipyrazole, having a K_i value of about 4.8 μM,[37] and by a series of 9-methyladenines, the most potent of which was N^6 cyclopentyl-9-methyladenine, having a K_i of 4.9 μM.[38]

V. RADIOLIGAND BINDING STUDIES

A. ADENINE NUCLEOTIDE RECEPTOR

Studies of binding to the adenine nucleotide receptor on intact platelets have concentrated, perhaps unwisely, on the agonist ADP, rather than an antagonist such as ATP. Radiolabeled ADP binding is complicated by its low affinity (which necessitates using a centrifugation assay), the variable amounts of radioactive suspending medium trapped by centrifuged platelets, the partial metabolism of ADP, the possible dilution of radiolabel by cold ADP donated by the release reaction, and the uptake of radioactive products arising from its dephosphorylation by platelet ectoenzymes,[12] Studies of ADP binding to platelet membranes generally employ a filtration assay, and are complicated by the exposure of large numbers of irrelevant nucleotide binding sites facing the interior of the platelet. These methodological difficulties explain in part the wide range of reported affinities (0.025 to 500 μM) and number of ADP receptors (0 to 32,800,000) per platelet, and the failure of displacers of binding to match their pharmacological activity at the ADP receptor.[12] When the more potent agonist 2-methylthio-ADP is used in an equilibrium binding assay, together with a more sophisticated methodology, saturable binding to intact platelets is detectable. Macfarlane et al.[39] found that β-[^{32}P]-2-methylthio-ADP bound to a single population of adenine nucleotide receptors, 400 to 1200 sites per platelet, with a K_D of 15 nM.[39] The kinetics of binding were too rapid to be measurable at 37°C, but at 23°C the second-order association rate constant (on-rate) was 3.5 × 10^6 M^{-1} s^{-1}, and the first-order dissociation rate (off-rate) was 0.024 s.[1] Binding of 2-methylthio-ADP was inhibited by ATP with a K_i of 7 μM, which is similar to the K_i obtained for antagonism by ATP of 2-methylthio-ADP-induced platelet aggregation. Analysis of bound radioactivity revealed that a small degree (13%) of phosphorylation of 2-methylthio-ADP to 2-methylthio-ATP had occurred, but no dephosphorylation to 2-methylthio-AMP and 2-methylthioadenosine was detected.[39]

In an attempt to circumvent problems associated with intact platelets, Jefferson and colleagues[23] employed human platelets which had been treated with formaldehyde to render them incapable of undergoing the release reaction and devoid of nucleotide-metabolizing ability, so as to enable steady-state ADP binding to them to be performed. In this study 2-[^{3}H]-ADP bound to about 160,000 high-affinity sites per fixed platelet, with a K_D of

$0.35~\mu M$, and was displaced potently by ATP with a K_i of $0.4~\mu M$, very weakly by AMP, and not at all by adenosine. In general the displacement of binding by analogs is as anticipated from their pharmacology at the ADP receptor, but the K_i for ATP is 50-fold lower than its pharmacological K_i $(20~\mu M)$.[23] Doubts must exist as to the identity of the binding site, and whether or not fixation really leaves the adenine nucleotide receptor unscathed, and although a later study demonstrated stereoselective displacement of bound ADP by a range of ADP agonists and antagonists to be in some accord with their pharmacology,[40] the study did not entirely remove these doubts.

Attempts to label covalently the ADP receptor on intact platelets using chemically reactive species have foundered on their extremely low affinity, instability, and/or lack of specificity for the adenine nucleotide receptor.[12] 2-[³H]-5′-Fluorosulfonylbenzoyladenosine (FSBA), which has been used extensively to label adenine nucleotide-utilizing enzymes, labels mainly one protein on intact platelets.[41] However, labeling is prevented by only very high (10 mM) concentrations of ATP, and also by adenosine, which is not in harmony with their pharmacology. FSBA is partly hydrolyzed to adenosine under conditions of its use, and is not a specific inhibitor of ADP-induced platelet aggregation, which makes its use in studies of adenine nucleotide receptor-mediated platelet function seem unwise. Attempts have been made to address these misgivings, however, including the use of 5′-fluorosulfonylbenzoylguanosine as a control for some of the nonspecific effects.[41] Photolysis of platelets in the presence of [β³²P]-2-azido-ADP failed to label covalently proteins relevant to the adenine nucleotide receptor, as incorporation of radioactivity was not prevented by excess ATP, probably because the azido group is orientated away from the receptor.[42]

Studies of ADP binding to membranes prepared from platelet homogenates are burdened with the need to prove that adenine nucleotide binding sites are on the outer membrane and are not merely acceptor sites or enzyme substrate sites for ADP or ATP, which seriously hampers acceptance of reports of isolated ADP binding proteins as being the adenine nucleotide receptor.[12] Lips et al.[43] found high-affinity reversible binding of [¹⁴C] ADP to human platelet membranes, with 206 pmol/mg protein, equivalent to 80,000 receptors per platelet, and a K_D of 1 μM. Bauvois et al.[44] also reports [¹⁴C] ADP binding, with a K_D of 0.5 μM and 400 pmol/mg protein, which approximates to 48,000 sites per platelet. Adler and Handin[45] solubilized and purified a 61-kDa protein which binds to ADP with a K_D of 0.38 μM, and approximates to about 10,000 sites per platelet.

B. ADENOSINE RECEPTOR

Radioligand binding of adenosine ligands to intact platelets has not been reported. Initial studies of agonist binding to platelet membranes employing [³H]-NECA detected a high-affinity site, with a K_D value of 160 nM and a B_{max} value of 8.4 pmol/mg protein, and a low-affinity site having a K_D value of 2.9 μM and B_{max} value of 33.4 pmol/mg protein. In subsequent studies by Lohse et al.,[46] the low-affinity component was removed by gel filtration and its lack of identity with the adenosine receptor was established. The remaining minor [³H]NECA binding component represented the major protein fraction, and binding of [³H]NECA to this was saturable, reversible, and GTP dependent, with a K_D value of 46 nM and a B_{max} of 0.51 pmol/mg protein, and the profile of displacement of binding by various adenosine receptor agonists and antagonists is in close agreement with their pharmacological profile at the A_2 adenosine receptor. Homologous desensitization of adenosine receptors in human platelets by prolonged treatment with 2-chloroadenosine correlated with a reduced capacity to stimulate adenylate cyclase with an approximately 50% reduction in the number of high-affinity [³H]NECA binding sites.[47]

Studies of antagonist binding to platelet membranes have employed radiolabeled XAC, a potent nonselective adenosine receptor antagonist. [³H]XAC binds in a saturable fashion to a homogeneous, noncooperative population of sites, having a K_D of 12 nM and a B_{max}

of 1.1 pmol/mg protein, a number near that obtained in the [³H]-NECA experiments.[36] Displacement of [³H]XAC binding by adenosine receptor antagonists is generally in good agreement with their pharmacology. In particular, theophylline displaced binding with a K_i value of 29 μM, compared to a K_i of about 25 μM for inhibition of stimulated adenylate cyclase in intact platelets, and XAC was not displaced by adenine nucleotides.[36]

VI. METABOLISM

A. ADENINE NUCLEOTIDES

Platelets possess ectophosphohydrolases (ectonucleotidases) capable of degrading ATP sequentially via ADP and AMP to adenosine.[12] Platelets also have an ectonucleoside diphosphate kinase (NDPK), which can convert [¹⁴C]ADP to [¹⁴C]ATP, and this NDPK activity is enhanced in the presence of phosphate donors such as UTP, and inhibited by AMP. NDPK activity in human platelet membranes is Mg^{2+} dependent, with a K_m value of 6 μM for ADP and 13 μM for ATP. Much of the NDPK is readily washed off platelet membranes, and the soluble enzyme has been purified from other ADP-binding proteins.[45] In radioligand binding studies of ADP analogs with whole platelets, analysis of bound radiolabel reveals conversion to corresponding ATP analogs, which illustrates a predominance of ecto-NDPK vs. ectonucleotidase activity.[39]

B. ADENOSINE

Adenosine is taken up by human platelets by a high-affinity system and by a low-affinity system.[48] Adenosine transported by the high-affinity system, with a K_m of about 10 μM and a V_{max} of about 800 pmol/min/10^9 platelets, is phosphorylated en route to AMP, which is further phosphorylated to ADP and ATP. This transport of adenosine is competitively inhibited by papaverine with a K_i of about 27 μM, by dipyridamole, and by 6-(4-nitrobenzyl)thioguanosine (NBTG). Adenosine transport via the low-affinity system with a K_m of about 10 mM, is competitively inhibited by adenine with a K_i of about 6 μM, and it arrives within the platelet unchanged to be either deaminated to inosine or phosphorylated to adenine nucleotides.[48] Some analogs of adenosine which inhibit platelet activation are also transported into the platelet. 2-Fluoroadenosine is taken up and phosphorylated to 2-fluoro-ATP, and similarly carbocyclic adenosine is phosphorylated to carbocyclic ATP, both nucleotides residing in the metabolic pool rather than in the dense granules.[28] 2-Chloroadenosine is not detectably transported, and, as expected, there is no direct correlation between inhibitory potency of adenosine analogs and their susceptibility to uptake.[28]

VII. THERAPEUTIC IMPLICATIONS AND SUMMARY

Adenosine and adenine nucleotides appear to play an important role in the cardiovascular system, where they are released into the blood following damage to vessel walls, stimulation of the endothelium, or conditions of hypoxia. Adenine nucleotides are rapidly dephosphorylated by ectoenzymes attached to the endothelium to products which can have differing or even opposing effects to the parent nucleotides. ATP itself inhibits ADP-induced platelet aggregation and adhesion, whereas ADP, the first product of dephosphorylation of ATP, activates platelets, but otherwise has the same effects on the cardiovascular system as ATP. AMP is generally inactive, while adenosine inhibits platelet function, though far less effectively *in vivo* than *in vitro*, has no effect on the endothelium, causes vasodilation by a direct effect on underlying smooth-muscle cells, and slows the heart directly and in a different way to ATP. The overall response of the cardiovascular system to adenosine and adenine nucleotides depends therefore on the results of a subtle interplay of pharmacology and enzymology. Such an interplay provides a potentially fertile ground for the design of drugs specific for individual actions of adenosine and of adenine nucleotides. We have found, using synthetic ATP analogs, that the receptors that mediate the various actions of ATP are different from each other and have designed agonists specific for each receptor subtype.[49] The design of specific modulators of the individual functions of adenosine and of adenine nucleotides could result in the development of novel drugs to treat hypertension, adjust platelet responsiveness, reduce the risk of thrombosis, and control the action of the heart.

ACKNOWLEDGMENT

We thank Ms. C. Lillie, Whitby Research, Inc., for preparation of the manuscript.

REFERENCES

1. **Haslam, R. J.,** Signal transduction in platelet activation, in *Thrombosis and Haemostasis,* Verstraete, M., Vermylen, J., Lijnen, H. R., and Arnout, J., Eds., Leuven University Press, Leuven, Belgium 1987, 147.
2. **Feinman, D. R. and Detwiler, T. C.,** Platelet secretion induced by divalent cation ionophores, *Nature,* 249, 172, 1974.
3. **Knight, D. E. and Scrutton, M. C.,** Direct evidence for a role for Ca^{2+} in amine storage granule secretion by human platelets, *Thromb. Res.,* 20, 437, 1980.
4. **MacIntyre, D. E., Shaw, A. M., Bushfield, M., MacMillan, L. J., and McNicol, A.,** Agonist-induced inositol phospholipid metabolism and Ca^{2+} flux in human platelet aggregation, in *Mechanisms of Stimulus-Response Coupling in Platelets,* Westwick, J., Scully, M. F., MacIntyre, D. E., and Kakkar, V. V., Eds., Plenum Press, New York, 1985, 127.
5. **Hallam, T. J. and Rink, T. J.,** Responses to adenosine diphosphate in human platelets loaded with the fluorescent calcium indicator quin 2, *J. Physiol.,* 386, 131, 1985.
6. **Sage, S. O., Merritt, J. E., Hallam, T. J., and Rink, T. J.,** Receptor mediated calcium entry in fura-2-loaded human platelets stimulated with ADP and thrombin, *Biochem. J.,* 258, 923, 1989.
7. **Mahaut-Smith, M. P., Sage, S. O., and Rink, T. J.,** Receptor-activated single channels in intact human platelets, *J. Biol. Chem.,* 265, 10479, 1990.
8. **Funder, J., Hershco, L., Rothstein, A., and Livne, A.,** Na^+/H^+ exchange and aggregation of human platelets activated by ADP: the exchange is not required for aggregation, *Biochim. Biophys. Acta,* 939, 425, 1988.
9. **Haslam, R. J. and Rosson, G. M.,** Effects of adenosine cyclic 3′,5′-monophosphate in human blood platelets in relation to adenosine incorporation and platelet aggregation, *Mol. Pharmacol.,* 11, 528, 1975.
10. **Haslam, R. J., Davidson, M. M. L., and Desjardins, J. V.,** Inhibition of adenylate cyclase by adenosine analogues in preparations of broken and intact human platelets, *Biochem. J.,* 176, 83, 1978.
11. **Londos, C. and Wolff, J.,** Two distinct adenosine-sensitive sites on adenylate cyclase, *Proc. Natl. Acad. Sci.,* 74, 5482, 1977.
12. **Haslam, R. J. and Cusack, N. J.,** Blood platelet receptor for ADP and for adenosine, in *Purinergic Receptors: Receptors and Recognition, Series B,* Vol. 12, Burnstock, G., Ed., Chapman and Hall, London, 1981, 223.
13. **Gough, G., Maguire, M. H., and Penglis, F.,** Analogues of adenosine 5′-diphosphate—new platelet aggregators, *Mol. Pharmacol.,* 8, 170, 1972.
14. **Cusack, N. J., Hickman, M. E., and Born, G. V. R.,** Effects of D- and L-enantiomers of adenosine, AMP and ADP and their 2-chloro- and 2-azido- analogues on human platelets, *Proc. R. Soc. London, Ser. B.,* 206, 139, 1979.
15. **Jefferson, J. R., Hunt, J. B., and Jamieson, G. A.,** Facile synthesis of 2-[(3-aminopropyl)thio]adenosine 5′-diphosphate: a key intermediate for the synthesis of molecular probes of adenosine 5′-diphosphate function, *J. Med. Chem.,* 30, 2013, 1987.
16. **Cusack, N. J. and Hourani, S. M. O.,** Competitive inhibition by adenosine 5′-triphosphate of the actions on human platelets of 2-chloroadenosine 5′-diphosphate, 2-azidoadenosine 5′-diphosphate, and 2-meththioadenosine 5′-diphosphate, *Br. J. Pharmacol.,* 77, 329, 1982.
17. **Cusack, N. J. and Hourani, S. M. O.,** Adenosine 5′-diphosphate antagonists and human platelets: no evidence that aggregation and inhibition of stimulated adenylate cyclase are mediated by different receptors, *Br. J. Pharmacol.,* 76, 221, 1982.

18. **Cusack, N. J. and Hourani, S. M. O.**, Actions and structure-activity relationships of purines on platelets, in *Purines: Pharmacology and Physiological Roles*, Stone, T. W., Ed., Macmillin, London, 1985, 163.

19. **Cusack, N. J. and Hourani, S. M. O.**, Partial agonist behaviour of adenosine 5'-O-(2-thiodiphosphate), *Br. J. Pharmacol.*, 73, 405, 1981.

20. **Cusack, N. J. and Hourani, S. M. O.**, Effects of Rp and Sp diastereoisomers of adenosine 5'-O-(1-thiodiphosphate) on human platelets, *Br. J. Pharmacol.*, 73, 409, 1981.

21. **Cusack, N. J. and Hourani, S. M. O.**, Differential inhibition by adenosine or by prostaglandin E_1 of human platelet aggregation induced by 5'-O-(1-thiodiphosphate) and adenosine 5'-O-(2-thiodiphosphate), *Br. J. Pharmacol.*, 75, 257, 1982.

22. **Macfarlane, D. E. and Mills, D. C. B.**, The effects of ATP on platelets: evidence against the central role of released ADP in primary aggregation, *Blood*, 46, 309, 1975.

23. **Jefferson, J. R., Harmon, J. T., and Jamieson, G. A.**, Identification of high affinity (K_d 0.35 μmol/L) and low affinity (K_d 7.9 μmol/L) platelet binding sites for ADP and competition by ADP analogues, *Blood*, 71, 110, 1988.

24. **Cusack, N. J. and Pettey, C. J.**, Effects of isopolar isosteric phosphonate analogues of adenosine 5'-diphosphate (ADP) on human platelets, in *Adenosine and Adenine Nucleotides: Physiology and Pharmacology*, Paton, D. M., Ed., Taylor & Francis, London, 1988, 287.

25. **Cusack, N. J. and Hourani, S. M. O.**, Specific but noncompetitive inhibition by 2-alkylthio analogues of adenosine 5'-monophosphate and adenosine 5'-triphosphate of human platelet aggregation by adenosine 5'-diphosphate, *Br. J. Pharmacol.*, 75, 297, 1982.

26. **Hourani, S. M. O., Welford, L. A., and Cusack, N. J.**, 2-MeS-AMP-PCP and human platelets: implications for the role of adenylate cyclase in ADP-induced aggregation?, *Br. J. Pharmacol.*, 87, 84, 1986.

27. **Antonini, I., Cristalli, G., Franchetti, P., Grifitani, M., Martelli, S., and Petrelli, F.**, Deaza analogues of adenosine as inhibitors of blood platelet aggregation, *J. Pharm. Sci.*, 73, 366, 1984.

28. **Agarwal, K. C. and Parks, R. E., Jr.**, Adenosine analogs and human platelets. II. Inhibition of ADP-induced aggregation by carbocyclic adenosine and imidazole ring modified analogs. Significance of alterations of the nucleotide pools, *Biochem. Pharmacol.*, 28, 501, 1979.

29. **Born, G. V. R., Haslam, R. J., Goldman, M., and Lowe, R. D.**, Comparative effectiveness of adenosine analogues as inhibitors of blood-platelet aggregation and as vasodilators in man, *Nature*, 205, 678, 1965.

30. **Kikugawa, K., Suehiro, H., and Ichino, M.**, Platelet aggregation inhibitors. 6. 2-Thioadenosine derivatives, *J. Med. Chem.*, 16, 1381, 1973.

31. **Kikugawa, K., Iizuka, K., and Ichino, M.**, Platelet aggregation inhibitors. 4. N^6-Substituted adenosines, *J. Med. Chem.*, 16, 358, 1973.

32. **Ukena, D., Böhme, E., and Schwabe, U.**, Effects of several 5'-carboxamide derivatives of adenosine on adenosine receptors of human platelets and rat fat cells, *N.-S. Arch. Pharmacol.*, 327, 36, 1984.

33. **Cusack, N. J. and Hourani, S. M. O.**, 5'-N-Ethylcarboxamidoadenosine: a potent inhibitor of human platelet aggregation, *Br. J. Pharmacol.*, 72, 443, 1981.

34. **Olsson, R. A., Kusachi, S., Thompson, R. D., Ukena, D., Padgett, W., and Daly, J. W.**, N^6-Substituted N-alkyladenosine-5'-uronamides: bifunctional ligands having recognition groups for A1 and A2 adenosine receptors, *J. Med. Chem.*, 29, 1683, 1986.

35. **Ukena, D., Shamin, M. T., Padgett, W., and Daly, J. W.**, Analogs of caffeine: antagonists with selectivity for A_2 adenosine receptors, *Life Sci.*, 39, 743, 1986.

36. **Ukena, D., Jacobson, K. A., Kirk, K. L., and Daly, J. W.**, A [^{3}H]amine congener of 1,3-dipropyl-8-phenylxanthine. A new radioligand for A_2 adenosine receptor of human platelets, *FEBS Lett.*, 199, 269, 1986.

37. **Peet, N. P., Dickerson, G. A., Abdallah, A. H., Daly, J. W., and Ukena, D.**, Benzo[1,2-c:5,4-c']dipyrazoles: non-xanthine adenosine antagonists, *J. Med. Chem.*, 31, 2034, 1988.

38. **Ukena, D., Padgett, W. L., Hong, O., Daly, J. W., and Olsson, R. A.**, N^6-Substituted 9-methyladenines: a new class of adenosine receptor antagonists, *FEBS Lett.*, 215, 203, 1987.

39. **Macfarlane, D. E., Srivastava, P. C., and Mills, D. C. B.**, 2-Methylthioadenosine [β^{32}P] diphosphate. An agonist and radioligand for the receptor that inhibits the accumulation of cyclic AMP in intact platelets, *J. Clin. Invest.*, 71, 420, 1983.

40. **Agrawal, A. K., Tandon, N. N., Greco, N. J., Cusack, N. J., and Jamieson, G. A.**, Evaluation of binding to fixed platelets of agonists and antagonists of ADP-induced aggregation, *Thromb. Haemostas.*, 62, 1103, 1989.

41. **Mills, D. C. B., Figures, W. R., Scearce, L. M., Stewart, G. J., Colman, R. F., and Colman, R. W.**, Two mechanisms for inhibition of ADP-induced platelet shape change by 5'-P-fluorosulfonylbenzoyladenosine, *J. Biol. Chem.*, 260, 8078, 1985.

42. **Macfarlane, D. E., Mills, D. C. B., and Srivastava, P. C.**, Binding of 2-azidoadenosine [β-^{32}P]diphosphate to the receptor on intact human blood platelets which inhibits adenylate cyclase, *Biochemistry*, 21, 544, 1982.

43. **Lips, J. P. M., Sixma, J. J., and Schiphorst, M. E.,** Binding of adenosine diphosphate to human blood platelets and to isolated blood platelet membranes, *Biochim. Biophys. Acta,* 628, 451, 1980.

44. **Bauvois, B., Legrand, C., and Caen, J. P.,** Interaction of adenosine and adenine nucleotides with the human platelet membrane. Further characterization of the ADP binding sites, *Haemostasis,* 9, 92, 1980.

45. **Adler, J. R. and Handin, R. L.,** Solubilization and characterization of a platelet membrane ADP-binding protein, *J. Biol. Chem.,* 254, 3866, 1979.

46. **Lohse, M. J., Elger, B., Linderborn-Fortinos, J., Klotz, K.-N., and Schwabe, U.,** Separation of solubilized A_2 adenosine receptors of human platelets from non-receptor [^{3}H]NECA binding sites by gel filtration, *N.-S. Arch. Pharmacol.,* 337, 64, 1988.

47. **Edwards, R. J., MacDermont, J., and Wilkins, A. J.,** Prostacyclin analogues reduce ADP-ribosylation of the α-subunit of the regulatory G_1-protein diminish adenosine (A_2) responsiveness of platelets, *Br. J. Pharmacol.,* 90, 501, 1987.

48. **Sixma, J. J., Lips, J. P. M., Trieschnigg, A. M. C., and Holmsen, H.,** Transport and metabolism of adenosine in human blood platelets, *Biochim. Biophys. Acta,* 443, 33, 1976.

49. **Cusack, N. J. and Hourani, S. M. O.,** Subtypes of P_2 purinoceptors: studies using analogues of ATP, in *Biological Actions of Extracellular ATP,* Dubyak, G. R. and Fedan, J. S., Eds., Ann. N.Y. Acad. Sci., Vol. 603, N.Y. Academy of Science, New York, 1990, 172.

Chapter 10

PURINES AND INFLAMMATION: NEUTROPHILS POSSESS P$_1$ and P$_2$ PURINE RECEPTORS

Bruce N. Cronstein

TABLE OF CONTENTS

I. INTRODUCTION

The effects of adenosine and adenine nucleotides on cells of the cardiovascular and central nervous systems have been known for many years. In contrast, the effects of purines on leukocyte function have only recently been appreciated. Moreover, the roles of purines in modulation of inflammation, the primary process in which neutrophils and other leukocytes are involved, are only now becoming clear. This review will briefly summarize the function of neutrophils, the effects of purines on neutrophil function, and the potential effects of purines at inflammatory sites.

II. THE ROLE OF NEUTROPHILS IN HOST DEFENSE AND INFLAMMATION

Neutrophils are highly specialized cells whose primary function is the phagocytosis and destruction of microorganisms and other noxious agents. In addition to their role in host defense, neutrophils are commonly present at sites of immunologically mediated tissue injury, e.g., synovial fluid in rheumatoid arthritis, and it has recently been appreciated that they mediate many of the events which occur at foci of immunologically induced acute inflammation.[1-5] More recent studies suggest that neutrophils also mediate injury following nonimmunologic injury, e.g, myocardial infarction.[6,7]

To arrive at inflammatory sites neutrophils first adhere to vascular endothelium, then follow a trail of chemoattractants (a large number of agents including activated complement components, cytokines, leukotrienes, and bacterial products) to the inflammatory site. Having arrived at the inflammatory site, neutrophils engulf or phagocytose bacteria, cellular debris, immune complexes, etc. Upon phagocytosis or stimulation with appropriate soluble agents, neutrophils undergo a ''respiratory burst''. The resulting products of the oxidative metabolism of neutrophils include superoxide anion (O_2^-) and hydrogen peroxide (H_2O_2, reviewed by Babior).[8] Both O_2^- and hydrogen peroxide serve as the substrate for further reactions to produce various potent bactericidal agents such as hypochlorite and hydroxyl radical. Hydrogen peroxide has been implicated in neutrophil-mediated tumor killing, as a fungicidal agent, and, more importantly for inflammatory diseases, in neutrophil-mediated toxicity to normal cultured vascular endothelium (reviewed by Clark).[9] Indeed, several groups have reported that toxic oxygen metabolites released by neutrophils are responsible for cell damage mediated by neutrophils both *in vivo* and *in vitro*.

III. RELEASE OF ADENOSINE AND ADENINE NUCLEOTIDES AT INFLAMMATORY SITES

To be useful physiologic regulators of neutrophil function (and inflammation), adenine nucleotides and adenosine must be released at appropriate loci following appropriate stimuli. At sites of vascular injury platelets aggregate and release their granular contents. Purine nucleotides (ATP and ADP) are among the compounds present within platelet granules which are released after platelet aggregation. Adenine nucleotides are also released by dead or dying cells undergoing lysis. Pearson and Gordon[10] have demonstrated that the adenine nucleotides released by platelets or other cells are rapidly metabolized to adenosine by ectonucleotidases and phosphatases. Additionally, Ager and Gordon[11] have suggested that endothelial cells selectively increase their release of adenosine upon exposure to H_2O_2. Several recent studies have indicated that normal neutrophils also release adenosine, probably as a result of ATP metabolism.[12,13] It has long been known that ischemic canine myocardium releases significant concentrations of adenosine (cf. Olsson).[14] Others have demonstrated that rapid cardiac pacing to angina also causes increased adenosine release from myocar-

TABLE 1
Effects of Adenosine on Human
Neutrophils

A_1 Receptor	A_2 Receptor
Promote chemotaxis	Inhibit O_2^-
Promote phagocytosis	Inhibit phagocytosis
Promote adherence	Inhibit adherence

dium.[15,16] Thus, adenine nucleotides may be released at sites of thrombosis or cell lysis whereas adenosine, in contrast, is released from intact and injured cells.

IV. THE EFFECT OF ATP ON NEUTROPHILS

While investigating the effect of platelet secretory products on neutrophil function, Ward and co-workers[17] demonstrated that platelets release ATP which affects neutrophil function. Whereas ATP does not stimulate neutrophils to generate O_2^- or other oxygen metabolites directly, ATP primes neutrophils for increased production of O_2^- after appropriate stimulation.[17,18] Since neither metabolism of ATP nor uptake of this nucleotide by neutrophils is required for ATP to prime the neutrophil, it has been concluded that ATP primes neutrophils by occupying specific receptors on the neutrophil surface.[17-20] ATP also stimulates neutrophil-neutrophil aggregation and neutrophil adhesion to albumin-coated latex beads by a mechanism which is associated with upregulation of CD11b/CD18 on the surface of the neutrophil.[21]

Current understanding of "priming" is far from complete; however, recent studies indicate that occupancy of ATP receptors on the neutrophil stimulates the generation of a number of intracellular intermediates. Thus ATP stimulates increases in $[Ca^{++}]$,[17-19,22] and metabolism of membrane phospholipids by phospholipase C, with generation of potentially important signal molecules such as inositol triphosphate by a pertussis toxin-sensitive pathway.[22] The role of these intermediates in priming the neutrophil remains to be demonstrated. In other studies, Walker et al.[20] showed that occupancy of neutrophil ATP receptors does not prime neutrophils by increasing the number or affinity of receptors for other stimuli.

The concentrations of ATP required to modulate neutrophil function are higher than those encountered in most extracellular fluids (1 to 100 μM). The high concentration of ATP required to prime neutrophil function suggests that ATP is a relevant intercellular signal only at sites where there is cellular lysis or at sites of platelet release of ATP.

V. THE EFFECT OF ADENOSINE ON HUMAN NEUTROPHILS

Initial studies of the effects of adenosine on human neutrophils (see Table 1) indicated that adenosine inhibits stimulated O_2^- generation without significantly affecting stimulated degranulation or aggregation.[12] Moreover, these early studies demonstrated that adenosine modulated neutrophil behavior by acting at a site on the external surface of the cell.[12] It was rapidly established by work in my laboratory and, subsequently, that of others that adenosine modulated stimulated neutrophil function by occupying a specific adenosine A_2 receptor on the surface of the neutrophil.[23-27] Unlike many other inhibitors of neutrophil function, adenosine is not only response specific (inhibiting O_2^- generation, but not degranulation), but stimulus specific; adenosine inhibits release of oxygen metabolites stimulated by the surrogate bacterial chemoattractant n-formyl-methionyl-leucyl-phenylalanine (FMLP), the lectin concanavalin A, the activated complement component C5a (zymosan-activated serum), and C3b-coated particles (serum-treated zymosan particles). Adenosine does not inhibit release

TABLE 2

Adenosine Promotes Chemotaxis by Occupying A_1 Receptors, but Inhibits $O_2^{\div}$ Generation by Occupying A_2 Receptors

Agonist	IC_{50} for $O_2^{\div}$ generation	EC_{50} for chemotaxis
5'-N-Ethylcarboxamidoadenosine	17 nM	9 pM
N^6-Cyclopentyladenosine	>1 μM	2 pM
CV-1808	23 nM	11 pM

of $O_2^{\div}$ or hydrogen peroxide by neutrophils exposed to phorbol myristate acetate, an agent which bypasses membrane receptors to activate the neutrophil, and adenosine is only a poor inhibitor of $O_2^{\div}$ or H_2O_2 generation in response to immune complexes.[12,28] Interestingly, adenosine inhibits H_2O_2 generation by adherent neutrophils more completely than it inhibits H_2O_2 generation by neutrophils in suspension.[29] Adenosine and its analogs modulate other neutrophil functions in addition to inhibiting $O_2^{\div}$ generation. In a recent study we found that 2-chloroadenosine inhibits adherence of neutrophils to endothelial cells,[30] and preliminary studies indicate that inhibition of neutrophil adherence to endothelium is mediated by adenosine A_2 receptors. Roberts et al.[31] have demonstrated that the poorly metabolized adenosine analog 2-chloroadenosine inhibits recovery of human neutrophils from attack by terminal components of the complement cascade. Our early studies had demonstrated that adenosine does not affect stimulated neutrophil aggregation or degranulation,[12] findings confirmed by some,[32,33] but not all reports.[24,34] Thus, in general, adenosine occupies A_2 receptors on the surface of the neutrophil to selectively inhibit stimulated neutrophil function.

Whereas the presence and role of the adenosine A_2 receptor on neutrophils was established early, later studies indicated that neutrophils also possess A_1 receptors. Rose et al.[35] first demonstrated that adenosine promoted neutrophil chemotaxis stimulated by either C5a (zymosan-treated plasma) or the surrogate bacterial chemoattractant FMLP. In this study the order of potency for promotion of chemotaxis was identical to that for inhibition of $O_2^{\div}$ generation (5'N-ethylcarboxamidoadenosine > N^6-phenylisopropyladenosine > adenosine), but the effective concentrations were significantly lower than those required to inhibit $O_2^{\div}$ generation (see Table 2). In a subsequent study, using other more specific ligands, Cronstein et al.[36] demonstrated that the effect of adenosine and its analogs on chemotaxis was mediated by occupancy of adenosine A_1 receptors.

VI. STIMULUS-RESPONSE COUPLING AT NEUTROPHIL ADENOSINE RECEPTORS

The mechanism by which adenosine affects neutrophil function is at present not completely understood. Similar to adenosine A_2 receptors on many other cell types, adenosine and its agonists promote accumulation of cAMP in neutrophils, but only in the presence of a phosphodiesterase inhibitor.[27,37] Moreover, occupancy of adenosine receptors enhances the rise in cAMP which follows stimulation of the neutrophil with FMLP.[37,38] Whether or not cAMP is a relevant second messenger in the neutrophil is in doubt for several reasons. First, other agents which also promote accumulation of cAMP in the neutrophil (e.g., prostaglandin E_2) inhibit not only $O_2^{\div}$ generation, but degranulation and aggregation as well. Second, adenosine deaminase reverses the effect of adenosine on neutrophil function despite preincubation of neutrophils with adenosine in the presence of a phosphodiesterase inhibitor. Thirdly, the phosphodiesterase inhibitor RO-20-1729 does not increase the effect of adenosine on neutrophil function.[37] The effect of A_2 receptor occupancy on other intracellular messengers for neutrophil function has also been examined. Grinstein and Furuya[33] have recently reported that adenosine does not affect the function of the Na^+/H^+ port. Pasini and co-workers[39] have reported that adenosine inhibits $O_2^{\dot{-}}$ generation by inhibiting cellular Ca^{++} uptake. In contrast, Cronstein et al.[37] found that adenosine inhibits FMLP-stimulated $O_2^{\dot{-}}$ generation even in the absence of extracellular Ca^{++}, a finding confirmed by Ward et al.[19] Moreover, several laboratories have now demonstrated that adenosine does not modulate the increment in $[Ca^{++}]_i$ which follows stimulation,[19,34,37] a finding at odds with the observations of Nielson and Vestal.[27] Occupancy of adenosine A_2 receptors also increases plasma membrane viscosity in the neutrophil without markedly affecting the affinity or number of chemoattractant (FMLP) receptors on the neutrophil surface.[20,40]

Recent studies suggest an alternative mechanism whereby adenosine receptor occupancy inhibits $O_2^{\dot{-}}$ generation by stimulated neutrophils. Jesaitis et al.[41,42] have presented data to indicate that following ligation, chemoattractant receptors rapidly associate with the cytoskeleton in "heavy" domains of the cytoskeleton (Figure 1). Once the chemoattractant receptors become associated with the cytoskeleton they become physically separated from the G proteins required for further signal propagation in the neutrophil.[43] Occupancy of adenosine A_2 receptors promotes association of occupied chemoattractant receptors with the cytoskeleton.[36] These data indicate, then, that adenosine receptor occupancy inhibits $O_2^{\dot{-}}$ generation by increasing the rate at which chemoattractant receptors become incapable of propagating the intracellular signals required to maintain $O_2^{\dot{-}}$ generation.

In contrast to A_2 receptors, adenosine A_1 receptors are linked to G proteins by which they modulate cellular function in other tissues.[44-54] Thus, it is not surprising that occupancy of neutrophil A_1 receptors promotes neutrophil chemotaxis by a mechanism which is inhibited by pertussis toxin and therefore dependent upon intact G proteins.[36] Unexpectedly, adenosine A_1 receptors also appear to depend upon intact microtubules, since both colchicine and vinblastine completely abrogate the effect of A_1 receptor occupancy on chemotaxis. Adenosine A_1 receptors may require intact microtubules in order to maintain their appropriate conformation or for stimulus transduction, since microtubules are not required for neutrophil chemotaxis.

VII. PURINES ARE INFLAMMATORY AUTOCOIDS

The foregoing discussion suggests several important roles for adenosine and adenine nucleotides at injured or inflammatory sites. Adenine nucleotides are released intravascularly by platelets or endothelial cells and these extracellular nucleotides can then trigger neutrophils to adhere to the vascular endothelium. The neutrophils then migrate out of the vasculature towards the inflammatory or infectious stimulus through relatively acellular connective tissue. The low adenosine concentrations present in acellular connective tissue promote neutrophil chemotaxis. At the periphery of the inflammatory site, adenosine leaked from accumulated neutrophils or injured cells diminishes generation of toxic oxygen metabolites by stimulated neutrophils, thereby preventing further injury to already damaged tissues. Once at the center of the inflammatory or infected focus, adenine nucleotides released from dying cells or bacteria prime neutrophils for more efficient bacterial killing.

ACTIVATION OF THE NEUTROPHIL

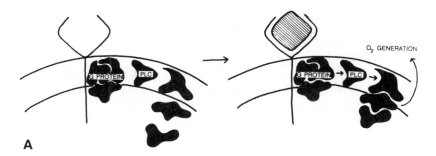

ACTIVATION OF THE NEUTROPHIL
EFFECT OF ADENOSINE A2
RECEPTOR LIGATION

FIGURE 1.   (A) Neutrophils are activated upon binding of ligands, such as chemoattractants, to specific receptors on their surface. Upon activation of the neutrophil there is assembly of G proteins, resulting in activation of phospholipases (and other intracellular messengers) with assembly of the oxidase system on the surface of the neutrophil. (B) After activation of the neutrophil, there is rapid association of bound chemoattractant receptors with the "cytoskeleton" (xxx), resulting in dissociation of bound chemoattractant receptors from stimulus-transduction mechanisms, such as G proteins. The dissociation of bound chemoattractant receptors from stimulus-transduction machinery leads to disassembly of the oxidase system. Occupancy of adenosine receptors is associated with more rapid and complete association of bound chemoattractant receptors with the cytoskeleton and, ultimately, more rapid disassembly of the neutrophil oxidase.

REFERENCES

1. **DeShazo, C. V., McGrade, M. T., Henson, P. M., and Cochrane, C. G.,** The effect of complement depletion on neutrophil migration in acute immunologic arthritis, *J. Immunol.,* 108, 1414, 1972.
2. **DeShazo, C. V., McGrade, M. T., and Henson, P. M.,** Acute immunologic arthritis in rabbits, *J. Immunol.,* 51, 50, 1972.
3. **Humphrey, J. H.,** The mechanism of Arthus reactions. I. The role of polymorphonuclear leucocytes and other factors in reversed passive Arthus reactions in rabbits, *Br. J. Exp. Pathol.,* 36, 268, 1955.
4. **Parrish, W. E.,** Effects of neutrophils on tissues. Experiments on the Arthus reaction, the flare phenomenon, and post-phagocytic release of lysosomal enzymes, *Br. J. Dermatol.,* 81, 28, 1969.
5. **Stetson, C. A.,** Similarities in the mechanisms determining the Arthus and Schwartzmann phenomena, *J. Exp. Med.,* 94, 347, 1951.

6. **Romson, J. L., Hook, B. G., Kunkel, S. L., Abrams, G. D., Schork, M. A., and Lucchesi, B. R.,** Reduction of the extent of ischemic myocardial injury by neutrophil depletion in the dog, *Circulation*, 67, 1016, 1983.

7. **Mullane, K. M., Read, N., Salmon, J. A., and Moncada, S.,** Role of leukocytes in acute myocardial infarction in anesthetized dogs: relationship to myocardial salvage by antiinflammatory drugs, *J. Pharmacol. Exp. Ther.*, 228, 510, 1984.

8. **Babior, B. M.,** The respiratory burst of phagocytes, *J. Clin. Invest.*, 73, 599, 1984.

9. **Clark, R. A.,** Extracellular effects of the myeloperoxidase-hydrogen peroxide-halide system, *Adv. Inf. Res.*, 5, 107, 1983.

10. **Pearson, J. D. and Gordon, J. L.,** Vascular endothelial and smooth muscle cells in culture selectively release adenine nucleotides, *Nature*, 281, 384, 1979.

11. **Ager, A. and Gordon, J. L.,** Differential effects of hydrogen peroxide on indices of endothelial function, *J. Exp. Med.*, 159, 592, 1985.

12. **Cronstein, B. N., Kramer, S. B., Weissmann, G., and Hirschhorn, R.,** Adenosine, a physiological modulator of superoxide anion generation by human neutrophils, *J. Exp. Med.*, 158, 1160, 1983.

13. **Newby, A. C., Holmquist, C. A., Illingworth, J., and Pearson, J. D.,** *Biochem. J.*, 214, 317, 1983.

14. **Olsson, R. A.,** Changes in content of purine nucleoside in canine myocardium during coronary occlusion, *Circ. Res.*, 26, 301, 1970.

15. **Fox, A. C., Reed, G. E., Glassman, E., Kaltman, A. J., and Silk, B.,** Release of adenosine from human hearts during angina induced by rapid atrial pacing, *J. Clin. Invest.*, 53, 1447, 1974.

16. **Fox, A. C., Reed, G. E., Meilman, H., and Silk, B. B.,** Release of nucleosides from canine and human hearts as an index of prior ischemia, *Am. J. Cardiol.*, 43, 52, 1979.

17. **Ward, P. A., Cunningham, T. W., McCulloch, K. K., Phan, S. H., Powell, J., and Johnson, K. J.,** Platelet enhancement of O_2^- responses in stimulated human neutrophils. Identification of platelet factor as adenine nucleotide, *Lab. Invest.*, 58, 37, 1988.

18. **Ward, P. A., Cunningham, T. W., McCulloch, K. K., and Johnson, K. J.,** Regulatory effects of adenosine and adenine nucleotides on oxygen radical responses of neutrophils, *Lab. Invest.*, 58, 438, 1988.

19. **Ward, P. A., Cunningham, T. W., Walker, B. A. M., and Johnson, K. J.,** Differing calcium requirements for regulatory effects of ATP, ATPγS and adenosine on O_2^- responses of human neutrophils, *Biochem. Biophys. Res. Commun.*, 154, 746, 1988.

20. **Walker, B. A. M., Cunningham, T. W., Freyer, D. R., Todd, R. F., III, Johnson, K. J., and Ward, P. A.,** Regulation of superoxide responses of human neutrophils by adenine compounds. Independence of requirement for cytoplasmic granules, *Lab. Invest.*, 61, 515, 1989.

21. **Freyer, D. R., Boxer, L. A., Axtell, R. A., and Todd, R. F., III,** Stimulation of human neutrophil adhesive properties by adenine nucleotides, *J. Immunol.*, 141, 580, 1988.

22. **Dubyak, G. R., Cowen, D. S., and Lazarus, H. M.,** Activation of the inositol phospholipid signaling system by receptors for extracellular ATP in human neutrophils, monocytes and neutrophil/monocyte progenitor cells, *Ann. N.Y. Acad. Sci.*, 551, 218, 1988.

23. **Cronstein, B. N., Rosenstein, E. D., Kramer, S. B., Weissmann, G., and Hirschhorn, R.,** Adenosine; a physiologic modulator of superoxide anion generation by human neutrophils. Adenosine acts via an A2 receptor on human neutrophils, *J. Immunol.*, 135, 1366, 1985.

24. **Schmeichel, C. J. and Thomas, L. L.,** Methylxanthine bronchodilators potentiate multiple human neutrophil functions, *J. Immunol.*, 138, 1896, 1987.

25. **Schrier, D. J. and Imre, K. M.,** The effects of adenosine agonists on human neutrophil function, *J. Immunol.*, 137, 3284, 1986.

26. **Roberts, P. A., Newby, A. C., Hallett, M. B., and Campbell, A. K.,** Inhibition by adenosine of reactive oxygen metabolite production by human polymorphonuclear leucocytes, *Biochem. J.*, 227, 669, 1985.

27. **Nielson, C. P. and Vestal, R. E.,** Effects of adenosine on polymorphonuclear leucocyte function, cyclic 3',5'-adenosine monophosphate, and intracellular calcium, *Br. J. Pharmacol.*, 97, 882, 1989.

28. **Cronstein, B. N., Kubersky, S. M., Weissmann, G., and Hirschhorn, R.,** Engagement of adenosine receptors inhibits hydrogen peroxide (H_2O_2) release by activated human neutrophils, *Clin. Immunol. Immunopathol.*, 42, 76, 1987.

29. **de la Harpe, J. and Nathan, C. F.,** Adenosine regulates the respiratory burst of cytokine-triggered human neutrophils adherent to biologic surfaces, *J. Immunol.*, 143, 596, 1989.

30. **Cronstein, B. N., Levin, R. I., Belanoff, J., Weissmann, G., and Hirschhorn, R.,** Adenosine, an endogenous inhibitor of neutrophil-mediated injury to endothelial cells, *J. Clin. Invest.*, 78, 760, 1986.

31. **Roberts, P. A., Morgan, B. P., and Campbell, A. K.,** 2-Chloroadenosine inhibits complement-induced reactive oxygen metabolite production and recovery of human polymorphonuclear leukocytes attacked by complement, *Biochem. Biophys. Res. Commun.*, 126, 692, 1985.

32. **Marone, G., Thomas, L., and Lichtenstein, L.,** The role of agonists that activate adenylate cyclase in the control of cAMP metabolism and enzyme release by human polymorphonuclear leukocytes, *J. Immunol.*, 125, 2277, 1980.

33. **Grinstein, S. and Furuya, W.,** Cytoplasmic pH regulation in activated human neutrophils, effects of adenosine and pertussis toxin on Na^+/H^+ exchange and metabolic acidification, *Biochim. Biophys. Acta,* 889, 301, 1986.

34. **Skubitz, K. M., Wickham, N. W., and Hammerschmidt, D. E.,** Endogenous and exogenous adenosine inhibit granulocyte aggregation without altering the associated rise in intracellular calcium concentration, *Blood,* 72, 29, 1988.

35. **Rose, F. R., Hirschhorn, R., Weissmann, G., and Cronstein, B. N.,** Adenosine promotes neutrophil chemotaxis, *J. Exp. Med.,* 167, 1186, 1988.

36. **Cronstein, B. N., Duguma, L., Nicholls, D., Hutchison, A., and Williams, M.,** The adenosine/neutrophil paradox resolved. Human neutrophils possess both A_1 and A_2 receptors which promote chemotaxis and inhibit O_2^- generation, respectively, *J. Clin. Invest.,* 85, 1750, 1990.

37. **Cronstein, B. N., Kramer, S. B., Rosenstein, E. D., Korchak, H. M., Weissmann, G., and Hirschhorn, R.,** Occupancy of adenosine receptors raises cyclic AMP alone and in synergy with occupancy of chemoattractant receptors and inhibits membrane depolarization, *Biochem. J.,* 252, 709, 1988.

38. **Iannone, M. A., Wolberg, G., and Zimmerman, T. P.,** Chemotactic peptide induces cAMP elevation in human neutrophils by amplification of the adenylate cyclase response to endogenously produced adenosine, *J. Biol. Chem.,* 264, 20177, 1989.

39. **Pasini, F. L., Capecchi, P. L., Orrico, A., Ceccatelli, L, and Di Perri, T.,** Adenosine inhibits polymorphonuclear leukocyte in vitro activation, a possible role as an endogenous calcium entry blocker, *J. Immunopharmacol.,* 7, 203, 1985.

40. **Cronstein, B. N., Rose, F. R., and Pugliese, C.,** Adenosine, a cytoprotective autocoid, effects of adenosine on neutrophil plasma membrane viscosity and chemoattractant receptor display, *Biochim. Biophys. Acta,* 987, 176, 1989.

41. **Jesaitis, A. J., Naemura, J. R., Sklar, L. A., Cochrane, C. G., and Painter, R. G.,** Rapid modulation of N-formyl chemotactic peptide receptors on the surface of human granulocytes, formation of high affinity ligand-receptor complexes in transient association with cytoskeleton, *J. Cell Biol.,* 98, 1378, 1984.

42. **Jesaitis, A. J., Tolley, J. O., and Allen, R. A.,** Receptor-cytoskeleton interactions and membrane traffic may regulate chemoattractant-induced superoxide production in human granulocytes, *J. Biol. Chem.,* 261, 13662, 1986.

43. **Jesaitis, A. J., Bokoch, G. M., Tolley, J. O., and Allen, R. A.,** Lateral segregation of neutrophil chemotactic receptors into actin- and fodrin-rich plasma membrane microdomains depleted in guanyl nucleotide regulatory proteins, *J. Cell Biol.,* 107, 921, 1988.

44. **Ramkumar, V. and Stiles, G. L.,** Reciprocal modulation of agonist and antagonist binding to A1 adenosine receptors by guanine nucleotides is mediated via a pertussis toxin-sensitive G protein, *J. Pharmacol. Exp. Ther.,* 246, 1194, 1988.

45. **Parsons, W. J., Ramkumar, V., and Stiles, G. L.,** Isobutylmethylxanthine stimulates adenylate cyclase by blocking the inhibitory regulatory protein, Gi, *Mol. Pharmacol.,* 34, 37, 1988.

46. **Ramkumar, V. and Stiles, G. L.,** A novel site of action of a high affinity A1 adenosine receptor antagonist, *Biochem. Biophys. Res. Commun.,* 153, 939, 1988.

47. **Monaco, L., DeManno, D. A., Martin, M. W., and Conti, M.,** Adenosine inhibition of the hormonal response in the Sertoli cell is reversed by pertussis toxin, *Endocrinology,* 122, 2692, 1988.

48. **Green, A.,** Adenosine receptor down-regulation and insulin resistance following prolonged incubation of adipocytes within an A1 adenosine receptor agonist, *J. Biol. Chem.,* 262, 15702, 1987.

49. **Trussell, L. O. and Jackson, M. B.,** Dependence of an adenosine-activated potassium current on a GTP-binding protein in mammalian central neurons, *J. Neurosci.,* 7, 3306, 1987.

50. **Arend, L. J., Sonnenburg, W. K., Smith, W. L., and Spielman, W. S.,** A1 and A2 adenosine receptors in rabbit cortical collecting tubule cells. Modulation of hormone-stimulated cAMP, *J. Clin. Invest.,* 79, 710, 1987.

51. **Rossi, N. F., Churchill, P. C., and Churchill, M. C.,** Pertussis toxin reverses adenosine receptor-mediated inhibition of renin secretion in rat renal cortical slices, *Life Sci.,* 40, 481, 1987.

52. **Parsons, W. J. and Stiles, G. L.,** Heterologous desensitization of the inhibitory A1 adenosine receptor-adenylate cyclase system in rat adipocytes. Regulation of both Ns and Ni, *J. Biol. Chem.,* 262, 841, 1987.

53. **Berman, M. I., Thomas, C. G., Jr., and Nayfeh, S. N.,** Inhibition of thyrotropin-stimulated adenosine 3′,5′-monophosphate formation in rat thyroid cells by an adenosine analog. Evidence that the inhibition is mediated by the putative inhibitory guanine nucleotide regulatory protein, *J. Cyclic Nucleotide Protein Phosphor. Res.,* 11, 99, 1986.

54. **Garcia Sainz, J. A. and Torner, M. L.,** Rat fat-cells have three types of adenosine receptors (Ra, Ri and P). Differential effects of pertussis toxin, *Biochem. J.,* 232, 439, 1985.

Chapter 11

ADENOSINE AND IMMUNE SYSTEM FUNCTION

Teresa S. Priebe and J. Arly Nelson

TABLE OF CONTENTS

I. INTRODUCTION

The discovery that certain immunodeficiency diseases are related to deficiency of the catabolic enzymes adenosine deaminase (ADA)[1] or purine nucleoside phosphorylase[2] stimulated considerable interest in searching for possible immunomodulatory roles of adenosine. Lymphocytes appear to be extremely sensitive to adenosine and, in normal individuals, relatively high ADA activity in lymphoid tissue[3] appears to maintain low extracellular levels of adenosine (~ 0.1 μM),[4] permitting lymphocyte survival. In addition, lymphocytes have been shown to be capable of secreting ADA into the extracellular compartment.[5] Adenosine may arise from either intracellular or extracellular metabolism. Intracellularly, adenosine can originate from ATP degradation or S-adenosyl-homocysteine hydrolysis. Studies by Barankiewicz and Cohen[6,7] demonstrated that adenosine is not accumulated under normal physiological conditions. Any adenosine generated intracellularly or transported into the cell can undergo the following metabolic fates. Adenosine kinase (AK, Km 2 to 4 μM) is capable of phosphorylating low concentrations of adenosine to AMP. At higher concentrations of adenosine, accumulation of S-adenosyl-homocysteine can occur as a result of the equilibrium kinetics of the S-adenosyl-homocysteine hydrolase. Finally, adenosine can be catabolized by ADA, an enzyme having low affinity (Km ~ 50 μM), but high capacity for this substrate, leading to the formation of inosine. Intracellular catabolism of ATP was found to be different in T and B lymphoblasts.[8] In T lymphoblasts, degradation of ATP occurs mainly via AMP deamination due to relatively low AMP-5'-nucleotidase activity. High activities of ADA and AK in these cells prevent adenosine accumulation and its possible release into the extracellular space. In contrast, B lymphoblasts catabolize ATP via AMP dephosphorylation, and because of low ADA and AK activities the adenosine is released from these cells. Extracellular adenosine is generated from ATP or other nucleotides by actions of cell-surface ectonucleotidases. Ectonucleotidase activity has been detected on B lymphocytes and B lymphoblastoid cell lines, but not on T lymphocytes or T cell lines.[9] These findings suggest that B cells are major producers of extracellular adenosine in lymphoid tissues. In contrast, T lymphocytes, but not B lymphocytes, were found to express an extracellular adenosine receptor, suggesting that adenosine might play a role in T and B cell communication.

Different concentrations of extracellular adenosine might produce differential, and possibly opposite effects as described below. Many actions attributed to adenosine appear to be mediated via specific cell-surface receptors linked to adenylyl cyclase. In micromolar concentrations, exogenous adenosine increases intracellular cyclic AMP (cAMP) levels by interacting with an extracellular receptor classified as subtype A_2, stimulatory for adenylyl cyclase. Nanomolar concentrations of adenosine may bind to high-affinity A_1 receptors, inhibitory for adenylyl cyclase. These extracellular actions are inhibited by methylxanthines, but not by nucleoside transport inhibitors. Adenosine receptors similar to those initially described in brain[10] and platelets[11] have now been found on a variety of lymphoid cells, viz, T lymphocytes, cytotoxic cells, monocytes, mast cells, basophils, and neutrophils. Alternatively, high (i.e., millimolar) concentrations of adenosine and its ribose-modified analogs can activate an intracellular P-site that inhibits adenylyl cyclase. In contrast to the A_1 and A_2 receptors, P-site agonist activities are not antagonized by methylxanthines.[12] The P-site is present in human lymphocytes and polymorphonuclear cells,[13] however, its biological significance in these cells is unknown.

The above described features of adenosine metabolism and action are characteristic of a hormone. For example, it is released from tissues at low concentrations, permitting its local or distal effects, and many of its actions can be attributed to interaction with specific cellular receptors. Finally, as with epinephrine, the actions of adenosine can be rapidly terminated by reuptake (often by carrier-mediated systems), by rephosphorylation to AMP, or by deamination. Deamination to form inosine is considered a pathway toward degradation, since further metabolism of inosine by purine nucleoside phosphorylase and hypoxanthine by xanthine oxidase yields the end product of human purine metabolism, uric acid. Genetic deficiencies of ADA or of purine nucleoside phosphorylase are associated with severe combined immunodeficiency or a selective T cell deficiency, respectively. In the absence of the deaminase, adenosine and deoxyadenosine accumulate and are thought to mediate the immunodeficiency. These experiments of Nature suggest roles for adenosine in modulating normal immune function. This chapter reviews the known actions of adenosine on cells that mediate immune response.

II. LYMPHOCYTES

A. ADENOSINE INHIBITION OF LECTIN-STIMULATED PROLIFERATIVE RESPONSE

Adenosine and some other naturally occurring nucleosides appear to be especially toxic to lymphocytes. The antiproliferative properties of adenosine on lymphocytes have been recognized since 1970 when Hirschhorn et al.[14] demonstrated that adenosine can inhibit DNA synthesis in phytohemagglutinin (PHA)-stimulated human lymphocytes. The inhibitory action of adenosine was potentiated when the stimulated lymphocytes were cultured in medium supplemented with ADA-deficient horse serum.[15] ADA activity increased significantly (up to threefold) within 24 h following exposure of human peripheral lymphocytes to PHA, and this elevation preceded DNA synthesis. A subsequent drop to 44% of resting values occurred at 72 h while AK activity was not changed.[16]

The ADA inhibitor coformycin (1 μM) does not impair mature T or B lymphocyte function; however, coformycin does inhibit the maturation of precursor cells, particularly T cells.[17] This finding suggests a physiological role for ADA and indicates the importance of adenosine degradation in proliferating lymphocytes, a population that appears much more vulnerable to adenosine than are resting lymphocytes. The mechanism by which adenosine inhibits cell proliferation is still not totally clear. There is evidence that adenosine can cause a decrease in the intracellular content of phosphoribosyl pyrophosphate (PRPP), a common precursor for purine and pyrimidine nucleotide biosynthesis. The importance of the de novo pathway of purine biosynthesis for proliferating lymphocytes is supported by the findings that the cellular content of PRPP increases within minutes after PHA stimulation.[18] Several observations suggest that adenosine inhibits pyrimidine biosynthesis de novo, perhaps as a consequence of its effect on PRPP pools. Addition of adenosine leads to accumulation of orotic acid, depletion of the pyrimidine nucleotide pool, and expansion of the adenine nucleotide pool. This toxicity due to pyrimidine starvation can be reversed by the addition of uridine.[19] Toxicity not reversible by uridine occurs when the intracellular content of adenosine is very high. The equilibrium coefficient of S-adenosylhomocysteine hydrolase favors the formaton of S-adenosylhomocysteine, a potent inhibitor of S-adenosylmethionine-mediated methylation reactions. Thus, high levels of adenosine lead to accumulaton of S-adenosylhomocysteine and arrest of lymphocyte blastogenesis. Adenosine concentrations that inhibit DNA and protein synthesis in human lymphocytes by 50% decrease DNA methylation by 10 to 15%; reduction of methylation by 50% is associated with 85% inhibition of DNA and protein synthesis.[20]

Marone et al.[21] and Schwartz et al.[22] independently postulated the presence of adenosine receptor(s) on human peripheral blood lymphocytes. Concentrations of adenosine encompassing the physiological range (0.01 to 10 μM) were shown to produce dose-dependent elevations of intracellular cAMP in human lymphocytes.[21] The increase in cAMP content was rapid and the higher concentrations of adenosine elevated the cyclic nucleotide level by 100% during the first 5 min of incubation.[22] The nucleoside transport inhibitor, dipyridamole, did not prevent the increase in cAMP.[21] This result suggested an extracellular site of action for adenosine. Theophylline and 3-isobutyl-methylxanthinel (IBMX), methylxanthines commonly employed as phosphodiesterase inhibitors, did not potentiate the effects of adenosine. The antagonism of adenosine action by these methylxanthins may relate to their interaction at an extracellular receptor site.[21] In these studies, several criteria for the presence of extracellular adenosine receptor(s) linked to adenylyl cyclase on human lymphocytes have been met: (1) adenosine is an effective stimulator of cAMP production at low concentrations; (2) inhibition of adenosine transport does not prevent the induced cAMP accumulation; (3) methylxanthines, adenosine receptor antagonists, prevent adenosine effects; (4) the ribose moiety of adenosine is required for cAMP elevation, i.e., adenine had no effect; and (5) effects of adenosine on adenylyl cyclase are independent from other adenylyl cyclase activators such as norepinephrine.[22] Adenosine was found to be less effective as a stimulator of adenylyl cyclase in human lymphocytes than other endogenous activators, i.e., a concentration of 10 μM caused a 4-fold elevation of cAMP, whereas prostaglandin E_2 produced a 10- to 12-fold elevation of cAMP.[23]

The above described observations employing unfractionated lymphocytes make it difficult to judge whether or not there exists a subpopulation of lymphocytes toward which adenosine acts preferentially. Sandberg and Fredholm[24] described inhibitory effects of adenosine on guinea pig and rat thymocyte proliferation *in vitro* due to the accumulation of cAMP.[25] The use of a variety of adenosine receptor agonists demonstrated the following order of potency: NECA > 2-ClAdo > PIA > CHA characteristic of an A_2-type adenosine receptor. Further studies by Dinjens et al.[26] demonstrated the presence of an A_2 type of receptor on human peripheral blood T lymphocytes and thymocytes.

B. INHIBITION OF E ROSETTE FORMATION

Adenosine and other agents that increase cAMP levels block the ability of human T lymphocytes to form rosettes with sheep red blood cels. Exposure of E rosette-forming cells (mature T lymphocytes) to 50 to 100 μM adenosine identifies two distinct cell subsets. The major subset is resistant to adenosine; the sensitive subset (not capable of forming rosettes) represents less than 10% of the E rosette-forming cells.[27] Similar effects were produced by theophylline.[28] The theophylline-resistant subpopulation bears the receptor for the Fc portion of IgM and expresses the helper phenotype (75.1% OKT4$^+$ and 20.5% OKT8$^+$), whereas the theophylline-sensitive population bears the receptor for IgG and expresses a suppressor phenotype (21.3% OKT4$^+$ and 53.5% OKT8$^+$).[29] Inhibition of rosette formation by adenosine and theophylline were found to be associated with the elevation of intracellular cAMP content.

C. ADENOSINE AS AN INDUCER OF NEW SUPPRESSOR ACTIVITY

Human T lymphocytes expressing receptors for the Fc portion of IgG (RFcγ) function as suppressors for B cell differentiation into antibody-producing plasma cells, whereas cells bearing receptors for IgM (RFcμ) have characteristics of helper cells supporting B cell differentiation.[30] Incubation of unfractionated peripheral blood lymphocytes with adenosine increased the number of cells with surface receptors Fcγ for IgG in the absence of concomitant cell division. Adenine did not increase the percentage of cells expressing the Fcγ suppressor phenotype and the adenosine effect was abrogated by theophylline, suggesting that the observed phenomenon is receptor mediated.[31] Subsequent exposure of theophylline-resistant (T helper) and theophylline-sensitive (T suppressor) cell subsets to adenosine demonstrated that induction of RFcγ suppressor phenotype was restricted to the resistant cell subpopulation. Induction of the suppressor phenotype paralleled the loss of helper function for B cell differentiation.[31,32]

The early (5-min) decline of OKT4$^+$ phenotype expression was associated with adenosine-induced cAMP elevation.[34] Later changes (5 to 30 min) into the suppressor (OKT8$^+$)

phenotype and development of immunosuppressive activity were found to require transport of adenosine into the cells. At these times cAMP was decreased.[33] A similar induction of new suppressor activity was achieved by treatment of the theophylline-resistant subpopulation with the cAMP elevating agent, impromidine, an H_2 histamine receptor agonist.[31] The expression of a new suppressor phenotype was relatively stable, remaining unchanged for 24 h after the removal of adenosine.[32] Adenosine induced a significant loss (65%) of T helper activity for B cell differentiation into plasma cells that produce IgM and IgG. In contrast, helper activity was maintained in cells exposed to IBMX, nitrobenzylthioinosine or 2-chloroadenosine, suggesting that stimulation of extracellular receptor was not sufficient to induce suppressor activity. Development of new suppressor activity seems to require adenosine interaction with a cell-surface receptor (A_2 type) and also entry of adenosine into the cell. Adenosine modulation of surface antigen expression in T cells was selective and did not involve other receptors, i.e., T3, T5, or Ia.[29]

D. INHIBITION OF INTERLEUKIN-2 SYNTHESIS

Adenosine and the adenosine receptor agonist 2-chloroadenosine were found to inhibit interleukin-2 (IL-2) production and murine T cell proliferation[34] by acting through the A_2-type adenosine receptor that is stimulatory for adenylyl cyclase.[28] This effect was also observed in human T cells.[35] Different treatments leading to enhancement of intracellular cAMP concentration, i.e., direct stimulation of adenylyl cyclase by forskolin, activation of the stimulatory G protein by cholera toxin, use of the physiological agonists prostaglandins type E, or treatment with the permeant cAMP analog dibutyryl cAMP result in significant reduction of IL-2 synthesis.[36] Adenosine stimulation of the A_2-type receptor enhances cAMP production and occupancy of intracellular cAMP receptors, resulting in lowering of the steady-state content of IL-2 mRNA and IL-2 synthesis.[37] Averill et al.[35] presented findings that also support a role for the adenylyl cyclase-cAMP-protein kinase A pathway for IL-2 production. Adenosine, like prostanglandins and histamine, can be considered an autocoid interacting with cell-surface receptors and inducing suppressor activity.[35] On the other hand, IL-2 was also reported to inhibit adenylyl cyclase activity by activation of protein kinase C, with resultant reduction of cAMP levels.[38,39] This reduction leads to inactivation of protein kinase A, promotion of transcription and translation, and eventually to cell proliferation.[40]

E. REGULATION OF CAPPING BY ADENOSINE

Capping of surface molecules represents a basic signal which triggers lymphocyte activation. It has been proposed that adenosine regulates capping by activating a cAMP-dependent pathway. Certain microtubule-associated proteins (MAP) were found to be substrates for cAMP-dependent protein kinase.[41] Phosphorylation of MAP results in impaired microtubule interaction with actin filaments, the result of which is an increase in capping. Exposure of human T lymphocytes to 10 μM adenosine resulted in a marked increase of capped T cells; T3 increased from 58 to 80%; T4 increased from 34 to 79%; and T8 increased from 33 to 70%.[42] Inosine and adenine were not effective, and pretreatment of T cells with 100 μM theophylline antagonized the adenosine effects. The phosphodiesterase inhibitor RO-20-1724 potentiated the effect of 10 nM adenosine. Enhancement of capping similar to that produced by adenosine was observed with the cAMP derivative, 8-BrcAMP.[42] These data suggest that regulation of capping by adenosine occurs via an adenosine receptor and the cAMP pathway. Photoaffinity labeling demonstrated that adenosine increases occupancy of intracellular cAMP receptors.[43]

F. ADENOSINE AND PROGRAMMED CELL LYSIS

Recent reports indicate that adenosine may play an important role in the process of apoptosis or programmed cell death.[44] Apoptosis is involved in the programmed elimination of cells during thymus development. Adenosine receptor-mediated accumulation of cAMP caused T lymphocyte cell death through internucleosomal DNA cleavage, characteristic of the apoptosis process related to the activation of nuclear endonucleases.[45]

III. CYTOTOXIC CELLS

A. T-CYTOTOXIC LYMPHOCYTES

Both T-cytotoxic lymphocyte (CTL) cells and natural killer (NK) cells share basically similar lytic mechanisms; however, they are of distinct origin. Human CTL bear the T3 antigen typical of T cells, they are T8 positive and MHC class I restricted. Unlike NK cells, development of T cytotoxicity requires cell proliferation, and this activity is directed against targets bearing correct MHC product and antigen, generally viral-infected cells. Similar to NK cells, CTL are sensitive to a variety of agents capable of elevating intracellular cAMP, the consequence of which is inhibition of lytic activity. The fundamental work of Wolberg et al.[46] demonstrated that adenosine is a potent inhibitor of murine CTL. Adenosine inhibited CTL activity against EL-4 tumor targets in a dose-dependent fashion. The use of an ADA inhibitor potentiated the adenosine effect twofold. Inhibition of CTL correlated with the increased intracellular cAMP content, strongly suggesting that the observed phenomenon may be receptor mediated. Further studies[47] using the adenosine analog 2-chloroadenosine, a selective agonist for adenosine A_2-type receptors, supported the hypothesis that an adenosine receptor occurs on murine CTL.

B. NATURAL KILLER CELLS (NK)

NK cells exhibit spontaneous cytotoxic activity (without prior sensitization) against a variety of syngeneic and allogeneic tumor cells, viral-infected cells, and certain undifferentiated normal cell types. NK cell activity is almost exclusively restricted to cells with large granular lymphocyte (LGL) morphology that represent approximately 10% of total human peripheral blood lymphocytes. NK cells differ from CTL by not expressing the T3 antigen in humans or Lyt-2,3 in mice. Also, NK cell activity is not restricted by the major histocompatibility complex (MHC). NK cells have several regulatory functions, i.e., suppressor effects on antibody production by B cells and accessory activity for the development of CTL. Natural toxicity is sensitive to a variety of biological modifiers. Among these, interferon and IL-2 stimulate NK cell proliferation and enhance lytic activity. Diverse pharmacological agents that elevate cAMP, such as prostaglandin E_2, cholera toxin, isoproterenol, or treatment of NK cells with the membrane-permeable cAMP analog dibutyryl cAMP lead to inhibition of NK cell activity.[48,49] Treatment of murine NK cells with adenosine (10 μM) plus the ADA inhibitor deoxycoformycin (1 μM) results in complete inhibition of NK lytic activity, whereas 1 μM of adenosine plus 1 μM of the ADA inhibitor caused 50% inhibition. These effects were antagonized by the adenosine receptor antagonist, 1,3-dipropyl-8-phenylxanthine amine congener (XAC), 1 μM, but were not antagonized by the nucleoside transport inhibitor, p-nitrobenzyl-6-thioinosine-5'-monophosphate (NBTIP), 10 μM (Figure 1),[50] suggesting the existence of an adenosine receptor on NK cells. In order to characterize the nature of this putative adenosine receptor, we investigated the influence of selected adenosine receptor agonists on NK cell activity. Adenosine receptor A_2 agonists (stimulatory for adenylyl cyclase) were anticipated to inhibit NK cell function, whereas Ado receptor A_1 agonists (inhibitory for cyclase) might stimulate this particular immune response. PIA and CPA (N^6-cyclopentyladenosine), representing selective Ado A_1 receptor agonists, were less potent inhibitors of NK lytic activity than were the A_2 receptor agonists NECA and 2-phenylaminoadenosine (PAA) (Figure 2). The order of potency for inhibition of mouse NK cell activity was NECA, PAA > Ado > PIA, CPA, consistent with inhibition due to agonist activity at an adenosine A_2-type receptor. In each case the adenosine receptor antagonist

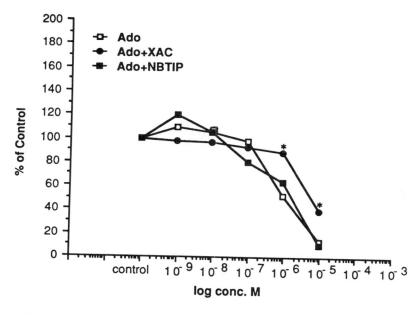

FIGURE 1. Inhibition of murine NK cell lysis by adenosine. Mouse spleen cells were incubated with various concentrations of adenosine in the presence of deoxycoformycin (1 μM). Adenosine and the ADA inhibitor were added at the beginning of the 4-h lytic assay. When present, the concentrations of XAC and NBTIP used were 10^{-6} and 10^{-5} M, respectively. (Modified from Priebe, T., Platsoucas, C., and Nelson, J. A., *Cancer Res.*, 50, 4328, 1990. With permission.)

XAC caused partial reversal, whereas the nucleoside transport inhibitor NBTIP did not change the effects produced by these agonists. These functional studies demonstrate that adenosine receptor agonists modify murine NK cell activity in a manner anticipated from known effects of cAMP on NK lytic activity.

There is no available information about adenosine receptors on human NK cells. Fredholm et al.[51] reported that adenosine (100 μM) had no effect on human NK cell activity from partially purified (nylon wool passage) lymphocytes. Moreover, this concentration of the nucleoside did not elevate cAMP levels. A study published by Nishida et al.[52] showed only minor inhibition of human NK cell activity by the A_2 receptor agonist 2-chloroadenosine. Also, preliminary experiments from our laboratory showed no inhibition of human NK cell lytic function by the A_2 agonist NECA at high concentrations (10^{-5} M).[53] Difficulties in purification and the relatively small percentage of NK cells in human peripheral blood make it difficult to study pure populations of normal human NK cells. As an alternative, we used a cloned human NK cell line (NK 3.3, kindly provided by Dr. Jackie Kornbluth, Department of Medicine, University of Arkansas for Medical Sciences).[54] The NK 3.3 cells also did not exhibit sensitivity toward NECA; however, dibutyryl cAMP (10^{-3} M) caused 100% inhibition and the direct stimulator of adenylyl cyclase, forskolin (10^{-4} M), produced 70% inhibition, consistent with the known inhibitory effects of cAMP on NK cell functon.

Binding experiments using the ligands [³H]NECA (A_2 receptor agonist) and [³H]R-PIA (A_1 receptor agonist) and NK 3.3 cell membranes failed to demonstrate specific binding of the A_1 ligand and only low-affinity ($K_D = 163$ nM), high-capacity binding of NECA, the A_2 agonist. Because NECA has affinity for both A_1 and A_2 receptors, the recently introduced, highly selective A_2 agonist, [³H] CGS 21680, was also employed. CGS 21680 did not demonstrate specific binding to NK 3.3 membranes.[55] The [³H]NECA binding results obtained with this human NK cell line is probably a nonfunctional NECA binding site, as

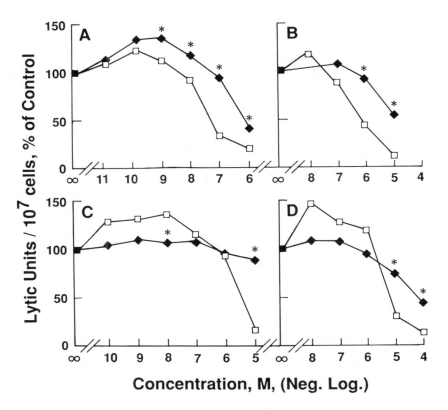

FIGURE 2.   Inhibition and stimulation of murine NK cell activity by adenosine receptor agonists. The agonists were added to the cultures at the beginning of the 4-h lytic assay. A, NECA; B, PAA; C, PIA; D, CPA. Asterisks, values significantly different from that obtained in the absence of XAC ($p \leq 0.05$, n $\geq$3). □, adenosine receptor agonist alone; ♦, plus XAC, 10^{-6} *M*. (Reproduced from Priebe, T., Platsoucas, C., and Nelson, J. A., *Cancer Res.*, 50, 4328, 1990. With permission.)

suggested earlier for rat striatum by Bruns et al.[56] The above findings suggest that murine and human NK cells differ substantially with regard to the presence of extracellular adenosine receptors.

IV. MONONUCLEAR PHAGOCYTES, MONOCYTES, AND MACROPHAGES

The major function of monocytes is phagocytosis. Another important function of mononuclear phagocytes is synthesis of complement components, especially in sites of inflammation. Adenosine was found to inhibit the synthesis of second complement (C2) by human monocytes.[57] The adenine nucleotides AMP, ADP, and ATP also inhibited C2 production when used alone. α-β-Methylene ADP, an inhibitor of ATPase and 5'-nucleotidase, completely prevented the inhibition, suggesting that the effect was due to the degradation of these nucleotides by ectoenzymes and adenosine formation. Agents capable of increasing intracellular levels of cAMP also decreased the production of complement components.[58] Results employing the adenosine receptor agonists NECA and R-PIA suggested that the action was mediated by adenosine receptors of the A_2 type.[59] Tissue macrophages and monocytes also produce tumor necrosis factor (TNF), which causes hemorrhagic necrosis of some tumors *in vivo* and tumor cells *in vitro*. TNF stimulates IL-1 production and inhibits lipoproteinase activity. The role of adenosine in regulating these macrophage activities has not been explored; however, by analogy to other adenylyl cyclase activating agents, similar effects involving an adenosine receptor can be predicted. Prostaglandin E_2 was reported to depress TNF and subsequently inhibit monocyte IL-1 expression by increasing the intracellular content of cAMP.[60]

Adenosine was also found to inhibit the maturation of human blood monocytes to macrophages *in vitro*.[61] Adenosine appears to inhibit human monocyte chemotaxis due to accumulation of intracellular *S*-adenosyl-homocysteine, a competitive inhibitor of *S*-adenosyl-L-methionine methylation.[62] Purine metabolism has not been studied extensively in this particular type of cell. However, it has been shown that adenosine is predominantly deaminated in activated and nonactivated macrophages with resultant formation of uric acid. Activation of macrophages does not change the deamination of adenosine markedly, the phosphorolysis of inosine, or hypoxanthine oxidation; however, ATP synthesis is elevated several fold.[62,63] It seems conceivable that the observed high catabolic activity of adenosine in macrophages is related to its biological function of phagocytosis.

V. ADENOSINE IN ALLERGIC RESPONSE

A. BASOPHILS

Basophils, short-lived circulating granulocytes, and mast cells residing in tissues function as effector cells in mediating inflammatory response, particularly that initiated by hypersensitivity reactions. Granules in these cells contain preformed mediators such as histamine, heparin, eosinophil chemotactic factor (ECF), and neutrophil chemotactic factor, while other mediators are synthesized *de novo*, i.e., leukotriene, C4, D4, and E4, formerly known as "slow-reacting substances of anaphylaxis" (SRS-A). Activation of degranulation and subsequent mediator release may be triggered by antigen cross linking to IgE molecules bound to the cell-surface Fc receptor.

Adenosine, at physiological concentrations, has been shown to inhibit histamine release from immunologically induced human basophils.[64] The response was dose dependent and inhibition was associated with elevation of the intracellular cAMP content.[65] The nucleoside transport inhibitor dipyridamole (10 μM) did not impair this effect of adenosine, whereas theophylline was a competitive antagonist. Further, adenine was not an effective inhibitor of histamine release. Studies with structurally modified adenosine analogs S-PIA, R-PIA, and NECA regarding inhibition and stimulation of IgE-dependent release of histamine from basophils indicated that these responses were mediated by an adenosine receptor of the A_2 subtype, stimulatory for adenylyl cyclase.[66,67] Marone et al.[68] extended these observations by demonstrating that activation of the A_2-type adenosine receptor leads to inhibition of *de novo* synthesis of leukotriene C4 (LTC$_4$) in human basophils. Modulation of newly synthesized LTC$_4$ was more sensitive to adenosine than was the release of the preformed mediator histamine, suggesting participation of adenosine in the regulation of arachidonic acid metabolism.[69] Interestingly, preincubation of human basophil leukocytes with adenosine before immunological stimulation resulted in inhibition of mediator release, whereas addition of the nucleoside after stimulation led to mild potentiation.[66]

Hughes et al.[70] proposed two distinct mechanisms by which adenosine exerts its inhibitory effect on histamine secretion from basophils. The first is suggested to be adenosine A_2-type receptor mediated, mimicked by NECA, antagonized by 8-phenyltheophylline, and associated with cAMP generation. The second mechanism requires prolonged exposure to the nucleoside and apparently involves intracellular events, resulting in the elevation of *S*-adenosyl-homocysteine. This action can be mimicked by 3-deazadenosine and is antagonized by dipyridamole, but is not antagonized by 8-phenyltheophylline.

B. MAST CELLS

Mast cells reside in tissues, particularly connective tissue and mucosa lining the respiratory and digestive tracts. Antigen-bound IgE antibodies trigger degranulation of mast cells, leading to the release of mediators.

Adenosine exerts opposite effects on mast cells compared to those observed with basophils. In human mast cells, adenosine and adenosine analogs potentiate histamine and LTC_4 release. Although an increased intracellular cAMP level was associated with decreased histamine secretion produced by adenosine in basophils, both inhibition and potentiation of mediator release occurred in mast cells by mechanisms suggested to result from adenosine A_2 receptor stimulation.[71,72] Murine bone marrow mast cells indeed exhibit an order of potency typical for A_2 receptor agonists, i.e., NECA > ADO > R-PIA.[73]

Marquardt and Wasserman[74] demonstrated specific binding of [^{3}H] adenosine to rat serosal mast cell membranes with a K_D of 28 nM. Competition studies indicated that 2-chloroadenosine had an affinity similar to that of adenosine, whereas NECA and R-PIA were 80 to 120 times less potent competitors. Changes of cAMP levels in resting mast cells exposed to adenosine and the analogs NECA and R-PIA showed an unusual potency order, i.e., adenosine $\simeq$ NECA > R-PIA = S-PIA. Studies of peritoneal rat mast cells have also shown an atypical agonist potency order, S-PIA > NECA = Ado > R-PIA.[75] Moreover, adenosine-mediated enhancement of histamine release was not blocked by the adenosine receptor antagonist 8-phenyltheophylline. Thus, cell-surface receptors that might mediate the effect cannot yet be classified as A_1 or A_2 type. Therefore, enhancement of histamine release may represent a novel cell-surface action of adenosine which might be independent of its effects on adenylyl cyclase.[75] Marone et al.[68] found that NECA (A_2 receptor agonist) and R-PIA (A_1 agonist) potentiate IgE-mediated release of histamine from human lung mast cells. Neither theophylline nor dipyridamole was capable of suppressing the adenosine-induced histamine release. Thus, the properties of putative mast cell adenosine receptors are not typical for those already described. It is plausible that these adenosine receptor subtypes are similar to those described in brain.[76]

VI. CONCLUSIONS

Adenosine appears to influence the immune system in several ways as a result of its effects on lymphoid cells. Interaction with cell-surface receptors seems to be a major mechanism by which adenosine and its analogs exert their effects. In the majority of lymphoid cells, the receptor appears to be of the A_2 type, i.e., low affinity, stimulatory for adenylyl cyclase. Occupancy of this receptor by agonists leads to increased cAMP production, increased occupancy of intracellular cAMP receptors, and activation of protein kinase A. Activation of the adenylyl cyclase-cAMP-protein kinase A pathway leads to down regulation of both the amplification and effector phases of the immune response resulting in immunosuppression.[40] Inhibition of proliferative responses by adenosine and other activators of cAMP synthesis can be generally attributed to enhancement of suppressor activity, lack of IL-2 production, and possible other growth factors. Suppression of lymphocyte-mediated cytotoxicity and NK cell activity and enhancement of the release of mediators from mast cells via extracellular receptors also suggest that this nucleoside acts as a ubiquitous regulator of immune response. As pointed out in previous reviews of adenosine and the immune system,[77,78] adenosine release in areas of ischemia may also have anti-inflammatory action due to its inhibition of superoxide anion formation by neutrophils. The effects of adenosine on neutrophil function are discussed in Chapter 10 of this text by Dr. B. N. Cronstein.

The potential importance of deoxyadenosine elevation in mediating the severe combined immunodeficiency associated with ADA deficiency has been widely accepted. The possible down regulation of putative adenosine receptors on cells of the immune system due to elevated adenosine may also play a role in this disease. Clearly, the various types of adenosine receptors that appear to occur on these important cells may permit selective physiological functions for adenosine in the regulation of immune responses.

ACKNOWLEDGMENTS

Results from the authors' laboratory were obtained through the assistance of The National Institutes of Health, Grant CA-28034 from the National Cancer Institute and Grant HD-13951 from the National Institute for Child Health and Human Development.

REFERENCES

1. **Giblett, E. R., Anderson, J. E., Cohen, F., Pollara, B., and Meuwissen, J. H.,** Adenosine-deaminase deficiency in two patients with severely impaired cellular immunity, *Lancet,* 2, 1067, 1972.
2. **Giblett, E. R., Ammann, A. J., Wara, D. W., Sandman, R., and Diamond, L. K.,** Nucleoside-phosphorylase deficiency in a child with severely defective T-cell immunity and normal B-cell immunity, *Lancet,* 1, 1010, 1975.
3. **Carson, D. A., Kaye, J., and Seegmiller, J. E.,** Lymphospecific toxicity in adenosine deaminase deficiency and purine nucleoside phosphorylase deficiency: possible role of nucleoside kinase(s), *Proc. Natl. Acad. Sci. U.S.A.,* 74, 5677, 1977.
4. **Mills, G. C., Schmalstieg, F. C., Trimmer, K. B., Goldman, A. S., and Goldblum, R. M.,** Purine metabolism in adenosine deaminase deficiency, *Proc. Natl. Acad. Sci. U.S.A.,* 73, 2867, 1976.
5. **Strauss, P. R.,** Murine lymphocytes and lymphocyte cell lines secrete adenosine deaminase, *Adv. Exp. Med. Biol.,* 195B, 275, 1986.
6. **Barankiewicz, J. and Cohen, A.,** Nucleotide catabolism and nucleoside cycles in human thymocytes, *Biochem. J.,* 219, 197, 1984.
7. **Barankiewicz, J. and Cohen, A.,** Evidence for distinct catabolic pathways of adenosine ribonucleotides and deoxyribonucleotides in human T lymphoblastoid cells, *J. Biol. Chem.,* 259, 15178, 1984.
8. **Barankiewicz, J., Cohen, A., and Gruber, H.,** Extracellular and intracellular formaton of adenosine, in *Purines in Cellular Signaling. Targets for New Drugs,* Jacobson, K. A., Daly, J. W., and Manganiello, V., Eds., Springer-Verlag, New York, 1990, 379.
9. **Barankiewicz, J., Dosch, H-M., and Cohen, A.,** Extracellular nucleotide catabolism in human B and T lymphocytes, *J. Biol. Chem.,* 263, 7094, 1988.
10. **Sattin, A. and Rall, T. W.,** The effect of adenosine and adenine nucleotides on the cyclic adenosine 3′5′-phosphate content of guinea pig cerebral cortex slices, *Mol. Pharmacol.,* 6, 13, 1970.
11. **Haslam, R. J. and Rosson, G. M.,** Effects of adenosine on levels of adenosine cyclic 3′5′monophosphate in human blood platelets in relation to adenosine incorporation and platelet aggregation, *Mol. Pharmacol.,* 11, 528, 1975.
12. **Londos, C. and Wolff, J.,** Two distinct adenosine-sensitive sites on adenylate cyclase, *Proc. Natl. Acad. Sci. U.S.A.,* 74, 5482, 1977.
13. **Marone, G., Vigorita, S., Triggiani, M., and Condorelli, M.,** Adenosine receptors on human lymphocytes, *Adv. Exp. Med. Biol.,* 195B, 1, 1986.
14. **Hirschhorn, R., Grossman, J., and Weissmann, G.,** Effect of cyclic 3′,5′-adenosine monophosphate and theophylline on lymphocyte transformation, *Proc. Soc. Exp. Biol. Med.,* 133, 1361, 1970.
15. **Hovi, T., Smyth, J. F., Allison, A. C., and Williams, S. C.,** Role of adenosine deaminase in lymphocyte proliferation, *Clin. Exp. Immunol.,* 23, 395, 1976.
16. **Snyder, F. F., Mendelsohn, J., and Seegmiller, J. E.,** Adenosine metabolism in phytohaemagglutinin-stimulated human lymphocytes, *J. Clin. Invest.,* 58, 654, 1976.
17. **Ballet, J. J., Insel, R., Merler, E., and Rosen, F. S.,** Inhibition of maturation of human precursor lymphocytes by coformycin an inhibitor of the enzyme adenosine deaminase, *J. Exp. Med.,* 143, 1271, 1976.
18. **Hovi, T., Allison, A. C., and Allsop, J.,** Rapid increase of phosphoribosyl pyrophosphate concentration after mitogenic stimulation of lymphocytes, *FEBS Lett.,* 55, 291, 1975.
19. **Green, H. and Chan, T. S.,** Pyrimidine starvation induced by adenosine in fibroblasts and lymphoid cells: role of adenosine deaminase, *Science,* 182, 836, 1973.
20. **Johnston, J. M. and Kredich, N. M.,** Inhibition of methylation by adenosine in adenosine deaminase-inhibited, phytohaemagglutinin-stimulated human lymphocytes, *J. Immunol.,* 123, 97, 1979.
21. **Marone, G., Plaut, M., and Lichtenstein, L. M.,** Characterization of a specific adenosine receptor on human lymphocytes, *J. Immunol.,* 121, 2153, 1978.

22. **Schwartz, A. L., Stern, R. C., and Polmar, S. H.,** Demonstration of an adenosine receptor on human lymphocytes in vitro and its possible role in the adenosine deaminase-deficient form of severe combined immunodeficiency, *Clin. Immunol. Immunopathol.,* 9, 499, 1978.

23. **Marone, G., Thomas, L. L., and Lichtenstein, L. M.,** The role of agonists that activate adenylate cyclase in the control of cAMP metabolism and enzyme release by human polymorphonuclear leukocytes, *J. Immunol.,* 125, 2277, 1980.

24. **Sandberg, G. and Fredholm, B. B.,** Regulation of thymocyte proliferation: effects of L-alanine, adenosine and cyclic AMP in vitro, *Thymus,* 3, 63, 1981.

25. **Fredholm, B. B. and Sandberg, G.,** Inhibition by xanthine derivatives of adenosine receptor stimulated cyclic adenosine 3′,5′-monophosphate accumulation in rat and guinea-pig thymocytes, *Br. J. Pharmacol.,* 80, 639, 1983.

26. **Dinjens, W. N. M., van Doorn, R., van Laarhoven, J. P. R. M., Ross, D., Zeijlemaker, W. P., and de Bruijn, C. H. M. M.,** Adenosine receptors on human T lymphocytes and human thymocytes, *Adv. Exp. Med. Biol.,* 195B, 1, 1986.

27. **Bessler, H., Djaldetti, M., Kupfer, B., and Moroz, C.,** Two T lymphocyte subpopulations isolated from human peripheral blood following ''in vitro'' treatment with adenosine, *Biomedicine,* 32, 66, 1980.

28. **Limatibul, S., Shore, A., Dosch, H. M., and Gelfand, E. W.,** Theophylline modulation of E-rosette formation. An indicator of T-cell maturation, *Clin. Exp. Immunol.,* 33, 503, 1978.

29. **Birch, R. E., Rosenthal, A. K., and Polmar, S. H.,** Pharmacological modification of immunoregulatory T lymphocytes. II. Modulation of T lymphocyte cell surface characteristics, *Clin. Exp. Immunol.,* 48, 231, 1982.

30. **Moretta, L., Webb, S. R., Grossi, C. E., Lydyard, P. M., and Cooper, M. D.,** Functional analysis of two human T-cell subpopulations. Help and suppression of B-cell responses by T-cell bearing receptors for IgM (Tμ) or IgG (Tγ), *J. Exp. Med.,* 146, 184, 1977.

31. **Birch, R. E. and Polmar, S. H.,** Pharmacological modification of immunoregulatory T lymphoyctes. I. Effect of adenosine, H_1 and H_2 histamine agonists upon T lymphocyte regulation of B lymphocyte differentiation in vitro, *Clin. Exp. Immunol.,* 48, 218, 1982.

32. **Birch, R. E. and Polmar, S. H.,** Induction of Fcγ receptors on a subpopulation of human T lymphocytes by adenosine and impromidine, an H_2-histamine agonist, *Cell. Immunol.,* 57, 455, 1981.

33. **Birch, R. E. and Polmar, S. H.,** Adenosine induced immunosuppression: the role of the adenosine receptor-adenylate cyclase interaction in the alteration of T-lymphocyte surface phenotype and immunoregulatory function, *Int. J. Immunopharmacol.,* 8, 329, 1986.

34. **Dos Reis, G. A., Nobrega, A. F., and Paes de Carvalho, R.,** Purinergic modulation of T-lymphocyte activation: differential susceptibility of distinct activation steps and correlation with intracellular 3′,5′-cyclic adenosine monophosphate accumulation, *Cell. Immunol.,* 101, 213, 1986.

35. **Averill, L. E., Stein, R. L., and Kammer, G. M.,** Control of human T-lymphocyte interleukin-2 production by cAMP-dependent pathway, *Cell. Immunol.,* 115, 88, 1988.

36. **Didier, M., Aussel, C., Ferrua, B., and Fehlmann, M.,** Regulation of interleukin 2 synthesis by cAMP in human T cells, *J. Immunol.,* 139, 1179, 1987.

37. **Cohen, M. B. and Glazer, R. I.,** Inhibition of interleukin-2 messenger RNA in mouse lymphocytes by 2′-deoxycoformycin and adenosine metabolites, *Ann. N.Y. Acad. Sci.,* 451, 180, 1985.

38. **Beckner, S. K. and Farrar, W. L.,** Interleukin 2 modulation of adenylate cyclase. Potential role of protein kinase C, *J. Biol. Chem.,* 261, 3043, 1986.

39. **Beckner, S. K. and Farrar, W. L.,** Inhibition of adenylate cyclase by IL 2 in human T lymphocytes is mediated by protein kinase C, *Biochem. Biophys. Res. Commun.,* 145, 176, 1987.

40. **Kammer, G. M.,** The adenylate cyclase-cAMP-protein kinase A pathway and regulation of the immune response, *Immunol. Today,* 9, 222, 1988.

41. **Sloboda, R. D., Rudolph, S. A., Rosenbaum, J. L., and Greengard, P.,** Cyclic AMP-dependent endogenous phosphorylation of a micro-tubule-associated protein, *Proc. Natl. Acad., Sci. U.S.A.,* 72, 177, 1975.

42. **Kammer, G. M. and Rudolph, S. A.,** Regulation of human T lymphocyte surface antigen mobility by purinergic receptor, *J. Immunol.,* 133, 3298, 1984.

43. **Kammer, G. M., Boehm, C. A., and Rudolph, S. A.,** Role of adenylate cyclase in human T-lymphocyte surface antigen capping, *Cell. Immunol.,* 101, 251, 1986.

44. **Kizaki, H., Shimada, H., Ohsaka, F., and Sakurada, T.,** Adenosine, deoxyadenosine and deoxyguanosine induce DNA cleavage in mouse thymocytes, *J. Immunol.,* 141, 1652, 1988.

45. **Kizaki, H., Suzuki, K., Tadakuma, T., and Ishimura, Y.,** Adenosine receptor-mediated accumulation of cyclic AMP-induced T-lymphocyte death through internucleosomal DNA cleavage, *J. Biol. Chem.,* 265, 5280, 1990.

46. **Wolberg, G., Zimmerman, T. P., Hiemstra, K., Winston, M., and Chu, L. C.,** Adenosine inhibition of lymphocyte-mediated cytolysis: possible role of cyclic adenosine monophosphate, *Science,* 187, 957, 1975.

47. **Wolberg, G., Zimmerman, T. P., Duncan, G. S., Singer, K. H., and Elion, G. B.,** Inhibition of lymphocyte-mediated cytolysis by adenosine analogs: biochemical studies concerning mechanism of action, *Biochem. Pharmacol.*, 27, 1487, 1978.

48. **Ullberg, M., Jondal, M., Lanefelt, F., and Fredholm, B. B.,** Inhibition of human NK cell cytotoxicity by induction of cyclic AMP depends on impaired target cell recognition, *Scand, J. Immunol.*, 17, 365, 1983.

49. **Hiserodt, J. C., Britvan, L. J., and Targan, S. R.,** Differential effects of various pharmacological agents on the cytolytic reaction mechanism of the human natural killer lymphocyte: further resolution of programming for lysis and KCIL into discrete stages, *J. Immunol.*, 129, 2266, 1982.

50. **Priebe, T., Platsoucas, C., and Nelson, J. A.,** Adenosine receptors and modulation of natural killer cell activity by purine nucleosides, *Cancer Res.*, 50, 4328, 1990.

51. **Fredholm, B. B., Jondal, M., Lanefelt, F., and Ng, J.,** Effect of 5'-methylthioadenosine, 3-deazaadenosine, and related compounds on human natural killer cell activity, *Scand. J. Immunol.*, 20, 511, 1984.

52. **Nishida, Y., Kamatani, N., Morito, T., and Miyamoto, T.,** Differential inhibition of lymphocyte function by 2-chloroadenosine, *Int. J. Immunopharmacol.*, 6, 325, 1984.

53. **Priebe, T. and Nelson, J. A.,** Unpublished data.

54. **Leiden, J. M., Gottesdiener, K. M., Quertermous, T., Coury, L., Bray, R. A., Gottschalk, L., Gebel, H., Seidman, J. G., Strominger, J. L., Landay, A. L., and Kornbluth, J.,** T-cell receptor gene rearrangement and expression in human natural killer cells: natural killer activity is not dependent on the rearrangement and expression of T-cell receptor α, β, or γ genes, *Immunogenetics*, 27, 231, 1988.

55. **Priebe, T., Sapsowitz, S. H., and Nelson, J. A.,** Unpublished data.

56. **Bruns, R. F., Lu, G. H., and Pugsley, T. A.,** Characterization of the A_2 adenosine receptor labeled by [^{3}H]NECA in rat striatal membranes, *Mol. Pharmacol.*, 29, 331, 1986.

57. **Lappin, D. and Whaley, K.,** Cyclic AMP mediated modulation of the production of the second component of human complement by monocytes, *Int. Arch. Allergy Appl. Immunol.*, 65, 85, 1981.

58. **Lappin, D. and Whaley, K.,** Control of monocyte C2 production by cyclic AMP, *Immunology*, 49, 625, 1983.

59. **Lappin, D. and Whaley, K.,** Adenosine A_2 receptors on human monocytes modulate C2 production, *Clin. Exp. Immunol.*, 57, 454, 1984.

60. **Knudsen, P. J., Dinarello, C. A., and Strom, T. B.,** Prostaglandins posttranscriptionally inhibit monocyte expression of interleukin 1 activity by increasing intracellular cyclic adenosine monophosphate, *J. Immunol.*, 137, 3189, 1986.

61. **Fischer, D., Van der Weyden, M. B., Snyderman, R., and Kelley, W. N.,** A role for adenosine deaminase in human monocyte maturation, *J. Clin. Invest.*, 58, 399, 1976.

62. **Pike, M. C., Kredich, N. M., and Snyderman, R.,** Requirement of S'-adenosyl-L-methionine-mediated methylation for human monocyte chemotaxis, *Proc. Natl. Acad. Sci. U.S.A.*, 75, 3928, 1978.

63. **Barankiewicz, J. and Cohen, A.,** Purine metabolism in rat macrophages, *Adv. Exp. Med. Biol.*, 165B, 227, 1984.

64. **Marone, G., Findlay, S. R., and Lichtenstein, L. M.,** Adenosine receptor on human basophils: modulation of histamine release, *J. Immunol.*, 123, 1473, 1979.

65. **Hughes, P. J., Holgate, S. T., Roath, S., and Church, M. K.,** The relationship between cyclic AMP changes and histamine release from basophil-rich human leukocytes, *Biochem. Pharmacol.*, 32, 2557, 1983.

66. **Church, M. K., Holgate, S. T., and Hughes, P. J.,** Adenosine inhibits and potentiates IgE-dependent histamine release from human basophils by an A_2-receptor mediated mechanism, *J. Pharmacol.*, 80, 719, 1983.

67. **Marone, G., Vigorita, S., Antonelli, C., Torella, G., Genovese, A., and Condorelli, M.,** Evidence for adenosine A_2/R_a receptor on human basophils, *Life Sci.*, 36, 339, 1985.

68. **Marone, G., Triggiani, M., Kagey-Sobotka, A., Lichtenstein, L., and Condorelli, M.,** Adenosine receptor on human basophils and lung mast cells, *Adv. Exp. Med. Biol.*, 195B, 35, 1986.

69. **Peachell, P. T., Lichtenstein, L. M., and Schleimer, R. P.,** Inhibition by adenosine of histamine and leukotriene release from human basophils, *Biochem. Pharmacol.*, 38, 1717, 1989.

70. **Hughes, P. J., Benyon, R. C., and Church, M. K.,** Adenosine inhibits immunoglobulin E-dependent histamine secretion from human basophil leukocytes by two independent mechanisms, *J. Pharmacol. Exp. Ther.*, 242, 1064, 1987.

71. **Hughes, P. J., Holgate, S. T., and Church, M. K.,** Adenosine inhibits and potentiates IgE dependent histamine release from human lung mast cells by an A_2-purino-receptor mediated mechanism, *Biochem. Pharmacol.*, 23, 3847, 1984.

72. **Holgate, S. T., Lewis, R. A., and Austen, K. F.,** Role of adenylate cyclase in immunological release of mediators from rat mast cells: agonist and antagonist effects of purine and ribose modified adenosine analogues, *Proc. Natl. Acad. Sci. U.S.A.*, 77, 6800, 1980.

73. **Marquardt, D. L., Walker, L. L., and Wasserman, S. I.,** Adenosine receptors on mouse bone marrow-derived mast cells. Functional significance and regulation by aminophylline, *J. Immunol.*, 133, 932, 1984.

74. **Marquardt, D. L. and Wasserman, S. I.**, [³H] Adenosine binding to rat mast cells—pharmacologic and functional characterization, *Agents Actions,* 16, 453, 1985.

75. **Church, M. K. and Hughes, P. J.**, Adenosine potentiates immunological histamine release from rat mast cells by a novel cyclic-AMP-independent cell surface action, *Br. J. Pharmacol.,* 85, 3, 1985.

76. **Murphy, K. M. M. and Snyder, S. H.**, Heterogeneity of adenosine A₁ receptor binding in brain tissue, *Mol. Pharmacol.,* 22, 250, 1982.

77. **Polmar, S. H., Fernandez-Mejia, C., and Birch, R. E.**, Adenosine receptors: immunologic aspects, in *Adenosine Receptors,* Cooper, D. M. F., and Londos, C., Eds., Alan R. Liss, New York, 1988, 97.

78. **Gilbertsen, R. B.**, Adenosine and adenosine receptors in immune function. Minireview and meeting report, *Agents Actions,* 22, 94, 1987.

Chapter 12

PURINERGIC REGULATION OF TRANSMITTER RELEASE

J. Alexandre Ribeiro

TABLE OF CONTENTS

I. INTRODUCTION

Most of the work to be discussed in this review has been performed on rat or frog neuromuscular junctions, where adenosine has only presynaptic effects and transmitter release can be measured electrophysiologically, by recording endplate potentials, or neurochemically, by recording [³H]-acetylcholine (ACh) release. This chapter is circumscribed to evidence that adenosine or adenosine analogs modify transmitter release, and does not review evidence that the intact adenosine 5′-triphosphate (ATP) molecule reduces neurotransmitter release (see Chapter 24 of the present volume).

It is known that endogenous adenosine is present at synapses. At the neuromuscular junction of the frog at least 40 to 50% of endogenous adenosine results from metabolism of adenine nucleotides.[1] In the frog neuromuscular junction adenosine is inactivated by uptake,[1] whereas in the rat neuromuscular junction both uptake and deamination contribute to the inactivation of extracellular adenosine.[2] It has been possible to establish the following steps involved in the metabolism of extracellular ATP at the innervated frog sartorius muscle: ATP can sequentially break down into adenosine 5′-monophosphate (AMP). Degradation of AMP might occur through ecto-5′-nucleotidase, producing adenosine, or through ecto-5′-AMP-deaminase, producing IMP.[3] When degradation of AMP is prevented by α,β-methylene adenosine 5′-diphosphate (AOPCP), an inhibitor of 5′-nucleotidase, and coformycin, an inhibitor of AMP-deaminase, adenine nucleotides can accumulate extracellularly upon nerve stimulation; about one third were formed when muscle contraction was prevented with tubocurarine, which suggests that its origin is from nerve endings.[4] Presynaptic release of ATP, measured with luciferin-luciferase at the rat neuromuscular junction, has been described by Silinsky.[5]

II. ADENOSINE AND TRANSMITTER RELEASE IN THE PERIPHERAL NERVOUS SYSTEM

A. IN THE NEUROMUSCULAR JUNCTION

The first extensive study about the characterization of the effect of adenosine on neurotransmitter release, now a classic in the adenosine area, was published about 20 years ago.[6,7] This study, performed on the innervated rat diaphragm, demonstrated that adenosine: (1) inhibits the evoked release of ACh measured as a reduction in the quantal content of the evoked endplate potentials (EPPs), an effect antagonized by theophylline; (2) inhibits the spontaneous release of ACh measured as the reduction in the frequency of miniature endplate potentials (MEPPs); this effect can be observed either in the presence of extracellular calcium or in the absence of calcium and in the presence of EDTA in hyperosmotic medium;[8] (3) causes as a maximal inhibition of transmitter release usually about 50% of its initial (preadenosine) value; (4) does not produce tachyphylaxis; (5) has only a presynaptic effect, since adenosine does not change the mean amplitude of MEPPs. The finding that at the neuromuscular junction adenosine has only presynaptic effects makes this preparation a very interesting model to study the effect of adenosine on neurotransmitter release. Other advantages of the use of the neuromuscular junction are (1) there is only one type of transmitter, and its chemical nature is well known; (2) it is easy to obtain intracellular recordings from the postsynaptic region; (3) in the absence of postsynaptic modifications, the response recorded from one cell is a function of the output of transmitter by only one nerve terminal; (4) it is possible to distinguish between pre- and postsynaptic effects (see Ribeiro and Sebastião).[9]

In their paper Ginsborg and Hirst[7] wrote: "Further experiments, for which time was unfortunately not available are required to establish the magnitude of the effect of adenosine in very low calcium concentrations". My colleagues and I did continue this work and, since

Silinsky and Hubbard[10] published that ATP is released from the rat phrenic nerve endings, it seemed of interest to investigate the effect of ATP on transmitter release from motor nerve endings. The results of these studies[11,12] demonstrated that ATP inhibits the quantal content of EPPs in the same proportion as does adenosine, and that this occurs either in mammalian (rat) or amphibian (frog) neuromuscular junctions. Interpreting these results with those described by Silinsky and Hubbard,[10] it was postulated[11] that if ATP and ACh are released together, with the same time course, from the same region of the nerve ending, the concentration of ATP could transiently be of the same order as the concentration that inhibits ACh release, if thus being conceivable that released ATP could contribute to the depression of the response to nerve stimulation with high frequencies and high output of transmitter (see Ribeiro and Walker[11] and Ribeiro[13]). The comparison of the inhibitory effects of adenosine and ATP on ACh release in both rat and frog showed that these substances caused similar reductions in transmitter release and that in the presence of a supramaximal concentration of adenosine, ATP does not cause further inhibition in the maximal effect of adenosine.[12] Studies performed later[1] revealed that the inhibitory effect of ATP depends on its previous hydrolysis to adenosine, since the effect of ATP is prevented by the 5′-nucleotidase inhibitor, α,β-methylene ADP, which prevents the formation of adenosine from ATP. Furthermore, the antagonism caused by the adenosine receptor antagonist, 1,3-dipropyl-8-cyclopentylxanthine (DPCPX), in relation to the inhibitory effects of adenosine or ATP on neuromuscular transmission, or the potentiation of these effects by the adenosine uptake blocker, dipyridamole, are of similar magnitude.[14] These findings indicate that the presynaptic inhibitory purinoceptor at motor nerve endings belongs to the P_1 family of purinoceptors.[15]

The inhibitory effect of adenosine on ACh release from motor nerve endings has also been observed at the mouse diaphragm[16] and at the rat diaphragm stimulated by K^+ (15 mM),[17] where adenosine reduces the occurrence of MEPPs. Inhibition of neuromuscular transmission *in vivo* has also been observed.[18]

1. Presynaptic Inhibitory Adenosine Receptor

After clear characterization of the inhibitory effect of adenosine on ACh release, studies on the type of adenosine receptor involved in this effect showed initially that this receptor is an R type, since it needs the ribosidic part of the adenosine molecule intact to be activated.[19] Substances with action on the P-site (intracellular), such as 2′-deoxyadenosine, did not inhibit transmitter release.[19]

Further characterization of the adenosine receptor, using as a preparation the innervated frog sartorius muscle, was reported[20] and in contradistinction to the initial proposal to classify adenosine receptors[21,22] into A_1 (with R-phenylisopropyladenosine [R-PIA] > 2-chloroadenosine [CADO] > 5′-N-ethylcarboxamide adenosine [NECA]) and A_2 (with NECA > CADO > R-PIA) subclasses, the agonist profile of the presynaptic inhibitory adenosine receptor revealed that R-PIA and NECA were equipotent and both more potent than CADO. This kind of agonist profile did not allow us to classify this inhibitory adenosine receptor into the A_1 or A_2 adenosine receptor families.[20] Recent observations[23] on the nature of this receptor with the use of the adenosine receptor antagonist, DPCPX, demonstrated that the K_i value for this antagonist is 35 nM, which is quite different from the K_i value obtained for the A_1 adenosine receptor in binding studies with [^{3}H]-DPCPX in nervous tissue, in which the K_i is lower than 0.5 nM[24,25] and in relation to the A_2 adenosine receptor, also in nervous tissue, where the K_i for DPCPX is 330 nM.[24,25] The K_i value obtained in the frog neuromuscular junction is highly suggestive of the existence of a third receptor (A_3) for adenosine (see also review by Ribeiro and Sebastião).[26] K_i values for DPCPX of similar magnitude (8 to 24 nM) have been recently described[27] for the adenosine receptor involved in the modulation of phosphatidylinositol turnover in guinea pig and mouse cortical slices.

At the neuromuscular junction of the rat[2] the adenosine receptor agonist profile is the same as that described for the frog neuromuscular junction. Both frog and rat neuromuscular junctions, used to define the adenosine receptor agonist profiles, were pretreated either with high magnesium, in which the safety margin of neuromuscular transmission was inhibited presynaptically, or pretreated with tubocurarine, in which the preparations were inhibited postsynaptically. Considering that these treatments could influence the adenosine receptor agonist profile, experiments were performed in which the potency of the agonists was studied on [³H]-ACh release from the phrenic nerve endings in the absence of pre- or postsynaptic initial inhibitions. The agonist profile obtained in these conditions remained the same (R-PIA = NECA > CADO)[28] as that already observed in pretreated preparations.[2] However, the affinity of DPCPX for the presynaptic inhibitory adenosine receptor in the rat neuromuscular junction[29] is similar ($K_i = 0.54$ nM) to that described for the A_1 adenosine receptor in binding studies, as well as similar to that found for the inhibitory adenosine receptor mediating inhibition of population spikes ($K_i = 0.45$ nM) at the rat hippocampal slices.[29]

Other studies on the potency of adenosine receptor agonists at the neuromuscular junction showed that both R-PIA and NECA reduced the frequency of evoked MEPPs in the mouse diaphragm.[30] Measuring the frequency of MEPPs in the frog, Barry[31] found R-PIA more potent than NECA and a great stereoselectivity for the PIA isomers, suggesting an A_1 profile. Since spontaneous and evoked release of ACh might be affected by adenosine through different mechanisms,[26] the discrepancies between the agonist profiles found in relation to the adenosine-induced inhibition of evoked and spontaneous transmitter might result from different adenosine-operated mechanisms.

2. Presynaptic Excitatory Adenosine Receptor

Besides the existence of presynaptic inhibitory adenosine receptors, it was recently shown[28] that motor nerve endings possess excitatory adenosine receptors susceptible to being activated by A_2 adenosine agonists such as NECA, when applied in the presence of DPCPX (to silence the A_1 receptors), or by CGS 21680C. These agonists increase evoked [³H]-ACh release, an effect antagonized by PD 115,199, but not by low nanomolar concentrations of DPCPX.[4] The same nerve ending has both inhibitory and excitatory receptors, since in the same endplate R-PIA inhibits and CGS 21680C increases EPPs amplitude.[4] The possibility that excitatory adenosine receptors could be activated by adenosine is suggested by the finding that theophylline in low concentrations (<50 μM) inhibits MEPP frequency in hyperosmotic solutions.[32]

3. Transducing System of the Presynaptic Inhibitory Adenosine Receptor

The transducing system of the adenosine receptor operating inhibition of transmitter release is not as yet clear, but one of the most frequently advanced hypotheses is that adenosine could reduce evoked transmitter release by interacting with a calcium channel[26,33] of the N type.[34] However, the way adenosine affects the calcium channel and/or the calcium needed at the neurotransmitter release site is still very intriguing (see Silinsky et al.),[35] since the manner of coupling between Ca^{2+} and the releasing site is still a mystery.

Adenosine is more effective in decreasing neuromuscular transmission at the rat diaphragm neuromuscular junction at 37°C than at room temperature.[36,37] At the frog neuromuscular junction, decreasing the temperature also attenuates the effect of adenosine on evoked transmitter release.[38] This influence of temperature might suggest that a second messenger system is involved in the mechanism of action of adenosine at nerve terminals.

Pertussis toxin prevents the inhibitory effect of adenosine in both the rat[39] and the frog[35] neuromuscular junctions, indicating that a G protein (G_i, G_p, or G_o) is involved in the transducing system operated by the inhibitory presynaptic adenosine receptor. This transducing system appears not to be related to adenylate cyclase (and thus to G_i), since manipulations of adenylate cyclase activity with the activator, forskolin, or the inhibitor, MDL 12,330A, do not modify the inhibitory effect of adenosine on EPP amplitude.[40]

Some indirect evidence suggests that adenosine might inhibit ACh release by interfering with phosphatidylinositol turnover. Pharmacological studies with the phorbol ester active on protein kinase C, 4β-phorbol,12,13-diacetate, demonstrated that the inhibitory effects of adenosine and CADO are attenuated in the presence of this substance, whereas the action of the adenosine receptor antagonist, 8-phenyltheophylline, is not affected.[40] As the inhibitor of protein kinase C, polymyxin B, does not modify the effect of adenosine, the interaction of adenosine with protein kinase C seems to be indirect and could be through a receptor coupled to phospholipase C, since the excitatory effect of lithium ions (which inhibit breakdown of inositol phosphates) on neurotransmitter release is attenuated by adenosine or its analog, CADO.[40]

4. Transducing System of the Presynaptic Excitatory Adenosine Receptor

The transducing system of the presynaptic excitatory adenosine receptor is probably an adenylate cyclase/G_s protein coupling and increases in cyclic AMP accumulation. Experimental evidence suggesting that this could be the case is (1) the A_2 nature of the receptor, which in many preparations is positively coupled to adenylate cyclase; (2) the observations that cAMP analogs, such as dibutyryl cAMP, increase ACh release and the amplitude of EPPs in both rat[41,42] and frog[14] neuromuscular junctions, and that (3) forskolin, an activator of the catalytic subunit of adenylate cyclase, increases transmitter release and EPP amplitude in both frog[40] and rat[4] neuromuscular junctions.

B. IN THE ELECTRIC ORGAN OF TORPEDO

In this model of nicotinic transmission, adenosine decreases the amplitude of the electrical discharge in the electric organ of Torpedo,[43] as well as decreases the release of [^{14}C]-ACh evoked by electrical stimulation, but not that triggered by K^+ (80 mM).[44]

C. IN THE SMOOTH MUSCLE
1. Cholinergic System

Adenosine decreases the electrically evoked release of ACh in the guinea pig ileum determined either by bioassay[45,46] or by gas chromatography-mass spectrometry.[47] The amplitude of the nerve-evoked contractions of the guinea pig ileum is also reduced by adenosine, this nucleoside being without effect on the sensitivity of the smooth-muscle cells to exogenously applied ACh.[45,48,49] The inhibitory effect of adenosine on nerve-evoked contractions of the guinea pig ileum[49] or on the release of ACh evoked by electrical stimulation[45,47] is more evident at lower frequencies of stimulation.

In the guinea pig ileum myenteric plexus synaptosomes, adenosine decreases the K^+-evoked release of ^{3}H-ACh,[50,51] but is less effective than when ACh release is evoked either by nicotinic stimulation with dimethyl-4-phenylpiperazinum[50] or by electrical stimulation.[51]

2. Noradrenergic System

Decrease in [^{3}H]-noradrenaline (NA) release evoked by electrical stimulation in the presence of adenosine has been shown to occur in guinea pig vas deferens,[52] rat vas deferens,[53,54] rat mesenteric artery,[55] rat portal vein,[54,56] and human fallopian tube.[57] The release of [^{3}H]-NA induced by K^+ in the canine blood vessels[58] and rat vas deferens[54] is also decreased by adenosine, an effect which is inversely related to the concentration of K^+ used to stimulate the release.[54] The electrically evoked contractions of isolated vas deferens,[53,54] dog saphenous vein,[58,59] and rabbit portal vein[60] are decreased by adenosine at concentrations which do not affect the response to exogenously added NA. As in the guinea pig ileum, the higher the frequency of stimulation, the smaller the inhibitory effect of adenosine on nerve-evoked contractions of the veins[58] or vas deferens.[54] Studies performed *in vivo* showed that adenosine decreases the contractile response of the dog saphenous vein to lumbar sympathetic stimulation, an effect which is also more evident with low frequency of stimulation.[58] Adenosine also depresses constrictor responses of the rat mesenteric arteries to sympathetic nerve stimulation.[61]

D. IN THE HEART

Decrease in electrically evoked release of NA by adenosine has been demonstrated in the heart of rabbits,[62] rats, and guinea pigs.[54] *In vivo,* adenosine impairs the tachycardia induced in dogs by stimulation of cardiac sympathetic nerves without affecting the positive chronotropic effect of NA.[63]

E. IN THE KIDNEY, SPLEEN, ADIPOSE TISSUE, SUBMANDIBULAR GLAND, AND ADRENAL MEDULLA

In all these tissues adenosine decreases NA release. This is the case for the rabbit kidney,[52] cat splenic nerve,[64] dog adipose tissue,[52] and rat salivary gland.[54] At the adrenal medulla, ATP and adenosine added to a pulsatile stimulation together with ACh were found to inhibit catecholamine release.[65,66]

F. IN THE GANGLIA

Stable adenosine analogs reversibly reduce the amplitude of the nicotinic fast excitatory postsynaptic potential in the rat superior cervical ganglion; the postsynaptic response to exogenous cholinergic agonists is not affected by the adenosine analogs, suggesting that their effects are predominantly presynaptic.[67] In the same preparation adenosine also has postsynaptic effects, since it facilitates ganglionic transmission during repetitive stimulaton of the preganglionic nerve.[68]

III. ADENOSINE AND TRANSMITTER RELEASE IN THE CENTRAL NERVOUS SYSTEM

In the central nervous system adenosine has both pre- and postsynaptic actions,[69,70] as well as nonsynaptic actions.[71] However, microelectrophysiological studies alone (recording from the cell body of neurons) hardly distinguish the synaptic structure where the nucleoside acts. The release studies with radiolabeled neurotransmitters are also not completely conclusive. As pointed out by Dunwiddie,[72] with the exception of the synaptosomal experiments, the finding that adenosine decreases transmitter release does not necessarily mean that the nucleoside has a direct effect upon the nerve terminal. Postsynaptic and/or nonsynaptic inhibitory effects of adenosine also result in an apparent decrease in the amount of transmitter released by the affected neuron. Nevertheless, there is a consensus that the predominant effect of adenosine in the central nervous system, responsible for the inhibition of synaptic transmission, is presynaptic. A brief description of the studies reporting inhibition of transmitter release in the central nervous system follows.

A. CHOLINERGIC SYSTEM

Adenosine has been shown to decrease the release of [³H]-ACh or [¹⁴C]-ACh from electrically stimulated rabbit hippocampus,[73] K⁺-depolarized rat striatum,[74] and veratridine-depolarized cholinergic synaptosomes purified by immunoaffinity from the rat striatum.[75] Decrease in endogenous ACh release, quantified by bioassay, from electrically stimulated rat brain slices has also been reported.[76] Adenosine also significantly decreases the electrically evoked release of [³H]-ACh from the rabbit caudate nucleus, but its effect is smaller and requires higher concentrations than to reduce ACh release from the rabbit hippocampus,[73] where the potency of the adenosine receptor agonists is similar to that observed at the frog neuromuscular junction.[20] It is possible that the lower sensitivity of the caudate nucleus is related to the higher frequency of stimulation used.

The turnover rate of ACh in the rat cortex and hippocampus is decreased after intracerebroventricular[77] or intracisternal[78] injections of the stable adenosine analog CADO.

B. NORADRENERGIC SYSTEM

Inhibition in the electrically evoked release of [³H]-NA in the central nervous system has been described by several authors. In the rat hippocampus[79] and rabbit hippocampus,[80] adenosine and its analogs have relative potencies similar to those observed in relation to the inhibition of the release of ACh.[73] In K⁺-stimulated slices of the rat brain neocortex it was observed that adenosine decreases [³H]-NA release with a maximal effect of about 35% at 100 μM.[81] Also using K⁺-evoked neurotransmitter release, Ebstein and Daly[82] described that adenosine and its analogs reduce ³H-NA release from a guinea pig brain vesicular preparation. This effect requires higher concentrations of adenosine or its analogs than those needed to inhibit the electrically evoked [³H]-NA release described by Jackisch et al.,[80] and is mediated by an adenosine receptor with a different agonist profile (for a comparison of the relative potency of the adenosine analogs see Ribeiro and Sebastião).[26]

C. DOPAMINERGIC SYSTEM

Reduction in the release of [³H]-dopamine (DA) by adenosine has been found in K⁺ (26 mM)-stimulated slices of the rat striatum[74] and in synaptosomes of the rat basal ganglia in which release of [³H]-DA was evoked by K⁺ (15 mM).[83] The effect observed by Harms et al.[74] was rather small compared with that described by Michaelis et al.[83] This discrepancy might be attributed to the different concentrations of K⁺ used to evoke the release of DA, since the inhibitory effect of the stable adenosine analog, CADO, is smaller the higher the K⁺ concentration in the bath.[83] This might explain why, when an even higher K⁺ concentration (43.4 mM) is used to stimulate the guinea pig brain vesicular preparation, only CADO causes inhibition of [³H]-DA release.[82] In the presence of this high K⁺ concentration, the other adenosine analogs, R-PIA, S-PIA, and 5′-cyclopropylcarboxamide adenosine, cause facilitation of [³H]-DA release.[82] Adenosine analogs can inhibit and facilitate striatal *in vivo* DA release measured with the microdialysis technique,[84] suggesting that both inhibitory and excitatory receptors are present in the rat striatum. Besides reducing DA release, the adenosine analogs R-PIA and NECA, administered intraperitoneally, also decrease the synthesis of DA in rat striatal nerve terminals.[85]

D. SEROTONERGIC SYSTEM

Adenosine slightly decreases the K⁺-elicited release of [³H]-serotonin (5-HT) in the rat corpus striatum, an effect which is theophylline insensitive.[74] Studies on [³H]-5-HT release have also been performed in the rabbit hippocampus[86] where it was demonstrated that adenosine and its analogs clearly decrease the electrically evoked release of [³H]-5-HT through a xanthine-sensitive process, similar to that found in the same preparation in relation to the inhibition of the release of ACh[73] or NA.[80] In the rabbit caudate nucleus, adenosine analogs did not inhibit [³H]-5-HT release evoked either electrically or by K⁺ (25 mM).[87]

E. GLUTAMATE

The electrically evoked release of glutamate from the rat hippocampus, determined by mass spectrometry, is decreased by adenosine.[88] Adenosine has also been shown to inhibit the K⁺-evoked release of endogenously synthesized ³H-glutamate from the rat dentate gyrus.[89]

F. ASPARTATE

Decreases in endogenous aspartate release from the electrically stimulated rat hippocampus[88] and in [3H]-aspartate release from K^+-stimulated rat hippocampal slices[90] after application of adenosine have been reported. Adenosine and adenosine analogs inhibit glutamate-stimulated [3H]-aspartate release from cerebellar granule cells.[91]

G. GAMMA-AMINOBUTYRIC ACID

In the rat cerebral cortical slices, adenosine decreases the K^+ (36 mM)-evoked release of gamma-aminobutyric acid (GABA),[92] but in the K^+ (25 mM)-stimulated slices of the rat dentate gyrus, CADO had no effect.[89]

H. TRANSDUCING SYSTEM OF THE PRESYNAPTIC INHIBITORY ADENOSINE RECEPTOR AT THE CENTRAL NERVOUS SYSTEM

Studies to discover how adenosine inhibits synaptic transmission at the central synapses have been carried out by several groups, and a number of difficulties have been found in particular related to the presynaptic mechanisms involving changes in calcium conductance and the relation between the effects on calcium and the transducing systems operated. Most of the difficulties in explaining the mechanism of action of adenosine on transmitter release reside in aspects, not yet resolved, about the *modus faciendi* of calcium affecting the releasing site for transmitter release (see Section II.A.3).

In cultured mouse sensory neurones, CADO reduces the N-calcium current, an effect abolished by pertussis toxin treatment, but probably not related to modulation of adenylate cyclase activity.[34] Also, no clear correlation appears to exist between the presynaptic inhibitory effect of adenosine in the hippocampus and modulation of adenylate cyclase activity.[93]

Evidence that the presynaptic adenosine receptor in the central nervous system is coupled to a G protein was provided firstly in cerebellar neurons in primary culture where pertussis toxin prevents the inhibitory effect of R-PIA on glutamate release.[94] In contrast, *in vivo* pertussis toxin treatment failed to block the presynaptic effect, but not the postsynaptic effect of adenosine analogs on rat hippocampal slices.[93] The presynaptic effect is, however, antagonized by N-ethylmaleimide,[95] which also inactivates G proteins.

IV. EFFECTS OF ADENOSINE ON SPONTANEOUS NEUROTRANSMITTER RELEASE

Besides decreasing the evoked release of the transmitter, adenosine and adenosine analogs reduce the spontaneous release of ACh from motor nerve terminals, quantified as a decrease in the frequency of MEPPs, an effect observed either in rat,[7,8,12,37] mouse,[30] or frog[12,32,96,97] neuromuscular junctions. The hypothesis that the inhibitory effect of adenosine on spontaneous release of ACh at the neuromuscular junction is mediated by a mechanism different from that responsible for the inhibitory effect of the nucleoside on evoked release was put forward on the basis that at the neuromuscular junction, little or no correlation appears to exist between the inhibitory effects of adenosine on spontaneous release and evoked release of the transmitter when both effects are measured from the same endplate.[26]

At the nerve endings that innervate the smooth muscle, for example, in ileal synaptosomes, Reese and Cooper[50] reported that adenosine decreases the basal release of [3H]-ACh, but Shinozuka et al.,[51] using the same preparation, did not detect any effect. In the longitudinal muscle strip of the guinea pig ileum, the resting release of ACh, quantified by bioassay, was significantly reduced by adenosine.[45]

At the central nervous system, adenosine inhibits the resting release of ACh from rat cortical slices,[76] but in the rabbit hippocampus adenosine and its stable analogs failed to

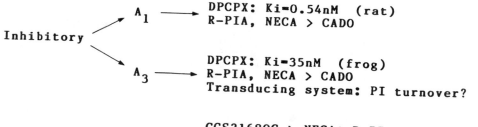

FIGURE 1. Presynaptic adenosine receptors on the motor nerve endings. PI: phosphoinositides; A. C.: adenylate cyclase. See text for the key of other abbreviations.

reduce the basal output of [³H]-ACh.[73] Also, the basal release of ³H-NA from rabbit hippocampus[80] or rat brain neocortex[81] is not affected by adenosine or its stable analogs.

V. CONCLUDING REMARKS

In summary, adenosine decreases the release of excitatory transmitters in many different preparations, either from the central or the peripheral nervous systems. In relation to the inhibitory transmitters, such as GABA, some controversy exists. Since the global effect of adenosine on neurons is inhibitory, it should be expected that the release of inhibitory transmitters is not affected, or if it is, it will be less affected than the release of excitatory neurotransmitters. It is possible that the inhibition of the release of inhibitory transmitters by adenosine in brain slices results from depression of the excitatory input to the inhibitory neurons.

Adenosine and its analogs also increase transmitter release in both the peripheral and the central nervous systems. The physiological significance of this effect remains to be established, since the manipulation of endogenous adenosine affects mainly the tonic inhibition exerted by the nucleoside.

The adenosine receptors involved in the inhibitory effects at the motor nerve endings are of the A_1 subtype in mammalian and A_3 subtype in amphibian. The excitatory effect is likely to be mediated by an A_2 adenosine receptor subtype. Figure 1 summarizes the characteristics of the presynaptic adenosine receptors detected with pharmacological studies on the motor nerve terminals.

The transducing systems for the A_2 adenosine receptors could be adenylate cyclase/cyclic AMP, and for the A_1 and A_3 receptors one does not know, but changes of Ca^{2+} conductance and/or phosphoinositide turnover have been postulated.

The effects of adenosine on K^+-induced release of neurotransmitters are small compared with the effects of adenosine on electrically evoked neurotransmitter release. This discrepancy might result from different mechanisms of transmitter release activated by K^+ and electrical stimulation, each of them having different sensitivities to adenosine.

ACKNOWLEDGMENT

My thanks to Dr. A. M. Sebastião for reading the manuscript.

REFERENCES

1. **Ribeiro, J. A. and Sebastião, A. M.,** On the role, inactivation and origin of endogenous adenosine at the frog neuromuscular junction, *J. Physiol. (London),* 384, 571, 1987.
2. **Sebastião, A. M. and Ribeiro, J. A.,** On the adenosine receptor and adenosine inactivation at the rat diaphragm neuromuscular junction, *Br. J. Pharmacol.,* 94, 109, 1988.
3. **Cunha, R. A., Sebastião, A. M., and Ribeiro, J. A.,** Parallel formation of IMP and adenosine from adenine nucleotides at the innervated frog sartorius muscle, *Br. J. Pharmacol.,* 98, 855P, 1989.
4. **Ribeiro, J. A., Correia-de-Sá, P., Cunha, R. A., Oliveira, J. C., and Sebastião, A. M.,** Role of adenosine at the neuromuscular junction: effects mediated by presynaptic inhibitory and excitatory adenosine receptors, in *Proc. 4th Int. Symp. on Adenosine and Adenine Nucleotides,* Imai, I., Ed., Elsevier, Tokyo, in press.
5. **Silinsky, E. M.,** On the association between transmitter secretion and the release of adenine nucleotides from mammalian motor nerve terminals, *J. Physiol. (London),* 247, 145, 1975.
6. **Ginsborg, B. L. and Hirst, G. D. S.,** Cyclic AMP, transmitter release and the effect of adenosine on neuromuscular transmission, *Nature New Biol.,* 232, 63, 1971.
7. **Ginsborg, B. L. and Hirst, G. D. S.,** The effect of adenosine on the release of the transmitter from the phrenic nerve of the rat, *J. Physiol. (London),* 224, 629, 1972.
8. **Ribeiro, J. A. and Dominguez, M. L.,** Mechanisms of depression of neuromuscular transmission by ATP and adenosine, *J. Physiol. (Paris),* 74, 491, 1978.
9. **Ribeiro, J. A. and Sebastião, A. M.,** Purinergic modulation of neurotransmitter release in the peripheral and central nervous systems, in *Presynaptic Regulation of Neurotransmitter Release,* Feigenbaum, J. and Hanani, M., Eds., Freund Publishing House, London, in press.
10. **Silinsky, E. M. and Hubbard, J. I.,** Release of ATP from rat motor nerve terminals, *Nature,* 243, 404, 1973.
11. **Ribeiro, J. A. and Walker, J.,** Action of adenosine triphosphate on endplate potentials recorded from muscle fibres of the rat-diaphragm and frog sartorius, *Br. J. Pharmacol.,* 49, 724, 1973.
12. **Ribeiro, J. A. and Walker, J.,** The effects of adenosine triphosphate and adenosine diphosphate on transmission at the rat and frog neuromuscular junctions, *Br. J. Pharmacol.,* 54, 213, 1975.
13. **Ribeiro, J. A.,** Purinergic modulation of transmitter release, *J. Theor. Biol.,* 80, 259, 1979.
14. **Sebastião, A. M. and Ribeiro, J. A.,** unpublished data.
15. **Burnstock, G.,** A basis for distinguishing two types of purinergic receptor, in *Cell Membrane Receptors for Drugs and Hormones: A Multidisciplinary Approach,* Straub, R. W. and Bollis, L., Eds., Raven Press, New York, 1978, 107.
16. **Chiou, L. C., Hong, S. J., and Chang, C. C.,** Does endogenous adenosine modulate the release of acetylcholine from motor nerve during single and repetitive stimulations in the mouse diaphragm?, *Jpn. J. Pharmacol.,* 44, 373, 1987.
17. **Miyamoto, M. D. and Breckenridge, B. McL.,** A cyclic adenosine monophosphate link in the catecholamine enhancement of transmitter release at the neuromuscular junction, *J. Gen. Physiol.,* 63, 609, 1974.
18. **Standaert, F. G., Dretchen, K. L., Skirboll, L. R., and Morgenroth, V. H., III,** A role of cyclic nucleotides in neuromuscular transmission, *J. Pharmacol. Exp. Ther.,* 199, 553, 1976.
19. **Silinsky, E. M.,** Evidence for specific adenosine receptors at cholinergic nerve endings, *Br. J. Pharmacol.,* 71, 191, 1980.
20. **Ribeiro, J. A. and Sebastião, A. M.,** On the type of receptor involved in the inhibitory action of adenosine at the neuromuscular junction, *Br. J. Pharmacol.,* 84, 911, 1985.
21. **Van Calker, D., Muller, M., and Hamprecht, B.,** Adenosine regulates via two different types of receptors, the accumulation of cyclic AMP in cultured brain cells, *J. Neurochem.,* 33, 999, 1979.
22. **Londos, C., Cooper, D. M. F., and Wolff, J.,** Subclasses of external adenosine receptors, *Proc. Natl. Acad. Sci., U.S.A.,* 77, 2551, 1980.
23. **Sebastião, A. M. and Ribeiro, J. A.,** 1,3,8- and 1,3,7-substituted xanthines: relative potency as adenosine receptor antagonists at the frog neuromuscular junction, *Br. J. Pharmacol.,* 96, 211, 1989.
24. **Bruns, R. F., Fergus, J. H., Badger, E. W., Bristol, J. A., Santay, L. A., Hartman, J. D., Hays, S. J., and Huang, C. C.,** Binding of the A_1-selective adenosine antagonist 8-cyclopentyl-1,3-dipropylxanthine to rat brain membranes, *Naunyn-Schmiedeberg's Arch. Pharmacol.,* 335, 59, 1987.
25. **Lohse, M. J., Klotz, K.-N., Lindenborn-Fotinos, J., Reddington, M., Schwabe, U., and Olsson, R. A.,** 8-Cyclopentyl-1,3-dipropylxanthine (DPCPX)—a selective high affinity antagonist radioligand for A_1 adenosine receptors, *Naunyn-Schmiedeberg's Arch. Pharmacol.,* 336, 204, 1987.
26. **Ribeiro, J. A. and Sebastião, A. M.,** Adenosine receptors and calcium: basis for proposing a third (A_3) adenosine receptor, *Prog. Neurobiol.,* 26, 179, 1986.

27. **Alexander, S. P. H., Kendall, D. A., and Hill, S. J.,** Differences in the adenosine receptors modulating inositol phosphates and cyclic AMP accumulation in mammalian cerebral cortex. *Br. J. Pharmacol.,* 98, 1241, 1989.

28. **Correia-de-Sá, P., Fougo, J. L., Sebastião, A. M., and Ribeiro, J. A.,** Evidence that phrenic nerve endings possess inhibitory and excitatory adenosine receptors, *Br. J. Pharmacol.,* 99, 227P, 1990.

29. **Sebastião, A. M., Stone T. W., and Ribeiro, J. A.,** The inhibitory adenosine receptor at the neuromuscular junction and hippocampus of the rat: antagonsim by 1,3,8-substituted xanthines, *Br. J. Pharmacol.,* 101, 453, 1990.

30. **Singh, Y. N., Dryden, W. F., and Chen, H.,** The inhibitory effects of some adenosine analogues on transmitter release at the mammalian neuromuscular junction, *Can. J. Physiol. Pharmacol.,* 64, 1446, 1986.

31. **Barry, S. R.,** Adenosine depresses spontaneous transmitter release from motor nerve terminals by acting at an A_1-like receptor, *Life Sci.,* 46, 1389, 1990.

32. **Barry, S. R.,** Dual effects of theophylline on spontaneous transmitter release from motor nerve terminals, *J. Neurosci.,* 8, 4427, 1988.

33. **Van der Kloot, W.,** The kinetics of quantal release during endplate currents at the frog neuromuscular junction, *J. Physiol. (London),* 402, 605, 1988.

34. **Gross, R. A., MacDonald, R. L., and Ryan-Jastrow, T.,** 2-Chloroadenosine reduces the N calcium current of cultured mouse sensory neurones in a pertussis toxin-sensitive manner, *J. Physiol. (London),* 411, 585, 1989.

35. **Silinsky, E. M., Solsona, C. S., Hirsh, J. K., and Hunt, J. M.,** Calcium-dependent acetylcholine secretion: influence of adenosine, in *Adenosine Receptors in the Nervous System,* Ribeiro, J. A., Ed., Taylor & Francis, London, 1989, 141.

36. **Ribeiro, J. A.,** The decrease of neuromuscular transmission by adenosine depends on previous neuromuscular depression, *Arch. Int. Pharmacodyn.,* 255, 59, 1982.

37. **Buckle, P. J. and Spence, I.,** The actions of adenosine and some analogues on evoked and potassium stimulated release at skeletal and autonomic neuromuscular junctions, *Naunyn-Schmiedeberg's Arch. Pharmacol.,* 319, 130, 1982.

38. **Silinsky, E. M. and Hirsh, J. K.,** The effect of reduced temperature on the inhibitory action of adenosine and magnesium ion at frog motor nerve terminals, *Br. J. Pharmacol.,* 93, 839, 1988.

39. **Silinsky, E. M., Solsona, C., and Hirsh, J. K.,** Pertussis toxin prevents the inhibitory effect of adenosine and unmasks adenosine-induced excitation of mammalian motor nerve endings, *Br. J. Pharmacol.,* 97, 16, 1989.

40. **Sebastião, A. M. and Ribeiro, J. A.,** Interactions between adenosine and phorbol esters or lithium at the frog neuromuscular junction. *Br. J. Pharmacol.,* 100, 55, 1990.

41. **Goldberg, A. L. and Singer, J. J.,** Evidence for a role of cyclic AMP in neuromuscular transmission, *Proc. Natl. Acad. Sci., U.S.A.,* 64, 134, 1969.

42. **Wilson, D. F.,** The effects of dibutyryl cyclic adenosine 3',5'-monophosphate, theophylline and aminophylline on neuromuscular transmission in the rat, *J. Pharmacol. Exp. Ther.,* 188, 447, 1974.

43. **Israel, M., Lesbats, B., Manaranche, R., Marsal, J., Mastour-Franchon, P., and Meunier, F. M.,** Related changes in amounts of ACh and ATP in resting and active *Torpedo* nerve electroplaque synapses, *J. Neurochem.,* 28, 1259, 1977.

44. **Israel, M., Lesbats, B., Manaranche, R., Meunier, F. M., and Franchon, P.,** Retrograde inhibition of transmitter release by ATP, *J. Neurochem.,* 34, 923, 1980.

45. **Vizi, E. S. and Knoll, J.,** The inhibitory effect of adenosine and related nucleotides on the release of acetylcholine, *Neuroscience,* 1, 391, 1976.

46. **Hayashi, E., Mori, M., Yamada, S., and Kunitomo, M.,** Effects of purine compounds on cholinergic nerves. Specificity of adenosine and related compounds on acetylcholine release in electrically stimulated guinea pig ileum, *Eur. J. Pharmacol.,* 48, 297, 1978.

47. **Gustafsson, L., Hedqvist, P., Fredholm, B. B., and Lundgren, G.,** Inhibition of acetylcholine release in the guinea pig ileum by adenosine, *Acta Physiol. Scand.,* 104, 469, 1978.

48. **Sawynok, J. and Jhamandas, K. H.,** Inhibition of acetylcholine release from cholinergic nerves by adenosine, adenine nucleotides and morphine: antagonism by theophylline, *J. Pharmacol. Exp. Ther.,* 197, 379, 1976.

49. **Hayashi, E., Maeda, T., and Shinozuka, K.,** Adenosine and dipyridamole: actions and interactions on the contractile response of guinea-pig ileum to high frequency electrical field stimulation, *Br. J. Pharmacol.,* 84, 765, 1985.

50. **Reese, J. H. and Cooper, J. R.,** Modulation of the release of acetylcholine from ileal synaptosomes by adenosine and adenosine 5'-triphosphate, *J. Pharmacol. Exp. Ther.,* 223, 612, 1982.

51. **Shinozuka, K., Maeda, T., and Hayashi, E.,** Effects of adenosine on ^{45}Ca uptake and ^3H-acetylcholine release in synaptosomal preparation from guinea-pig ileum myenteric plexus, *Eur. J. Pharmacol.,* 113, 417, 1985.

52. **Hedqvist, P. and Fredholm, B. B.,** Effects of adenosine on adrenergic neurotransmission; prejunctional inhibition and postjunctional enhancement, *Naunyn-Schmiedeberg's Arch. Pharmacol.,* 293, 217, 1976.
53. **Clanachan, A. S., Johns, A., and Paton, D. M.,** Presynaptic inhibitory actions of adenine nucleotides and adenosine on neurotransmission in the rat vas deferens, *Neuroscience,* 2, 597, 1977.
54. **Wakade, A. R. and Wakade, T. D.,** Inhibition of noradrenaline release by adenosine, *J. Physiol. (London),* 282, 35, 1978.
55. **Kubo, T. and Su, C.,** Effects of adenosine on ^{3}H-norepinephrine release from perfused mesenteric arteries of SHR and renal hypertensive rats, *Eur. J. Pharmacol.,* 87, 349, 1983.
56. **Enero, M. A. and Saidman, B. Q.,** Possible feed-back inhibition of noradrenaline release by purine compounds, *Naunyn-Schmiedeberg's Arch. Pharmacol.,* 297, 39, 1977.
57. **Wiklund, N. P., Samuelson, U. E., and Brundin, J.,** Adenosine modulation of adrenergic neurotransmission in the human fallopian tube, *Eur. J. Pharmacol.,* 123, 11, 1986.
58. **Verhaeghe, R. H., Vanhoutte, P. M., and Shepherd, J. T.,** Inhibition of sympathetic neurotransmission in canine blood vessels by adenosine and adenine nucleotides, *Circ. Res.,* 40, 208, 1977.
59. **De Mey, J., Burnstock, G., and Vanhoutte, P. M.,** Modulation of the evoked release of noradrenaline in canine saphenous vein via presynaptic receptors for adenosine but not ATP, *Eur. J. Pharmacol.,* 55, 401, 1979.
60. **Su, C.,** Purinergic inhibition of adrenergic transmission in rabbit blood vessels, *J. Pharmacol. Exp. Ther.,* 204, 351, 1978.
61. **Kamikawa, Y., Cline, W. H., Jr, and Su, C.,** Diminished purinergic modulation of the vascular adrenergic neurotransmission in spontaneously hypertensive rats, *Eur. J. Pharmacol.,* 66, 347, 1980.
62. **Hedqvist, P. and Fredholm, B. B.,** Inhibitory effect of adenosine on adrenergic neuroeffector transmission in the rabbit heart, *Acta Physiol. Scand.,* 105, 120, 1979.
63. **Lokhandwala, M. F.,** Inhibition of cardiac sympathetic neurotransmission by adenosine, *Eur. J. Pharmacol.,* 60, 353, 1979.
64. **Mueller, A. L., Mosimann, W. F., and Weiner, N.,** Effects of adenosine on neurally mediated norepinephrine release from the cat spleen, *Eur. J. Pharmacol.,* 53, 329, 1979.
65. **Kumakura, K., Sasakawa, N., Oizumi, N., Kobayashi, K., and Kuroda, Y.,** Modulation of catecholamine release by ATP and its metabolites in isolated adrenal chromaffin cells: a study with cell bed closed perfusion system, *Adv. Biosci.,* 36, 79, 1982.
66. **Chern, Y-J., Herrera, M., Kao, L. S., and Westhead, E. W.,** Inhibitions of catecholamine release from bovine chromaffin cells by adenine nucleotides and adenosine, *J. Neurochem.,* 48, 1573, 1987.
67. **Henon, B. K. and McAfee, D. A.,** The ionic basis of adenosine receptor actions on post-ganglionic neurones in the rat, *J. Physiol. (London),* 336, 607, 1983.
68. **Henon, B. K. and McAfee, D. A.,** Modulation of calcium currents by adenosine receptors on mammalian sympathetic neurons, in *Regulatory Function of Adenosine,* Berne, R. M., Rall, T. W., and Rubio, R., Eds, Martinus Nijhoff, The Hague, 1983, 455.
69. **Siggins, G. R. and Schubert, P.,** Adenosine depression of hippocampal neurons in vitro: an intracellular study of dose-dependent actions on synaptic and membrane potentials, *Neurosci. Lett.,* 23, 55, 1981.
70. **Segal, M.,** Intracellular analysis of a postsynaptic action of adenosine in the rat hippocampus, *Eur. J. Pharmacol.,* 79, 193, 1982.
71. **Schubert, P. and Lee, K. S.** Non-synaptic modulation of repetitive firing by adenosine is antagonized by 4-aminopyridine in a rat hippocampal slice, *Neurosci. Lett.,* 67, 334, 1986.
72. **Dunwiddie, T. V.,** The physiological role of adenosine in the central nervous system, *Int. Rev. Neurobiol.,* 27, 63, 1985.
73. **Jackisch, R., Strittmatter, H., Kasakov, L., and Hertting, G.,** Endogenous adenosine as a modulator of hippocampal acetylcholine release, *Naunyn-Schmiedeberg's Arch. Pharmacol.,* 327, 319, 1984.
74. **Harms, H. H., Wardeh, G., and Mulder, A. H.,** Effects of adenosine on depolarization-induced release of various radiolabelled neurotransmitters from slices of rat corpus striatum, *Neuropharmacology,* 18, 577, 1979.
75. **Richardson, P. J., Brown, S. J., Bailes, E. M., and Luzio, J. P.,** Ectoenzymes control adenosine modulation of immunoisolated cholinergic synapses, *Nature,* 327, 232, 1987.
76. **Pedata, F., Antonelli, T., Lambertini, L., Beani, L., and Pepeu, G.,** Effect of adenosine, adenosine triphosphate, adenosine deaminase, dipyridamole and aminophylline on acetylcholine release from electrically-stimulated brain slices, *Neuropharmacology,* 22, 609, 1983.
77. **Murray, T. F., Blaker, W. D., Cheney, D. L., and Costa, E.,** Inhibition of acetylcholine turnover rate in rat hippocampus and cortex by intraventricular injection of adenosine analogs, *J. Pharmacol. Exp. Ther.,* 222, 550, 1982.
78. **Haubrich, D. R., Williams, M., and Yarbrough, G. G.,** 2-Chloroadenosine inhibits brain acetylcholine turnover in vivo, *Can. J. Physiol. Pharmacol.,* 59, 1196, 1981.
79. **Jonzon, B. and Fredholm, B. B.,** Adenosine receptor mediated inhibition of noradrenaline release from slices of the rat hippocampus, *Life Sci.,* 35, 1971, 1984.

80. **Jackisch, R., Fehr, R., and Hertting, G.,** Adenosine: an endogenous modulator of hippocampal nora-drenaline release, *Neuropharmacology,* 24, 499, 1985.
81. **Harms, H. H., Wardeh, G., and Mulder, A. H.,** Adenosine modulates depolarization-induced release of ^{3}H-noradrenaline from slices or rat brain neocortex, *Eur. J. Pharmacol.,* 49, 305, 1978.
82. **Ebstein, R. P. and Daly, J. W.,** Release of norepinephrine and dopamine from brain vesicular preparations: effects of adenosine analogues, *Cell. Mol. Neurobiol.,* 2, 193, 1982.
83. **Michaelis, M. L., Michaelis, E. K., and Myers, S. L.,** Adenosine modulation of synaptosomal dopamine release, *Life Sci.,* 24, 2083, 1979.
84. **Zetterstrom, T. and Fillenz, M.,** Adenosine agonists can both inhibit and enhance in vivo striatal dopamine release, *Eur. J. Pharmacol.,* 180, 137, 1990.
85. **Myers, S. and Pugsley, T. A.,** Decrease in rat striatal dopamine synthesis and metabolism in vivo by metabolically stable adenosine receptor agonists, *Brain Res.,* 375, 193, 1986.
86. **Feuerstein, T. J., Hertting, G., and Jackisch, R.,** Modulation of hippocampal serotonin (5-HT) release by endogenous adenosine, *Eur. J. Pharmacol.,* 107, 233, 1985.
87. **Feuerstein, T. J., Bar, K. I., and Lucking, C. H.,** Activation of A_1 adenosine receptors decreases the release of serotonin in the rabbit hippocampus, but not in the caudate nucleus, *Naunyn-Schmiedeberg's Arch. Pharmacol.,* 338, 664, 1988.
88. **Corradetti, R., Lo Conte, G., Moroni, F., Passani, M. B., and Pepeu, G.,** Adenosine decreases aspartate and glutamate release from rat hippocampal slices, *Eur. J. Pharmacol.,* 104, 19, 1984.
89. **Dolphin, A. C. and Archer, E. R.,** An adenosine agonist inhibits and a cyclic AMP analogue enhances the release of glutamate but not GABA from slices of rat dentate gyrus, *Neurosci. Lett.,* 43, 49, 1983.
90. **Bowker, H. M. and Chapman, A. G.,** Adenosine analogues. The temperature-dependence of the anti-convulsant effect and inhibition of ^{3}H-D-aspartate release, *Biochem. Pharmacol.,* 35, 2949, 1986.
91. **Drejer, J., Frandsen, A., Honoré, T., and Schousboe, A.,** Adenosine inhibits glutamate stimulated [^{3}H]D-aspartate release from cerebellar granule cells, *Neurochem. Int.,* 11, 77, 1987.
92. **Hollins, C. and Stone, T. W.,** Adenosine inhibition of γ-aminobutyric acid release from slices of rat cerebral cortex, *Br. J. Pharmacol.,* 69, 107, 1980.
93. **Fredholm, B. B., Fastbom, J., Dunér-Engstrom, M., Hu, P.-S., Van der Ploeg, I., and Dunwiddie, T. V.,** Mechanism(s) of inhibition of transmitter release by adenosine receptor activation, in *Adenosine Receptors in the Nervous System,* Ribeiro, J. A., Ed., Taylor & Francis, London, 1989, 123.
94. **Dolphin, A. C. and Prestwich, S. A.,** Pertussis toxin reverses adenosine inhibition of neuronal glutamate release, *Nature,* 316, 148, 1985.
95. **Dunér-Engstrom, M. and Fredholm, B. B.,** Evidence that prejunctional adenosine receptors regulating acetylcholine release from rat hippocampal slices are linked to an N-ethylmaleimide-sensitive G-protein, but not to adenylate cyclase or dihydropyridine-sensitive Ca^{2+}-channels, *Acta Physiol. Scand.,* 134, 119, 1988.
96. **Branisteanu, D. D., Haulica, I. D., Proca, B., and Nhue, B. G.,** Adenosine effects upon transmitter release parameters in the Mg^{2+}-paralyzed neuromuscular junction of frog, *Naunyn-Schmiedeberg's Arch. Pharmacol.,* 308, 273, 1979.
97. **Silinsky, E. M.,** On the mechanism by which adenosine receptor activation inhibits the release of acetyl-choline from motor nerve endings, *J. Physiol. (London),* 346, 243, 1984.

IV. Adenosine and Blood Flow in Vascular Beds

Chapter 13

EFFECT OF ADENOSINE ON THE CORONARY CIRCULATION

Michael G. Collis

TABLE OF CONTENTS

I. INTRODUCTION

The early studies of Anrep[1] and of Drury and Szent-Gyorgyi[2] established that adenosine is a potent coronary vasodilator. This vasodilator action has since been confirmed by numerous other investigators in most species, including man.[3] The potency of adenosine as a coronary vasodilator is illustrated in the guinea pig isolated heart,[4] where a detectable coronary vasodilation is evoked by 10^{-8} M of the purine, with maximal dilation occuring at 5×10^{-6} M (Figure 1). The dilator effect of adenosine is not homogeneous, being more pronounced on small coronary arteries and arterioles (less than 0.5 mm O. D.) than on the larger coronary arteries.[5,6] Submaximal infusion rates of adenosine administered *in vivo* preferentially dilate the subendocardium, whilst maximally effective concentrations increase endocardial and epicardial blood flow to a similar extent.[7,8] Intramyocardial small-vessel volume is not increased by adenosine infusion[5] and this may indicate a lack of vasodilator effect on precapillary sphincters.

II. MECHANISM OF ACTION

The vasodilator effect of adenosine on coronary vessels could potentially be mediated via an interaction of the purine with the endothelial cells or with the medial smooth-muscle cells of the coronary vessels. Furchgott[9] has shown that a number of vasoactive agents can release a diffusable substance from the vascular endothelium which can evoke vascular relaxation. It has been suggested that this substance, termed endothelial-derived relaxing factor (EDRF), is nitric oxide.[10] An involvement of EDRF in mediating a vasodilator response can be implied when removal of the endothelium abolishes the dilator response, though care must be taken to distinguish the effects of EDRF from the effects of other potential vasodilator substances from the endothelium, like prostacyclin. In pig, human, bovine, and monkey large coronary arteries, however, endothelial removal has no effect on adenosine-evoked vasodilator responses.[11-14] Removal of the endothelium from the dog coronary artery[15] has been shown to attenuate, but not to abolish, adenosine-induced relaxation. The situation may be different in the guinea pig heart, where intra-arterial adenosine can cause coronary dilation at concentrations which are insufficient to cross the endothelial layer and reach the vascular smooth muscle.[16] At present it is not clear whether this phenomenon in the guinea pig[15] is the result of a species difference or is due to an endothelial mediation of relaxation in coronary arterioles, which are only accessible to adenosine when it is administered intraluminally into the intact vascular system.

Since removal of the endothelium has little effect on the coronary vasoactivity of adenosine in most species, the dilator effect must result primarily from an interaction with the vascular smooth-muscle cells in the arterial wall. The major site of action of the purine on smooth muscle appears to be at a cell-surface receptor. Action at a cell surface has been implied because compounds that block nucleoside transport and consequently reduce access of adenosine into the cell enhance the vasoactivity of low concentrations of the purine,[17] and because the coronary vasodilator action of adenosine is retained when it is chemically linked to large stachyose molecules which prevent permeation into the cell.[18] The effects of adenosine at the cell-surface receptor are antagonized by alkylxanthines[12,17,19] such as theophylline and 8-phenyltheophylline; thus, the receptor involved can be classified as P_1 according to some of the criteria laid down by Burnstock.[20] Cell-surface adenosine receptors can be divided into two types. This subdivision was originally based on the observation of Van Calker et al.[21] that adenosine could either stimulate or inhibit adenylate cyclase activity in a brain cell culture. Adenosine was found to be more potent than its analog N^6-phenylisopropyladenosine (PIA) as a stimulant of the enzyme, whereas the reverse order of potency pertained for the inhibitory effect. The inhibitory receptor was tentatively named A_1 and the

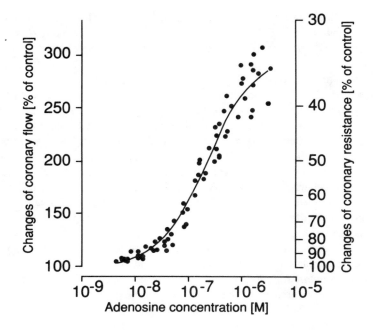

FIGURE 1.   Dose-response relation for changes of coronary flow or coronary resistance caused by exogenous adenosine. Perfusion pressure was maintained at 60 cmH$_2$O when changes in flow were monitored. Coronary flow was kept constant when perfusion pressure was monitored. During perfusion with constant volume, the initial flow was adjusted to yield a perfusion pressure of 60 cmH$_2$O. Resistance was then calculated. (From Schrader, J., Haddy, F. J., and Gerlach, E., *Pfluegers Arch., 369*, 1, 1977. With permission.)

stimulatory receptor A$_2$. The structure-activity relations for a number of adenosine analogs with the putative A$_1$ and A$_2$ receptors have been elucidated by their ability to displace radioactive ligands and by their effects on adenylate cyclase activity.[22-25] Based on the results of these studies, the A$_1$ receptor has been characterized by the high binding affinity or potency of the N^6-substituted adenosine analogs R-N^6-phenylisopropyladenosine (R-PIA), N^6-cyclohexyladenosine (CHA), and N^6-cyclopentyladenosine (CPA). The 2-substituted analog 2-chloradenosine and 5'-substituted 5'-N-cyclopropylcarboxamide (NCPCA) and 5'-N-ethylcarboxamideadenosine (NECA) have slightly lower potencies or binding affinities than R-PIA and CHA. Two adenosine analogs which show very low affinity for A$_1$ receptors are 2-phenylaminoadenosine (CV1808) and 2-[p-(2 carboxyethyl) phenethylamino -5'-N-ethylcarboxamide] (CGS 21680).[26] In addition, the A$_1$ receptor exhibits marked stereoselectivity for the diastereoisomers of PIA, R-PIA being 50 to 100 times more potent than S-PIA.[27] A different order of potency pertains at the A$_2$ receptor. The 5'-substituted analogs NECA, CGS 21680, and NCPCA are significantly more potent than the N^6-substituted analogs R-PIA, S-PIA, CPA, and CHA. 2-Chloroadenosine and CV1808 are intermediate in potency between the 5'- and the N^6-substituted analogs. In contrast to the stereoselectivity of the A$_1$ receptor for the R isomer of PIA, the difference in potency between the R and the S isomers at the A$_2$ receptor is only threefold to fivefold.

Antagonists with selectivity between the A$_1$ and the A$_2$ receptors have not been easy to find. However, 8-cyclopentyl-1,3,dipropylxanthine exhibits clear selectivity for A$_1$ receptors in all systems examined to date.[27,28] The triazoloquinazoline CGS 15943A has been suggested to be an A$_2$-selective antagonist;[29] however, studies in isolated tissues such as the guinea pig aorta and trachea, which possess A$_2$ receptors, have not confirmed this selectivity.[20]

Studies using adenosine analogs in isolated coronary arteries have demonstrated an order of agonist potency with NECA=CGS 21680 > 2-chloroadenosine=CV1808 > R-PIA=CHA=CPA. In addition, there has been little stereoselectivity shown for R-PIA. These studies, which have been performed on coronary vessels from rabbits, dogs, pigs, cows, and humans, indicate the presence of an A_2 receptor subtype.[11-14,30-32] Studies in which the coronary flow in rat isolated perfused heart has been used as an indicator of vascular resistance have also shown an order of potency of adenosine analogs which is indicative of an A_2 receptor and which correlates well with binding affinities at A_2 receptors in rat central nervous system tissue.[33] Because of the lack of an antagonist with clear A_2 selectivity, the characterization of coronary adenosine receptors with antagonists is in its infancy. CGS 15943A has been reported to have a very high affinity on dog isolated coronary arteries ($pA_2 = 10.8$).[34] However, the A_2 selectivity of this nonxanthine antagonist has not been confirmed and therefore results achieved with it must be regarded with caution.[28]

Since the consensus of opinion favors an A_2 adenosine receptor site on coronary smooth-muscle cells, an increase in cyclic adenosine 5'-monophosphate (cAMP) levels should be expected as a consequence of receptor activation. Published evidence on the involvement of cAMP in the vasedilator action of adenosine is, however, equivocal. Studies of cAMP levels in cultured vascular smooth-muscle cells and in homogenized brain microvessels have demonstrated an enhancement in response to low micromolar concentrations of adenosine or 2-chloroadenosine.[35,36] The significance of these elevations is difficult to assess since vascular contractile function was not assessed in these studies. In studies using large coronary vessels, attempts have been made to correlate alterations in cAMP levels with vascular relaxation. Herlihy et al.[37] could not detect an increase in cAMP levels in porcine coronary arteries exposed to concentrations of adenosine that evoked vascular relaxation. Some correlation between cAMP levels and vascular relaxation was reported in porcine and bovine coronary arteries by Kukovetz et al.[38] However, the increases in cAMP content were only noted at concentrations of adenosine in excess of $10^{-5}\ M$, whereas relaxation could be evoked by lower concentrations of the purine.

In a recent study using rabbit femoral artery, Cassis et al.[39] have demonstrated a good correlation between increased cAMP content and vasodilation evoked by 10^{-7} to $10^{-5}\ M$ adenosine. These increases in cAMP concentration preceded the relaxation and had returned to the control level before relaxation was maximal (after about 4 min). The transient nature of this increase in cAMP could explain the inability of Herlihy et al.[37] to detect it as they assayed for the nucleotide 4 to 18 min after the addition of adenosine. Thus, it appears that if cAMP is involved as a mediator of the vasodilator effect of low concentrations of adenosine, then its role is probably transient. It is tempting to speculate that the large and prolonged increases in cAMP concentration evoked by high concentrations of the purine[37,38] could be related to an additional intracellular action of the purine[40] or of its breakdown product, inosine (see later).

Recent studies have shown that the relaxation of bovine coronary arteries evoked by adenosine analogs are inhibited by *N*-ethylmaleimide and by cholera toxin, which implies a role for a GTP-dependent regulatory proteins (possibly Gs) in the response, since these inhibitory agents alkylate the agonist-receptor-Gs complex or ADP-ribosylate Gs, respectively.[41]

The effects of adenosine receptor activation on ion fluxes have not been studied in great detail in coronary arteries. Bovine coronary rings in calcium-free solution still partially relax to the purine, which suggests that its mechanism of action involves effects on both extra- and intracellular calcium fluxes.[42] Electrophysiological studies have shown a moderate hyperpolarization of large bovine coronary arteries to adenosine receptor activation.[42] In small, but not in large, canine coronary arteries, adenosine can block action potentials provoked by tetraethylammonium.[6]

Adenosine is rapidly metabolized in most cells to inosine. This deaminated metabolite also exhibits vasoactivity in coronary vessels, albeit with much reduced potency when compared to adenosine.[44] The site of action of inosine does not appear to be the cell-surface adenosine receptor, as its effects are not blocked by adenosine receptor antagonists.[44] The blockade of inosine-evoked responses by inhibitors of purine transport suggest that it acts at an intracellular site in the vascular smooth muscle.[40] The second-messenger systems involved in the coronary vasodilator effect of inosine are not known.

III. PHYSIOLOGICAL SIGNIFICANCE

The potent coronary dilator action of adenosine has led Berne[45] to propose that this nucleoside is the mediator of metabolically controlled coronary vasomotion. In the heart, as in all other vascular beds except the kidney, an increase in tissue metabolism or a reduction in blood supply leads to vasodilation, which is mediated via local mechanisms. Adenosine appears to be a good candidate for an endogenous regulator of coronary blood flow, since it is released from the myocardium when the oxygen supply/oxygen demand ratio decreases.[45,46] Thus, ischemia or hypoxia have been shown to elevate myocardial adenosine levels, indicating an increased production of the purine when oxygen supply is inadequate.[47] Furthermore, enhanced adenosine levels are seen when cardiac rate or force are increased, maneuvers that increase myocardial oxygen demand.[48] It has also been observed that cardiac adenosine levels change during the cardiac cycle, being increased in systole, an observation which is compatible with a role for adenosine in linking cardiac metabolism to coronary vasomotor tone.[49]

A considerable amount of research has been undertaken to challenge Berne's hypothesis. One experimental approach has involved the use of potentiating agents. If adenosine plays an important endogenous role in regulating coronary blood flow, then agents that potentiate its effects should be vasoactive and should potentiate metabolically linked vasodilation. The coronary vasodilator drug dipyridamole reduces the influx of adenosine into various cell types, particularly red blood cells and endothelial cells, by blocking the carrier-mediated diffusion of the purine.[50,51] This effect of the drug should enhance the interstitial concentration of adenosine and potentiate its vasodilator action. Consistent with this idea, dipyridamole has been shown to enhance the coronary vasodilator effects of exogenous adenosine.[52] There are discrepant results, however, concerning the potentiating effects of dipyridamole on coronary reactive hyperemia,[53] a response which might be mediated by endogenous adenosine. One factor which complicates the use of dipyridamole to test the adenosine hypothesis is that the drug can also inhibit the efflux of adenosine from myocardial cells.[54] Given this dual effect of dipyridamole, it becomes difficult to interpret its overall effect on a coronary vascular response.

Another approach to testing Berne's hypothesis has been the use of agents that destroy adenosine or block its receptors. Adenosine deaminase, a large-molecular weight enzyme derived from the calf intestinal mucosa, has been used to test the hypothesis that adenosine is an important regulator of coronary vascular tone. If adenosine has a vasodilator role, then it should be attenuated by adenosine deaminase, which converts it to inosine, which has low vasoactivity. It is assumed that the enzyme, administered intra-arterially, can penetrate capillary walls and reach the interstitial spaces where it can deaminate the adenosine.

In the heart, administration of adenosine deaminase into the coronary arteries does appear to allow access to the interstitial fluid, since it can be recovered from cardiac lymph.[55] Experiments using this enzyme have not revealed an effect on resting coronary flow nor on the autoregulatory response to reductions in perfusion pressure.[55-60] The reactive hyperemia following a short period of coronary artery occlusion or hypoxia is reduced, but not abolished by the enzyme.[57,61,62] Coronary vasodilation due to increased myocardial activity evoked by β-adrenoceptor stimulation is also reduced, but not abolished by adenosine deaminase.[63]

The adenosine receptor antagonist theophylline or its ethylene-diamine aminophylline have also been used to test the hypothesis that adenosine is involved in the regulation of vascular tone. Aminophylline has been shown to block the coronary and systemic effects of exogenous adenosine.[64] If the adenosine hypothesis of coronary flow control is correct, then the vasodilation evoked by hypoxia, by ischemia, and by increased myocardial work should be attenuated by aminophylline or theophylline. The results of a number of studies, however, do not give a consistent pattern. The xanthines have been shown to have no effect on resting coronary flow[65] or on the coronary dilator response to hypoxia.[66-68] The coronary vasodilator response to cardiac pacing at 150/min in the dog is reduced by aminophylline.[65] However, that evoked by noradrenaline or isoprenaline stimulation of cardiac function is not affected.[19,68] In man, doses of theophylline which block the coronary vasodilator response to dipyridamole have little effect on coronary blood flow at rest or on exercise.[69]

The effects of aminophylline or theophylline on coronary hyperemia following the release of an occlusion are somewhat less controversial. Most investigators have demonstrated some attenuation of the dilator response.[70-72] One group, however, has observed little effect.[73] In general, an effect of these xanthines is more pronounced after occlusions of long than of short duration.[74] The reduction in the reactive hyperemia is generally in the order of 20 to 25%.

It can be argued that theophylline (or aminophylline) is not the most appropriate agent with which to evaluate the role of adenosine in the control of coronary blood flow. Theophylline is of low potency as an adenosine antagonist ($pA_2 = 5$)[75] and therefore there is the potential for the antagonism to be overcome by high local concentrations of the purine. Theophylline also has a positive inotropic and chronotropic effect (probably due to inhibition of cAMP phosphodiesterase) which could cause compensatory increases in coronary flow.[75] There is also recent evidence to suggest that theophylline enhances adenosine release from the myocardium in the presence of β-adrenergic stimulation,[19] an effect that would obviously complicate the interpretation of the results obtained with this xanthine. The advent of more potent and specific adenosine receptor antagonists such as 8-phenyltheophylline,[53,76] which lacks an effect on phosphodiesterase enzymes[77] and which is not a positive inotropic agent,[75] has allowed a less equivocal investigation of the role of the purine in the control of blood flow in the coronary circulation.

Studies using 8-phenyltheophylline have revealed an attenuation of reactive coronary hyperemia following a 1-min period of coronary ligation in the anesthetized dog[78] (Figure 2). In the conscious dog, Bache et al.[79] have also demonstrated a reduction in coronary reactive hyperemia, but found no effect on coronary dilation during exercise. In neither of these studies did 8-phenyltheophylline alter resting coronary blood flow. A recent study in the dog has also shown that 8-phenyltheophylline abolishes coronary dilation evoked by systemic hypoxia when coronary flow is held constant. The physiological significance of this effect is not clear because of the unusual experimental conditions employed.[80]

IV. CONCLUSIONS

Adenosine interacts with an A_2 adenosine receptor located on coronary artery smooth muscle to evoke relaxation and vasodilation. A number of studies have been performed to investigate the role of adenosine as a physiological mediator of changes in coronary arterial tone. The results obtained with antagonists at the adenosine receptor and with adenosine deaminase have not supported a role for adenosine in the maintenance of resting coronary tone or in autoregulatory responses. There is, however, consistent evidence in favor of a role for the purine as one of the mediators of reactive hyperemia following occlusion and reperfusion of the coronary arteries. On the other hand, there is little evidence to support a role of adenosine in exercise-induced coronary vasodilation. There may be a role for the

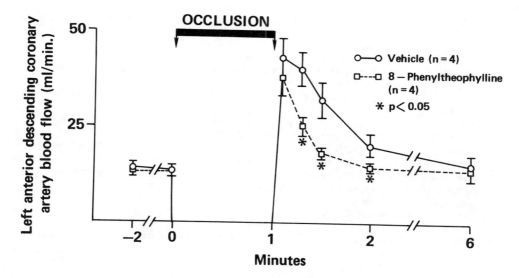

FIGURE 2.   Effect of 8-phenyltheophylline (5mg/kg, i.v.) on left anterior descending coronary artery flow, before and after occlusion of the artery for 1 min. Coronary perfusion pressure, heart rate, and segmental contractile function were not altered by 8-phenyltheophylline. (From Collis, M. G., in *Adenosine and Adenine Nucleotides*, Paton, D. M., Ed., Taylor & Francis, London, 1988. With permission.)

purine in vasodilation induced by hypoxia, but this requires further investigation. The majority of the studies whose results have suggested that adenosine may play a role in some aspects of coronary vascular control have been performed in the dog. Further studies, especially those using potent and selective adenosine antagonists in noncanine species, should be performed in the future in order to fully evaluate Berne's hypothesis that adenosine is an important physiological modulator of coronary vascular tone.

REFERENCES

1. **Anrep, G. V.,** The regulation of the coronary circulation, *Physiol. Rev.*, 6, 596, 1926.
2. **Drury, A. N. and Szent-Gyorgyi, A.,** The physiological activity of adenine compounds with special reference to their action upon the mammalian heart, *J. Physiol. (London)*, 68, 213, 1929.
3. **Watt, A. H., Penny, W. J., Singh, H., Routledge, P. A., and Henderson, A. H.,** Adenosine causes transient dilatation of coronary arteries in man, *Br. J. Clin. Pharmacol.*, 24, 665, 1987.
4. **Schrader, J., Haddy, F. J., and Gerlach, E.,** Release of adenosine, inosine and hypoxanthine from the isolated guinea pig heart during hypoxia, flow-autoregulation and reactive hyperemia, *Pfluegers Arch.*, 369, 1, 1977.
5. **Crystal, G. J., Downey, F. H., and Bashour, F. A.,** Small vessel and total coronary blood volume during intracoronary adenosine, *Am. J. Physiol.*, 241, H194, 1982.
6. **Harder, D. R., Belardinelli, L., Sperelakis, N., Rubio, R., and Berne, R. M.,** Differential effects of adenosine and nitroglycerin on the action potentials of large and small coronary arteries, *Circ. Res.*, 44, 176, 1979.
7. **Rembrant, J. C., Boyd, L. M., Watkinson, W. P., and Greenfield, J. C.,** Effect of adenosine on transmural myocardial blood flow distribution in the awake dog, *Am. J. Physiol.*, 239, H7, 1980.
8. **Bache, R. J. and Cobb, F. R.,** Effect of maximal coronary vasodilation on transmural myocardial perfusion during tachycardia in the awake dog, *Circ. Res.*, 41, 648, 1977.
9. **Furchgott, R. F.,** The role of endothelium in the responses of vascular smooth muscle to drugs, *Annu. Rev. Pharmacol. Toxicol.*, 24, 175, 1984.

10. **Palmer, R. M. J., Fennige, A. G., and Moncada, S.,** Nitric oxide release accounts for the biological activity of endothelium-derived relaxing factor, *Nature (London),* 327, 524, 1987.

11. **Ramagopal, M. V., Chitwood, J. R. W., and Mustafa, S. J.,** Evidence for an A_2 adenosine receptor in human coronary arteries, *Eur. J. Pharmacol.,* 151, 438, 1988.

12. **Mustafa, S. J. and Askar, A. O.,** Evidence suggesting an Ra-type adenosine receptor in bovine coronary arteries, *J. Pharmacol. Exp. Ther.,* 232, 49, 1985.

13. **Toda, N.,** Isolated human coronary arteries in response to vasoconstrictor substances, *Am. J. Physiol.,* 245, H937, 1983.

14. **King, A. D., Milavec, Krizman, M., and Muller-Schweinitzer, E.,** Characterization of the adenosine receptor in purine coronary arteries, *Br. J. Pharmacol.,* 100, 483, 1990.

15. **Rubanyi, G. and Vanhoutte, P. M.,** Endothelium-removal decreases relaxations of canine coronary arteries caused by beta-adrenergic agonists and adenosine, *J. Cardiovasc. Pharmacol.,* 7, 139, 1985.

16. **Newman, W. H., Becker, B. F., Heier, M., Nees, S., and Gerlach, E.,** Endothelium-mediated coronary dilatation by adenosine does not depend on endothelial adenylate cyclase activation: studies in isolated guinea-pig hearts, *Pfluegers Arch.,* 43, 1, 1988.

17. **Collis, M. G. and Brown, C. M.,** Adenosine relaxes the aorta by interacting with an A_2 receptor and an intracellular site, *Eur. J. Pharmacol.,* 96, 61, 1983.

18. **Olsson, R. A., Davis, C. J., Khouri, E. M., and Patterson, R. E.,** Evidence for an adenosine receptor on the surface of dog coronary myocytes, *Circ. Res.,* 39, 93, 1976.

19. **McKenzie, J. E., Steffen, R. P., and Haddy, F. J.,** Effect of theophylline on adenosine production in the canine myocardium, *Am. J. Physiol.,* 252, 204, 1987.

20. **Burnstock, G.,** A basis for distinguishing two types of purinergic receptor, in *Cell Membrane Receptors for Drugs and hormones: A Multi-Disciplinary Approach,* Straub, R. W. and Bolis, L., Eds., Raven Press, New York, 1978, 107.

21. **Van Calker, D., Muller, M., and Hamprecht, B.,** Adenosine regulates via two different types of receptors, the accumulation of cyclic AMP in cultured brain cells, *J. Neurochem.,* 33, 999, 1979.

22. **Trost, T. and Stock, K.,** Effects of adenosine derivatives on cAMP accumulation and lipolysis in rat adipocytes and on adenylate cyclase in adipocyte plasma membranes, *Naunyn-Schmiedeberg's Arch. Pharmacol.,* 299, 33, 1977.

23. **Bruns, R. F.,** Adenosine receptor activation in human fibroblasts: nucleoside agonists and antagonists, *Can. J. Physiol. Pharmacol.,* 58, 673, 1980.

24. **Bruns, R. F., Daly, J. W., and Snyder, S. H.,** Adenosine receptors in brain membranes: binding of N6-cyclohexyl[3H] adenosine and 1,3 diethyl-8-[3H] phenylxanthine, *Proc. Nat. Acad. Sci. U.S.A.,* 77, 5547, 1980.

25. **Londos, C., Cooper, D. M., and Wolff, J.,** Subclasses of external adenosine receptors, *Proc. Nat. Acad. Sci. U.S.A.,* 77, 2551, 1980.

26. **Hutchison, A. J., Webb, R. L., Oei, H. H., Ghai, G. R., Zimmerman, M. B., and Williams, M.,** CGS 21680C, an A_2 selective adenosine receptor agonist with preferential hypotensive activity, *J. Pharmacol. Exp. Ther.,* 251, 47, 1989.

27. **Collis, M. G., Stoggall, S. M., and Martin, F. M.,** Apparent affinity of 1,3-dipropyl-8-cyclopentylxanthine for adenosine A_1 and A_2 receptors in isolated tissues from guinea-pigs, *Br. J. Pharmacol.,* 97, 1274, 1989.

28. **Collis, M. G.,** Adenosine receptor sub-types in isolated tissues: antagonist studies, in *Purines in Cellular Signalling,* Jacobson, K. A., Daly, J. W., and Manganiello, V., Eds., Springer-Verlag, 1990, 48.

29. **Williams, M., Francis, J., Ghai, G., Braunwalder, A., Psychoyos, S., Stone, G. A., and Cash, W. D.,** Biochemical characterization of the triazoloquinazoline, CGS 15943, a novel non-xanthine adenosine antagonist, *J. Pharmacol. Exp. Ther.,* 241, 415, 1987.

30. **Kusachi, S., Thompson, R. D., and Olsson, R. A.,** Ligand selectivity of dog coronary adenosine receptor resembles that of adenylate cyclase stimulatory (Ra) receptors, *J. Pharmacol. Exp. Ther.,* 227, 316, 1983.

31. **Nakazawa, M. and Mustafa, S. J.,** Effects of adenosine and calcium entry blockers on 3,4-diaminopyridine-induced rythmic contractions in dog coronary artery, *Eur. J. Pharmacol.,* 149, 345, 1988.

32. **Odwarara, S., Kurahashi, K., Usui, H., Taniguchi, T., and Fujiwara, M.,** Relaxations of isolated rabbit coronary artery by purine derivatives: A_2 adenosine receptors, *J. Cardiovasc. Pharmacol.,* 8, 567, 1986.

33. **Hamilton, H. W., Taylor, M. D., Steffen, R. F., Haleen, S. J., and Bruns, R. F.,** Correlation of adenosine receptor affinities and cardiovascular activity, *Life Sci.,* 41, 2295, 1987.

34. **Ghai, G., Francis, J. E., Williams, M., Dotson, R. A., Hopkins, M. F., Cote, D. T., Goodman, F. R., and Zimmerman, M. B.,** Pharmacological characterization of CGS 15943A: a novel non-xanthine adenosine antagonist, *J. Pharmacol. Exp. Ther.,* 242, 784, 1987.

35. **Palmer, G. G. and Ghai, G.,** Adenosine receptors in capillaries and pia-arachnoid of rat cerebral cortex, *Eur. J. Pharmacol.,* 81, 129, 1982.

36. **Anand-Srivastava, M. B. and Franks, D. J.,** Stimulation of adenylate cyclase by adenosine and other agonists in mesenteric artery smooth muscle cells in culture, *Life Sci.,* 37, 857, 1985.

37. **Herlihy, J. T., Bockman, E. L., Berne, R. M., and Rubio, R.,** Adenosine relaxation of isolated vascular smooth muscle, *Am. J. Physiol.,* 230, 1239, 1976.
38. **Kukovetz, W. R., Poch, G., Holzmann, S., Wurm, A., and Rinner, I.,** Role of cyclic nucleotides in adenosine-mediated regulation of coronary flow, *Adv. Cyclic Nucleotide Res.,* 9, 397, 1987.
39. **Cassis, L. A., Loeb, A. L., and Peach, M. J.,** Mechanisms of adenosine- and ATP-induced relaxation in the rabbit femoral artery: role of the endothelium and cyclic nucleotides, in *Topics and Perspectives in Adenosine Research,* Gerlach, E. and Becker, B. F., Eds., Springer-Verlag, Heidelberg, 1986, 486.
40. **Collis, M. G., Palmer, D. B., and Baxter, G. S.,** Evidence that the intracellular effects of adenosine in the guinea-pig aorta are mediated by inosine, *Eur. J. Pharmacol.,* 121, 141, 1986.
41. **Sabouni, M. H., Cushing, D. J., and Mustafa, S. J.,** Adenosine receptor-mediated relaxation in coronary artery: evidence for a guanyl nucleotide-binding regulatory protein involvement, *J. Pharmacol. Exp. Ther.,* 251, 943, 1989.
42. **Ramagopal, M. V., Nakazawa, M., and Mustafa, S. J.,** Relaxing effects of adenosine in coronary artery in calcium-free medium, *Eur. J. Pharmacol.,* 159, 33, 1989.
43. **Sabouni, M. H., Hargittai, P. T., Lieberman, E. M., and Mustafa, S. J.,** Evidence for adenosine receptor-mediated hyperpolarization in coronary smooth muscle, *Am. J. Physiol.,* 257, H1750, 1989.
44. **Jones, C. E., Mayer, L. R., Smith, E. E., and Hurst, T. W.,** Relaxation of the isolated coronary artery by inosine: noninvolvement of the adenosine receptor, *J. Cardiovasc. Pharmacol.,* 3, 612, 1981.
45. **Berne, R. M.,** Cardiac nucleotides in hypoxia: possible role in regulation of coronary blood flow, *Am. J. Physiol.,* 204, 317, 1963.
46. **Berne, R. M.,** The role of adenosine in the regulation of coronary blood flow, *Circ. Res.,* 47, 807, 1980.
47. **Bacchus, A. N., Ely, S. W., Knabb, R. M., Rubio, R., and Berne, R. M.,** Adenosine and coronary blood flow in conscious dogs during normal physiologic stimuli, *Am. J. Physiol.,* 243, H628, 1982.
48. **Miller, W. L., Belardinelli, L., Bacchus, A., Foley, D. H., Rubio, R., and Berne, R. M.,** Canine myocardial adenosine and lactate production, oxygen consumption, and coronary flow during stellate ganglia stimulation, *Circ. Res.,* 45, 708, 1979.
49. **Thompson, C. I., Rubio, R., and Berne, R. M.,** Changes in adenosine and glycogen phosphorylase activity during the cardiac cycle, *Am. J. Physiol.,* 238, H389, 1980.
50. **Bunag, R. D., Douglas, C. R., Imai, S., and Berne, R. M.,** Influence of a pyrimidopyridine derivative on deamination of adenosine by blood, *Circ. Res.,* 15, 83, 1964.
51. **Pearson, J. D., Carleton, J. S., Hutchings, A., and Gordon, J. L.,** Uptake and metabolism of adenosine by pig aortic endothelial and smooth-muscle cell in culture, *Biochem. J.,* 170, 265, 1978.
52. **Hashimoto, K., Kumayura, S., and Tanemura, I.,** Mode of action of adenine, uridine and cytidine nucleotides and 2,6-bis (diethanol-amino)-4,8-dipiperidino-pyrimidino (5,4-d) pyrimidine on the coronary, renal and femoral arteries, *Arzneim. Forsch.,* 14, 1252, 1964.
53. **Collis, M. G.,** The vasodilator role of adenosine, *Pharmacol. Ther.,* 41, 143, 1989.
54. **Kubler, W., Spieckermann, P. G., and Bretschneider, H. J.,** Influence of dipyrimidamole (Persantin) on myocardial adenosine metabolism, *J. Mol. Cell Cardiol.,* 1, 23, 1970.
55. **Kroll, K. and Feigl, E. O.,** Adenosine is unimportant in controlling coronary blood flow in unstressed dog hearts, *Am. J. Physiol.,* 249, H1176, 1985.
56. **Hanley, F. L., Grattan, M. T., and Stevens, M. B.,** Role of adenosine in coronary autoregulation, *Am. J. Physiol.,* 250, H558, 1986.
57. **Saito, D., Steinhart, C. R., Nixon, D. G., and Olsson, R. A.,** Intracoronary adenosine deaminase reduces canine myocardial reactive hyperaemia, *Circ. Res.,* 49, 1262, 1981.
58. **Dole, W. P., Yamada, N., Bishop, V. S., and Olsson, R. A.,** Role of adenosine in coronary blood flow regulation after reductions in perfusion pressure, *Circ. Res.,* 56, 517, 1985.
59. **Gewirtz, H., Brautigan, D. L., Olsson, R. A., Brown, P., and Most, A.,** Role of adenosine in the maintenance of coronary vasodilation distal to a severe coronary stenosis, *Circ. Res.,* 53, 42, 1983.
60. **Gewirtz, H., Olsson, R. A., and Most, A. S.,** Role of adenosine in mediating coronary autoregulation under basal conditions, *Circulation,* 70 (Suppl. II), 14, 1984.
61. **Merrill, G. F., Downey, H. F., and Jones, C. E.,** Adenosine deaminase attenuates canine coronary vasodilation during systemic hypoxia, *Am. J. Physiol.,* 250, H579, 1986.
62. **Rubio, R., Knabb, R. M., Ely, S. W., and Berne, R. M.,** 50% Decrease in the coronary flow response to hypoxia caused by ''micro'' adenosine deaminase, *Fed. Proc., Fed. Am. Soc. Exp. Biol.,* 41, 1599, 1982.
63. **Merrill, G. F., Jones, C. E., and Downey, H. F.,** Adenosine deaminase attenuates norepinephrine-induced coronary functional hyperaemia, *Physiologist,* 28, 340, 1985.
64. **Afonso, S. and O'Brien, G. S.,** Inhibition of cardiovascular, metabolic and haemodynamic effects of adenosine by aminophylline, *Am. J. Physiol.,* 219, 1672, 1985.
65. **Lammerant, T. and Becsei, I.,** Inhibition of pacing-induced coronary dilation by aminophylline, *Cardiovasc. Res.,* 9, 532, 1975.

66. **Afonso, S., Ansfield, T. J., Berndt, B., and Rowe, G. G.,** Coronary vasodilator responses to hypoxia before and after aminophylline, *J. Physiol. (London),* 221, 589, 1972.
67. **Wadsworth, R. M.,** The effects of aminophylline on the increased myocardial blood flow produced by systemic hypoxia or by coronary artery occlusion, *Eur. J. Pharmacol.,* 20, 130, 1972.
68. **Jones, C. E., Hurst, T. W., and Randall, J. R.,** Effect of aminophylline on coronary functional hyperaemia and myocardial adenosine, *Am. J. Physiol.,* 243, H480, 1982.
69. **Edlund, A., Sollevi, A., and Wennmalm, A.,** The role of adenosine and prostacyclin in coronary flow regulation in healthy man, *Acta Physiol. Scand.,* 135, 39, 1989.
70. **Radford, M. J., McHale, P. A., Sadick, N., Schwartz, G. G., and Greenfield, J. C.,** Effect of aminophylline on coronary reactive and functional hyperaemic response in conscious dogs, *Cardiovasc. Res.,* 18, 377, 1984.
71. **Giles, R. W. and Wilken, D. E. L.,** Reactive hyperaemia in the dog heart: interrelations between adenosine, ATP, and aminophylline and the effect of indomethacin, *Cardiovasc. Res.,* 11, 113, 1977.
72. **Curnish, R. R., Berne, R. M., and Rubio, R.,** Effect of aminophylline on myocardial reactive hyperaemia, *Proc. Soc. Exp. Biol. Med.,* 141, 593, 1972.
73. **Bittar, N. and Pauly, T. J.,** Myocardial reactive hyperaemia responses in the dog after aminophylline and lidoflazine, *Am. J. Physiol.,* 220, 812, 1971.
74. **Schutz, W., Zimpfer, M., and Raberger, G.,** Effect of aminophylline on coronary reactive hyperaemia following brief and long occlusion periods, *Cardiovasc. Res.,* 11, 507, 1977.
75. **Collis, M. G., Keddie, J. R., and Torr, S. R.,** Evidence that the positive inotropic effects of the alkylxanthines are not due to adenosine receptor blockade, *Br. J. Pharmacol.,* 81, 401, 1984.
76. **Collis, M. G., Palmer, D. B., and Saville, V. L.,** Comparison of the potency of 8-phenyltheophylline as an antagonist at A1 and A2 adenosine receptors in atria and aorta from the guinea-pig, *J. Pharm. Pharmacol.,* 37, 278, 1985.
77. **Smellie, F. W., Davis, C. W., Daly, J. W., and Wells, J. N.,** Alkylxanthines: inhibition of adenosine-elicited accumulation of cyclic AMP in brain slices and of brain phosphodiesterase activity, *Life Sci.,* 24, 2475, 1979.
78. **Collis, M. G.,** Cardiac and renal effects of adenosine antagonists in vivo, in *Physiology and Pharmacology of Adenosine and Adenine Nucleotides,* Paton, D. M., Ed., Taylor & Francis, London, 1988, 259.
79. **Bache, R. J., Dai, X-Z., Schwartz, J. S., and Homans, D. C.,** Role of adenosine in coronary vasodilation during exercise, *Circ. Res.,* 62, 846, 1988.
80. **Wei, H. M., Kang, Y. H., and Merrill, G. F.,** Canine coronary vasodepressor responses to hypoxia are abolished by 8-phenyltheophylline, *Am. J. Physiol.,* 245, H1043, 1989.

Chapter 14

PURINERGIC CONTROL OF SKELETAL MUSCLE BLOOD FLOW

Mark W. Gorman, Sharon S. Kelley, Lana Kaiser, and Harvey V. Sparks, Jr.

TABLE OF CONTENTS

I. INTRODUCTION

The most striking aspect of the skeletal muscle circulation is the hyperemia that accompanies exercise, during which flow increases of up to tenfold are possible. The mechanism of exercise hyperemia has been an active area of research for more than 100 years. Because of the excellent match between muscle oxygen consumption and blood flow[1] and because exercise hyperemia persists in a chronically denervated muscle, the hypothesis arose very early that exercise hyperemia results from the release of a substance from the myocytes which causes vasodilation in proportion to the workload.[2] The thrust of research in this area has been to determine the identity of this substance.

Any potential mediator of exercise hyperemia must satisfy certain criteria:

1.   When infused intra-arterially, it must be capable of causing vasodilation equivalent to that during exercise.
2.   Either the mediator or the enzymes necessary for its formation must be present in skeletal muscle.
3.   A mechanism must be present for its rapid removal following cessation of exercise.
4.   During exercise, its concentration and time course at the resistance vessels must be sufficient to cause the observed vasodilation.
5.   Inhibitors or potentiators of its actions must have the appropriate effects on exercise hyperemia.

Exercise hyperemia results from an orchestration of several mediators. For example, potassium ion and plasma osmolarity are both elevated to vasoactive levels at the onset of exercise, but not during steady-state exercise.[3] The mechanisms initiating exercise hyperemia are therefore not the same as those maintaining it. In this brief review we will concentrate on adenosine and ATP as potential mediators of steady-state exercise hyperemia.

II. ADENOSINE

A. VASODILATOR PROPERTIES

Adenosine has been known as a vasodilator in skeletal muscle since the experiments of Drury and Szent-Gyorgi in 1929.[4] It was proposed, but rejected, as a mediator of exercise hyperemia as early as 1935.[5] This rejection of adenosine was based on the observation that the vasodilator activity of reperfused venous blood persisted after standing for 20 to 30 min, during which time it was thought that all of the adenosine would be destroyed. Since we now know that formed elements in standing blood are capable of releasing adenosine as well as taking it up[6] and that most of the vasodilator activity of venous blood is due to changes in blood gases and pH,[7] this rejection of adenosine was premature. Interest in adenosine was rekindled following the proposals of Berne[8] and Gerlach et al.[9] that it is the link between metabolism and blood flow in the heart.

B. FORMATION

Although adenosine can be measured in acid extracts of rapidly frozen muscle, the overwhelming majority of this adenosine is bound to protein and not available for vasoregulation (see Chapter 5 of this volume). Thus, vasoactive adenosine must be newly synthesized. Adenosine is formed by the action of the enzyme 5′-nucleotidase on AMP. 5′-nucleotidase activity has been found in skeletal muscle of many species, so criterion number 2 above is satisfied at least superficially. However, it is also essential that the enzyme be in the right location and have the right kinetics. Much of the 5′-nucleotidase activity in skeletal muscle is in the form of a plasma membrane-bound ectoenzyme. It was therefore proposed that this enzyme acted upon intracellular AMP to simultaneously form adenosine and transport it into the interstitial fluid.[10,11] Strong evidence against this mechanism comes from the observation that α,β-methylene ADP (AOPCP), an inhibitor of ecto-5'-nucleotidase, does not reduce adenosine release during free-flow exercise hyperemia.[12] Because of similar observations in the heart[13] it is currently proposed that adenosine is formed in the myocyte cytosol and transported into the interstitium via a nucleoside carrier in the plasma membrane. Cytosolic forms of 5'-nucleotidase have been isolated from almost every organ, including rat skeletal muscle, with varying kinetics and preferences for AMP vs. IMP.[14-17]

An additional consideration is that 5'-nucleotidase must compete for AMP with AMP deaminase. The product of AMP deaminase, IMP, will also decrease adenosine formation since it competes for 5'-nucleotidase with AMP. In extracts from dog skeletal muscle, AMP deaminase activity is 300-fold higher than 5'-nucleotidase activity.[18] This ratio by itself is probably meaningless since the assay conditions do not reflect enzyme activities *in vivo*. The ratio is much lower in dog myocardium, and in skeletal muscle depends on fiber type.[18] In summary, it appears that skeletal muscle does possess the enzymatic machinery needed for adenosine formation, but the conditions for its formation are less favorable than in the heart.

The presence of ectonucleotidases on the outer membrane of myocytes and endothelial cells means that adenosine can also be produced from nucleotides released into the extracellular space. These nucleotides might come from nerves, formed elements, or myocytes. In this review we will treat this scenario as part of the ATP hypothesis. Thus, the adenosine hypothesis refers here to the situation in which adenosine is formed within the myocytes, transported into the interstitial fluid, and bound to vascular smooth-muscle receptors, causing vasodilation.

What is the stimulus for increased adenosine formation during exercise? The most viable hypothesis is that adenosine formation is stimulated by a decrease in the O_2 supply/demand ratio.[19] Since venous PO_2 decreases during exercise hyperemia even under free-flow conditions, this indicates a decreased tissue PO_2 and a decreased supply/demand ratio. Furthermore, reducing the flow to a well-perfused exercising muscle dramatically increases tissue adenosine content.[20,21] The precise chemical signal linking O_2 supply/demand ratio to adenosine formation remains uncertain. This could be as simple as an increase in substrate (AMP) concentration, but might also involve activation of 5'-nucleotidase by decreased ATP, increased ADP, decreased creatine phosphate, increased Mg^{++}, or decreased intracellular pH.[14-17]

C. DESTRUCTION

Any putative mediator of exercise hyperemia must have a mechanism for rapid removal, since the decrease in flow following cessation of contraction is quite rapid, especially under free-flow conditions. This criterion is easily satisfied for adenosine. Adenosine can be taken up by myocytes or endothelial cells and rephosphorylated, deaminated to nonvasoactive inosine by interstitial adenosine deaminase (ADA), or simply washed out by blood flow. Both experimental studies and model simulations suggest that flow washout by itself is too slow, and that cellular uptake is the major route of adenosine removal.[22,26]

D. ADENOSINE CONCENTRATION DURING EXERCISE

Is adenosine released from exercising muscle? This question has been examined over the last 25 years in increasingly physiological preparations. Due to the difficulty in measuring adenosine, early studies concentrated on measurements of tissue adenosine content. It was first established that adenosine content increases during contractions under conditions of either ischemia or low constant flow.[20,21,23,24] During free-flow exercise, tissue adenosine does not increase in dog anterior calf muscles,[21] but does increase in dog gracilis.[25] Since most tissue adenosine is bound to protein, plasma adenosine concentration is a better index of interstitial adenosine than tissue adenosine content. Adenosine *release* (flow times venous-arterial concentration difference) increases during free-flow exercise in both dog gracilis and hindlimb muscles.[12,26,27] Mathematical modeling suggests that venous adenosine *concentration* is a better index of interstitial concentration than is adenosine release, assuming that arterial adenosine concentration is relatively constant.[27] In the dog hindlimb, venous adenosine concentration increases during 6-Hz exercise, but not during 3-Hz exercise.[12,27] In dog gracilis, venous adenosine increases during 4-Hz exercise with high constant flow.[26] We can therefore conclude that interstitial adenosine concentration increases under at least some exercise conditions.

E. IS ADENOSINE RESPONSIBLE FOR EXERCISE HYPEREMIA?

A more important question is whether or not the increased adenosine concentration is sufficient to account for steady-state exercise hyperemia. This is the most difficult criterion for adenosine or any other putative mediator to meet. It is also the most difficult question experimentally. Approaches so far have included treatment of muscles with adenosine antagonists, ADA, or potentiators of adenosine action.

1. Receptor Antagonists

Several studies have used the adenosine receptor antagonists aminophylline, theophylline, or 8-phenyltheophylline, which should decrease exercise vasodilation to the extent that adenosine is responsible. The results obtained depend on exercise intensity and on the perfusion conditions. Tabaie et al.[28] found decreased vasodilation with theophylline in dog gracilis, but under restricted flow conditions. In the same muscle with free flow and 0.75- to 5-Hz stimulation for 90 s, Honig and Frierson[29] found no effect with theophylline. They attribute the earlier result of Tabaie et al.[28] to the use of norepinephrine to eliminate vasodilation caused by theophylline. Thompson et al.[27] found no effect of aminophylline during 30 min of free-flow, 3-Hz contractions in dog hindlimb. Koch et al.[30] have reported a similar negative result with aminophylline and with 8-phenyltheophylline in conscious dogs during 10 to 15 min of treadmill exercise. One possible explanation for these negative findings is that adenosine receptor competitive antagonists can be overwhelmed by increased adenosine concentration during exercise, or by a stimulatory effect of the adenosine antagonist on adenosine release.[31] Thompson et al.[27] ruled out both of these possibilities by demonstrating that aminophylline blocked the response to adenosine infused during 3-Hz exercise, and that adenosine release rate and venous concentration were unchanged by aminophylline.

Similar studies have been carried out in microcirculatory preparations, again with varying results depending on species and exercise conditions. Theophylline reduced postexercise vasodilation in hamster cremaster muscles stimulated at 5 Hz,[32] but aminophylline was ineffective in rat cremaster stimulated at 2 Hz.[33] In the latter study, 2-Hz contractions produced a vasodilation approximating the maximum achievable with exogenous adenosine.

2. Adenosine Deaminase

Pretreatment with ADA should also reduce exercise hyperemia if adenosine is responsible. Applied topically to the hamster cremaster, ADA reduced estimated postexercise flow by 22% following 2-Hz stimulation and by 44% following 10-Hz stimulation.[32] In dog gracilis, ADA infusion increased vascular resistance during free-flow, 4-Hz contractions by 11%.[34] During moderate treadmill exercise in rats, however, Klabunde et al.[35] found no effect of ADA on skeletal muscle flow. Again, there appears to be a correlation between the effectiveness of ADA and exercise intensity.

3. Potentiation of Adenosine

Potentiators of adenosine action have also been used to test the adenosine hypothesis.

Dipyridamole inhibits adenosine uptake and erythro-9-(2-hydroxy-3-nonyl) adenine hydrochloride (EHNA) inhibits ADA. Both agents should increase interstitial fluid adenosine concentration. Because the effects of adenosine will be magnified, a positive result with these agents indicates that adenosine is present, but does not necessarily mean that it is contributing to vasodilation under normal conditions. Dipyridamole and EHNA increase postcontraction vasodilation following maximal tetanic contractions.[36] However, dipyridamole is without effect when tension is limited to 20% of maximum, unless arterial inflow is restricted.[37] Honig and Frierson[29] also saw no effect of dipyridamole on twitch exercise in dog gracilis. Laughlin et al.[38] examined the effect of dipyridamole on forelimb muscle blood flow during treadmill exercise in miniature swine. During exercise at 70% maximum VO_2, dipyridamole decreased resistance only in heart, diaphragm, and the medial head of the triceps muscle. During maximal VO_2 exercise, dipyridamole lowered arterial pressure and decreased vascular resistance in all muscles studied. They concluded that only slow-twitch oxidative skeletal muscle fibers and diaphragm produce adenosine when adequately perfused, but all muscles are capable of producing adenosine during intense exercise associated with flow restriction.

4. Adenosine Desensitization

Hester et al.[39] employed an unusual test of the adenosine hypothesis. A very high concentration of adenosine was continuously infused in the dog gracilis muscle. This initially increased blood flow sevenfold, but after 1 to 3 h the flow returned to control levels. At that time, during continued adenosine infusion, free-flow exercise hyperemia during 5-s periods of contraction was normal. Presumably, any adenosine released by the muscle would be dwarfed by the extremely high exogenous adenosine concentration. This result is a severe challenge to the adenosine hypothesis for the initiation of exercise hyperemia. Whether or not the result can be extended to steady-state exercise is unknown.

5. Conclusions

The pattern that emerges from these varied tests of the adenosine hypothesis is that adenosine probably does contribute to steady-state exercise hyperemia when flow is restricted, or under free-flow conditions when the exercise level is intense. During submaximal free-flow exercise, however, adenosine does not appear to play a role. It should be emphasized that most experimental preparations for studying exercise hyperemia involve exercise intensities and patterns that are unphysiologic. Usually, electrical stimulation is used to induce maximal isometric contractions, either twitch or tetanic. Such voluntary contractions are rare, and it would be even less likely to repeat them voluntarily at 1 to 10 Hz for extended periods. With this in mind, the three treadmill exercise studies cited earlier assume increased importance.[30,35,38] All three studies indicated little or no role for adenosine in voluntary exercise hyperemia.

III. ADENOSINE TRIPHOSPHATE (ATP)

A. BACKGROUND

ATP is an even more potent vasodilator than adenosine in skeletal muscle, and has been suggested as the mediator of exercise hyperemia.[40] It is present in high concentration within all cell types and can be rapidly degraded via plasma membrane nucleotidases. Thus, it satisfies the first three of the criteria for a metabolic mediator. ATP has been measured in the venous effluent of exercising muscle in man, possibly in concentratons sufficient to account for exercise hyperemia.[41] However, evaluating the true role of ATP in exercise hyperemia has been hampered by the lack of a clear understanding on the source of the ATP found in venous effluent. This ATP could come from myocytes, nerves, or formed elements of blood.

B. MECHANISM OF ACTION

It has become apparent in recent years that the vasodilator effects of ATP may be entirely mediated by endothelial cells, and that the direct effect of ATP on vascular smooth muscle is vasoconstriction.[42,43] This is both good news and bad news for proponents of the ATP hypothesis of exercise hyperemia. It is good news in that production of ATP by formed elements, long cited as a criticism of ATP measurements from exercising muscle, is converted from a weakness to a strength. Intravascular ATP should be a much more effective vasodilator than interstitial ATP. Endothelial-dependent vasodilation is bad news for the ATP hypothesis in that ATP must now be considered a double-edged sword, capable of causing net vaso-constriction if it is released from either myocytes or nerves on the adventitial surface of blood vessels. Thus the source of ATP from exercising muscle has become an even more important consideration than previously thought.

One form of the ATP hypothesis that can probably be ruled out is that extracellular ATP released during exercise is degraded by ectonucleotidases to AMP or adenosine, which then cause the vasodilation. Since methylxanthines antagonize responses to both adenosine and AMP, the negative results with these agents argue strongly against this hypothesis as well as the adenosine hypothesis.[27,29,30] If ATP is causing exercise vasodilation, it must be doing so as ATP or ADP rather than as AMP or adenosine.

C. PHARMACOLOGY

The methylxanthine adenosine antagonists are much less effective against ATP and ADP. Based on this and on differences in agonist potencies, Burnstock[44] proposed a clas-sification for purine receptors. P_1 receptors are antagonized by methylxanthines and are most sensitive to adenosine and least sensitive to ATP. P_2 receptors are not antagonized by methylxanthines and are most sensitive to ATP and least sensitive to adenosine. The P_2 receptors have been subdivided to account for endothelial-dependent ATP dilation.[43] P_{2x} receptors are located on vascular smooth muscle and cause vasoconstriction, while P_{2y} receptors on endothelial cells cause vasodilation. P_{2y} receptors also differ from P_{2x} receptors in their sensitivity to different ATP analogs.[43]

An ideal pharmacologic test of the ATP hypothesis in exercise hyperemia would be to administer a selective antagonist of P_{2y} receptors. Short of this, a selective P_2 agonist would be very useful. The lack of selective antagonists has made it very difficult to evaluate this hypothesis. Recently, some new candidates have arisen as selective P_2 antagonists: anta-zoline, arylazido aminopropionyl ATP (ANAPP$_3$), and ATP desensitization. Antazoline inhibited the dilator responses to ATP and to hypothalamic stimulation in rabbit hindlimb.[45] It also inhibited the responses to histamine and adenosine, but separate experiments with either antihistamines or aminophylline showed no effect on vasodilation induced by hypo-thalamic stimulation. Antazoline might therefore be useful as a P_2 antagonist. ATP desen-sitization refers to the tachyphylaxis induced by continuous ATP infusion. ANAPP$_3$ is a photoaffinity-labeled ATP molecule that is activated in the presence of ultraviolet or visible light to bind covalently with ATP receptors.[46,47] Nonphotolyzed ANAPP$_3$ competitively inhibits ATP-induced contractions in guineapig vas deferens[47] and ATP-induced dilation in isolated rabbit femoral arteries.[48] In rat aorta, photolyzed ANAPP$_3$ had no effect on ATP-induced relaxation when endothelium was intact, but inhibited the contractile response to ATP analogs when endothelium was removed.[49] These results suggest that photolyzed ANAPP$_3$ is a selective antagonist for contractile P_{2x} receptors, while nonphotolyzed ANAPP$_3$ antag-onizes both P_{2x} and P_{2y} receptors.

D. EXPERIMENTS WITH ATP ANTAGONISTS

1. Preparation

We have examined the three agents above in a pharmacologic test of the ATP hypothesis

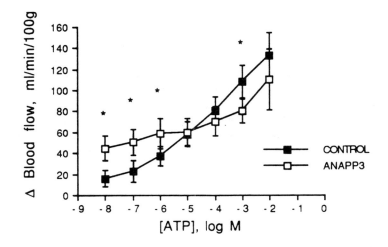

FIGURE 1. ATP dose-response curves before (■) and after (□) ANAPP$_3$. ATP was injected as a 0.3-ml bolus in the concentrations indicated. Δ Blood flow represents the maximum flow above baseline in response to each concentration. Values are mean ± SEM (n = 5, except at 10^{-5} M, where n = 4). * p <0.05 ANAPP$_3$ vs. control.

in exercising canine skeletal muscle. Dog gastrocnemius-plantaris muscles were vascularly isolated and perfused at constant pressure via a servo-controlled pump. Twitch contractions were elicited by supramaximal stimuli to the sciatic nerve at 1.5 to 6 Hz. Isometric force, arterial-venous oxygen content difference, and flow were measured continuously. The effectiveness of these putative P$_2$ antagonists was measured by injections or infusions of ATP and other vasodilators before and after treatment.

2. Antazoline and ATP Desensitization

Antazoline was infused intra-arterially at 4 mg/kg over 15 min. It significantly depressed the vasodilator responses to ATP, but also to adenosine and isoproterenol. Arterial infusions of ATP (ATP desensitization) reduced the vasodilator response to ATP by about 75%, but also reduced resting flow and the responses to adenosine and isoproterenol. Thus neither of these agents was a selective P$_2$ receptor antagonist in this preparation.

3. ANAPP$_3$

ANAPP$_3$ was infused in arterial concentrations of 20 to 130 μM. This dose enhanced the vasodilator response to low doses of ATP, but depressed the response to higher doses (Figure 1). ANAPP$_3$ did not reduce responsiveness to either adenosine or acetylcholine. Thus, ANAPP$_3$ appears to be a selective antagonist of P$_2$ receptors. When exercise was performed in the presence of ANAPP$_3$, we found that for a given oxygen consumption, flow was higher in the presence of ANAPP$_3$ (Figure 2). This suggests that ATP actually limits exercise vasodilation. When the effects of ANAPP$_3$ on ATP-induced dilation and on exercise hyperemia are taken together, a novel hypothesis emerges for the role of ATP in exercise hyperemia.

4. Conclusions

The ATP dose-response data suggest that the observed dilator response to exogenous ATP is really the algebraic sum of an endothelial-mediated (P$_{2y}$ receptor) vasodilation and a vascular smooth-muscle-mediated (P$_{2x}$ receptor) vasoconstriction. ANAPP$_3$ appears to shift the dose-response curve for P$_{2x}$ receptors far more than that of P$_{2y}$ receptors. Thus at low doses of ATP we observe an uncontested P$_{2y}$ vasodilation (higher flow). At higher concen-

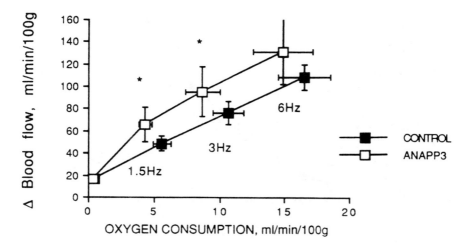

FIGURE 2.   Exercise hyperemia before (■) and after (□) ANAPP$_3$ treatment. Measurements were made during steady-state conditions at rest, and during 1.5-, 3-, and 6-Hz twitch stimulation. Values are means ± SEM. * Indicates a significant difference between oxygen consumption before and after ANAPP$_3$. None of the Δ blood flow changes were statistically significant.

trations of ATP, the ANAPP$_3$ blockade of P$_{2x}$ receptors is overcome, but not the P$_{2y}$-receptor blockade, resulting in lower flows relative to control.

Based on these data, we propose that ATP is released into the interstitial space, probably as a cotransmitter from either motor or sympathetic neurons, and possibly from skeletal muscle cells. ATP release from perivascular nerves causes a direct vascular smooth-muscle constrictor effect via P$_{2x}$ receptors. When ANAPP$_3$ is given, it antagonizes this direct vasoconstrictor effect, and so skeletal muscle blood flow is higher for a given oxygen consumption. We conclude from this particular series of experiments that ATP is unlikely to mediate exercise vasodilation. On the other hand, it may play an important role in limiting blood flow during exercise. Other experiments we have performed suggest that it may limit peak reactive hyperemia as well. At present we do not know whether these responses represent an adaptive effect with an unforseen benefit to the organism or if it simply reflects an inappropriate response to ATP.

IV. SUMMARY

The evidence presented here suggests that neither adenosine nor ATP is necessary for steady-state exercise hyperemia. There may be other situations, though, where adenosine and/or ATP contribute to skeletal muscle blood flow. Adenosine probably contributes under restricted flow conditions and during hypoxia. Both adenosine and ATP modulate α-adrenergic vasoconstriction by inhibiting norepinephrine release from sympathetic neurons.[50] During exercise, ATP released from nerves and/or myocytes may limit exercise hyperemia. ATP release from aggregating platelets may very well induce endothelial-dependent vasodilation. In cases of cell injury, leakage of intracellular ATP could cause either vasoconstriction or vasodilation, depending on the proximity of the source to vascular smooth muscle and endothelial cells.

ACKNOWLEDGMENTS

Supported by U.S. Public Health Service grants HL24232, HL01842, and HL25779. The authors thank Dr. Jack Barclay for his assistance with the ATP experiments.

REFERENCES

1. **Kramer, K., Obal, F., and Quensel, W.,** Untersuchungen uber den Muskelstoffwechseldes warmbluters. III. Mittilung. Die saverstoffaufnahmedes Muskels wahrend ryhthmischer Tatigkeit, *Pfluegers Arch. Gesamte Physiol., Menschen Tiere,* 241, 717, 1939.

2. **Gaskell, W. H.,** On the changes of the blood stream in muscle through stimulation of their nerves, *J. Anat.,* 11, 360, 1877.

3. **Shepherd, J. T.,** Circulation to skeletal muscle, in *Handbook of Physiology, Section 2, Volume III,* Shepherd, J. T. and Abboud, F. M., Eds., American Physiological Society, Bethesda, MD, 1983, 319.

4. **Drury, A. N. and Szent-Gyorgyi, A.,** The physiological activity of adenine compounds with especial reference to their action upon the mammalian heart, *J. Physiol. (London),* 68, 213, 1929.

5. **Anrep, G. V. and von Saalfeld, E.,** The blood flow through skeletal muscle in relation to its contraction, *J. Physiol. (London),* 85, 375, 1935.

6. **Klabunde, R. E. and Althouse, D. G.,** Adenosine metabolism in dog whole blood: effect of dipyridamole, *Life Sci.,* 28, 2631, 1981.

7. **Stowe, D. F., Owen, T. L., Anderson, D. K., Haddy, F. J., and Scott, J. B.,** Interaction of O_2 and CO_2 in sustained exercise hyperemia of canine skeletal muscle, *Am. J. Physiol.,* 229, 28, 1975.

8. **Berne, R. M.,** Cardiac nucleotides in hypoxia: possible role in regulation of coronary blood flow, *Am. J. Physiol.,* 204, 317, 1963.

9. **Gerlach, E., Deuticke, B., and Dreisbach, R. H.,** Der Nucleotid-Abbau im Herzmuskel bei Sauerstoffmangel und seine mogliche Bedeutung fur die Coronardurchblutung, *Naturwissenschaften,* 50, 228, 1963.

10. **Rubio, R., Berne, R. M., and Dobson, J. G.,** Sites of adenosine producton in cardiac and skeletal muscle, *Am. J. Physiol.,* 225, 938, 1973.

11. **Frick, G. P. and Lowenstein, J. M.,** Vectorial production of adenosine by 5'-nucleotidase in the perfused rat heart, *J. Biol. Chem.,* 253, 1240, 1978.

12. **Fuchs, B. D., Gorman, M. W., and Sparks, H. V.,** Adenosine release into venous plasma during free flow exercise, *Proc. Soc. Exp. Biol. Med.,* 181, 364, 1986.

13. **Schutz, W., Schrader, J., and Gerlach, E.,** Different sites of adenosine formation in the heart, *Am. J. Physiol.,* 240, H963, 1981.

14. **Collinson, A. R., Peuhkurinen, K. J., and Lowenstein, J. M.,** Regulation and function of 5'-nucleotidases, in *Topics and Perspectives in Adenosine Research,* Gerlach, E. and Becker, B. F., Eds., Springer-Verlag, Berlin, 1987, 133.

15. **Newby, A. C., Worku, Y., and Meghji, P.,** Critical evaluation of the role of ecto- and cytosolic 5'-nucleotidase in adenosine formation, in *Topics and Perspectives in Adenosine Research,* Gerlach, E. and Becker, B. F., Eds., Springer-Verlag, Berlin, 1987, 155.

16. **Itoh, R. and Ozasa, H.,** Regulation of rat heart cytosol 5'-nucleotidase by adenylate energy charge, *Biochem. J.,* 235, 847, 1986.

17. **Spychala, J., Madrid-Marina, V., Nowak, P. J., and Fox, I. H.,** AMP and IMP dephosphorylation by soluble high and low Km 5'-nucleotidase, *Am. J. Physiol.,* 256, E386, 1989.

18. **Bockman, E. L. and McKenzie, J. E.,** Tissue adenosine content in active soleus and gracilis muscles of cats, *Am. J. Physiol.,* 244, H552, 1983.

19. **Sparks, H. V. and Bardenheuer, H.,** Regulation of adenosine formation by the heart, *Circ. Res.,* 58, 193, 1986.

20. **Bockman, E. L., Berne, R. M., and Rubio, R.,** Release of adenosine and lack of release of ATP from contracting skeletal muscle, *Pfluegers Arch.,* 355, 229, 1975.

21. **Phair, R. D. and Sparks, H. V.,** Adenosine content of skeletal muscle during active hyperemia and ischemic contraction, *Am. J. Physiol.,* 237, H1, 1979.

22. **Mohrman, D. E. and Sparks, H. V.,** Resistance and venous oxygen dynamics during sinusoidal exercise of dog skeletal muscle, *Circ. Res.,* 33, 337, 1973.

23. **Dobson, J. G., Rubio, R., and Berne, R. M.,** Role of adenine nucleotides, adenosine, and inorganic phosphate in the regulation of skeletal muscle blood flow, *Circ. Res.,* 29, 375, 1971.

24. **Belloni, F. L., Phair, R. D., and Sparks, H. V.,** The role of adenosine in prolonged vasodilation following flow-restricted exercise of canine skeletal muscle, *Circ. Res.,* 44, 759, 1979.

25. **Steffen, R. P., McKenzie, J. E., Bockman, E. L., and Haddy, F. J.,** Changes in dog gracilis muscle adenosine during exercise and acetate infusion, *Am. J. Physiol.,* 244, H387, 1983.

26. **Ballard, H. J., Cotterrell, D., and Karim, F.,** Appearance of adenosine in venous blood from the contracting gracilis muscle and its role in vasodilation in the dog, *J. Physiol. (London),* 387, 401, 1987.

27. **Thompson, L. P., Gorman, M. W., and Sparks, H. V.,** Aminophylline and interstitial adenosine during sustained exercise hyperemia, *Am. J. Physiol.,* 251, H1232, 1986.

28. **Tabaie, H. M., Scott, J. B., and Haddy, F. J.,** Reduction of exercise dilation by theophylline, *Proc. Soc. Exp. Biol. Med.,* 154, 93, 1977.

29. **Honig, C. R. and Frierson, J. L.,** Role of adenosine in exercise vasodilation of dog gracilis muscle, *Am. J. Physiol.,* 238, 703, 1980.

30. **Koch, L. G., Britton, S. L., and Metting, P. J.,** Adenosine is not essential for exercise hyperaemia in the hindlimb in conscious dogs, *J. Physiol. (London),* 429, 63, 1990.

31. **McKenzie, J. E., Steffen, R. P., and Haddy, F. J.,** Effect of theophylline on adenosine production in the canine myocardium, *Am. J. Physiol.,* 252, H204, 1987.

32. **Proctor, K. G.,** Reduction of contraction-induced arteriolar vasodilation by adenosine deaminase or theophylline, *Am. J. Physiol.,* 247, H195, 1984.

33. **Mohrman, D. E. and Heller, L. J.,** Effect of aminophylline on adenosine and exercise dilation of rat cremaster arterioles, *Am. J. Physiol.,* 246, H592, 1984.

34. **Karim, F. and Goonewardene, I. P.,** The effect of adenosine deaminase on free-flow exercise hyperaemia, *J. Vasc. Med. Biol.,* 1 (Abstr.), 180, 1989.

35. **Klabunde, R. E., Laughlin, M. H., and Armstrong, R. B.,** Systemic adenosine deaminase administration does not reduce active hyperemia in running rats, *J. Appl. Physiol.,* 64, 108, 1988.

36. **Kille, J. M. and Klabunde, R. E.,** Adenosine as a mediator of postcontraction hyperemia in dog gracilis muscle, *Am. J. Physiol.,* 246, 274, 1984.

37. **Klabunde, R. E.,** Conditions for dipyridamole potentiation of skeletal muscle active hyperemia, *Am. J. Physiol.,* 250, H62, 1986.

38. **Laughlin, M. H., Klabunde, R. E., Delp, M. D., and Armstrong, R. B.,** Effects of dipyridamole on muscle blood flow in exercising miniature swine, *Am. J. Physiol.,* 257, H1507, 1989.

39. **Hester, R. L., Guyton, A. C., and Barber, B. J.,** Reactive and exercise hyperemia during high levels of adenosine infusion, *Am. J. Physiol.,* 243, H181, 1982.

40. **Forrester, T.,** Release of adenosine triphosphate from active skeletal muscle, *J. Physiol. (London),* 186, 107P, 1966.

41. **Forrester, T.,** An estimate of adenosine triphosphate release into the venous effluent from exercising human forearm muscle, *J. Physiol. (London),* 224, 611, 1972.

42. **DeMey, J. G., Claeys, M., and Vanhoutte, P. M.,** Endothelium-dependent inhibitory effects of acetylcholine, adenosine triphosphate, thrombin and arachidonic acid in the canine femoral artery, *J. Pharmacol. Exp. Ther.,* 222, 166, 1982.

43. **Burnstock, G. and Kennedy, C.,** A dual function for adenosine 5'-triphosphate in the regulaton of vascular tone, *Circ. Res.,* 58, 319, 1986.

44. **Burnstock, G.,** A basis for distinguishing two types of purinergic receptor, in *Cell Membrane Receptors for Drugs and Hormones: A Multidisciplinary Approach,* Straub, R. W. and Bolis, L., Eds., Raven Press, New York, 1978, 107.

45. **Shimada, S. G. and Stitt, J. T.,** An analysis of the purinergic component of active muscle vasodilation obtained by electrical stimulation of the hypothalamus in rabbits, *Br. J. Pharmacol.,* 83, 577, 1984.

46. **Fedan, J. S., Hogaboom, G. K., O'Donnell, J. P., and Westfall, D. P.,** Use of photoaffinity labels as P_2-purinoceptor antagonists, in *Methods In Pharmacology,* Vol. 6, Paton, D. M., Ed., Plenum Press, New York, 1985, 279.

47. **Hogaboom, G. K., O'Donnell, J. P., and Fedan, J. S.,** Purinergic receptors: photoaffinity analog of adenosine triphosphate is a specific adenosine triphosphate antagonist, *Science,* 208, 1273, 1980.

48. **Cassis, L. A., Loeb, A. L., and Peach, M. J.,** Mechanisms of adenosine- and ATP-induced relaxation in rabbit femoral artery: role of the endothelium and cyclic nucleotides, in *Topics and Perspectives in Adenosine Research,* Gerlach, E. and Becker, B. F., Eds., Springer-Verlag, Berlin, 1987, 486.

49. **White, T. D., Chaudry, A., Vohra, M. M., Webb, D., and Leslie, R. A.,** Characteristics of P_2 (nucleotide) receptors mediating contraction and relaxation of rat aortic strips: possible physiological relevance, *Eur. J. Pharmacol.,* 118, 37, 1985.

50. **Verhaeghe, R. H., Vanhoutte, P. M., and Shepherd, J. T.,** Inhibition of sympathetic neurotransmission in canine blood vessels by adenosine and adenine nucleotides, *Circ. Res.,* 40, 208, 1977.

Chapter 15

ADENOSINE AND THE REGULATION OF CEREBRAL BLOOD FLOW

David G. L. Van Wylen, Veronica M. Sciotti, and H. Richard Winn

TABLE OF CONTENTS

I. INTRODUCTION

Although it has been recognized for 100 years that the cerebral circulation is regulated by local metabolic mechanisms,[1] it is only since the mid-1970s that adenosine has been considered to be involved in the metabolic regulation of cerebral blood flow (CBF).[2] The mechanism by which adenosine is proposed to play a role in the regulaton of CBF is similar to that in the heart and other organs; when there is a mismatch between local blood flow and local metabolism, such that an area of the brain receives an inadequate oxygen or substrate supply for a given level of activity, adenosine release from the brain cells into the surrounding interstitial fluid is enhanced. In the interstitial fluid, adenosine dilates the blood vessels supplying the region, thereby increasing oxygen and substrate supply and compensating for the discrepancy between blood flow and metabolism.

At the outset of this chapter, it is important to understand the context in which adenosine participates in CBF regulation. Blood flow in the brain, perhaps more so than any other organ, is regulated primarily by intrinsic mechanisms. This self-regulation is, in part, related to the fact that the brain is subdivided into countless regions which are capable of independent physiological activation. For example, the motor cortex is a separate entity from the sensory cortex, and within the motor cortex, hand function is distinct from leg activity. Within the hand cortex, the thumb region is distinguishable from the regions which control the other digits. The redundancy of blood supply into the circle of Willis via the carotid and vertebral arterial systems ensures that metabolic and myogenic adjustments can compensate for this highly diverse physiology and protect against occlusion of one or, in many species, two major inflow arteries. Furthermore, the blood-brain barrier insulates the brain from vasoactive substances in the systemic circulation. Finally, external neural influences seem minimally involved in the regulation of CBF, while several intrinsic neural networks have been described which may modulate or act in concert with chemically mediated metabolic adjustments in CBF.[3]

It is also important to appreciate that adenosine is only one of several chemical mediators which are likely involved in the metabolic adjustments of the cerebral circulation; potassium and hydrogen ions have also received substantial experimental support as mediators of CBF.[3] The role of adenosine in the regulation of CBF must therefore be viewed as one piece of a multifactored metabolic control puzzle which is placed in a unique and complex environment designed to ensure adequate and appropriate CBF.

This chapter summarizes the evidence for and against a role for adenosine in the regulation of CBF. First, background information pertaining to adenosine action on the cerebral vasculature and pertinent aspects of adenosine metabolism will be addressed. Second, the physiological contribution of adenosine to the maintenance of resting cerebrovascular tone and the role of adenosine in the circulatory adjustments to decreased oxygen supply or increased brain activity will be discussed. Finally, the overall role of adenosine in the regulation of cerebral function and some unresolved issues regarding adenosine and CBF regulation will be discussed.

II. BACKGROUND

A. CEREBROVASCULAR EFFECTS OF ADENOSINE

The principal prerequisite for participation of adenosine in the metabolic regulation of CBF is that it must be a cerebral vasodilator. The vasodilatory characteristics of adenosine have been demonstrated in isolated cerebral blood vessels,[4-7] from the pial vessel response to topically applied adenosine,[2,8-11] from the intraparenchymal blood flow response to locally administered adenosine,[12,13] and despite an apparent blood-brain barrier for adenosine,[2,14-16] from the global CBF response to intravascular adenosine.[17-22] Furthermore, it has been established with adenosine analogs that, as in most other vascular beds, the cerebrovascular relaxation induced by adenosine is mediated by the interaction with what is currently classified as the A_2 adenosine receptor.[5,6,13,23-25] Thus, adenosine fulfills the initial criteria for a metabolic mediator of having an appropriate mechanism by which to alter CBF.

B. PRODUCTON AND METABOLISM OF ADENOSINE IN THE BRAIN

The detailed aspects of adenosine production and metabolism are discussed elsewhere in this book and have been recently reviewed by Phillis.[26] However, it should be emphasized that for adenosine to participate as a regulator of CBF, there must exist biochemical pathways to rapidly produce and metabolize adenosine in response to changes in metabolic requirements or substrate availability. The primary source of adenosine in the brain is likely the dephosphorylation of adenosine monophosphate (AMP), a reaction catalyzed by 5'-nucleotidase, a generalized term for a group of enzymes with differing cellular and subcellular locations distributed throughout the brain, but concentrated in the plasma membrane of the glial footpads which surround blood vessels.[27] This location of 5'-nucleotidase is important, since it catalyzes the formation of adenosine proximal to the site of action on the vasculature. Other sources of adenosine production in the brain include the extracellular degradation of adenosine triphosphate (ATP) released from nerve terminals and the cytoplasmic metabolism of S-adenosylhomocysteine.[26,28]

The enzymes involved in the adenosine degradation pathway (adenosine deaminase, nucleoside phosphorylase, and xanthine oxidase) are located primarily in astrocytes, glial cells, and capillary endothelial cells,[26,29-31] suggesting that little degradative activity occurs in neurons. However, adenosine kinase, the enzyme which rephosphorylates adenosine to AMP, is found in neurons, indicating that adenosine taken up into neurons is preferentially reincorporated into the adenine nucleotide pool.[32]

III. ADENOSINE AND CEREBRAL BLOOD FLOW REGULATION

A. CONTRIBUTION TO RESTING CEREBROVASCULAR TONE

Adenosine may exert a vasodilatory influence on cerebral blood vessels during normoxic (i.e., resting) conditions, but evidence of its action in this setting is difficult to assess. The majority of experimental approaches designed to address the role of adenosine in the maintenance of resting tone have examined the cerebrovascular response to adenosine receptor antagonists. If adenosine contributes to resting cerebrovascular tone, then inhibition of adenosine action should induce vasoconstriction. However, increases,[11,15,16,33-35] decreases,[20,35-40] or no changes[11,15,16,22,33,35,41-46] in resting CBF or pial vessel diameter have been observed following the administration of adenosine receptor antagonists. Despite these contradictory findings, several patterns emerge. First, dose-response studies indicate that vasodilatory responses to the adenosine receptor antagonist theophylline, whether indicated by increased CBF[15] or increased pial vessel diameter,[11,35] are induced only by high concentrations of theophylline, and are probably due to the inhibitory effect of theophylline on cyclic AMP phosphodiesterase. Second, support for a contribution of adenosine to resting tone comes only from studies in which CBF or pial vessel changes were determined following systemic administration of theophylline or caffeine; several investigators have been unable to detect a decrease in resting pial vessel diameter with topically applied theophylline.[11,33-35] Third, when decreases in resting CBF are detected following systemic theophylline administration, they generally amount to approximately 20% reductions in CBF.[20,35-40]

Another experimental approach has been to determine if inhibition of adenosine deaminase or adenosine uptake into cells increases resting CBF. Dipyridamole, an inhibitor of adenosine uptake, increases resting CBF in the rabbit,[19,20] but not in the dog[47] or the rat.[38,43] Since dipyridamole induces moderate systemic hypotension, calculated cerebrovascular resistance decreases in all cases. However, constant CBF during dipyridamole-induced hypotension can be explained by an autoregulatory response, thus obscuring the interpretation of the dipyridamole response in dogs and rats. Inhibitors of adenosine deaminase also fail to significantly increase resting CBF in the rat.[45]

The contribution of adenosine to resting cerebrovascular tone is further confused by the wide range of estimates of basal interstitial fluid adenosine levels and the technical limitations inherent in attempts to obtain estimates of interstitial adenosine concentration. Extrapolation from the adenosine content of rapidly frozen brain tissue yields estimates of approximately 2 μM,[48] but these estimates may be unreliable due to the uncertain extent of adenosine binding within cells. Cerebrospinal fluid adenosine levels range from 0.02 to 0.6 μM,[26,49] but considering the rapidity of adenosine production and metabolism, it is possible that cerebrospinal fluid adenosine concentration differs from intraparenchymal interstitial fluid adenosine concentration. Studies using the brain microdialysis technique to sample interstitial fluid lead to estimates of intraparenchymal interstitial fluid adenosine of 0.5 to 2.0 μM,[50-55] but the microdialysis technique has been criticized as giving high estimates of interstitial adenosine because of tissue trauma associated with implantation.[26] Cortical cup adenosine levels are approximately 40 nM,[56] but this technique requires exposure of the brain surface from which measurements are made. Furthermore, the volume of the sample placed in the cortical cup is relatively large compared to the surface area of brain sampled. A high volume to surface area ratio has recently been suggested to contribute to low estimates of cardiac interstitial fluid adenosine levels.[57]

Accurate determination of resting interstitial fluid adenosine levels is important to the assessment of the contribution of adenosine to resting tone, since the pial vessel vasodilatory response to topically applied adenosine suggests that interstitial fluid adenosine in the 10^{-8} M range would have a minimal influence on resting tone, while interstitial fluid adenosine in the 10^{-6} M range would exert a significant vasodilatory influence (approximately 20% dilation) on resting cerebrovascular tone.[8-11] However, some caution must be used in this extrapolation, since intraparenchymal vessels appear to have a different response to topically applied adenosine than pial vessels.[12]

In summary, while adenosine does appear to have some influence on resting cerebrovascular tone, resting interstitial fluid adenosine concentration and the impact of this level on resting tone have yet to be determined.

B. CONTRIBUTION TO CEREBRAL VASODILATION DURING DECREASED OXYGEN SUPPLY

1. Hypoxia

Reductions of arterial oxygen content evoke well-described and potent increases in CBF. Several lines of evidence support a role of adenosine in hypoxia-induced cerebral hyperemia. First, increases in brain adenosine levels accompany reductions in P_aO_2. This has been demonstrated in brain tissue,[58,59] interstitial fluid,[53-56] and cerebrospinal fluid.[41,53] Winn et al.,[59] using the freeze-blow technique to rapidly freeze brain tissue in the rat, observed that brain adenosine content increased sevenfold from a normoxic level of 0.36 nmol/g when P_aO_2 was reduced to 30 mmHg. At a P_aO_2 of approximately 50 mmHg, brain adenosine content tended to increase, but this increase did not reach statistical significance. These investigators also demonstrated that during acute hypoxia, induced by 100% N_2 inhalation, brain adenosine content increased within 30 s after the onset of hypoxia. Using the brain-microdialysis technique, Zetterstrom et al.,[55] Van Wylen et al.,[54] and Park et al.[53] have demonstrated increases in cerebral interstitial fluid adenosine levels during hypoxia, as has Phillis et al.[56] using the cortical cup technique. Despite different estimates of basal interstitial fluid adenosine levels with the brain-microdialysis technique[50-55] and the cortical cup technique,[56] quantitatively similar increases are seen during hypoxia with both methods. Van

Wylen et al.[54] estimated a 10-fold increase in interstitial fluid adenosine levels during 10 min of sustained hypoxemia at a P_aO_2 of 42 mmHg with brain microdialysis, while Phillis et al.[56] report 4.5- to 16-fold increases in cortical perfusate adenosine over a similar hypoxemic range. It should be noted that these studies[54,56] were performed in anesthetized rats, which become hypotensive and hypocapnic during hypoxia. Therefore, some of the increase in interstitial adenosine seen during hypoxia may be a result of the combined influences of hypoxemia, hypotension, and hypocapnia on oxygen supply to the brain.

The second line of evidence which supports a role for adenosine during hypoxic hyperemia comes from studies using adenosine receptor antagonists. Theophylline and caffeine, adenosine receptor antagonists, have been shown to attenuate or completely block the increase in CBF during moderate ($P_aO_2 = 40$ to 60 mmHg) and severe ($P_aO_2 < 35$ mmHg) hypoxia.[15,16,35,39,41,43] This has been shown in several animal models (rat,[16,35,43] dog,[15] rabbit,[39] and piglet),[41] using a variety of CBF measurement techniques (venous outflow,[15,35,43] microspheres,[16,35,41] and mass spectrometry),[39] and during acute[35,43] or sustained[15,16,35,39,41] hypoxia. Although the majority of the studies were performed in anesthetized animals, Pinard et al.[39] recently demonstrated in the conscious rabbit a 30 to 40% reduction in the hyperemic response to moderate hypoxia (P_aO_2 46 to 50 mmHg) with theophylline in three deep brain structures (caudate nucleus, thalamus, and hippocampus). Similar results were obtained in five adult humans during exposure to hypoxia (arterial O_2 saturation $= 80\%$) after aminophylline treatment.[36]

Another approach to the determination of the role of adenosine in the cerebrovascular response to hypoxia has been the attempt to potentiate the CBF response with inhibitors of adenosine deaminase or inhibitors of adenosine uptake, which presumably potentiate CBF due to enhanced interstitial fluid adenosine levels. Phillis et al.[45] demonstrated that inhibition of either adenosine deaminase or adenosine uptake potentiated the CBF response to brief (24- to 45-s) hypoxic challenges. Furthermore, hypoxia in the presence of deoxycoformycin, an inhibitor of adenosine deaminase, was associated with an enhanced increase in interstitial fluid adenosine.[60]

Despite an abundance of evidence supporting the involvement of adenosine in the vasodilatory response to hypoxia, conflicting results have been obtained from several laboratories. McPhee and Maxwell[42] were unable to determine an effect of theophylline on CBF in the newborn piglet during reduction of P_aO_2 to 23 mmHg, a result which is in contrast with the attenuation of hypoxic hyperemia obtained in the piglet by Laudignon et al.[41] using the more specific adenosine receptor antagonist 8-phenyltheophylline. There are also conflicting data from studies of the cerebrocortical circulation. Both Haller and Kuschinsky[34] and Dora[33] have been unable to detect an effect of either topical or systemic theophylline on the cerebrocortical response to hypoxia in cats. This is in contrast to Morii et al.,[35] who observed an attenuation of pial vessel dilation during hypoxia in rats treated systemically with theophylline. Likewise, topical cerebrocortical adenosine deaminase treatment has produced conflicting results. Wei and Kontos[61] were able to attenuate hypoxia-induced pial vessel dilation with adenosine deaminase superfusion, but Dora et al.[62] were unable to detect an effect of topical adenosine deaminase on cerebrocortical vascular volume during hypoxia. Although there are many technical aspects which may account for these differences, such as the method of cerebrocortical observation (closed cranial window vs. mineral oil-covered brain) or the detection method (direct microscopic observation vs. surface fluororeflectometry), these discrepant results remain essentially unresolved.

In summary, the weight of evidence suggests that adenosine participates in the cerebrovascular adjustments to the reduction of oxygen supply associated with hypoxemia.

2. Autoregulation and Severe Hypotension
Cerebral autoregulation refers to the ability of the cerebral resistance vessels to dilate

in response to a decrease in cerebral perfusion pressure and to constrict in response to an increase in cerebral perfusion pressure. The result of cerebral autoregulation is that CBF is maintained relatively constant over a wide range of arterial blood pressures. According to the adenosine hypothesis, hypotension transiently decreases CBF, thus decreasing the oxygen supply to the brain. In response to the tissue hypoxia, adenosine is released from the brain cells into the surrounding interstitial fluid, where it dilates the resistance vessels and restores CBF to near control levels. Thus, if adenosine is involved in the cerebral vasodilation which accompanies hypotension, increases in brain adenosine content or cerebral interstitial fluid adenosine concentration would be expected as blood pressure is reduced. Van Wylen et al.,[46] using the brain-microdialysis technique, estimated changes in cerebral interstitial fluid adenosine levels in rats which were hemorrhaged to reduce mean arterial blood pressure (MABP) from control levels (MABP = 101 mmHg) to 80, 70, 60, 50, 40, and 30 mmHg. Cerebral autoregulation was evidenced by an absence of significant changes in CBF until MABP was decreased to 60 mmHg. However, interstitial fluid adenosine did not increase until MABP was decreased to 50 mmHg. At 30 mmHg, there was a fivefold increase in dialysate adenosine concentration. A similar observation was made by Park et al.[63] who, using microdialysis in the newborn piglet, demonstrated that interstitial fluid adenosine levels were not increased at a MABP of 60 mmHg, but increased progressively at pressures of 50 mmHg and below. The changes in interstitial fluid adenosine during hypotension are consistent with data on changes in brain tissue adenosine content. Rubio et al.[58] observed a 4-fold increase in rat brain adenosine content at a blood pressure of 45 mmHg, while Winn et al.[64] showed that, when compared to brain tissue adenosine levels at 135 mmHg, brain adenosine was not significantly increased at 107 mmHg, but doubled at 72 mmHg, and increased 5.6-fold at 45 mmHg. Taken together, these studies indicate that brain adenosine levels change minimally over a large portion of the autoregulatory range. However, progressive and substantial increases in brain adenosine are seen below blood pressures of 50 to 70 mmHg. This would suggest that adenosine is not essential for the cerebral autoregulatory response to moderate changes in arterial blood pressure. However, adenosine may contribute to the decrease in cerebral vascular resistance observed at arterial pressures at the lower end of the autoregulatory range and down to arterial pressures of 30 to 40 mmHg, where maximal vasodilation is observed. Consistent with this, Kontos and Wei[65] demonstrated a reduction of pial vessel vasodilation in response to a reduction of arterial pressure to 58 mmHg in the presence of topically applied adenosine deaminase.

In contrast, Phillis and Delong[44] have shown that adenosine receptor blockade with caffeine did not alter the CBF response to reductions of MABP to 40 mmHg. Furthermore, cortical cup estimates of changes in adenosine during hypotension demonstrated no change in cortical perfusate adenosine when blood pressure was lowered to 46 mmHg.[56] However, since the artificial cerebrospinal fluid place in the cortical cup contained oxygen, the stimulus for enhanced adenosine production may be reduced in this setting. Park et al.[53] showed that the presence of elevated oxygen in the artificial cerebrospinal fluid which was used to perfuse the microdialysis probe virtually eliminated the increase in interstitial fluid adenosine seen during reduced arterial pressure; interstitial adenosine increased only at a MABP of 20 mmHg in the presence of locally supplied oxygen.

In summary, the available evidence suggests a possible role for adenosine in cerebral autoregulation, but its importance is unclear. In light of the observations of Wahl and Kuschinsky,[66] who were unable to detect changes in perivascular hydrogen or potassium ions over the MABP range of 60 to 200 mmHg, the overall contribution of metabolic mechanisms to cerebral autoregulation is questionable. It is possible that cerebral autoregulation is mediated primarily by a myogenic mechanism, and that metabolic factors come into play only when a myogenically induced vasoconstriction would be inappropriate for the existing state of metabolism or when the oxygen supply/oxygen demand ratio is reduced.

3. Ischemia

There is little doubt that cerebral ischemia is a powerful stimulus for increased cerebral adenosine production. This has been shown by rapid increases in brain tissue adenosine content following the onset of global cerebral ischemia.[2,48,67,68] Winn et al.[48] demonstrated a 2.4-fold increase in brain tissue adenosine content only 5 s after aortic transection in the rat. Increases in interstitial fluid adenosine during cerebral ischemia have also been shown by both brain microdialysis and cortical cup techniques.[50,51,54,56,69,70] Obviously, the increase in brain adenosine during total or severe ischemia has little to do with blood flow regulation; however, the increase in brain adenosine during moderate reductions in CBF induced by hypotension suggests that adenosine is involved in the vasodilatory response during moderate ischemia.

If reperfusion is established following cerebral ischemia, a well-described reactive hyperemia ensues. Sciotti et al.[70] recently demonstrated a temporal relationship between CBF and interstitial fluid adenosine levels during reperfusion following a 10-min period of cerebral ischemia; dialysate adenosine levels were increased almost 30-fold during the reactive hyperemic phase (3.5-fold increase in CBF) and then declined parallel to the restoration of normal CBF. This suggests that adenosine which accumulates in the interstitial fluid during ischemia participates in reactive hyperemia following transient ischemia.

Recently, considerable interest has developed in the role of adenosine in the protection of ischemic brain.[26] Although the effect of adenosine on presynaptic calcium permeability and neurotransmitter release is dealt with in detail elsewhere in this book, suffice it to say that the marked elevation in interstitial adenosine during ischemia may be beneficial to the brain not only because of its effect on CBF, but also because of its inhibitory effects on excitotoxic neurotransmitter release.

4. Hypocapnia

Hypocapnia exerts a well-known and powerful vasoconstrictive effect on the cerebral circulation. From the standpoint of a role for adenosine in matching blood flow to the metabolic demands of the brain, it is interesting to speculate if adenosine opposes hypocapnia-induced cerebral vasoconstriction. Brain tissue and interstitial fluid adenosine are increased during hypocapnia,[46, 58] and Ibayashi et al.[71] have recently demonstrated that the hypocapnia-induced decrease in CBF is exaggerated in the presence of adenosine receptor blockade and reduced by inhibition of adenosine uptake. Incidentally, although many investigators have been unable to detect an effect of adenosine receptor blockers on sustained hypercapnia-induced cerebral hyperemia,[15,16,35] Phillis and Delong[38] recently reported that adenosine receptor blockade with caffeine decreased the peak CBF response to a brief hypercapnic challenge, while inhibition of adenosine uptake enhanced the response.

C. CONTRIBUTION TO CEREBRAL VASODILATION DURING INCREASED BRAIN ACTIVITY

Although the coupling between cerebral activity and CBF involves many factors,[3] there is substantial evidence that adenosine participates in the hyperemic response to increased neuronal activity. Accelerated formation of adenosine by the brain has been observed during seizures[72-74] and electrical stimulation.[75] Winn et al.[74] demonstrated a sevenfold increase in brain adenosine levels after 10 s of bicuculline-induced seizure activity in rats, a response that was augmented by reductions of P_aO_2 or arterial blood pressure. Similar results were obtained by Schrader et al.,[73] who demonstrated that brain adenosine content increased within 15 s after the beginning of seizure activity and remained elevated for 20 min. Increases in interstitial fluid adenosine have been seen during increased brain activity as well. Using brain microdialysis, Park et al.[72] observed increases in CBF and interstitial fluid adenosine levels in newborn piglets subjected to bicuculline-induced seizures, while Van Wylen et al.[75] reported graded increases in interstitial adenosine in response to progressive increases in the frequency of localized electrical stimulation in the rat.

The most compelling evidence for a role of adenosine in seizure-induced hyperemia comes from recent studies in which effects of adenosine are either attenuated or enhanced via pharmacological manipulations of adenosine action. Van Wylen and Moffe[76] used microdialysis to locally infuse kainic acid (an agonist of the excitatory neurotransmitter glutamate) into the caudate nucleus of rats, and demonstrated 64 and 236% increases in dialysate adenosine and local CBF, respectively. This hyperemia was attenuated by adenosine receptor blockade with 8-(p-sulfophenyl)theophylline and was associated with a further increase in interstitial adenosine. Pinard et al.[40] determined the effect of systemic theophylline treatment on CBF, tissue PO_2, and tissue histology during kainic acid-induced seizures of unanesthetized rats. Theophylline administration resulted in a delay and an overall reduction of the hyperemic response to kainic acid. Interestingly, theophylline treatment resulted in a greater decrease in tissue PO_2 during seizures and increased neuronal cell vulnerability to kainic acid-induced damage. In a recent study Ko et al.,[77] using a closed cranial window technique, observed a 38% increase in pial vessel diameter in the contralateral somatosensory cortex in response to sciatic nerve stimulation.[78] This response was augmented by dipyridamole (62% dilation) and attenuated by theophylline (18% dilation). These findings are in contrast to the findings of Dora et al.,[33,62] who detected minimal or no effects of topically applied theophylline or adenosine deaminase on the cerebrocortical response to brain activation.

IV. ADENOSINE AND THE REGULATION OF CEREBRAL FUNCTION

This review has focused on the role of adenosine in the regulation of CBF. However, it is important to recognize that this is only one of several actions of adenosine in the brain. Based on the influences of adenosine in the heart, Newby et al.[79] coined the term "retaliatory metabolite" to describe the multiple effects of adenosine. This term is appropriate for the brain as well, as the actions of adenosine on the vasculature and neurons seem designed to retaliate against reductions in the energy supply/demand ratio, inducing vascular and neural responses which reapproximate cerebral energy balance and protect the brain from cellular damage which could ensue in the face of maintained energy imbalance.[26,28] An awareness of the many actions of adenosine in the brain is important in studies using pharmacological manipulation of adenosine action or metabolism to study CBF regulation, since these drugs alter other aspects of cerebral function as well.

V. UNRESOLVED ISSUES

Despite increasing information about the role of adenosine in the brain, there are significant issues which remain unresolved. First, interstitial fluid adenosine concentration and the relationship between the changes in interstitial adenosine and changes in CBF need to be conclusively determined. To address this issue, consideration must be given to the interactions of adenosine not only with other known mediators, but also with inosine, the metabolite of adenosine which has recently been shown to potentiate the vascular effects of adenosine.[77,80] In addition, further work is required to obtain a better understanding of the coordinated vascular response to adenosine. For example, Segal and Duling[81] have shown that vasodilation can be propagated upstream from a primary site of vasodilation to the larger upstream vessels. Therefore, the endogenous adenosine levels necessary to alter local CBF may be less than that predicted from studies of exogenous adenosine application to pial arterioles. Furthermore, Hester[82] recently described the venular-arteriolar diffusion of adenosine, which would provide an additional mechanism whereby adenosine could send a signal to cause dilation of the more proximal arterioles.

A second unresolved issue is the cellular (neuron, glial, endothelial, etc.) and chemical (ATP, AMP, S-adenosyl-homocysteine) sources of adenosine in the brain. Addressing this issue will require resolution of the site and regulation of adenosine production and a further understanding of the complex actions of adenosine in the modulation of neuronal activity.

A third unresolved issue is the regional differences of adenosine action in the brain. Since adenosine receptors and enzymes involved with adenosine production and metabolism are heterogeneously distributed throughout the brain, it is likely that the role of adenosine in CBF regulation differs on a regional basis as well.

Finally, there will likely be continued study of the role of adenosine as a cerebroprotective agent during pathological processes such as ischemia, seizures, and migraine. This important avenue of research may lead to therapeutic uses of adenosine, adenosine agonists, or agents which enhance the levels of endogenous adenosine in the brain.

REFERENCES

1. **Roy, C. S. and Sherrington, C. S.,** On the regulation of the blood supply to the brain, *J. Physiol.,* 11, 85, 1890.
2. **Berne, R. M., Rubio, R., and Curnish, R. R.,** Release of adenosine from ischemic brain. Effect on cerebral vascular resistance and incorporation into cerebral adenine nucleotides, *Circ. Res.,* 35, 262, 1974.
3. **Wahl, M.,** Local chemical,, neural, and humoral regulation of cerebrovascular resistance vessels, *J. Cardiovasc. Pharmacol.,* 7 (Suppl. 3), S36, 1985.
4. **Buyniski, J. P. and Rapela, C. E.,** Cerebral and renal vascular smooth muscle responses to adenosine, *Am. J. Physiol.,* 217, 1660, 1969.
5. **Edvinsson, L. and Fredholm, B. B.,** Characterization of adenosine receptors in isolated cerebral arteries of cat, *Br. J. Pharmacol.,* 80, 631, 1983.
6. **McBean, D. E., Harper, A. M., and Rudolphi, K. A.,** Effects of adenosine and its analogues on porcine basilar arteries: are only A_2 receptors involved? *J. Cereb. Blood Flow Metab.,* 8, 40, 1988.
7. **Toda, N.,** The action of vasodilating drugs on isolated basilar, coronary and mesenteric arteries of the dog, *J. Pharmacol. Exp. Ther.,* 191, 139, 1974.
8. **Dora, E.,** Effect of adenosine and its stabile analogue 2-chloroadenosine on cerebrocortical microcirculation and NAD/NADH redox state, *Pfluegers Arch.,* 404, 208, 1985.
9. **Gregory, P. C., Boisvert, D. P. J., and Harper, A. M.,** Adenosine response on pial arteries, influence of CO_2 and blood pressure, *Pfluegers Arch.,* 386, 187, 1980.
10. **Morii, S., Ngai, A. C., and Winn, H. R.,** Reactivity of rat pial arterioles and venules to adenosine and carbon dioxide: with detailed descriptions of the closed cranial window technique in rats, *J. Cereb. Blood Flow Metab.,* 6, 34, 1986.
11. **Wahl, M. and Kuschinsky, W.,** The dilatory action of adenosine on pial arteries of cats and its inhibition by theophylline, *Pfluegers Arch.,* 362, 55, 1976.
12. **Livermore, P. and Mitchell, G.,** Adenosine causes dilation and constriction of hypothalamic blood vessels, *J. Cereb. Blood Flow Metab.,* 3, 529, 1983.
13. **Van Wylen, D. G. L., Park, T. S., Rubio, R., and Berne, R. M.,** The effect of local infusion of adenosine and adenosine analogues on local cerebral blood flow, *J. Cereb. Blood Flow Metab.,* 9, 556, 1989.
14. **Boarini, D. J., Kassel, N. F., Sprowell, J. A., and Olin, J.,** Intravertebral artery adenosine fails to alter cerebral blood flow in the dog, *Stroke,* 15, 1057, 1984.
15. **Emerson, T. E. and Raymond, R. M.,** Involvement of adenosine in cerebral hypoxic hyperemia in the dog, *Am. J. Physiol.,* 241, H134, 1981.
16. **Hoffman, W. E., Albrecht, R. F., and Miletich, D. J.,** The role of adenosine in CBF increases during hypoxia in young vs. aged rats, *Stroke,* 15, 124, 1984.
17. **Anwar, M. and Weiss, H. R.,** Adenosine and cerebral capillary perfusion and blood flow during middle cerebral artery occlusion, *Am. J. Physiol.,* 257, H1656, 1989.
18. **Forrester, T., Harper, A. M., MacKenzie, E. T., and Thomson, E. M.,** Effect of adenosine triphosphate and some derivatives on cerebral blood flow and metabolism, *J. Physiol.,* 296, 343, 1979.
19. **Heistad, D. D., Marcus, M. L., Gourley, K. K., and Busija, D. W.,** Effect of adenosine and dipyridamole on cerebral blood flow, *Am. J. Physiol.,* 240, H775, 1981.

20. **Puiroud, S., Pinard, E., and Seylaz, J.,** Dynamic cerebral and systemic circulatory effects of adenosine, theophylline, and dipyridamole, *Brain Res.,* 453, 287, 1988.

21. **Sollevi, A., Ericson, K., Eriksson, L., Lindqvist, C., Largerkranser, M., and Stone-Elander, S.,** Effect of adenosine on human cerebral blood flow as determined by positron emission tomography, *J. Cereb. Blood Flow Metab.,* 7, 673, 1987.

22. **Torregrosa, G., Terrasa, J. C., Salom, J. B., Miranda, F. J., Campos, V., and Alborch, E.,** P_1-purinoceptors in the cerebrovascular bed on the goat in vivo, *Eur. J. Pharmacol.,* 149, 17, 1988.

23. **Hardebo, J. E., Kahrstrom, J., and Owman, C.,** P_1- and P_2-purine receptors in brain circulation, *Eur. J. Pharmacol.,* 144, 343, 1987.

24. **Ibayashi, S., Hgai, A. C., Meno, J. R., and Winn, H. R.,** Effects of topical adenosine analogs and forskolin on rat pial arterioles *in vivo, J. Cereb. Blood Flow Metab.,* 11, 72, 1990.

25. **Kalaria, R. N. and Harik, S. I.,** Adenosine receptors of cerebral microvessels and choroid plexus, *J. Cereb. Blood Flow Metab.,* 6, 463, 1986.

26. **Phillis, J. W.,** Adenosine in the control of the cerebral circulation, *Cerebrovasc. Brain Metab. Rev.,* 1, 26, 1989.

27. **Kreutzberg, G. W., Barron, K. D., and Schubert, P.,** Cytochemical localization of 5′-nucleotidase in glial plasma membranes, *Brain Res.,* 158, 247, 1978.

28. **Dunwiddie, T. V.,** The physiological role of adenosine in the central nervous system, *Int. Rev. Neurobiol.,* 27, 63, 1985.

29. **Mistry, G. and Drummond, G. I.,** Adenosine metabolism in microvessels from heart and brain, *J. Mol. Cell. Cardiol.,* 18, 13, 1986.

30. **Schrader, W. P., West, C. A., and Strominger, N. L.,** Localization of adenosine deaminase and adenosine deaminase complexing protein in rabbit brain, *J. Histochem. Cytochem.,* 35, 443, 1987.

31. **Yamamoto, T., Geiger, J. D., Daddona, P. E., and Nagy, J. I.,** Subcellular, regional, and immuno-histological localization of adenosine deaminase in various species, *Brain Res. Bull.,* 19, 473, 1987.

32. **Winn, H. R., Park, T. S., Curnish, R. R., Rubio, R., and Berne, R. M.,** Incorporation of adenosine and its metabolites into brain nucleotides, *Am. J. Physiol.,* 239, H212, 1980.

33. **Dora, E.,** Effect of theophylline treatment on the functional hyperemic and hypoxic responses of cerebro-cortical microcirculation, *Acta Physiol. Hung.,* 68, 183, 1986.

34. **Haller, C. and Kuschinsky, W.,** Moderate hypoxia: reactivity of pial arteries and local effects of theo-phylline, *J. Appl. Physiol.,* 63, 2208, 1987.

35. **Morii, S., Ngai, A. C., Ko, K. R., and Winn, H. R.,** Role of adenosine in regulation of cerebral blood flow: effects of theophylline during normoxia and hypoxia, *Am. J. Physiol.,* 253, H165, 1987.

36. **Bowton, D. L., Haddon, W. S., Prough, D. S., Adair, N., Alford, P. T., and Stump, D. A.,** Theophylline effect on the cerebral blood flow response to hypoxemia, *Chest,* 94, 371, 1988.

37. **Grome, J. J. and Stefanovich, V.,** Differential effects of methylxanthines on local cerebral blood flow and glucose utlization in the conscious rat, *Naunyn-Schmiedeberg's Arch. Pharmacol.,* 333, 172, 1986.

38. **Phillis, J. W. and Delong, R. E.,** An involvement of adenosine in cerebral blood flow regulation during hypercapnia, *Gen. Pharmacol.,* 18, 133, 1987.

39. **Pinard, E., Puiroud, S., and Seylaz, J.,** Role of adenosine in cerebral hypoxic hyperemia in the un-anesthetized rabbit, *Brain Res.,* 481, 124, 1989.

40. **Pinard, E., Riche, D., Puiroud, S., and Seylaz, J.,** Theophylline reduces cerebral hyperaemia and enhances brain damage induced by seizures, *Brain Res.,* 511, 303, 1990.

41. **Laudignon, N., Farri, E., Beharry, K., Rex, J., and Aranda, J. V.,** Influence of adenosine on cerebral blood flow during hypoxic hypoxia in the newborn piglet, *J. Appl. Physiol.,* 68, 1534, 1990.

42. **McPhee, A. J. and Maxwell, G. M.,** The effect of theophylline on regional cerebral blood flow responses to hypoxia in newborn piglets, *Pediatr. Res.,* 21, 573, 1987.

43. **Phillis, J. W., Preston, G., and DeLong, R. E.,** Effects of anoxia on cerebral blood flow in the rat brain: evidence for a role of adenosine in autoregulation, *J. Cereb. Blood Flow Metab.,* 4, 586, 1984.

44. **Phillis, J. W. and Delong, R. E.,** The role of adenosine in cerebral vascular regulation during reductions in perfusion pressure, *J. Pharm. Pharmacol.,* 38, 460, 1986.

45. **Phillis, J. W., DeLong, R. E., and Towner, J. K.,** Adenosine deaminase inhibitors enhance cerebral anoxic hyperemia in the rat, *J. Cereb. Blood Flow Metab.,* 5, 295, 1985.

46. **Van Wylen, D. G. L., Park, T. S., Rubio, R., and Berne, R. M.,** Cerebral blood flow and interstitial fluid adenosine during hemorrhagic hypotension, *Am. J. Physiol.,* 255, H1211, 1988.

47. **Boarini, D. J., Kassell, N. F., Olin, J. J., and Sprowell, J. A.,** The effect of intravenous dipyridamole on the cerebral and systemic circulations of the dog, *Stroke,* 13, 842, 1982.

48. **Winn, H. R., Rubio, R., and Berne, R. M.,** Brain adenosine production in the rat during 60 seconds of ischemia, *Circ. Res.,* 45, 486, 1979.

49. **Laudignon, N., Beharry, K., Rex, J., and Aranda, J. V.,** Effect of adenosine on total and regional cerebral blood flow of the newborn piglet, *J. Cereb. Blood Flow Metab.,* 10, 392, 1990.

50. **Hagberg, H., Andersson, P., Lacarewicz, J., Jacobson, I., Butcher, S., and Sandberg, M.,** Extracellular adenosine, inosine, hypoxanthine, and xanthine in relation to tissue nucleotides and purines in rat striatum during transient ischemia, *J. Neurochem.*, 49, 227, 1987.

51. **Hillered, L., Hallstrom, A., Segersvard, S., Persson, L., and Ungerstedt, U.,** Dynamics of extracellular metabolites in the striatum after middle cerebral artery occlusion in the rat monitored by intracerebral microdialysis, *J. Cereb. Blood Flow Metab.*, 9, 607, 1989.

52. **Ngai, A. C., Meno, J. R., Kay, K., Ko, K. O., and Winn, H. R.,** Interstitial levels of adenosine in rat cerebral cortex, *J. Cereb. Blood Flow Metab.*, 7 (Suppl. 1), S256, 1987.

53. **Park, T. S., Van Wylen, D. G. L., Rubio, R., and Berne, R. M.,** Increased brain interstitial adenosine concentration during hypoxia in the newborn piglet, *J. Cereb. Blood Flow Metab.*, 7, 178, 1987.

54. **Van Wylen, D. G. L., Park, T. S., Rubio, R., and Berne, R. M.,** Increases in cerebral interstitial fluid adenosine concentration during hypoxia, local potassium infusion, and ischemia, *J. Cereb. Blood Flow Metab.*, 6, 522, 1986.

55. **Zetterstrom, T., Vernet, L., Ungerstedt, U., Tossman, U., Jonzon, B., and Fredholm, B. B.,** Purine levels in the intact brain. Studies with an implanted perfused hollow fibre, *Neurosci. Lett.*, 29, 111, 1982.

56. **Phillis, J. W., Walter, G. A., O'Regan, M. H., and Stair, R. E.,** Increases in cerebral cortical perfusate adenosine and inosine concentrations during hypoxia and ischemia, *J. Cereb. Blood Flow Metab.*, 7, 679, 1987.

57. **Kaiser, D. M., Rubio, R., Gidday, J., and Berne, R.,** Heterogeneity and sampling volume dependence of epicardial adenosine (ADO) concentrations, *FASEB J.*, 4, A852, 1990.

58. **Rubio, R., Berne, R. M., Bockman, E. L., and Curnish, R. R.,** Relationship between adenosine concentration and oxygen supply in rat brain, *Am. J. Physiol.*, 228, 1896, 1975.

59. **Winn, H. R., Rubio, R., and Berne, R. M.,** Brain adenosine concentration during hypoxia in rat, *Am. J. Physiol.*, 241, H235, 1981.

60. **Phillis, J. W., O'Regan, M. H., and Walter, G. A.,** Effects of deoxycoformycin on adenosine, inosine, hypoxanthine, xanthine, and uric acid release from the hypoxemic rat cerebral cortex, *J. Cereb. Blood Flow Metab.*, 8, 733, 1988.

61. **Wei, P. and Kontos, H. A.,** Role of adenosine in cerebral arteriolar dilation from arterial hypoxia, *J. Cereb. Blood Flow Metab.*, 1 (Suppl. 1), S395, 1981.

62. **Dora, E., Koller, A., and Kovach, A. G. B.,** Effect of topical adenosine deaminase treatment on the functional hyperemic and hypoxic responses of cerebrocortical microcirculation, *J. Cereb. Blood Flow Metab.*, 4, 447, 1984.

63. **Park, T. S., Van Wylen, D. G. L., Rubio, R., and Berne, R. M.,** Brain interstitial adenosine and sagittal sinus blood flow during systemic hypotension in piglet, *J. Cereb. Blood Flow Metab.*, 8, 822, 1988.

64. **Winn, H. R., Welsh, J. E., Rubio, R., and Berne, R. M.,** Brain adenosine production during sustained alteration in systemic blood pressure, *Am. J. Physiol.*, 239, H636, 1980.

65. **Kontos, H. A. and Wei, E. P.,** Oxygen-dependent mechanisms in cerebral autoregulation, *Ann. Biomed. Eng.*, 13, 329, 1985.

66. **Wahl M. and Kuschinsky, W.,** Unimportance of perivascular H^+ and K^+ activities for the adjustment of pial arterial diameter during changes of arterial blood pressure in cats, *Pfluegers Arch.*, 382, 203, 1979.

67. **Nordstrom, C. H., Rehncrona, S., Siesjo, B. K., and Westerberg, E.,** Adenosine in rat cerebral cortex: its determination, normal values and correlation to AMP and cyclic AMP during shortlasting ischemia, *Acta Physiol. Scand.*, 101, 63, 1977.

68. **Hsu, S. S., Meno, J., and Winn, H. R.,** Influence of hyperglycemia on cerebral adenosine production during ischemia and reperfusion, *Am. J. Physiol.*, in press.

69. **Park, T. S. and Gidday, J. M.,** Effect of dipyridamole on cerebral extracellular adenosine level in vivo, *J. Cereb. Blood Flow Metab.*, 10, 424, 1990.

70. **Sciotti, V., Litchmore, T., and Van Wylen, D. G. L.,** Interstitial fluid (ISF) purine metabolite levels during and after cerebral ischemia (ISC), *FASEB J.*, 4, A1095, 1990.

71. **Ibayashi, S., Ngai, A. C., Meno, J. R., and Winn, H. R.,** The effects of dipyridamole and theophylline on rat pial vessels during hypocarbia, *J. Cereb. Blood Flow Metab.*, 8, 829, 1988.

72. **Park, T. S., Van Wylen, D. G. L., Rubio, R., and Berne, R. M.,** Interstitial fluid adenosine and cerebral blood flow during bicuculline-seizures in newborn piglet, *J. Cereb. Blood Flow Metab.*, 7, 633, 1987.

73. **Schrader, J., Wahl, M., Kuschinsky, W., and Kreutzberg, G. W.,** Increase of adenosine content in cerebral cortex of the cat during bicuculline-induced seizure, *Pfluegers Arch.*, 387, 245, 1980.

74. **Winn, H. R., Welsh, J. E., Rubio, R., and Berne, R. M.,** Changes in brain adenosine during bicuculline-induced seizures in rats. Effects of hypoxia and altered systemic blood pressure, *Circ. Res.*, 47, 568, 1980.

75. **Van Wylen, D. G. L., Park, T. S., Rubio, R., and Berne, R. M.,** Increases in brain dialysate adenosine (ADO), inosine (INO), and hypoxanthine (HYPO) during local brain stimulation, *Physiologist*, 30, 180, 1987.

76. **Van Wylen, D. G. L. and Moffe, A. L.,** Attenuation of kainic acid (KA) induced increases in cerebral blood flow (CBF) with adenosine (ADO) receptor blockade, *Physiologist,* 31, A219, 1988.

77. **Ko, K. R., Ngai, A. C., and Winn, H. R.,** The role of adenosine in the regulation of regional cerebral blood flow in sensory cortex, *Am. J. Physiol.,* 259, H1703, 1990.

78. **Ngai, A. C., Ko, K. R., Morii, S., and Winn, H. R.,** Effect of sciatic nerve stimulation on pial arterioles in rats, *Am. J. Physiol.,* 254, H133, 1988.

79. **Newby, A. C., Worku, Y., and Holmquist, C. A.,** Adenosine formation: evidence for a direct link with energy metabolism, *Adv. Myocardiol.,* 6, 273, 1985.

80. **Ngai, A. C., Monsen, M. R., Ibayashi, S., Ko, K. R., and Winn, H. R.,** Effect of inosine on pial arterioles: potentiation of adenosine-induced vasodilation, *Am. J. Physiol.,* 256, H603, 1989.

81. **Segal, S. S. and Duling, B. R.,** Communication between feed arteries and microvessels in hamster striated muscle: segmental vascular responses are functionally coordinated, *Circ. Res.,* 59, 283, 1986.

82. **Hester, R. L.,** Venular-arteriolar diffusion of adenosine in hamster cremaster microcirculation, *Am. J. Physiol.,* 258, H1918, 1990.

Chapter 16

ADENOSINE AND BLOOD FLOW THROUGH THE GASTROINTESTINAL TRACT

Kenneth G. Proctor and Bobbi Langkamp-Henken

TABLE OF CONTENTS

I. INTRODUCTION

Several extrinsic and intrinsic factors regulate blood flow to the gastrointestinal tract. The relative contribution of each component cannot be easily distinguished in the intact organism; but, in simplified experimental models, the gut regulates its blood flow in the absence of extrinsic influences, which attests to the importance of local mechanisms.[1-3] This intrinsic control can be attributed to the enteric nervous system,[4] a myogenic mechanism,[5] and a local metabolic mechanism.

In theory, changes in metabolism are coupled to changes in blood flow through the release of vasoactive materials into the extracellular space. Adenosine (ADO) is viewed as a signal for inadequate ATP levels that maintains a balance between oxygen supply and metabolic demand,[6-8] so it has received the greatest attention of all the potential metabolic mediators.

When considering the role of ADO in the regulation of gastrointestinal circulation, it is convenient to address the criteria defined in an earlier review on the role of ADO in the coronary circulation.[6] Briefly, there must be an endogenous metabolic pathway and a mechanism for delivering it to the extracellular space; the extracellular levels must vary with the oxygen supply to demand ratio; those endogenous levels must produce vasodilation with a time course and magnitude that are proportional to the metabolic stimulus; and drugs that attenuate, potentiate, or mimic its action should evoke corresponding changes in blood flow. Within this framework, even a casual review of earlier work reveals fundamental deficiencies in the tenets of the ADO hypothesis. Most important, oxygen extraction in the gastrointestinal circulation is low in basal conditions, so tissue oxygen requirements can be met by increases in capillary perfusion or by increases in oxygen extraction.[9] Accordingly, functional hyperemia and hypoxic vasodilation occur in the absence of appropriate PO_2 changes in the perivascular region of the flow-controlling arterioles.[10] Therefore, in contrast to the heart, metabolism and blood flow are only loosely coupled in the gastrointestinal tract. Thus, it is not surprising that ADO generally acts as a *modulator*, rather than as a *mediator*, of blood flow to the gastrointestinal tract.

II. POTENTIAL SOURCES OF ADO

Since intestinal epithelial cells cannot synthesize purines *de novo*,[11] purines from the diet are probably imported from the lumen via specific high-affinity nucleoside and nucleotide transport systems.[12-15] Dietary sources include organ meats, legumes, and shellfish; but only adenine is incorporated into cellular nucleotide pools in significant amounts.[16] Since rapidly proliferating cells, such as those in the intestinal epithelium or immune tissue, seem to have a special need for dietary nucleotides,[17] RNA has recently been added to nutritional support therapies for metabolically stressed patients.[11] Removal of dietary nucleotides, or administration of the antimetabolite 6-mercaptopurine, caused a dramatic decrease in total RNA and mRNA in the small intestine and colon, but not the liver, of adult rats, which suggests that these dietary components could control the synthesis of specific proteins.[18] Nevertheless, it is not clear whether or not dietary purines directly influence intestinal nucleoside pools.

Extracellular ADO can originate from the catabolism of extracellular ATP. For many years, it was recognized that low-frequency stimulation of splanchnic nerves evoked an atropine-resistant vasodilator response in the stomach, intestine, and mesenteric artery.[19-21] In addition, mechanical distention of the stomach or intestine elicits a local neural vasodilator reflex that is resistant to adrenergic and cholinergic antagonists and surgical denervation.[22] It is now well established that these responses are mediated by ATP released from nerve terminals, as well from the cytoplasm of several cell types.[23,24] Extracellular ATP is then locally metabolized to ADO by ectonucleotidases.[7,25]

The oldest idea is that ADO is generated inside the cell and then exported into the interstitium. It is usually assumed that AMP formed during the sequential dephosphorylation of ATP is the immediate intracellular precursor, but it is now known that as much as 90% of the ADO can be derived from the transmethylation pathway (e.g., in the heart) by the hydrolysis of S-adenosylhomocysteine.[26] The transmethylation pathway is oxygen insensitive, and it probably supplies a major fraction of the intracellular ADO in normal resting conditions.[8] On the other hand, during hypoxia and ischemia, the major route of ADO formation is probably via ATP breakdown.

In the gastrointestinal tract, the metabolic pathways have not been fully defined, but the ATP/ADP + AMP ratio suggests active kinases and nucleotidases.[27] In addition, the mucosa may have the highest adenosine deaminase activity of any tissue in the body.[28] AMP is probably degraded via 5'-nucleotidase to ADO, rather than via AMP deaminase to IMP.[29] In the heart, most of the ADO formed via the cytosolic nucleotidase or transmethylation pathway is reincorporated into the adenine nucleotide pool by the action of kinases, and only a small fraction escapes into the extracellular space.[8] In the intestine, the relative contributions from the membrane-bound ecto-5'-nucleotidase, the cytosolic 5'-nucleotidase, or the S-adenosyl-homocysteine pathway are unknown.

The levels of adenine nucleotides and nucleosides are similar to those in most tissues (e.g., see Sarau et al.).[27] Total ADO in resting jejunum-ileum of the cat averages 0.1 to 0.2 μM,[30] but these values cannot be equated with interstitial concentration based on data from the heart where a large fraction of the total ADO pool is bound to S-adenosyl-homocysteine, which does not freely exchange with the extracellular space.[31] ADO averages 50 to 60 nM in jejunal venous plasma from the dog,[32] but plasma values cannot be equated with interstitial values because uptake and metabolism by platelets, erythrocytes, and endothelium rapidly lower blood levels.[25,33-35] For example, endothelial cells can trap as much as 90% of infused ADO, even in micromolar concentrations.[36]

III. RESTING VASOMOTOR TONE

It is unlikely that ADO has any role in the control of resting vasomotor tone because methylxanthines (aminophylline, theophylline, or 8-phenyltheophylline), ADO uptake inhibitors (dipyridamole or nitrobenzylthioinosine), and adenosine deaminase have no effect on resting vasomotor tone in any species yet tested.[37-44]

Relatively high levels of ADO must be administered to evoke vasodilation, but the apparent insensitivity probably reflects rapid inactivation of the exogenous substrate. Therefore, plasma or suffusate concentrations cannot be equated with concentrations at the receptor level. In the autoperfused cat ileum, ADO caused dose-related increases (maximal effects at 40 μM plasma concentrations) in blood flow.[38] In the terminal ileum from the dog, i.a. ADO evoked dose-related increases in blood flow with a half maximal effect at 3.1 nmol/kg-min.[43] Maximal blood flow increases of 2.5-fold were observed at a calculated concentration of 1 μM i.a. in the dog jejunum-ileum.[9] Maximal increases were evoked by 30 μmol/kg-min i.a. (50 μM) in partially isolated jejunal segments in the same species.[41] In partially isolated jejunal segments from rats, topical application evoked maximal increases in arteriolar diameter (150% of control) at 250 μM.[40] In the partially isolated stomach from the same species, topical application evoked increases in submucosal arteriolar diameter (150% of control) at 1 mM.[37] All the responses were antagonized by methylxanthines.

The deaminase-resistant synthetic analog, 2-chloroadenosine (CADO), was 6 times more potent than ADO on a molar basis in the dog intestine.[44] In the rat intestine, compared to ADO, CADO was at least 10 times more potent and N-ethylcarboxamidoadenosine (NECA) was at least 100 times more potent; threshold responses were evoked in the submicromolar range with serosal application.[41] In contrast, with mucosal application, even 10 mM ADO or 100 μM NECA or CADO evoked only minimal responses in both the intestine[40,41] and stomach.[37] The most likely explanation is that the mucosa presents a physical and metabolic diffusion barrier that limits the passage of luminal purines into the interstitium.[41] Since the metabolically stable analogs evoke changes with interstitial concentrations in the micromolar range, it is relatively safe to assume that micromolar concentrations of endogenous ADO at the receptor level would evoke changes in blood flow similar to those evoked by various metabolic stimuli.

Although ADO causes dose-related increases in mesenteric artery blood flow, it decreases oxygen uptake and may shift blood flow away from the metabolically active mucosa.[9,38,45] Since mucosal and serosal arterioles originate from a common submucosal arteriole that branches off the mesenteric artery, changes in mesenteric artery flow do not necessarily reflect microcirculatory flow in the individual layers.[46] Thus, ADO might evoke nonuniform microcirculatory changes within the intestinal wall. Perhaps there are differences in distribution of A_1- or A_2-receptor subtypes or differences in the receptor density between the serosal and mucosal layers.

ADO-evoked vasodilation could be caused by at least three actions within the blood vessel wall:[23,33] release of endothelium-derived relaxing factor by a direct or indirect action on endothelial cells, relaxation of arterioles coupled to the activation of A_2 receptors on vascular smooth muscle, or inhibition of norepinephrine release from adrenergic nerve terminals coupled to A_1 receptors at a prejunctional site.

IV. AUTOREGULATION

There is a weak tendency to preserve blood flow with fluctuating perfusion pressure in the small intestine and liver, but no appreciable tendency in the stomach or colon. However, autoregulation of gastric microcirculatory blood flow is improved by sympathethic denervation; this effect is attributed either to a reduction in sympathetic tone and/or an increase in gastric oxygen demand.[47] In fasted cats[48] and fed dogs,[9] methylxanthines attenuated autoregulation in the superior mesenteric artery. However, in fasted dogs the response was not altered.[9] It is difficult to reconcile these observations with a physiologic role for ADO.

The contribution probably depends on the intensity of the metabolic stimulus or the fasted or fed state of the tissue, which influences arteriovenous oxygen difference.[9] Another possibility is that the intestinal vasculature has more than one receptor subtype: a high-affinity A_1 receptor that mediates vasoconstriction at low concentrations and a low-affinity A_2 receptor that mediates vasodilation at higher concentrations. For example, biphasic responses evoked by ADO can be attributed to A_1 and A_2 receptors in some vasculatures, such as the kidney and skin.[49,50] This possibility deserves consideration because the gastrointestinal tract, like the skin, is endowed with a rich purinergic innervation.[22] At least two subtypes of purinoceptors (on the endothelium and vascular smooth muscle) subserve vasoconstriction or vasodilation.[51] In any case, these hypotheses probably cannot be tested with ADO per se because it is a nonselective receptor agonist, as well as a metabolic substrate. It may be more appropriate to use the new classes of potent antagonists that have minimal nonspecific actions and the novel receptor-selective agonists that are metabolically inert.[52]

Sympathetic nerve activation or norepinephrine infusion causes an intense, but only transient, vasoconstriction followed by a poststimulation hyperemia that has been attributed to the accumulation of metabolic vasodilators.[53] In fasted cats, exogenous ADO inhibited the vasoconstrictions, but 8-phenyltheophylline had no effect, which ruled out a contribution of endogenous ADO coreleased or released subsequent to stimulation.[54] In piglets, the "autoregulatory escape" was inhibited by adenosine deaminase, but the poststimulation hyperemic response was not altered.[55]

V. REACTIVE HYPEREMIA

Brief periods of arterial occlusion (1 to 2 min) promote the accumulation of vasoactive metabolites, which evoke a brisk reactive hyperemia upon release of the occlusion. The magnitude of the hyperemia correlated with the duration of occlusion and with blood ADO concentration,[30] but aminophylline reduced the peak hyperemia only 15% in cats.[37] In contrast, in the dog, theophylline decreased the duration of the response and the effect was proportional to the duration of arterial occlusion.[9]

These inconsistent results can also be explained by the postulate that ADO has minimal effects on local flow control unless severe stresses such as complete ischemia are imposed on the tissue.[9] Alternatively, the disparity could reflect heterogeneity within the intestinal wall. Reactive hyperemia after short occlusions is typically confined to the mucosa; the response is reflected in total tissue blood flow changes because the mucosa normally receives the largest fraction of mural blood flow.[45] If perfusion is redistributed to the muscularis in basal conditions, then total flow changes after mesenteric artery occlusion could be masked.

VI. ISCHEMIA-REPERFUSION INJURY

The intestine is unusually sensitive to ischemia. If the mesenteric artery is totally occluded for >5 min or partially occluded for >1 h, then released, a transient hyperemia is followed by a sustained, intense vasoconstriction or "no reflow phenomena". There are profound structural and functional derangements in the tissue, including villus atrophy, crypt degeneration, and breakdown of intestinal barrier function. The pathogenesis of this injury can be attributed to a burst of oxygen radicals generated by the action of xanthine oxidase on hypoxanthine, which accumulates during the ischemic degradation of ATP.[56] In addition to a toxic effect per se, the highly reactive oxyradicals promote the infiltration, adherence, and secretion of activated granulocytes, whose cytotoxins can further aggravate the microvascular and tissue injury.[57]

ADO dramatically attenuates ischemia-reperfusion injury by interfering with granulocyte function, by causing vasodilation directly or by inhibiting sympathetic vasoconstrictor tone, and (possibly) by restoring the ischemic purine deficit. These beneficial effects are seen even if ADO is administered postischemia. In a perfused segment of cat ileum subjected to 60 min of low-flow ischemia and 60 min of reperfusion, a continuous intra-arterial infusion of 2 μM ADO attenuated the increases in intestinal capillary permeability, oxyradical generation, granulocyte infiltration, and granulocyte extravasation.[57] In the partially isolated rat jejunum subjected to 5 min of total arterial occlusion and 60 min of reperfusion, topical application of 100 μM ADO for 30 min at the time of reperfusion (but not during ischemia) prevented the no-reflow phenomena and postischemic decrease in mucosal surface pH, reduced granulocyte infiltration, reduced histological damage, and reduced oxyradical generation.[58,59] This beneficial effect cannot be entirely attributed to a vasodilator property because other vasodilators at equieffective concentrations (prostacyclin or acetylcholine) did not achieve equal protection.[58] Nor could the protection be attributed solely to inhibiting oxyradical generation, because ADO reduced oxyradical production by granulocytes only 20%,[57] and because even a brief application of exogenous ADO was more effective than a continuous application of superoxide dismutase + catalase or vitamin E.[59] Endogenous ADO probably has a similar protective effect, because the injury was aggravated with 8-phenyltheophylline and attenuated with amino-imidazole carboxamide riboside (AICAR), a novel substance that increases endogenous ADO levels in ischemic tissues.[59] It seems likely that ADO could also preserve ATP levels in reperfused tissues.[7,8] In context, these observations support an emerging hypothesis that ADO is an endogenous, anti-inflammatory autacoid.[60]

VII. HORMONE-STIMULATED HYPEREMIA

Whereas caffeine is a well-known gastric secretagogue (see Debas),[61] ADO and its synthetic analogs inhibit the basal release of hydrogen ion and the volume of the secretions. The rank order of potency is R-phenylisopropyladenosine (R-PIA) = NECA > CADO > S-PIA, and the ID_{50} for NECA is 8 μg/kg.[62] Thus, endogenous ADO may be an important regulator of gastric secretory function in basal conditions. Additionally, the gastric acid secretory response evoked by histamine is mediated through A_1 receptors on the parietal cell,[63] and 8-phenyltheophylline inhibited both the pentagastrin-evoked gastric hyperemia and acid secretion.[37] Since stomach acid secretion is a highly energy-dependent process that is linearly correlated with oxygen consumption[64] and since ADO is a vasodilator in the stomach microcirculation,[37] it is logical to hypothesize that ADO generated during secretory activity could partially account for the gastric mucosal hyperemia.[37]

The evidence favoring a role for ADO in the gastric basal and stimulated secretory function is compelling,[62,63] but the evidence favoring a role in the metabolic hyperemia response[37] can be challenged, because there is no simple relationship between secretion and blood flow.[64] Indeed, acid output and microvascular flow velocity are poorly correlated in most conditions.[65] If oxygen consumption is limited, then blood flow and secretion change in parallel; otherwise, the variables change independently.[66] These characteristics can be expected if tissue oxygen requirements can be met either by increased extraction or by increased blood flow, in the stomach[64] as well as in the intestine.[10]

VIII. POSTPRANDIAL HYPEREMIA

Cardiac output, arterial blood pressure, heart rate, and organ blood flow change during the anticipation and ingestion of a meal. Thereafter, the increased blood flow is confined to the segment of the gastrointestinal tract that contains the partially digested food, which affirms that the response is locally mediated.[1-3,67,68] The hyperemia is influenced by a variety of regulatory pathways that depend on the constituents of the chyme,[69] for example, actively absorbed glucose acts primarily on the exchange vessels, whereas passively absorbed oleic acid acts primarily on the resistance vessels to increase oxygen delivery.[70] In addition to the hypermetabolic state of absorption, several of the absorbed nutrients are vasoactive per se, including fatty acids, carbohydrates, and the byproducts of protein digestion, but individual amino acids have minimal effects at physiological concentrations. Bile enhances the vasoactivity of most of these nutrients.[68] The hyperemia is mediated by several factors, including intrinsic nerves,[71] gastrointestinal hormones,[72] tissue hyperosmolarity,[73,74] arachidonic acid metabolites,[75-77] histamine,[78] and vasoactive byproducts of local metabolism, including ADO. The contribution of ADO (or any other substance derived from oxidative metabolism) to food-induced hyperemia probably depends on the arteriovenous oxygen difference, which depends on the experimental conditions and on the species. For example, in dogs, changes in oxygen consumption are correlated with blood flow, but the slope of the relationship depends on the initial oxygen extraction.[9] In contrast, in rats, resting intestinal oxygen consumption is two to three times higher than that in dogs or cats,[79] and increased blood flow is required to meet increments in oxygen demand.

In rats, the time course of mucosal tissue PO_2 decrements during glucose absorption parallels the increments in blood flow.[46] During absorption of glucose + oleic acid, the hyperemia was reduced, but not eliminated, by supramaximal doses of either adenosine deaminase or theophylline,[40] which suggests that ADO is one, but not the only, local factor with an important role. Other vasoactive factors include interstitial hyperosmolarity during glucose absorption,[76,74] and arachidonic acid metabolites[80] during fat, but not glucose, absorption.[74]

In dogs, the relative contributions of increased oxygen extraction or blood flow for delivering oxygen during absorption of predigested food fluctuate with time.[81] Theophylline had no effect on the increments in blood flow and oxygen utilization,[9] but it can be argued that theophylline is a relatively weak antagonist[82] whose actions could be complicated by motility.[42] Venous plasma ADO levels are transiently elevated during the initial phase of hyperemia.[32] Furthermore, dipyridamole enhanced[43] and the potent, specific antagonist, 8-phenyltheophylline, reduced, but did not eliminate, the response.[42] Nevertheless, even in the best case, ADO in only one factor mediating the food-induced hyperemia because histamine,[77] arachidonic acid metabolites,[75,76] and gastrointestinal hormones[72] also have important roles in the same experimental model.

Several previous studies have shown that ADO or its synthetic analogs evoke either minimal or negative vascular responses at the mucosal site of absorption.[37,38,40,41,45] It is difficult to reconcile these observations with the hypothesis that ADO is an important mediator of the mucosal hyperemia.

IX. SUMMARY

Due to the complexity and multiple interactions within control systems, and due to the fact that oxygen uptake can be provided by increments in blood flow or oxygen extraction in most conditions, there are relatively few situations where a metabolic mechanism linked to ADO dominates the other intrinsic or extrinsic factors in the regulation of gastrointestinal blood flow. Although ADO is an important local modulator of blood flow, its major role in the gastrointestinal tract may be in the regulation of neuronal activity, hormone secretion, and ischemia-reperfusion injury.

REFERENCES

1. **Granger, D. N. and Kvietys, P. R.,** The splanchnic circulation: intrinsic regulation, *Annu. Rev. Physiol.,* 43, 409, 1981.
2. **Granger, D. N., Kvietys, P. R., Premen, A. J., and Korthuis, R. J.,** Microcirculation of the intestinal mucosa, in *Handbook of Gastrointestinal Physiology,* Vol. 1, Wood, J. D., Ed., American Physiological Society, Rockville, MD, 1989, chap. 39.
3. **Shepherd, A. P.,** Local control of intestinal oxygenation and blood flow, *Annu. Rev. Physiol.,* 44, 13, 1982.
4. **Wood, J. D.,** Physiology of the enteric nervous system, in *Physiology of the Gastrointestinal Tract,* Johnson, L. R. et al., Eds., Raven Press, New York, 1987, chap. 3.
5. **Johnson, P. C.,** Autoregulation of blood flow, *Circ. Res.,* 59, 484, 1986.
6. **Berne, R. M.,** The role of adenosine in the regulation of coronary blood flow, *Circ. Res.,* 47, 807, 1980.
7. **Newby, A. C., Worku, Y., Meghji, P., Nakazawa, M., and Skladanowski, A. C.,** Adenosine: a retaliatory metabolite or not?, *News Physiol. Sci.,* 5, 67, 1990.
8. **Schrader, J.,** Adenosine: a homeostatic metabolite in cardiac energy metabolism, *Circulation,* 81, 389, 1990.
9. **Granger, H. J. and Norris, C. P.,** Intrinsic regulation of intestinal oxygenation in the anesthetized dog, *Am. J. Physiol.,* 238, H836, 1980.
10. **Granger, H. J. and Nyhof, R. A.,** Dynamics of intestinal oxygenation: interactions between oxygen supply and oxygen uptake, *Am. J. Physiol.,* 243, G91, 1982.
11. **Rudolph, F. B., Kulkarni, A. D., Fanslow, W. C., Pizzini, R. P., Kumar, S., and Van Buren, C. T.,** Role of RNA as a dietary source of pyrimidines and purines in immune function, *Nutrition,* 6, 45, 1990.
12. **Berlin, R. D. and Hawkins, R. A.,** Secretion of purines by the small intestine: transport mechanism, *Am. J. Physiol.,* 215, 942, 1968.
13. **Harms, V. and Stirling, C. E.,** Transport of purine nucleotides and nucleosides by *in vitro* rabbit ileum, *Am. J. Physiol.,* 233, E47, 1977.

14. **Kolassa, N., Schutzenberger, W. G., Wiener, H., and Turnheim, K.,** Active secretion of hypoxanthine and xanthine by guinea pig jejunum *in vitro, Am. J. Physiol.,* 238, G141, 1980.

15. **Kolassa, N., Stengg, R., and Turnheim, K.,** Adenosine uptake by the isolated epithelium of the guinea pig jejunum, *Can. J. Physiol. Pharmacol.,* 55, 1033, 1977.

16. **Clifford, A. J. and Story, D. L.,** Levels of purines in foods and their metabolic effects in rats, *J. Nutr.,* 106, 435, 1976.

17. **Pizzini, R. P., Kumar, S., Kulkarni, A. D., Rudolph, F. B., and Van Buren, C. T.,** Dietary nucleotides reverse malnutrition and starvation-induced immunosuppression, *Arch. Surg.,* 125, 86, 1990.

18. **Leleiko, N. S., Martin, B. A., Walsh, M., Kazlow, P., Rabinowitz, S., and Sterling, K.,** Tissue-specific gene expression results from a purine- and pyrimidine-free diet and 6-mercaptopurine in the rat small intestine and colon, *Gastroenterology,* 93, 1014, 1987.

19. **Hirst, G. D. S. and Neild, T. O.,** Evidence for two populations of excitatory receptors for noradrenaline on arteriolar smooth muscle, *Nature (London),* 283, 767, 1980.

20. **Holman, M. E. and Surprenant, A. M.,** An electrophysiological analysis of the effects of noradrenaline and α-receptor antagonists on neuromuscular transmission in mammalian muscular arteries, *Br. J. Pharmacol.,* 71, 651, 1980.

21. **Muramatsu, I.,** Evidence for sympathetic, purinergic transmission in the mesenteric artery of the dog, *Br. J. Pharmacol.,* 87, 478, 1986.

22. **Burnstock, G.,** Cholinergic and purinergic regulation of blood vessels, in *Handbook of Physiology, Section 2, The Cardiovascular System, Vol. II, Vascular Smooth Muscle,* American Physiological Society, Bethesda, MD, 1982, 567.

23. **Burnstock, G.,** Local control of blood pressure by purines, *Blood Vessels,* 24, 156, 1987.

24. **Burnstock, G. and Kennedy, C.,** A dual function of adenosine 5'-triphosphate in regulation of vascular tone: excitatory cotransmitter with noradrenaline from perivascular nerves and locally released inhibitory intravascular agent, *Circ. Res.,* 58, 319, 1985.

25. **Gordon, J. L.,** Extracellular ATP: effects, sources, and fate, *Biochem. J.,* 233, 309, 1986.

26. **Lloyd, H. G. E., Deussen, A., Wuppermann, H., and Schrader, J.,** The transmethylation pathway as a source for adenosine in the isolated guinea pig heart, *Biochem. J.,* 252, 489, 1988.

27. **Sarau, H. M., Foley, J., Moonsammy, G., Wiebelhaus, V. D., and Sachs, G.,** Metabolism of dog gastric mucosa: nucleotide levels in parietal cells, *J. Biol. Chem.,* 250, 8321, 1975.

28. **Dobbins, J. W., Laurenson, J. P., and Forrest, J. N.,** Adenosine and adenosine analogues stimulate adenosine cyclic 3'-5'-monophosphate dependent chloride secretion in the mammalian ileum, *J. Clin. Invest.,* 74, 929, 1985.

29. **Wilson, D. W. and Wilson, H. C.,** Studies *in vitro* of the digestion and absorption of purine ribonucleotides by the intestine, *J. Biol. Chem.,* 237, 1643, 1962.

30. **Mortillaro, N. A. and Mustafa, S. J.,** Possible role of adenosine in the development of post occlusion reactive hyperemia, *Fed. Proc., Fed. Am. Soc. Exp. Biol.,* 37 (Abstr.), 874, 1978.

31. **Olsson, R. A., Saito, D., and Steinhart, C. R.,** Compartmentalization of the adenosine pool of dog and rat hearts, *Circ. Res.,* 50, 617, 1982.

32. **Sawmiller, D. R. and Chou, C. C.,** Jejunal adenosine increases during food-induced jejunal hyperemia, *Am. J. Physiol.,* 258, G370, 1990.

33. **Collis, M. G.,** The vasodilator role of adenosine, *Pharmacol. Ther.,* 41, 143, 1989.

34. **Gerlach, E., Becker, B. F., and Nees, S.,** Formation of adenosine by vascular endothelium: a homeostatic and antithrombogenic mechanism?, in *Topics and Perspectives in Adenosine Research,* Gerlach, E. and Becker, B. F., Eds., Springer-Verlag, Berlin, 1987, 309.

35. **Pearson, J. D. and Gordon, J. L.,** Nucleotide metabolism by endothelium, *Annu. Rev. Physiol.,* 47, 617, 1985.

36. **Kroll, K., Kelm, M. K. M., Burrig, K. F., and Schrader, J.,** Transendothelial transport and metabolism of adenosine and inosine in the intact rat aorta, *Circ. Res.,* 64, 1147, 1989.

37. **Gerber, J. G. and Guth, P. H.,** Role of adenosine in the gastric blood flow response to pentagastrin in the rat, *J. Pharmacol. Exp. Ther.,* 251, 550, 1989.

38. **Granger, D. N., Valleau, J. D., Parker, R. E., Lane, R. S., and Taylor, A. E.,** Effects of adenosine on intestinal hemodynamics, oxygen delivery, and capillary fluid exchange, *Am. J. Physiol.,* 235, H707, 1978.

39. **Granger, H. J. and Norris, C. P.,** Role of adenosine in local control of intestinal circulation in the dog, *Circ. Res.,* 46, 764, 1980.

40. **Proctor, K. G.,** Possible role for adenosine in local regulation of absorptive hyperemia in rat intestine, *Circ. Res.,* 59, 474, 1986.

41. **Proctor, K. G.,** Intestinal arteriolar responses to mucosal and serosal application of adenosine analogs, *Circ. Res.,* 61, 187, 1987.

42. **Sawmiller, D. R. and Chou, C. C.,** Adenosine plays a role in food-induced jejunal hyperemia, *Am. J. Physiol.,* 255, 168, 1988.

43. **Sawmiller, D. R. and Chou, C. C.,** Dipyridamole enhances postprandial intestinal hyperemia, *FASEB J.,* 4 (Abstr.), A1188, 1990,

44. **Walus, K. M., Fondacaro, J. D., and Jacobson, E. D.,** Effects of adenosine and its derivatives on the canine intestinal vasculature, *Gastroenterology,* 81, 327, 1981.

45. **Shepherd, A. P., Riedel, G. L., Maxwell, L. C., and Kiel, J. W.,** Selective vasodilators redistribute intestinal blood flow and depress oxygen uptake, *Am. J. Physiol.,* 247, G377, 1984.

46. **Bohlen, H. G.,** Intestinal tissue PO_2 and microvascular responses during glucose exposure, *Am. J. Physiol.,* 238, H164, 1980.

47. **Holm-Rutili, L., Perry, M. A., and Granger, D. N.,** Autoregulation of gastric blood flow and oxygen uptake, *Am. J. Physiol.,* 241, G143, 1981.

48. **Lautt, W. W.,** Autoregulation of superior mesenteric artery is blocked by adenosine antagonism, *Can. J. Physiol. Pharmacol.,* 64, 1291, 1986.

49. **Stojanov, I. and Proctor, K. G.,** Pharmacological evidence for A1 and A2 adenosine receptors in the skin microcirculation, *Circ. Res.,* 65, 176, 1989.

50. **Stojanov, I. and Proctor, K. G.,** Temperature sensitive adenosine-mediated vasoconstriction in the skin microcirculation, *J. Pharmacol. Exp. Ther.,* 253, 1083, 1990.

51. **Ralevic, V. and Burnstock, G.,** Actions mediated by P_2 purinoceptor subtypes in the isolated perfused mesenteric bed of the rat, *Br. J. Pharmacl.,* 95, 637, 1988.

52. **Daly, J. W.,** Adenosine agonists and antagonists, in *Purines in Cellular Signalling: Targets for New Drugs,* Jacobson, K. A., et al., Eds., Springer-Verlag, New York, 1990, 3.

53. **Folkow, B., Lewis, D. H., Lundgren, O., Mellander, S., and Wallentin, I.,** The effect of graded vasoconstrictor fiber stimulation on the intestinal resistance and capacitance vessels, *Acta Physiol. Scand.,* 61, 445, 1964.

54. **Lautt, W. W., Lockhart, L. K., and Legare, D. J.,** Adenosine modulation of vasoconstrictor responses to stimulation of sympathetic nerves and norepinephrine infusion in the superior mesenteric artery of the cat, *Can. J. Physiol. Pharmacol.,* 66, 937, 1988.

55. **Crissinger, K. D., Kvietys, P. R., and Granger, D. N.,** Autoregulatory escape from norepinephrine infusion: roles of adenosine and histamine, *Am. J. Physiol.,* 254, G560, 1988.

56. **Granger, D. N.,** Role of xanthine oxidase and granulocytes in ischemia-reperfusion injury, *Am. J. Physiol.,* 255, H1269, 1988.

57. **Grisham, M. B., Hernandez, L. A., and Granger, D. N.,** Adenosine inhibits ischemia-reperfusion-induced leukocyte adherence and extravasation, *Am. J. Physiol.,* 257, H1334, 1989.

58. **Kaminski, P. M. and Proctor, K. G.,** Attenuation of no-reflow phenomenon, neutrophil activation, and reperfusion injury in intestinal microcirculation by topical adenosine, *Circ. Res.,* 65, 426, 1989.

59. **Kaminski, P. M. and Proctor, K. G.,** Actions of adenosine on nitro blue tetrazolium deposition and surface pH during intestinal reperfusion injury, *Circ. Res.,* 66, 1713, 1990.

60. **Cronstein, B. N., Angaw-Duguma, L., Nicholls, D., Hutchison, A., and Williams, M.,** Adenosine is an antiinflammatory autacoid: adenosine receptor occupancy promotes neutrophil chemotaxis and inhibits superoxide anion generation, in *Purines in Cellular Signalling: Targets for New Drugs,* Jacobson, K. A., et al., Eds., Springer-Verlag, New York, 1990, 114.

61. **Debas, H. T.,** Peripheral regulation of gastric acid secretion, in *Physiology of the Gastrointestinal Tract,* Johnson, L. R. et al., Eds., Raven Press, New York, 1987, chap. 31.

62. **Westerberg, V. S. and Geiger, J. D.,** Adenosine analogs inhibit gastric acid secretion, *Eur. J. Pharmacol.,* 160, 275, 1989.

63. **Gerber, J. G. and Payne, N. A.,** Endogenous adenosine modulates gastric acid secretion to histamine in canine parietal cells, *J. Pharmacol. Exp. Ther.,* 244, 190, 1988.

64. **Holm, L. and Perry, M. A.,** Role of blood flow in gastric acid secretion, *Am. J. Physiol.,* 254, G281, 1988.

65. **Holm-Rutili, L. and Obrink, K. J.,** Rat gastric mucosal microcirculation *in vivo, Am. J. Physiol.,* 248, G741, 1985.

66. **Holm-Rutili, L., and Berglindh, T.,** Pentagastrin and gastric mucosal blood flow, *Am. J. Physiol.,* 250, G575, 1986.

67. **Chou, C. C.,** Splanchnic and overall cardiovascular hemodynamics during eating and digestion, *Fed. Proc., Fed. Am. Soc. Exp. Biol.,* 42, 1658, 1983.

68. **Gallavan, R. H. and Chou, C. C.,** Possible mechanisms for the initiation and maintenance of postprandial intestinal hyperemia, *Am. J. Physiol.,* 249, G301, 1985.

69. **Chou, C. C., Kvietys, P., Post, J., and Sit, S. P.,** Constituents of chyme responsible for postprandial intestinal hyperemia, *Am. J. Physiol.,* 235, H677, 1978.

70. **Chou, C. C., Nyhof, R. A., Kvietys, P. R., Sit, S. P., and Gallavan, R. H.,** Regulation of jejunal blood flow and oxygenation during glucose and oleic acid absorption, *Am. J. Physiol.,* 249, G691, 1985.

71. **Eklund, S., Cassuto, J., Jodal, M., and Lundgren, O.,** The involvement of the enteric nervous system in the intestinal secretion evoked by cyclic adenosine 3', 5'-monophosphate, *Acta Physiol. Scand.,* 120, 311, 1984.

72. **Gallavan, R. H., Chen, M. H., Joffe, S. N., and Jacobson, E. D.,** Vasoactive intestinal polypeptide, cholecystokinin, glucagon, and bile-oleate-induced jejunal hyperemia, *Am. J. Physiol.,* 248, G208, 1985.

73. **Bohlen, H. G.,** Na^+-induced intestinal interstitial hyperosmolality and vascular responses during absorptive hyperemia, *Am. J. Physiol.,* 242, H785, 1982.

74. **Proctor, K. G.,** Contribution of hyperosmolality to glucose-induced intestinal hyperemia, *Am. J. Physiol.,* 248, G521, 1985.

75. **Mangino, M. J. and Chou, C. C.,** Arachidonic acid and postprandial intestinal hyperemia, *Am. J. Physiol.,* 246, G521, 1984.

76. **Mangino, M. J. and Chou, C. C.,** Thromboxane synthesis inhibition and postprandial intestinal hyperemia and oxygenation, *Am. J. Physiol.,* 250, G64, 1986.

77. **Proctor, K. G.,** Differential effect of cyclooxygenase inhibitors on absorptive hyperemia, *Am. J. Physiol.,* 249, H755, 1985.

78. **Chou, C. C. and Siregar, H.,** Role of histamine H_1 and H_2 receptors in postprandial intestinal hyperemia, *Am. J. Physiol.,* 243, G248, 1982.

79. **Anzueto, L., Benoit, J. N., and Granger, D. N.,** A rat model for studying the intestinal circulation, *Am. J. Physiol.,* 246, G56, 1984.

80. **Proctor, K. G., Falck, J. R., and Capdevila, J.,** Intestinal vasodilation by epoxyeicosatrienoic acids: arachidonic acid metabolites produced by a cytochrome P450 monoxygenase, *Circ. Res.,* 60, 50, 1987.

81. **Sit, S. P. and Chou, C. C.,** Time course of jejunal blood flow, O_2 uptake, and O_2 extraction during nutrient absorption, *Am. J. Physiol.,* 247, H395, 1984.

82. **Daly, J. W.,** Adenosine receptors: targets for future drugs, *J. Med. Chem.,* 25, 197, 1982.

Chapter 17

ADENOSINE-MEDIATED REGULATION OF HEPATIC BLOOD FLOW

W. Wayne Lautt

TABLE OF CONTENTS

The hepatic circulation is, by its very nature, a unique system with its dual blood flow, highly permeable sinusoids, and huge blood volume reservoir capacity. Recent reviews describe these various aspects in some detail.[1,2] Intrinsic regulation of hepatic blood flow operates through a mechanism that appears to be unique to the liver. The adenosine washout hypothesis is suggested to account for both types of intrinsic blood flow regulation.

I. BACKGROUND

The portal vein drains blood from the pancreas, spleen, intestines, stomach, and omentum, and provides two thirds of the total blood supply to the liver. This flow is not under control of the liver. The hepatic artery provides the remaining one third of flow and it is only through the hepatic artery that the liver can regulate its own blood flow.

The microvascular unit of the liver is the hepatic acinus, a cluster of parenchymal cells hanging like a berry on a vascular stalk. About 100,000 acini, each 2 mm in diameter, comprise the human liver. Blood is supplied to each acinus via terminal branches of the hepatic arteriole and portal venule. These vessels, along with a terminal bile ductule, lie in intimate contact within a small fluid space, the space of Mall. A limiting plate of hepatic cells surrounds this "portal triad" of vessels and separates the space of Mall from other fluid compartments.

A number of neural, humoral, and physical factors act as extrinsic influences on hepatic blood flow. Intrinsic regulation refers to mechanisms that regulate hepatic arterial flow exclusive of nerves and blood-borne vasoactive compounds. This area has recently undergone some major conceptual changes. Previous views were that the hepatic arterial flow was controlled by metabolic demands of the parenchymal cell mass; the hepatic artery showed mild autoregulation in response to altered perfusion pressure, and this autoregulation was myogenic in origin; altered portal perfusion caused an inverse change in hepatic arterial flow that was mediated by a myogenic mechanism or was secondary to altered oxygen supply. Recent data disproved these earlier concepts and are consistent with three hypotheses: (1) the hepatic artery is not controlled or affected by parenchymal metabolism; (2) the hepatic artery shows mild autoregulation that is dependent on a washout of locally produced adenosine; (3) the effect of portal flow on hepatic arterial resistance is mediated by washout of locally produced adenosine, which results in inverse changes in hepatic arterial flow. This latter mechanism, the hepatic arterial buffer response, is a major intrinsic control that buffers the impact of portal flow changes on total hepatic flow.

II. LACK OF METABOLIC CONTROL OF THE HEPATIC ARTERY

The data showing lack of subservience of the hepatic arterial flow to hepatic oxygen or metabolic requirements have been reviewed.[3] Alteration of hepatic oxygen supply by hemodilution or reducing oxygen content in arterial blood selectively perfusing the liver (unpublished observations) does not lead to the expected vasodilation of the hepatic artery. Alteration of oxygen demand by the use of dinitrophenol to increase oxygen demand, or SKF525A, to inhibit metabolism, similarly did not lead to the anticipated changes in arterial vascular tone. These observations are compatible with the morphology of the hepatic acinus where blood from the terminal branches of the hepatic arteriole and portal venule enter the center of the acinus and drain past 15 to 20 hepatocytes into the hepatic venules. Blood in adjacent sinusoids flows concurrently and there is no opportunity for metabolites that enter the hepatic sinusoids to diffuse back upstream to the arterial resistance vessels.

In these early studies that demonstrated a lack of metabolic control of the hepatic artery, one observation was consistently made, that is, as portal blood flow changed acutely within the experiment, hepatic arterial flow changed in the opposite direction.

III. HEPATIC ARTERIAL BUFFER RESPONSE

The earliest known studies reporting an effect of changes in portal perfusion on the hepatic arterial flow were by Betz in 1863 and Gad in 1873. The term hepatic arterial buffer response (HABR) was suggested for this intrinsic regulatory mechanism.[4] The mechanism of this response has remained elusive until recently. To study the mechanism it was essential to have a preparation that shows reproducible responses that could then be manipulated according to the hypothesis being tested. A suitable model was developed in 1985.[5] Hepatic arterial flow and superior mesenteric arterial flow were measured with electromagnetic flow meters after all other arteries supplying the portal flow were ligated. Although the responses produced were not quantitatively normal, the reproducibility of the response was well suited for pharmacological studies of the mechanism. The most consistent quantitative expression of the buffer response is the buffer capacity — the change in hepatic arterial flow divided by the change in portal flow expressed as a percentage. The full range of hepatic arterial dilation and constriction is able to be attained by producing very low and high portal flows, respectively.[6]

A. HEPATIC ARTERIAL BUFFER RESPONSE HYPOTHESIS

The HABR hypothesis proposes that adenosine is constantly secreted into the fluid of the space of Mall. The local concentrations of adenosine are regulated by the rate of washout into the terminal portal venule that lies in intimate contact with the hepatic arteriole. If portal blood flow decreases, less adenosine is washed away and the accumulated adenosine causes the hepatic arteriole to dilate, thus increasing arterial flow (Figure 1).

B. CRITERIA

The first criterion that adenosine must fulfill as a candidate for the intrinsic regulator of the HABR is to dilate the hepatic artery. Dose-response curves indicate that adenosine is an extremely effective dilator. A dose of 1 to 2 mg/min per kilogram body weight into the hepatic artery more than doubles hepatic arterial conductance without systemic hemodynamic effects due to recirculation.[7] The extreme lability of adenosine prevents calculations of local adenosine concentrations from the rates of infusion. However, pharmacodynamic analysis indicates that the maximal dilation obtainable with adenosine infusions is equivalent to an increase in conductance by 150%. The intraportal ED_{50} is roughly 0.2 mg/kg body weight.[7] Adenosine causes the most dramatic dilation of the hepatic artery reported with any drug or stimulus.[1]

If adenosine levels in the area of the hepatic arterial resistance vessels are controlled by washout into portal blood, it must be demonstrated that portal blood has access to the arterial resistance vessels. This criterion was shown by the fact that vasoactive compounds infused into the portal blood produced vascular effects on the hepatic arterioles. Adenosine infused into the portal vein produces a dilation of at least 50% the magnitude that is seen with the same infusion directly into the hepatic artery.[5] If adenosine can diffuse from portal blood to arterial resistance vessels, it is reasonable to assume that diffusion can also occur in the opposite direction.

A third criterion is that potentiators of exogenous adenosine effects should potentiate the buffer response. Dipyridamole blocked uptake of adenosine into cells[8] and potentiated the dilator effects of exogenous adenosine. The buffer response was also potentiated.[5]

A fourth criterion is that blockers of exogenous adenosine effects should inhibit the buffer response. This criterion has been fulfilled with 3-isobutyl-1-methylxanthine;[5] with the more consistent antagonist 8-phenyltheophylline;[7,9] and with 8-sulfophenyltheophylline.[10] These antagonists blocked the buffer response as well as the response to close arterial infusions of adenosine. Dose-response relationships showed that the capacity of the buffer

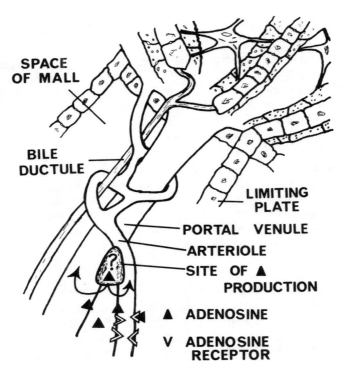

FIGURE 1.   Model for control of hepatic arterial flow by adenosine concentrations that are regulated by washout into portal blood vessels. Area shown is the portal triad contained within a limiting plate of parenchymal cells. Adenosine produced at a site in or near the hepatic artery can accumulate to produce dilation or can be washed away into portal venules. Reduced portal flow leads to accumulated adenosine and arterial dilation, thus accounting for the hepatic arterial buffer response. Similarly, washout into the hepatic artery accounts for arterial autoregulation. (From Lautt, W. W., Legare, D. J., and d'Almeida, M. S., *Am. J. Physiol.,* 248, H331, 1985. With permission.)

response and the dilator effect of exogenous adenosine were reduced in a parallel manner[7] (Figure 2). An important demonstration made in all of these studies is that the dilator effect of isoproterenol was not altered by the adenosine antagonists, thereby demonstrating the selective nature of the antagonist and avoiding the obvious potential pitfall that lack of an intrinsic dilator response could possibly be due to preparation deterioration.

Antagonists of the dilator effects of exogenous adenosine do not necessarily block endogenous adenosine-induced effects. Almost 50% inhibition of exogenous adenosine is seen before significant blockade of the HABR is detected,[7] as shown in Figure 2.

Aminophylline and theophylline produce profound systemic hemodynamic effects at doses needed to block the HABR.[4] 1,3-Dipropyl-8-*p*-sulfophenylxanthine is a very poor antagonist of the HABR, we suspect because it has a strong electrical charge at physiological pH and may therefore be excluded from the space of Mall.[27]

Caffeine has proven to be an effective antagonist of vascular adenosine receptors in the central nervous system.[11] However, it is a very weak antagonist in both the hepatic and intestinal circulations.[12] In the hepatic artery of the cat, caffeine is a noncompetitive antagonist of adenosine-induced vasodilation, and this antagonism is only seen at very high doses. Caffeine showed a maximal inhibition of 59% of the vasodilation induced by adenosine. Caffeine did not alter the response to isoproterenol. This indicates that adenosine may be

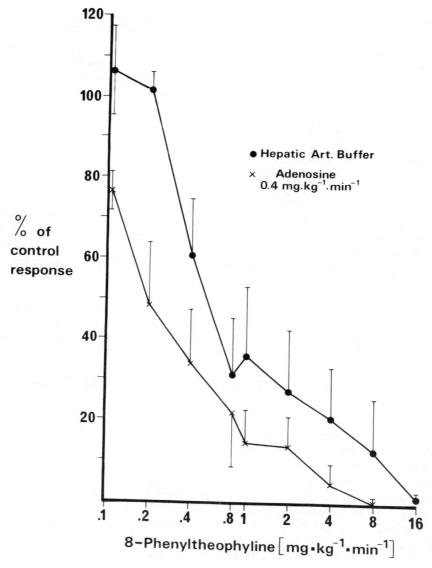

FIGURE 2. The progressive competitive antagonism of the hepatic arterial vasodilator effect of complete reduction of portal blood flow (the hepatic arterial buffer response) and the dilator effect of intraportal adenosine in the presence of progressive increases in intra-arterial dose of adenosine antagonist (8-phenyltheophylline). The control dilation (percent change in conductance) seen with the buffer response was classed as 100%. The first responses plotted were in the presence of 0.1 mg $\cdot$ kg^{-1} $\cdot$ min^{-1} of antagonist. Note the generally parallel depression of vasodilator effect of exogenous adenosine and the buffer response. Roughly three times the dose of 8-phenyltheophylline is needed to produce equivalent depression of the buffer response compared with the response to exogenous adenosine (n = 5, mean ± SE). (From Lautt, W. W. and Legare, D. J., *Can. J. Physiol. Pharmacol.*, 63, 717, 1985. With permission.)

producing dilation in the hepatic artery through two independent mechanisms, one of which (calcium dependent?) is antagonized by caffeine and the other (cAMP dependent?) is not blocked.

Overall, several criteria have been fulfilled to support the hypothesis that the local concentration of adenosine at the site of the hepatic arterial resistance vessels is controlled by washout into the portal vein. Alternate hypotheses have been considered and rejected.[4] Alternate hypotheses not involving adenosine can be eliminated based on the complete abolition of the buffer response by adenosine antagonists, although the hepatic artery shows normal vascular responses to nerve stimulation and infusions of vasoconstrictors and vasodilators.[7] (The development of this hypothesis with details of alternate hypotheses and implications can be followed in a series of reviews.)[2,3,4,13]

IV. AUTOREGULATION

Autoregulation is the response whereby, for example, an increase in arterial perfusion pressure results in constriction of the perfused artery. This has previously been assumed to be due to a myogenic response of the arterial smooth muscle to stretch. Recently, however, we have shown that, like the HABR where portal flow can wash adenosine from the area of the hepatic arterial resistance vessels, flow in the hepatic artery can also wash out this endogenous adenosine to completely account for autoregulation of the hepatic artery.[10] By this mechanism, increased arterial flow secondary to increased pressure results in a washout of adenosine from the space of Mall and leads to a constriction of the hepatic artery. The adenosine antagonists, 8-phenyltheophylline and 8-sulfophenyltheophylline, convert the typical pressure flow curve from one that is convex to the pressure axis to one that is linear. Dose-related decreases in autoregulation, buffer response, and response to exogenous adenosine are demonstrated.[10] The observation that the mechanism of autoregulation is compatible with a washout mechanism provides strong, but indirect support for a similar mechanism of the HABR.

V. UNTESTED ASPECTS OF THE HABR HYPOTHESIS

The site of adenosine production and the site of hepatic arterial resistance are not confirmed. These sites may be localized within the portal triad, enclosed within the limiting plate that surrounds the hepatic arteriole, portal venule, and bile ductule in the space of Mall. Although the data are consistent with the hypothesis that adenosine production is constant and local concentrations are regulated by washout, it is not clear if the basal production of adenosine can be altered by physiological factors. In addition, the metabolic pathway producing the adenosine is unclear.

The capacity of the HABR in the conscious animal has not been reported. Buffer capacities reported over the years vary greatly, depending on the methodology used. Isolated perfused livers[14] and livers perfused with arterial long-circuits[15] show weak buffer capacities. Anesthetized acute surgical preparations show a buffer capacity of about 25% in cats[4,5,16] and dogs.[17] Other studies done for different purposes where the arterial pressure is uncontrolled suggest buffer capacities of 61 to 100% in rats[18] and dogs[19,20] and 44% in anesthetized cats.[6] Problems of quantitation have been discussed in detail.[21]

VI. PHYSIOLOGICAL ROLES OF THE HABR

The role of the buffer response has not yet been studied extensively. The original hypothesis relating the HABR to liver function proposed that the hepatic artery, rather than being subservient to hepatic metabolic demands, was the guardian of humoral clearance rates.[13] Reduction of hepatic blood flow is known to decrease hepatic clearance of a variety of endogenous and exogenous compounds.[21,22] The HABR buffers changes in portal flow and tends to minimize the impact of these portal flow changes on hepatic clearances.

The HABR serves to selectively protect the oxygen supply to the liver in situations of low blood flow states. During hemorrhage the portal blood flow with its low oxygen content is reduced, thus activating the HABR and increasing the amount of well-oxygenated arterial blood perfusing the liver. This protective effect of the HABR is completely eliminated by adenosine receptor antagonists.[9]

VII. CLINICAL RELEVANCE

The lack of ability of the liver to regulate its blood flow in accordance with the metabolic activity of the parenchymal cells is not disadvantageous to the liver under normal physiological conditions because of the excess oxygen delivered and the hepatic capacity to increase oxygen extraction. However, a relatively hypoxic liver shows increased toxic effects of alcohol,[23] carbon tetrachloride,[24] and halothane.[25] The formation of active toxic metabolites may occur to a greater extent, and the ability of glutathione to detoxify such metabolites may be reduced when the $NAD^+/NADH$ ratio is altered by hypoxia. Could selective hepatic arterial vasodilation protect the liver by providing a more aerobic environment? If the production of adenosine in the space of Mall is not dependent upon the energy status of the hepatic parenchymal cells, it seems unlikely that adenosine is produced via breakdown of the adenine nucleotides. If adenosine is able to be produced by some alternate pathway, such as demethylation,[26] it seems possible that pharmacological intervention may be possible to produce selective dilation of the hepatic arterial blood supply to perhaps aid in liver regeneration and recovery from toxic reactions.

ACKNOWLEDGMENTS

This work has been supported by operating grants from the Manitoba Heart and Stroke Foundation. Collaborators for the projects include Dallas J. Legare, Waleed Ezzat, Mark d'Almeida, and Janet McQuaker. The manuscript was prepared by Karen Sanders.

REFERENCES

1. **Greenway, C. V. and Lautt, W. W.,** Hepatic circulation, in *Handbook of Physiology—The Gastrointestinal System I,* Vol. 1, Schultz, S. G., Wood, J. D., and Rauner, B. B., Eds., American Physiological Society, Oxford University Press, New York, 1989, 1519.
2. **Lautt, W. W. and Greenway, C. V.,** Conceptual review of the hepatic vascular bed, *Hepatology,* 7, 952, 1987.
3. **Lautt, W. W.,** Relationship between hepatic blood flow and overall metabolism: the hepatic arterial buffer response, *Fed. Proc., Fed. Am. Soc. Exp. Biol.,* 42, 1662, 1983.
4. **Lautt, W. W.,** Role and control of the hepatic artery, in *Hepatic Circulation in Health and Disease,* Lautt, W. W., Ed., Raven Press, New York, 1981, 203.
5. **Lautt, W. W., Legare, D. J., and d'Almeida, M. S.,** Adenosine as putative regulator of hepatic blood flow (the buffer response), *Am. J. Physiol.,* 248, H331, 1985.
6. **Lautt, W. W., Legare, D. J., and Ezzat, W. R.,** Quantitation of the hepatic arterial buffer response to graded changes in portal blood flow, *Gastroenterology,* 98, 1, 1990.
7. **Lautt, W. W. and Legare, D. J.,** The use of 8-phenyltheophylline as a competitive antagonist of adenosine and inhibitor of the intrinsic regulatory mechanism of the hepatic artery, *Can. J. Physiol. Pharmacol.,* 63, 717, 1985.
8. **Moritoki, H.,** Possible mechanism of potentiation of the action of adenosine by some vasodilators, in *Physiology and Pharmacology of Adenosine Derivatives,* Daly, J. W., Kuroda, Y., Phillis, J. W., Shimizu, H., and Ui, M., Eds., Raven Press, New York, 1983, 197.
9. **Lautt, W. W. and McQuaker, J. E.,** Maintenance of hepatic arterial blood flow during hemorrhage is mediated by adenosine, *Can. J. Physiol. Pharmacol.,* 67, 1023, 1989.
10. **Ezzat, W. R. and Lautt, W. W.,** An adenosine mediated mechanism of hepatic arterial pressure-flow autoregulation, *Am. J. Physiol.,* 252, H836, 1987.

11. **Phillis, J. W. and DeLong, R. E.**, An involvement of adenosine in cerebral blood flow regulation during hypercapnia, *Gen. Pharmacol.*, 18, 133, 1987.

12. **Lautt, W. W.**, Non-competitive antagonism of adenosine by caffeine on the hepatic and superior mesenteric arteries of anesthetized cats, *J. Pharmacol. Exp. Ther.*, 254, 400, 1990.

13. **Lautt, W. W.**, The hepatic artery: subservient to hepatic metabolism or guardian of normal hepatic clearance rates of humoral substances, *Gen. Pharmacol.*, 8, 73, 1977.

14. **Sato, T., Shirataka, M., Ikeda, N., and Grodins, F. S.**, Steady-state systems analysis of hepatic hemodynamics in the isolated perfused canine liver, *Am. J. Physiol.*, 233, R188, 1977.

15. **Hanson, K. M. and Johnson, P. C.**, Local control of hepatic arterial and portal venous flow in the dog, *Am. J. Physiol.*, 211, 712, 1966.

16. **Lautt, W. W. and Legare, D. J.**, Adenosine modulation of hepatic arterial but not portal venous constriction induced by sympathetic nerves, norepinephrine, angiotensin and vasopressin, *Can. J. Physiol. Pharmacol.*, 64, 449, 1986.

17. **Mathie, R. T. and Blumgart, L. H.**, The hepatic haemodynamic response to acute portal venous blood flow reductions in the dog, *Pfluegers Arch.*, 399, 223, 1983.

18. **Fernandez-Munoz, D., Caramelo, C., Santos, J. C., Blanchart, A., Hernando, L., and Lopez-Novoa, J. M.**, Systemic and splanchnic haemodynamic disturbances in conscious rats with experimental liver cirrhosis without ascites, *Am. J. Physiol.*, 249, G316, 1985.

19. **Groszmann, R. J., Blei, A. T., Kniaz, J. L., Storer, E. H., and Conn, H. O.**, Portal pressure reduction induced by partial mechanical obstruction of the superior mesenteric artery in the anaesthetized dog, *Gastroenterology*, 75, 187, 1978.

20. **Hughes, R. L., Mathie, R. T., Fitch, W., and Campbell, D.**, Liver blood flow and oxygen consumption during metabolic acidosis and alkalosis in the greyhound, *Clin. Sci.*, 60, 355, 1980.

21. **Lautt, W. W.**, Mechanism and role of intrinsic regulation of hepatic arterial blood flow: the hepatic arterial buffer response, *Am. J. Physiol.*, 249, G549, 1985.

22. **Messerli, F. H., Nowaczynski, W., Honda, M., Genest, J., Boucher, R., Kuchel, O., and Rojo-Ortega, J. M.**, Effects of angiotensin II on steroid metabolism and hepatic blood flow in man, *Circ. Res.*, 40, 204, 1977.

23. **Israel, Y., Kalant, H., Orrego, H., Khanna, J. M., Videla, L., and Phillips, J. M.**, Experimental alcohol-induced hepatic necrosis: suppresson by propylthiouracil, *Proc. Natl. Acad. Sci. U.S.A.*, 72, 1137, 1975.

24. **Shen, E. S., Garry, V. F., and Ander, M. W.**, Effect of hypoxia on carbon tetrachloride hepatotoxicity, *Biochem. Pharmacol.*, 31, 3787, 1982.

25. **McLain, G. E., Sipes, I. G., and Brown, B. B.**, An animal model of halothane hepatotoxicity, *Anesthesiology*, 51, 321, 1979.

26. **Lloyd, H. G. E. and Schrader, J.**, The importance of the transmethylation pathway for adenosine metabolism in the heart, in *Topics and Perspectives in Adenosine Research*, Gerlach E. and Becker, B. F., Eds., Springer-Verlag, Berlin, 1987, 199.

27. **d'Almeida, M. S., Zhang, Y., and Lautt, W. W.**, Unpublished observations.

Chapter 18

EFFECTS OF ADENOSINE AND ADENINE NUCLEOTIDES ON THE PULMONARY CIRCULATION

A. Hyman, B. Cai, C. Feng, Q. Hao, and H. Lippton

TABLE OF CONTENTS

I. INTRODUCTION

Although the pulmonary vascular bed is in series with the systemic bed, it propels the entire cardiac output through the lung at a pressure less than 20% of the systemic vascular pressure. By maintaining this low vascular pressure, the lung endeavors to preserve efficient gas exchange and concurrently protect itself against pulmonary edema. Thus, the responses of these vessels to a variety of physiologic and pharmacologic stimuli would be expected to be dissimilar to those in the systemic circulation. Initially, the pulmonary vascular bed was thought to respond as a passive conduit, conducting blood from the right to the left side of the heart by distending or puckering in response to changes in cardiac output, pulmonary blood volume, and left arterial pressure. Moreover, reactive hyperemia and autoregulation, as such, have not been identified in the lung vessels. Nonetheless, an enormous number of studies have confirmed the ability of pulmonary vessels to actively respond to physiologic stimuli such as hypoxia, changes in blood pH, and a host of humoral and exogenously administered vasoactive agents.

Since the first description of the effects of isolated crystalline AMP and of adenosine on the heart and circulation in 1929,[1] the role of adenosine and adenine nucleotides in the regulation of the heart and the systemic vasculature has received much attention, and a recent in-depth review has summarized our present knowledge of cardiovascular purinoceptors.[2] This original description[1] indicated that the response to either adenosine or its precursor, AMP, depends on the presence of both the 6-amino group on the purine base and on the ribose moiety. The duration of action is dependent on the metabolic activity in degradation. Drury and Szent-Gyorgyi[1] described in detail the effects on the heart, inducing at times heart block and at other times serious bradycardia, and even conversion of supraventricular arrhythmias, enhanced coronary blood flow, and intense peripheral vasodilation. Later, Burnstock[3] described certain visceral nerves that released ATP as a neurotransmitter and described the effects of released ATP in the presence of adrenergic and cholinergic blocking agents. This ATP response implicated specific receptors, which Burnstock identified as purinergic receptors and proposed criteria for their identification, both in neural tissue and in other organ systems.[4,5] The likelihood that ATP may act at receptors different from its metabolic breakdown product, adenosine, has been suggested for many years. In some vascular beds ATP is more potent than adenosine, for example, in constricting the pulmonary vascular bed[6,7] or dilating the coronary bed,[8,9] but in others, such as the renal, ATP induces vasodilation and adenosine induces vasoconstriction.[10]

A. PURINERGIC RECEPTOR SUBTYPES

Burnstock[4] classified purinoceptors into two general types on the basis of the natural ligands that they recognize. P_1 receptors recognize adenosine and possible AMP, whereas P_2 receptors recognize ATP and ADP. The classification is less than ideal pharmacologically in that it lacks a large variety of specific agonists and antagonists by which receptor responses can be characterized. Albeit, P_1 receptors are activated by adenosine and not by ATP. Further, these receptors are antagonized by methylxanthines and potentiated by inhibition of adenosine transport with dipyridamole. P_1 receptors are linked to adenylate cyclase and are not generally dependent on prostaglandin synthesis. Conversely, P_2 receptors recognize ATP, but not adenosine, are not antagonized by methylxanthines or potentiated by adenosine uptake inhibitors, and are not linked to adenylate cyclase but do activate prostaglandin production. Further, Van Calker et al.[11] designated P_1 receptors which inhibit adenylate cyclase as A_1, and those that stimulate adenylate cyclase as A_2 receptors. Common attributes of both A_1 and A_2 receptors include their location on the cell surface,[12] their coupling to adenylate cyclase through guanosine 5'-triphosphate-binding transduction proteins[13] and their susceptibility to competitive inhibition by alkylxanthines, such as caffeine and theophylline.[14,15] Strong evidence for two distinct receptor types is further indicated by the fact that A_1 receptors have a molecular mass of 35 to 38 Da, but A_2 receptors have a mass of 45 kDa. In addition, A_1 receptors activate inhibitory G protein, whereas A_2 receptors activate stimulatory G protein. Pertussis toxin ribosylates G_i protein and blocks A_1 receptor activity, whereas cholera toxin inhibits G_s protein and blocks A_2 receptor activity. Moreover, selective agonists and antagonists differ in selected species and organ systems.

Adenosine receptors which couple to effectors other than adenylate cyclase have also been identified. Most of these have the pharmacologic profile of the A_1 receptor type, but whether or not this represents a single receptor coupled to different transduction mechanisms, or whether there is molecular diversity of A_1 receptors is unknown. These A_1-type receptors have been identified as A_1-potassium receptors, which increase outward potassium conduction. G_i protein is implicated in their signal transduction, but cyclic nulceotides apparently play no role.[16] Other A_1-type receptors indirectly activate or antagonize phospholipase receptors. Further adenosine influences histamine (H_1) receptor-initiated hydrolysis of brain inositol phospholipids by phospholipase C.[17] The addition of the calcium ionophore A23187 or the alpha-1 adrenoceptor agonist phenylephrine to hamster brown fat cells stimulates the production of arachidonic acid and lysophospholipids, but adenosine deaminase potentiates the activity of these agonists. On the other hand, N^6-1-phenyl-2R propyladenosine, an A_1 agonist not subject to breakdown by adenosine deaminase, reverses the effect of the latter agent.[18] Additionally, other A_1 receptor analogs promote the concentration-dependent accumulation of cyclic guanosine 5′-monophosphate in cultures of aortic smooth-muscle cells, as well as activation of guanylate cyclase in partially purified plasma membranes of the aortic media.[19] Still other studies indicate the presence of A_1-calcium receptors, in which A_1 receptors are coupled to a calcium channel via G proteins, and these receptors have been shown to reduce the calcium component of nerve cell action potentials.[20] Another cyclic AMP (cAMP)-independent effect of adenosine is on an A_1-glucose receptor which adenosine stimulates to induce glucose transport in the adipocyte. Since this action is blocked by pertussis toxin, this activity appears to be linked to G_i protein.[21] These A_1-glucose receptors potentiate insulin-stimulated glucose uptake in heart muscle, but stimulation of these receptors overshadows the direct effect of adenosine to inhibit glucose uptake.

The effect of these cAMP-independent A_1 receptors on vascular smooth muscle has received little attention, and the effect in intact animals is largely unknown.

Burnstock[4] classified other receptors as P_2 receptors, based on their greater selectively for ATP than adenosine. Since ATP has contractile effects on some organ smooth muscle, but relaxant effects on others, Burnstock and Kennedy[22] suggested that P_2 receptors be subclassified as P_{2x}, indicating those having excitatory or contractile effects, whereas those with inhibitory or relaxant properties be identified as P_{2y} receptors. In general, ADP and ATP have similar activity at P_2 receptors in most blood vessels, but AMP is weak or inactive. Several different effector systems have been linked to P_2 receptor activation. Needleman and co-workers[23] first linked the relaxant effect of ATP to prostaglandin I_2 formation and release. Further, the ATP-induced contraction of other smooth muscle is inhibited by indomethacin suggesting activation of a constrictor-type prostaglandin.[5] In the cardiovascular system, ATP receptor stimulation of the piglet aorta induces prostaglandin release that is unrelated to relaxation, which raises the issue of whether or not the release of prostaglandins is always linked to the contractile or relaxant activity of ATP.[24] The lack of very selective agonists that are specific for P_{2y} and P_{2x} receptors, the difficulty in synthesis of ATP analogs, and the rapidity of hydrolysis of ATP and ADP to adenosine by ectophosphatases make complete characterization of P_2 receptors and subtypes difficult in the cardiovascular system.

II. PULMONARY VASCULAR RESPONSES

Although extensive investigation has identified the action of adenosine nucleotides on the heart and systemic circulation, the effects of these agents on the pulmonary vascular bed has received relatively little attention. This vascular bed is remote, and access for investigation has, for the most part, been from *in vitro* studies, or isolated lung perfusion studies. Only recently has attention been directed toward studying these nucleotides in the pulmonary vessels of intact, spontaneously breathing animals. Actively induced changes in pulmonary vascular tone are often difficult to assess from changes in calculated vascular resistance. Pulmonary vessels may be passively distended (decreased vascular resistance) by concurrent increases in cardiac output, i.e., pulmonary blood flow, or by increases in left atrial pressure, i.e., downstream pressure. On the other hand pulmonary vessels may be passively constricted (increased vascular resistance) by decreasing cardiac output and pulmonary blood flow and blood volume, or decreasing downstream pressure. Thus the concurrent activity of these purinergic agents on systemic vessels in other organs, and indeed on cardiac ionotrophy, is capable of passively altering vascular resistance in the intact lung and obscuring any direct effect of these agents on the intact pulmonary vessels.

Furthermore, the pulmonary vascular endothelium plays an important role in metabolic breakdown of circulating adenosine nucleotides. Although ATP was first thought to be an intracellular purine, it is now established that ATP does cross the cell membrane. Extracellular ATP has been detected in the blood from working muscles and after trauma, from aggregating platelets, from ischemia myocardium, and from stimulated sympathetic nerve terminals.[25] ATP in small doses is virtually all removed from the extracellular space in a single passage through the lung,[26] by an ectonucleotidase on the endothelial cells of the lung.[27] Indeed, subsequent electron microscopic cytochemical studies revealed that 5'-nucleotidase activity was localized to the luminar surface of the plasma membrane and was present mainly on caveolae.[28] Further, other ectopeptidases catabolize ADP, removing a platelet aggregating agent from plasma.[29] There are apparently three separate enzymes that sequentially catabolize ATP, ADP, AMP, and adenosine, and taken together with adenosine transport by endothelium, these ectonucleotidases provide a mechanism for regulating the concentration of circulating vasoactive purines.[30,31] Adenosine itself is also taken up by capillary endothelium in the pulmonary vascular beds, as well as skeletal and cardiac endothelium. Once reintroduced into the cell, it is primarily incorporated into the cellular adenine nucleotide pool, mainly that of vascular endothelium.[32] Although adenosine deaminase was thought to be a cytosolic enzyme, it is now clear that it is an ectoenzyme, reversibly bound to the cell exterior.[33] Thus, the major site of purine degradation is the microvascular endothelium, of which the lung contains the largest supply, and thus plays a major role in its regulation.

A. EFFECTS OF ATP

Early studies of the direct pulmonary vascular effects of adenosine nucleotides, described by Gaddum and Holtz,[7] concluded that small doses of adenosine produced dilatation and larger doses constriction. Additionally, adenylpyrophosphate produced both effects, and was more potent than adenosine or muscle adenylic acid. Subsequent experiments using isolated perfused lung lobes have identified only a pulmonary vasoconstrictor effect of ATP.[34-38] This vasoconstrictor response has been thought to have physiologic significance, since purines may be released in hypoxia and may contribute to this vasoconstrictor response.[39] Moreover, others have reported that ATP inhibits the hypoxic vasoconstrictor response.[40] In intact dogs with resting pulmonary vascular tone, infusions of ATP did not alter pulmonary vascular resistance.[41] Both pulmonary arterial pressure and cardiac output rose slightly, but resistance was unchanged. However, when tone was increased by an infusion of serotonin, the ATP infusion now decreased pulmonary artery pressure 15% without significantly changing cardiac output, suggesting vasodilation. This apparent divergence of experimental data was restudied, in part in this laboratory, in intact cats anesthetized with pentobarbital (35 mg/kg i.v.).[42] They were strapped in the supine position to a Phillips fluoroscopic table, and breathed spontaneously through a cuffed endotracheal tube. A specially designed 6F triple-lumen balloon perfusion catheter was passed under fluoroscopic guidance from the left external jugular vein into the arterial branch of the left lower lobe. After heparinization, the lobar artery was vascularly isolated by distention of the balloon cuff on the catheter. The lobe was perfused with blood withdrawn from the femoral artery by a Harvard 1210 peristaltic pump at a constant flow of 35 to 41 ml/min and pumped through the catheter to a port immediately beyond the balloon cuff. Perfusion pressure was measured through a downstream port of that catheter. Left atrial pressure was measured with a specially designed double-lumen 5F catheter placed transseptally under fluoroscopic guidance into the vein draining the left lower lobe. The second lumen on the transseptal catheter was used to carefully withdraw or infuse blood to maintain constant lobar venous pressure. Thus, since lobar blood flow and downstream pressure are held constant, any changes in lobar arterial perfusion pressure reflects a change in pulmonary vascular resistance. This technique has been previously described.[43] The effects of bolus injections of ATP and other adenine nucleotides were studied at resting pulmonary vascular tone (10 to 12 mmHg) and when tone was elevated to 35 to 38 mmHg by a continuous infusion of (15S) hydroxyl-11 alpha 9 alpha (epoxymethano) prosta-5Z, 13E, dienoic acid (U46619), a thromboxane mimic, at 30 to 410 ng/min into the lobar artery.[42] Under conditions of resting (low) tone (10 to 12 mmHg), bolus injections of ATP (30, 100, 300 μg) into the lobar artery induced a dose-related vasoconstrictor response (Figure 1A and B). On the other hand, when tone was increased to 35 mmHg in the same cat, bolus injections of ATP now induced a dose-related vasodilator response. Thus the vasoconstrictor responses described in earlier publications may have been observed under conditions of resting (low) vascular tone, but the vasodilator response described earlier under conditions of hypoxia were in the lungs at elevated vascular tone. The failure to observe a vasoconstrictor response to infusions of ATP at resting tone in dogs may have resulted from the passive effects of increasing flow, which would obscure this response.[41] Thus, the data indicate that at resting tone ATP induces vasoconstriction, and at elevated tone, vasodilation. The mechanisms by which changes in pulmonary vascular tone affect the response to ATP are not clear. Presumably, at resting tone the vasoconstrictor response is mediated in part by ATP stimulation of P_{2x} receptors in vascular smooth muscle. Support for this interpretation is found in the fact that stable analogs of ATP, β,γ-methylene ATP and α,β-methylene ATP, also induce pulmonary vasoconstriction and are not blocked by the adenosine antagonist BWA1433U.[42] In addition, this vasoconstrictor response is blunted by meclofenamate, suggesting that the injected bolus of ATP is metabolized to adenosine (which in turn activates a cyclooxygenase enzyme to produce prostaglandins; see Section II.C) before reaching the P_{2x} receptor.[43,44] Thus, part of the vasoconstrictor response results from adenosine, the ATP breakdown product acting at A_2 receptors. On the other hand, the vasodilator response to ATP at elevated tone may be mediated by several mechanisms. Vasodilation may result from stimulation of P_{2y} receptors, or A_1 receptors after metabolic degradation of ATP to adenosine. ATP stimulates P_{2y} receptors on vascular endothelium and induces release of endothelial-derived relaxing factor (EDRF). Indeed, in contracted isolated dog pulmonary arterial rings, the vasodilator response to ATP is endothelium dependent.[46,47] In addition to P_{2y} receptor activating EDRF release, recent data have shown that stimulation of P_{2y} receptors on endothelial cells from a variety of blood vessels leads to the release of prostacyclin, which may contribute to the pulmonary vasodilator response at elevated tone.[48-50] Endothelial cells from isolated rabbit pulmonary artery have been shown to release prostacyclin when exposed to ATP.[50] In the lungs of intact cats, however, cyclooxygenase blockade with meclofenamate did not affect the vasodilator

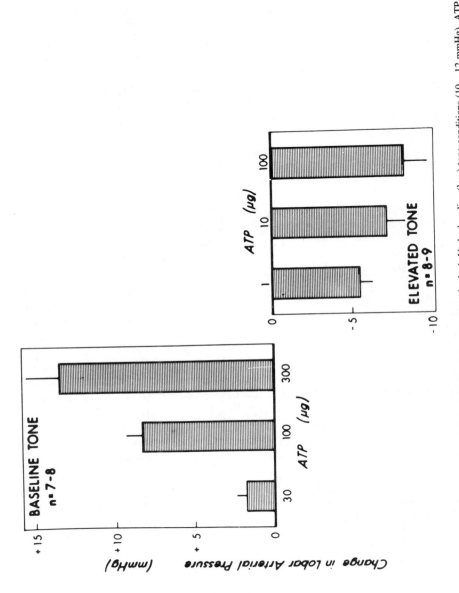

FIGURE 1A.   Influence of the level of tone on response to ATP in the feline pulmonary vascular bed. Under baseline (low) tone conditions (10—12 mmHg), ATP in doses of 30 to 300 μg caused a dose-related increase in lobar arterial pressure (left panel). However, when the level of tone is raised to 30 to 40 mmHg with U46619, ATP in doses of 1 to 100 μg caused dose-related decreases in lobar arterial pressure (right panel); n = number of experiments.

INFLUENCE OF PULMONARY VASOMOTOR TONE ON RESPONSES TO ATP

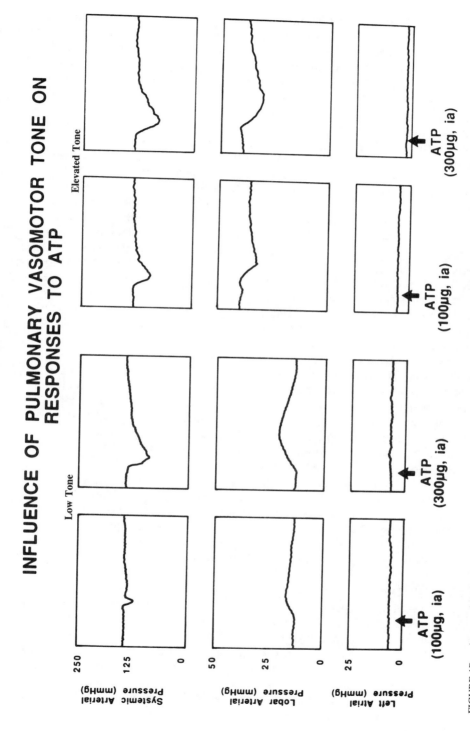

FIGURE 1B. At normal low pulmonary vascular tone, intralobar injections of ATP induce vasoconstriction (left panel), but at increased vascular tone, ATP induces vasodilation (right panel).

response to ATP.[42] The response is also not mediated by muscarinic, beta adrenergic, or histamine receptors.[42] Albeit the vasodilator effect reported from the release of prostacyclin is transient, but that of EDRF may persist for many minutes. Further evidence to suggest that two P_2 receptors or P_2 transduction mechanisms are involved has been published. The stable analogs of ATP, β,γ-methylene ATP and α,β-methylene ATP induce only vasconstriction at both resting tone and elevated tone.[43] Moreover, specific desensitization of P_{2x} receptors, mediating vasoconstriction in isolated, perfused rat lungs, uncovers a vasodilator response, suggesting masked P_{2y} receptor activity.[45] Despite considerable circumstantial evidence indicating that P_{2x} and P_{2y} receptors are different membrane proteins, the pharmacologic classification of these receptors rests on empirical criteria. Their identification rests for the most part, on the rank order of potency of ATP and its analogs.

ATP, like adenosine, may also influence pulmonary vascular tone by stimulating presynaptic receptors that modulate neurotransmission from sympathetic and parasympathetic nerve terminals.[51] Indeed, ATP is stored in sympathetic and cholinergic secretory granules, and is a cotransmitter upon sympathetic stimulation. Because of the rapid hydrolysis of ATP by ectophosphatases, it is not clear whether the modulation results from the corelease of ATP or its breakdown product, adenosine. Studies involving field stimulation of rabbit pulmonary artery have shown that ATP induces an initial presynaptic inhibition of norepinephrine release, thought to be due primarily to ADP-mediated, short-lasting prostaglandins of the E series.[52,53] The data suggest that the long-term inhibition is due to ATP and its metabolites, mainly adenosine, and involves a decreased entry of calcium into the adrenergic terminal with depolarization. On the other hand, the postjunctional effects of A_1 receptors serve to enhance the effect of norepinephrine released from the adrenergic terminal. It is likely that the postjunctional receptors have greater activity. This modulating effect may prove important in the pulmonary vascular bed, where the effects of sympathetic and parasympathetic nerve stimulation are tone dependent.[54-56]

ATP is present in millimolar concentration in most cells of the body, and any perturbation that releases ATP may lead to transient levels of the nucleotide in sufficient concentrations to stimulate P_2 receptors. Tissue trauma, vascular injury, and proteases such as thrombin and elastase may lead to release of these adenine nucleotides.[57] Moreover, ATP can be detected in venous effluent from organs or isolated tissue following ischemia-reperfusion,[58] nerve stimulation,[59] or acetylcholine infusions.[60] These concentrations of ATP may readily reach the pulmonary blood vessels and contribute to the regulation of pulmonary vascular resistance as ATP itself or as a degradation product. Moreover, ATP has greater activity than adenosine in lung vessels.[42] At low resting tone, ATP constricts pulmonary vessels and may serve to improve ventilation-perfusion coupling, whereas at elevated tone it can modulate the vasoconstrictor effect induced by other mechanisms. Since the vasodilator response is endothelium dependent, ATP may constrict the pulmonary vascular bed during conditions of acute lung injury with compromised endothelial function.

B. EFFECTS OF ADP

The effects of ADP on the pulmonary vascular bed are, in the main, similar to those of ATP. Furthermore, the activity of ATP and ADP at P_2 receptors is about equal. Unpublished data from this laboratory have indicated that bolus injections of ADP into the hemodynamically isolated lobe of intact cats induce pulmonary vasoconstrictor responses at low resting tone and vasodilation at elevated tone, as ATP does. However, ADP differs from the other purines in that it activates platelets to contract, aggregate, and secrete. On the contrary, ATP and adenosine have antiaggregatory properties. The main intracellular event resulting from ADP stimulation of platelets is increased intracellular calcium, induced receptor-operated calcium influx and, to varying extents, increased inositol phospholipid metabolism, activation of arachidonic acid metabolism by cyclooxygenase enzyme, and phosphorylation and dephosphorylation of cellular proteins.[61]

Although the effects of bolus injections involve exposure to a limited volume of blood and platelets, slow infusions of ADP permit the effects of platelet aggregation in the pulmonary vascular bed to be recognized as increased vascular resistance due to platelet (mechanical) obstruction of the lung microcirculation (Figure 2). Studies in dogs in this laboratory have shown that platelet aggregation is responsible in large measure for the pulmonary vasopressor response to ADP infusions.[62] In those experiments, a variety of techniques were used to remove the ADP-induced platelet aggregates from the blood before perfusing it into the hemodynamically isolated lobe. In each instance, after removal of platelet aggregates, infusion of ADP induced a vasodilator response in these experiments under conditions of elevated pulmonary vascular tone. Furthermore, the vasoconstrictor response to ADP was blunted by methysergide, suggesting that 5-hydroxytryptamine released from platelets might have contributed to the pulmonary vasoconstrictor response to ADP.

Additionally, the mediation of the platelet aggregation induced in response to ADP is different in different species. Although cyclooxygenase products are involved in the vasopressor responses in the perfused lobe of intact rabbits, they are not involved in aggregation (Figure 3).

C. EFFECTS OF ADENOSINE

The pulmonary vascular effects of adenosine have been the subject of more investigation than the purine nucleotides. Since it has been evident that adenosine may serve as an endogenous mediator of metabolic vasodilation in the heart and skeletal muscle, it has been suggested that the pulmonary vasoconstrictor response to hypoxia may be modulated by adenosine.[37] This hypothesis is supported by experiments in isolated perfused dog lungs exposed to 95% nitrogen and 5% carbon dioxide for periods up to 25 min. The overall pattern of response in the lung tissue was one of degradation of ATP, and accumulation of AMP, adenosine, and the degradation products inosine and hypoxanthine. In these experiments, the levels of adenosine in lung tissue increased about tenfold in 3 min of anoxia, whereas inosine and hypoxanthine, which are adenosine-degradation products, increased tenfold and sevenfold, respectively. These degradation products are formed mainly in the capillary endothelial cells that contain the degradative enzyme nucleotidase phosphorylase. This study further showed that adenosine and AMP dilated the pulmonary vessels under control normoxia conditions when these vessels are apparently at resting low tone. Further, adenosine infusions completely abolished the hypoxic vasoconstrictor response. Moreover, other studies have shown that circulating adenosine is efficiently taken up by the pulmonary vascular bed and this process is potently inhibited by the uptake blocker dipyridamole.[63] Following uptake, adenosine is incorporated into intracellular nucleotides, and at low perfusion rates little or none of the adenosine is returned after the first pass to the general circulation. At higher concentrations cellular uptake is saturable, and degradation products (especially inosine and hypoxanthine) are returned to the circulation. Further, the adenosine deaminase also may occur as an ectoenzyme on the pulmonary capillary endothelial membrane. This property of pulmonary endothelium may also contribute significantly to the maintenance of vascular homeostasis.

At low resting tone in the intact sheep and cat, adenosine has been shown to induce pulmonary vasoconstriction, which is blocked by cyclooxygenase inhibitors and a prostaglandin endoperoxide/thromboxane A_2 receptor antagonist.[42,44] Tachyphylaxis to the vasoconstrictor response to adenosine is readily demonstrated with infusions and it persists generally for 8 to 10 min. The mediation of this tachyphylaxis is not entirely clear, but generally exposure to agonists leads to an uncoupling of the receptor from the effector system, which may also be associated with an internalization of receptors from cell surfaces. Following exposure to R-phenylisopropyladenosine (R-PIA), an A_1 agonist, the ability of both R-PIA and prostaglandin E_1 to inhibit adenylate cyclase is diminished in rat adipocytes,

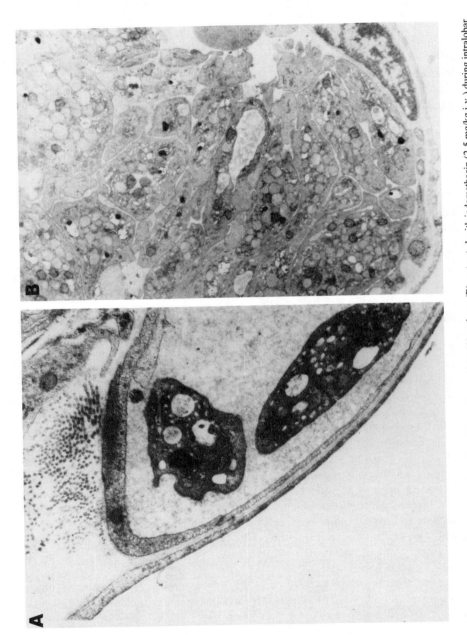

FIGURE 2.   Electron microscopic studies of lungs from rabbits (A) and cats (B) pretreated with indomethacin (2.5 mg/kg i.v.) during intralobar infusions of ADP. Although the arterioles of the rabbit lung have few platelet aggregates, the arterioles of the cat lung are severely obstructed with platelet aggregates.

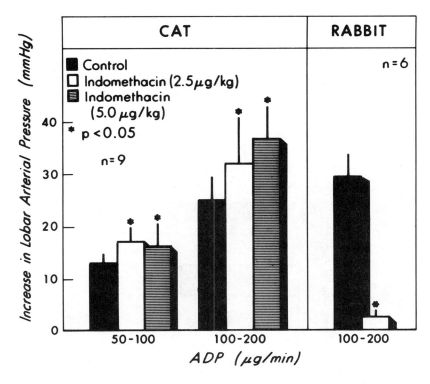

FIGURE 3. Influence of indomethacin on increases in lobar arterial pressure in response to intralobar infusion of ADP into the anesthetized cat and rabbit under conditions of constant pulmonary blood flow and left atrial pressure. Indomethacin was given at two doses to cats and at one dose to rabbits; n = number of animals and ≤ 0.05 was the criterion for statistical significance using paired Student test; * indicates statistically different than control.

suggesting heterologous desensitization.[64] The mechanism appears to relate to reversible block of G_i protein and increased sensitivity of G_s protein. The vasoconstrictor response is readily blocked by theophylline, suggesting that A_1 adenosine receptors in the lung activate cyclooxygenase enzyme to produce prostaglandins. Induction of vasoconstriction by purinergic P_1 receptor activation of vasopressor prostaglandins appears to be unique to the pulmonary vasculature, since P_1 receptor-induced prostaglandin formation is not recognized in the systemic vascular bed. On the other hand, purinergic P_2 receptor activation has been shown to release prostacyclin in a number of organs.[23] Both ATP and adenosine have antiaggregatory effects on platelets.

At elevated tone, induced by hypoxia in the isolated perfused ferret lung[65] and in the intact cat lung infused with U46619,[42] the thromboxane mimic, adenosine dilates the pulmonary vascular bed (Figure 4A and B). The response is competitively blocked by an A_2 receptor blocker, BWA1433U, and not by P_{2y} inhibitors. Unpublished data from this laboratory have also shown that this adenosine-induced vasodilation is not blocked by pretreatment of cats with pertussis toxin, suggesting that this vasodilator response is not G_i protein dependent. Since the vasodilator response is also not blocked by sulfonylureas or charybdotoxin, the respone is probably not mediated by activation of potassium ion channels. In addition, the pulmonary vasodilator response to adenosine is not altered by atropine, propranolol, cimetidine, and meclofenamate, suggesting adenosine dilates the feline lung independent of activation of muscarinic, beta adrenergic, and histaminergic$_2$ receptors as well as formation of cyclooxygenase products, including prostacyclin.[42] Initial studies in isolated pulmonary arterial rings from men[66] and dog[46] suggested that adenosine-induced relaxation

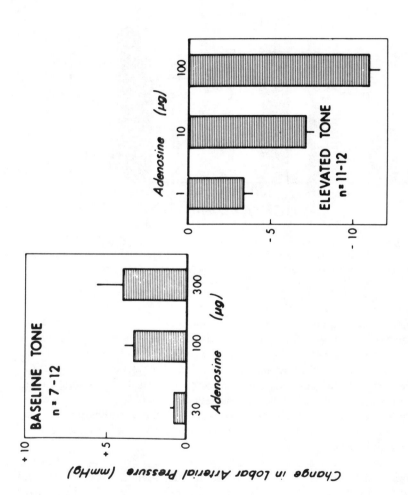

FIGURE 4A.   Influence of the level of tone on responses to adenosine in the feline pulmonary vascular bed. Under baseline (low) tone conditions adenosine in doses of 30 to 300 μg caused a dose-related increase in lobar arterial pressure (left panel). When the level of tone is raised to a high value (30 to 40 mmHg) with U46619, adenosine in doses of 1 to 100 μg caused a dose-related decrease in lobar arterial pressure (right panel); n = number of experiments.

INFLUENCE OF PULMONARY VASOMOTOR TONE ON RESPONSES TO ADENOSINE

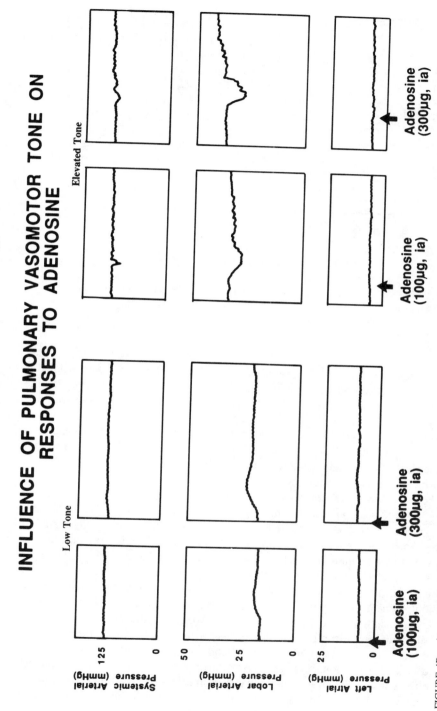

FIGURE 4B. At normal low pulmonary vascular tone, intralobar bolus injections of ADP induce vasoconstriction (left panel), but at increased vascular tone, ADP induces vasodilation (right panel).

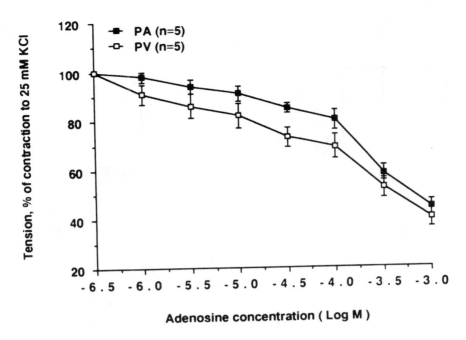

FIGURE 5.   Effects of adenosine on bovine intrapulmonary arterial (PA) and venous (PV) rings with an intact endothelium precontracted with potassium chloride (KCl, 25 m*M*). Cumulative dose-response curves were generated with vessels in oxygenated Krebs solution at 37°C and are expressed as percent (%) tension of contraction to KCl (25 m*M*); n = number of vessels from three different animals.

is not endothelial dependent, in contrast to the ATP relaxant response. Data from this laboratory indicate that adenosine has somewhat greater relaxant activity on isolated bovine pulmonary venous rings when compared to arterial segments (Figure 5). Moreover, the relaxant response to adenosine in isolated bovine pulmonary arterial and venous rings is competitively antagonized by BWA1433U, suggesting that arterial and venous segments possess substantial populations of A_2 receptors (Figures 6 and 7). Thus, adenosine retains its relaxant properties on pulmonary vessels despite change to the endothelial cell layer. In addition, adenosine may serve to modulate pulmonary vasoconstrictor responses in the lung regardless of whether or not the endothelium has been damaged by acute lung injury, since the pulmonary A_2 receptors are on vascular smooth muscle and are not endothelial dependent, in contrast to P_{2y} receptors. The ability of adenosine to relax pulmonary veins may represent a protective mechanism in the lung to reduce edema formation during pathophysiologic states, such as sepsis or pneumonia, which may accelerate purine nucleotide metabolism.

Two studies have characterized the hemodynamic profile of adenosine in man.[67,68] Although initial experiments demonstrated cardiac and systemic vasoactive properties of adenosine infusion,[67] recent work by Edlund et al.[68] reported the effects of adenosine infusion on the pulmonary circulation in conscious volunteers. Adenosine did not alter main pulmonary arterial and pulmonary capillary wedge pressures; however, cardiac output significantly increased and calculated pulmonary and systemic vascular resistances significantly decreased.[68] The lack of a pulmonary pressor effect in response to adenosine in man, but which is seen in the cat and the sheep, may have been masked by the direct and indirect effects of adenosine on cardiac performance. Alternatively, the pulmonary vasoconstrictor activity of adenosine may be species dependent and may be less in man.

Adenosine may also serve adversely in the pathogenesis of acute lung injury induced by reperfusion injury. This injury results from pathologic alterations in the metabolism of

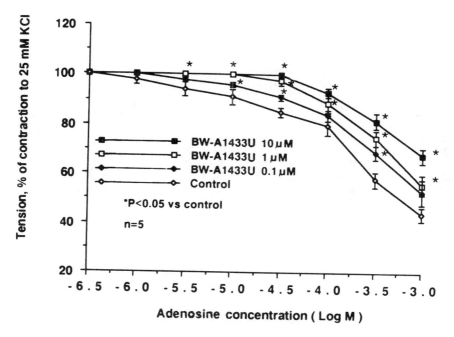

FIGURE 6. Influence of BW-A 1433U on decreases in tension in response to adenosine on isolated bovine intrapulmonary arterial rings with intact endothelium. Cumulative dose-response curves were generated in responses to adenosine in control vessels and in vessels pretreated for 15 min with BW-A 1433U (0.1 to 10 μM) in oxygenated Krebs solution at 37°C. Data are expressed as percent (%) tension of potassium chloride (KCl, 25 mM) and n = number of vessels from three different animals. The criterion for statistical significance was $p \leq 0.05$ using ANOVA, * indicates significantly different than control.

microvascular endothelial cells induced by ischemia and the influence of oxygen and neutrophils resulting from reperfusion. Adenosine released by hypoxic parenchymal cells undergoes enzymatic degradation to hypoxanthine, inducing superoxide formation through xanthine oxidase. The cascade of superoxide and hydroxyl radicals then leads to cell injury. Activated neutrophils under chemokinesis and chemotaxis release lysosomes and adhere to other neutrophils and endothelial cells. Although adenosine may serve as a source of the xanthine degradation products inducing lung injury, it also may mitigate this adverse effect through its direct vasodilating effect, platelet deaggregating effect, and inhibition of activated neutrophils. Investigation into the factors which balance these latter, likely beneficial, effects and the potential to form damaging oxygen radicals via xanthine metabolism is a fertile area of future research.

III. SUMMARY

Our understanding of the regulation of pulmonary vascular tone has significantly improved since the early concepts of a purely passive vascular bed. The direct effects of hypoxia, autonomic nervous regulation, and humoral agents on regulation of lung vascular tone have been clearly demonstrated. The possibility that adenine nucleotides may also serve as homeostatic regulators in the lung have been considered since the original observations of Drury and Szent-Gyorgyi[1] in 1929.

Classification of purinergic receptors has contributed heavily to our understanding of the contributions of these nucleotides to vascular homeostasis, and recent studies have begun to elucidate these effects in the pulmonary vascular bed. Though both ATP and adenosine induce vasoconstriction at low resting pulmonary vascular tone, the mechanisms responsible

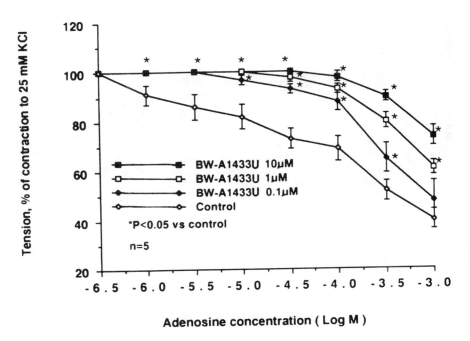

FIGURE 7.   Influence of BW-A 1433U on decreases in tension in response to adenosine on isolated bovine intrapulmonary venous rings with intact endothelium. Cumulative dose-response curves were generated in responses to adenosine in control vessels and in vessels pretreated for 15 min with BW-A 1433U (0.1 to 10 μM) in oxygenated Krebs solution at 37°C. Data are expressed as percent (%) tension of potassium chloride (KCl, 25 mM); n = number of vessels from three different animals. The criterion for statistical significance was $p \leq 0.05$ using ANOVA; * indicates significantly different than control.

for this response differ in that the adenosine response is mediated by A_1 receptor activation of constrictor prostanoids, but the ATP response is in part due to P_{2x} receptor activation on vascular smooth muscle, and in part due to its degradation to adenosine with resultant prostanoid production. The responses of these two purines is entirely reversed when vascular tone is elevated, both producing vasodilation. The mediation of these vasodilator responses is also different. Adenosine induces vasodilation directly by stimulating A_2 receptors on vascular smooth muscle, and in the pulmonary vascular bed is not endothelial dependent, whereas ATP-induced vasodilation is mediated by stimulation of P_{2y} receptors on endothelial cells leading to formation of EDRF, and is not prostanoid dependent. In the systemic vascular bed, much of the ATP-induced vasodilator response is mediated by P_{2y} receptor stimulation of prostacyclin formation, and responses to adenosine are not prostanoid dependent. On the other hand, in the pulmonary vascular bed, adenosine-induced vasoconstriction is mediated by prostanoids, but the ATP responses, mediated by P_{2x} and P_{2y} receptors, are not mediated by prostanoid. A more complete understanding of the receptor coupling, transduction, and effector mechanisms will lead to a better appreciation of tone dependency of these responses and the mechanisms by which they serve as homeostatic regulators. Thus, at low tone, these agents may induce vasoconstriction to improve ventilation-perfusion ratios, and may modulate the hypoxic vasoconstriction. In addition, since the vasodilator response to adenosine is not endothelium dependent, this nucleotide may modulate the vasoconstrictor response to lung injury. Improved understanding of receptor proteins, transduction mechanisms, and effector mechanisms may lead to improved specific agonists with greater potency for treating pulmonary hypertensive states.

REFERENCES

1. **Drury, A. H. and Szent-Gyorgyi, A.,** The physiologic activity of adenine compounds with especial reference to their action upon mammalian heart, *J. Physiol.,* 68, 213, 1929.
2. **Olsson, R. A. and Pearson, J. D.,** Cardiovascular purinoceptors, *Physiol. Rev.,* 70, 761, 1990.
3. **Burnstock, G.,** Neural nomenclature, *Nature,* 229, 282, 1971.
4. **Burnstock, G.,** A basis for distinguishing two types of purinergic receptors, in *Cell Membrane Receptors for Drugs and Hormones: A Multidisciplinary Approach,* Balis, L. and Staub, R. W., Eds., Raven Press, New York, 1978, 107.
5. **Burnstock, G. and Brown, C. M.,** An introduction to purinergic receptors, in *Purinergic Receptors and Recognition,* Burnstock, G., Ed., Chapman and Hall, London, 1981, 1.
6. **Emmelin, N. and Feldberg, W.,** Systemic effects of adenosine triphosphate, *Br. J. Pharmacol. Chemother.,* 3, 273, 1948.
7. **Gaddum, J. H. and Holtz, P.,** The localization of the action of drugs on the pulmonary vessels of dogs and cats, *J. Physiol.,* 77, 139, 1933.
8. **Green, H. N. and Stoner, H. B.,** *Biological Actions of Adenosine Nucleotides,* Lewis, London, 1950.
9. **Winburg, M. M., Papierski, D. H., Heumner, M. L., and Hemrourger, W. E.,** Coronary dilator action of adenine-ATP series, *J. Pharmacol. Exp. Ther.,* 109, 255, 1953.
10. **Sakai, K., Akima, M., and Nabata, H.,** A possible purinergic mechanism for reactive ischemia in isolated cross-circulated rat kidney, *J. Pharmacol.,* 29, 235, 1979.
11. **Van Calker, D., Muller, M., and Hamprecht, B.,** Adenosine regulates via two different receptors, the accumulation of cyclic AMP in cultured brain cells, *J. Neurochem.,* 33, 999, 1979.
12. **Olsson, R. A., Davis, C. J., Khouri, E. M., and Patterson, R. E.,** Evidence for adenosine receptors on the surface of dog coronary myocytes, *Circ. Res.,* 39, 93, 1976.
13. **Cooper, D. M., Londos, C., and Rodbell, M.,** Adenosine receptor mediated inhibition of rat cerebral cortical adenylate cyclase by GTP-dependent process, *Mol. Pharmacol.,* 18, 598, 1980.
14. **Alfonso, S. and O'Brien, G. S.,** Inhibition of cardiovascular metabolic and hemodynamic effects of adenosine by aminophylline, *Am. J. Physiol.,* 219, 1672, 1970.
15. **Bunger, R., Haddy, F. J., and Gerlach, E.,** Coronary responses to dilating substances and competitive inhibition by theophylline in isolated perfused guinea pig heart, *Pfluegers Arch.,* 358, 213, 1975.
16. **Bohm, M., Bruchna, R., Hackbonik, I., Hambitz, B., Linhart, R., Meyer, N., Schmidt, B., Schmitz, W., and Schalg, H.,** Adenosine inhibition of catecholamines-induced increase in force of contraction of guinea-pig atrial and ventricular heart preparations. Evidence against a cyclic AMP and cyclic GMP dependent effect, *J. Pharmacol. Exp. Ther.,* 230, 483, 1984.
17. **Hill, S. J. and Kendall, D. A.,** Studies on adenosine receptor mediating the augmentation of histamine induced inositol phospholipid hydrolysis in guinea pig cerebral cortex, *Br. J. Pharmacol.,* 91, 661, 1987.
18. **Schimmel, R. J. and Elliott, M. E.,** Adenosine inhibits phenylephrine activation of phospholipase A_2 in hamster brown adipocytes, *Biochem. Biophys. Res. Commun.,* 152, 886, 1988.
19. **Kurtz, A.,** Adenosine stimulates guanylate cyclase activity in vascular smooth muscle cells, *J. Biol. Chem.,* 262, 6296, 1987.
20. **Proctor, W. R. and Dunwiddie, T. V.,** Adenosine inhibits calcium spikes in hippocampal pyramidal neurons in vitro, *Neurosci. Lett.,* 35, 197, 1983.
21. **Londos, C.,** On multiple targets for fat cell receptors, in *Topics and Perspectives in Adenosine Research,* Gerlach, E. and Beckler, B. F., Eds., Springer-Verlag, Berlin, 1987, 239.
22. **Burnstock, G. and Kennedy, C.,** Is there a basis for distiguishing two types of P_1-purinoceptors?, *Gen. Pharmacol.,* 16, 433, 1985.
23. **Needleman, P., Minkes, M. S., and Douglas, J. R., Jr.,** Stimulation of prostaglandin biosynthesis by adenine nucleotides, *Circ. Res.,* 34, 455, 1974.
24. **Gordon, J. L. and Martin, W.,** Stimulation of endothelial prostacyclin production plays no role in endothelium-dependent relaxation of the pig aorta, *Br. J. Pharmacol.,* 80, 179, 1983.
25. **Gordon, J. L.,** Extracellular ATP: effects, sources and fats, *Biochem. J.,* 233, 309, 1986.
26. **Born, G. V. R. and Kratzer, M. A. A.,** Source of concentration of extracellular adenosine triphosphate during homeostasis in rats, rabbits and man, *J. Physiol.,* 354, 419, 1984.
27. **Ryan, J. W. and Smith, U.,** Metabolism of adenosine 5'-monophosphate during circulation through the lung, *Trans. Assoc. Am. Phys.,* 84, 297, 1971.
28. **Ryan, U. S.,** Structural bases for metabolic activity, *Annu. Rev. Physiol.,* 44, 223, 1982.
29. **Cooper, D. R., Lewis, G. P., Lieberman, G. E., Webb, H., and Westwick,** ADP metabolism in vascular tissue, a possible thrombo-regulatory mechanism, *Thromb. Res.,* 14, 901, 1979.
30. **Cusack, N. J., Pearson, J. D., and Gordon, J. L.,** Stereoselectivity of ectonucleotidases on vascular endothelial cells, *Biochem. J.,* 214, 975, 1983.
31. **Pearson, J. D. and Gordon, J. L.,** Nucleotide metabolism by endothelium, *Annu. Rev. Physiol.,* 47, 617, 1985.

32. **Lowenstein, J. M.,** Ammonia production in muscle and other tissue: the purine nucleotide cycle, *Physiol. Rev.,* 52, 382, 1972.

33. **Andy, R. J. and Kornfeld, R.,** The adenosine deaminase binding protein of human skin fibroblasts is located on cell surface, *J. Biol. Chem.,* 257, 7922, 1982.

34. **Evans, R. G., Forrester, T., and Mueller, H. S.,** Intravascular passage of adenosine triphosphate through the lung of a baboon, *J. Physiol.,* 276, 70P, 1978.

35. **Emmelin, N. and Feldberg, W.,** Systemic effects of adenosine triphosphate, *Br. J. Pharmacol. Chemother.,* 3, 273, 1948.

36. **Reeves, J. T., Joke, P., Joaquin, M., and Leathers, J. E.,** Pulmonary vascular obstruction following administration of high energy nucleotides, *J. Appl. Physiol.,* 22, 475, 1967.

37. **Mentzer, R. M., Rubio, R., and Berne, R. M.,** Release of adenosine by hypoxic canine lung tissue and its possible role in the pulmonary circulation, *Am. J. Physiol.,* 229, 1625, 1975.

38. **Green, H. N. and Stoner, H. B.,** *Biological Actions of Adenosine Nucleotides,* Lewis, London, 1950.

39. **Burnstock, G. and Kennedy, C.,** A dual function for adenosine 5'-triphosphate in the regulation of vascular tone, *Circ. Res.,* 58, 319, 1985.

40. **Bennmof, J. L., Fukuanago, A. F., and Trousdale, F. R.,** ATP inhibits hypoxic vasoconstriction, *Anesthesiology,* 57, A474, 1982.

41. **Rudolph, A. M., Kurland, M. D., Auld, P. A. M., and Paul, M.,** Effect of vasodilator drugs on normal and serotonin-constricted pulmonary vessels of the dog, *Am. J. Physiol.,* 197, 617, 1959.

42. **Neely, C. F., Kadowitz, P., Lippton, H., Neiman, M., and Hyman, A.,** Adenosine does not mediate the pulmonary vasodilator response of ATP in the feline pulmonary vascular bed, *J. Pharmacol. Exp. Ther.,* 250, 170, 1989.

43. **Hyman, A. L.,** The effects of bradykinin on the pulmonary veins, *J. Pharmacol. Exp. Ther.,* 161, 78, 1968.

44. **Biaggioni, I., King, L. S., Enayot, N., Robertson, D., and Newman, J. H.,** Adenosine produces pulmonary vasoconstriction in the sheep, *Circ. Res.,* 65, 1516, 1989.

45. **McCormack, D. G., Barnes, P. J., and Evans, J. W.,** Purinoceptors in the pulmonary circulation of the rat and their role in hypoxic vasoconstriction, *Br. J. Pharmacol.,* 98, 367, 1989.

46. **DeMey, J. G. and Vanhoutte, P. M.,** Heterogenous behavior of the canine arterial and venous wall. Importance of endothelium, *Circ. Res.,* 51, 439, 1982.

47. **Frank, G. W. and Bevan, J. A.,** Vasodilatation by adenosine-related nucleotides is reduced after endothelial destruction in basilar, lingual and pulmonary arteries, in *Regulatory Function of Adenosine,* Berne, R. M., Rall, T. W., and Rubio, R., Eds., Martinus Nijoff, Boston, 1983, 511.

48. **Needham, L., Cusack, N. J., Pearson, J. O., and Gordon, J. L.,** Characteristics of the P_2 purinoceptors that mediate endothelial prostacyclin production by pig endothelial cells, *Eur. J. Pharmacol.,* 134, 199, 1987.

49. **Pearson, J. D., Slakey, L. L., and Gordon, J. L.,** Stimulation of prostaglandin production through purinoceptors on cultured porcine endothelial cells, *Biochem. J.,* 214, 273, 1983.

50. **Boeynuems, J. M. and Garland, N.,** Stimulation of vascular prostacyclin synthesis by extracellular ADP and ATP, *Biochem. Biophys. Res. Commun.,* 112, 290, 1983.

51. **Gordon, J. L. and Martin, J. W.,** Stimulation of endothelial prostacyclin production plays no role in endothelium-dependent relaxation of the pig aorta, *Br. J. Pharmacol.,* 80, 179, 1983.

52. **Husted, W. E. and Nedergaard, O. A.,** Dual inhibitory action of ATP on adrenergic neuroeffector transmission in rabbit pulmonary artery, *Acta Pharmacol. Toxicol.,* 57, 204, 1985.

53. **Holton, F. A. and Holton, P.,** The possibility that ATP is a transmitter at sensory nerve endings, *J. Physiol.,* 119, 50P, 1953.

54. **Kadowitz, P. J. and Hyman, A. L.,** Effect of sympathetic nerve stimulation on pulmonary vascular resistance in the dog, *Circ. Res.,* 32, 221, 1973.

55. **Hyman, A. L., Lippton, H., and Kadowitz, P. J.,** Analysis of pulmonary vascular responses to sympathetic nerve stimulation under conditions of elevated tone, *Circ. Res.,* 67, 862, 1990.

56. **Nandiwada, P. A., Hyman, A. L., and Kadowitz, P. J.,** Pulmonary vasodilator responses to vagal stimulation and acetylcholine in the cat, *Circ. Res.,* 53, 86, 1983.

57. **Leroy, E. C., Ager, A., and Gordon, J. L.,** Effects on neutrophil elastase and other proteases on porcine aortic endothelial release of PGI_2 production, *J. Clin. Invest.,* 74, 1003, 1982.

58. **Forrester, T. and Lind, A. R.,** Identification of adenosine triphosphate in human plasma and concentration in venous effluent of forearm muscles, *J. Physiol.,* 204, 347, 1969.

59. **Fredholm, B. B., Hedqvist, P., Lindstrom, K., and Wennmalm, M.,** Release of nucleosides and nucleotides from rabbit heart by sympathetic nerve stimulation, *Acta Physiol. Scand.,* 116, 285, 1982.

60. **Schrader, J., Thompson, C. I., Hendlmayer, G., and Gerlach, E.,** Role of purines in acetylcholine induced coronary vasodilation, *J. Mol. Cell. Cardiol.,* 14, 427, 1982.

61. **Hallam, T. J. and Rink, T. J.,** Responses to adenosine diphosphate in human platelets loaded with fluorescent calcium indicator quin 2, *J. Physiol.,* 368, 131, 1985.

62. **Hyman, A. L., Woolverton, W. C., Pennington, D. G., and Jaques, W. E.,** Pulmonary vascular responses to adenosine diphosphate, *J. Pharmacol. Exp. Ther.*, 178, 549, 1971.

63. **Hellewell, P. G. and Pearson, J. D.,** Metabolism of circualting adenosine by porcine isolated perfused lung, *Circ. Res.*, 53, 1, 1983.

64. **Parson, W. J. and Stiles, G. L.,** Heterologous desensitization of the inhibitor A_1 adenosine receptor-adenylate cyclase system in the rat adipocyte, *J. Biol. Chem.*, 262, 841, 1987.

65. **Gottlieb, J. E., Peake, M. D., and Sylvester, J. T.,** Adenosine and hypoxic pulmonary vasodilation, *Am. J. Physiol.*, 246, H541, 1984.

66. **McCormack, D. G. and Barnes, P. J.,** Characterization of adenosine effects and receptors in human pulmonary arteries, *Am. Rev. Respir. Dis.*, (Abstract), 390, 1990.

67. **Biaggioni, I., Olafsson, B., Robertson, R. M., Hollister, A. S., and Robertson, D.,** Cardiovascular effects of adenosine in conscious man. Evidence for chemoreceptor activation, *Circ. Res.*, 61, 779, 1987.

68. **Edlund, A., Sollevi, A., and Lande, B.,** Haemodynamic and metabolic effects of infused adenosine in man, *Clin. Sci.*, 79, 131, 1990.

Chapter 19

ROLE OF ADENOSINE IN ANGIOGENESIS

Cynthia J. Meininger and Harris J. Granger

TABLE OF CONTENTS

I. INTRODUCTION

Angiogenesis, or neovascularization, denotes the growth of new capillary vessels from an established microvasculature following stimulation by various physiological or pathological processes. Angiogenesis is essentially an endothelial event. The development of a new blood vessel requires both that endothelial cells move to those areas where neovascularization takes place and that they proliferate to generate the extra cells necessary to form a vessel.[1] New capillaries arise from preexisting capillaries or small venules which lack smooth muscle.[2] The endothelial cells bud out from the existing vessel, forming a sprout protruding from the wall of that vessel. Lumen formation takes place within the growing sprouts, producing blind-end channels which then randomly anastomose with other sprouts to form a capillary loop through which blood begins to flow.

Neovascularization has become a subject of increasing interest in recent years because of the observation that growth of solid tumors is dependent upon angiogenesis.[3] Moreover, the persistence of several other pathologic conditions such as arthritis and diabetic retinopathy is associated with abnormal proliferation of the microvessels. On the other hand, angiogenesis plays a major role in the establishment and maintenance of normal function in physiologic processes such as embryogenesis, ovulation, and wound healing.

Tissue hypoxia or tissue ischemia is a common feature of many of the conditions in which neovascular growth is observed. When the metabolic requirements of the tissue chronically exceed the perfusion capabilities of the vessels, the vascular system responds by increasing the number of blood vessels. For example, increased vascularity is observed following the decrease in oxygen availability associated with chronic aerobic exercise,[4] prolonged electrical stimulation of skeletal muscle,[5] wound healing,[6,7] or direct exposure to a hypoxic environment.[8] In retinopathy of prematurity, a condition involving neovascularization of the retina, premature babies are placed into high-oxygen environments for a period of time and then returned to room air. The neovascularization that occurs has been proposed to be a response of the immature vascular system to a retina which has become hypoxic after the withdrawal of supplemental oxygen.[9] The vascular proliferation that occurs in diabetic retinopathy has also been attributed to local hypoxia.[10] Presumably, once a tissue becomes hypoxic, it elaborates angiogenic factors that induce neovascularization. Indeed, angiogenesis factors have been isolated from hypoxic tissues, such as the infarcted tissue of the myocardium[11] and macrophages exposed to low oxygen tensions.[7]

Adenosine is a low molecular weight metabolite released by hypoxic tissue. It is a potent vasodilator and increases blood flow in heart and skeletal muscle during periods of increased metabolic demand. In addition, capillary proliferation has been reported in heart and skeletal muscle after long-term administration of adenosine.[12] Although increased blood flow has been implicated as a physical factor involved in capillary growth, it is also possible that neovascularization in the intercellular spaces of tissues suffering from hypoxia may be due to direct chemical stimulation from locally increased concentrations of adenosine.

II. EFFECTS OF ADENOSINE ON ENDOTHELIAL CELL PROLIFERATION AND MIGRATION

A. ADENOSINE AS AN ENDOTHELIAL CELL MITOGEN AND CHEMOATTRACTANT

When bovine aortic or coronary venular endothelial cells are incubated in a low-oxygen environment (i.e., 10 to 20 mmHg), their rate of division is accelerated over that of control cells grown under standard tissue culture conditions (i.e., 140 to 150 mmHg).[13] This accelerated growth is not due to a direct effect of oxygen, but rather to the release of a mitogenic factor into the growth medium that acts in an autocrine manner to stimulate the division of the endothelial cells. This is evident by the fact that filtered medium conditioned by cells growing under hypoxic conditions can be used to stimulate the division of other endothelial cells grown under normoxic conditions.

Adenosine is released from endothelial cells under hypoxic conditions. At a concentration of 5.0 μM, adenosine stimulates the division of bovine aortic or coronary venular endothelial cells. This mitogenic action can be demonstrated by proliferation assays[13] as well as by measurements of tritiated thymidine incorporation.[22] Addition of the adenosine receptor blocker, 8-phenyltheophylline, blocks the stimulation of proliferation brought about by adenosine, conditioned medium, or hypoxic growth conditions. Thus, the stimulatory effect of hypoxia is apparently due to the release of adenosine from hypoxic cells.

In addition to its mitogenic effect, adenosine also has a chemotactic effect on endothelial cells. Coronary venular endothelial cells respond to a positive gradient of adenosine by migrating through the pores of a modified Boyden chamber in the direction of increasing adenosine concentration.[13] This chemotactic response is dose dependent.

B. NEOVASCULARIZATION IN THE CHICK CHORIOALLANTOIC MEMBRANE

Adenosine stimulates a dose-dependent increase in the vascular density of the chick chorioallantoic membrane (CAM), an *in vivo* model of angiogenesis.[14] This vasoproliferative effect is only evident when the adenosine is gradually released from a polymer and not when the adenosine is topically applied. While dipyridamole, a blocker of adenosine reuptake by endothelial and other cells, has little effect on the vascularity of the CAM by itself, it can enhance stimulation of vessel growth by adenosine. Dipyridamole potentiates the action of adenosine in tissues by inhibiting adenosine deaminase and blocking the uptake of adenosine. Presumably, the enhancement of the action of adenosine in the CAM by dipyridamole is due to decreased metabolism (i.e., prolonged action) of the adenosine.

C. STRUCTURAL ADAPTATION IN THE CHICK EMBRYO VASCULATURE

Structural adaptation of the blood vascular system to chronic hypoxia may involve increases in the number of blood vessels as well as an increase in the size of blood vessels. The role for adenosine in the development of the vascular system in the growing chick embryo has been investigated using vascular resistance measured after complete vasodilation as a functional estimate of vascularity.[15] This estimate is based on a whole-body perfusion technique in which the maximally dilated vasculature is perfused at constant pressures. Adenosine decreases whole-body structural vascular resistance of the embryo in a dose-dependent manner, reaching maximum effects at higher dosage levels. This decrease in resistance can be blocked using aminophylline, an adenosine receptor blocker. Since acute administration of adenosine has no effect on structural resistance, the decrease in whole-body resistance is due to structural changes in the vascular system, i.e., an increase in the number of perfused vessels or growth of existing vessels to a larger size. In the embryo, dipyridamole alone stimulates proliferation and/or growth of vessels, suggesting potentiation of the vasoproliferative actions of endogenous adenosine.

D. VASOPROLIFERATIVE EFFECTS OF ADENOSINE SECONDARY TO VASODILATION

Endothelial cells normally represent a very quiescent population of cells. However, their proliferation may be stimulated in areas of blood flow disturbance.[16] While much recent attention has been focused on the proliferation of endothelial cells in response to various angiogenic growth factors, these growth factors are released mainly under pathological situations, e.g., growth of solid tumors, wound healing, and diabetic angiopathy. In contrast, the growth of vessels occurring under normal physiological conditions, e.g., development, exposure to high-altitude hypoxia, and aerobic training, is associated with high blood flow. Thus, metabolic factors may be involved in vessel growth, but mechanical factors connected with increased blood flow may also influence endothelial cell proliferation.

Long-term administration of adenosine, in doses which cause marked increases in coronary and muscle blood flow, produces an increase in capillarity in both heart and skeletal muscle.[12] In addition, dipyridamole alone can stimulate vessel growth in the heart[18,19] and skeletal muscle[20] of adult rats. Some investigators[17] have argued that the angiogenic effect of adenosine and dipyridamole is nonspecific and is the result of mechanical factors associated with increased blood flow. In support of this view, chronic electrical stimulation of skeletal muscle leads to capillary growth. However, when blood flow is limited by ligation of the major vessel supplying the muscle, this capillary growth does not occur. Metabolic changes still occur in these stimulated muscles, yet angiogenesis is not evident. Other studies, however, clearly demonstrate vasoproliferation after chronic restriction of blood flow to muscle, supporting the notion that hyperperfusion is not a necessary condition for initiation of angiogenesis.

III. SIGNAL TRANSDUCTION IN THE PROLIFERATIVE RESPONSE TO ADENOSINE

Although the role of mechanical factors in modulating the behavior of endothelial cells cannot be ignored, direct studies of endothelial cell proliferation and chemotaxis in the absence of a flow stream clearly demonstrate the ability of adenosine to stimulate cell migration and division. In recent years, our laboratory has focused on the transmembrane signaling pathways responsible for initiating these processes in endothelial cells of coronary venules.

A. RECEPTOR INVOLVEMENT
Adenosine is known to bind to specific receptors on the external surface of endothelial cells which interact with the enzyme adenylate cyclase. Adenosine is also known to be taken up into endothelial cells where it is rapidly converted to intracellular nucleotides, predominantly adenosine triphosphate. Adenosine, therefore, could act by occupying a receptor and changing adenosine $3',5'$-cyclic monophosphate (cAMP) levels within the cell or it could be taken up into the endothelial cell where it contributes to the nucleotide pool.

When the chick CAM is stimulated with adenosine in the presence of isobutylmethylxanthine, an adenosine receptor blocker, no change in vascularity can be demonstrated.[14] When endothelial cells in culture are stimulated with adenosine in the presence of 8-phenyltheophylline, another adenosine receptor blocker, no stimulation of proliferation is observed.[13] In contrast, when adenosine uptake is prevented by utilizing polyadenylic acid, a high-molecular weight derivative of adenosine, proliferation of the endothelial cells can still be observed.[21] These data suggest that adenosine acts directly on endothelial cells by binding to specific receptors on the external surface of the cells.

B. ROLE OF ADENYLATE CYCLASE
The effects of adenosine in various tissues have been shown to be mediated via membrane-bound receptors which cause either inhibition (A_1 receptors) or stimulation (A_2 receptors) of adenylate cyclase. A third class of adenosine receptors is found at an intracellular site (P receptors) and mediates only inhibition of adenylate cyclase. The extracellular adenosine receptors are now being defined operationally by the relative potencies of synthetic agonist analogs that favor one or the other subtype. The majority of this work was done using brain slices or brain cell membranes. At the A_1 adenosine receptor the N^6-substituted adenosine analog, R-N^6-phenylisopropyladenosine (R-PIA) is more potent than adenosine

or 2-chloroadenosine, which in turn are more potent than the 5'-carboxamides of adenosine, such as 5'-*N*-ethylcarboxamide adenosine (NECA). At the A_2 adenosine receptor this order of agonist potency is reversed (i.e., NECA is most potent and R-PIA is least potent).

The identity of adenosine receptors on endothelial cells has not been clearly established by comparing the relative potency of the putative specific analogs. When coronary venular endothelial cells in culture are stimulated with adenosine analogs showing selectivity in the brain for A_1 and A_2 receptors, they respond in an identical manner. A dose-dependent stimulation of endothelial cell proliferation can be seen with either type of analog.[21] This suggests that either cAMP does not mediate the response to adenosine or these analogs are not selective in coronary venular endothelial cells.

The adenosine analog 2',5'-dideoxyadenosine (DDA) blocks the catalytic action of adenylate cyclase without binding to surface receptors. The addition of DDA to cultures of coronary venular endothelial cells blocks the stimulation of proliferation brought about by adenosine, NECA, or R-PIA, suggesting that cAMP does indeed play a role in the proliferative response to adenosine or an adenosine analog. The level of cAMP within endothelial cells treated with these analogs can be measured directly via radioimmunoassay and increases with either NECA or R-PIA. Thus, it would appear that the proliferative response to adenosine or an adenosine analog is positively coupled to the activation of adenylate cyclase with its subsequent rise in cAMP. From a functional point of view, A_2 receptors appear to predominate in the angiogenic process.

C. ROLE OF G PROTEINS

Adenosine receptors are coupled to adenylate cyclase via guanosine 5'-triphosphate-binding proteins (i.e., G proteins). These G proteins are heterotrimers with subunits designated α, β, and γ. Differences in the α-subunits serve to distinguish the various G proteins, whereas the β- and γ-subunits appear very similar between the different receptors. The α-subunits contain a single high-affinity binding site for GTP and possess the GTP-hydrolyzing activity that is crucial for the action of the G proteins. This subunit also contains the site for NAD-dependent ADP-ribosylation catalyzed by bacterial toxins.

The identity of the G protein associated with adenylate cyclase determines whether cAMP will rise or fall in the cell. The α-subunit of G_s can be ADP ribosylated by cholera toxin, resulting in continuous activation of adenylate cyclase. When endothelial cells are pretreated with cholera toxin (1 ng/ml) for 2 h they are stimulated to proliferate with or without adenosine. The finding supports the hypothesis that a rise in cAMP is a necessary signal to initiate endothelial cell proliferation. NAD-dependent ADP ribosylation of the α-subunit of G_i with pertussis toxin inactivates this G protein by impairing its ability to interact with the adenosine receptor. Thus, pertussis toxin should stimulate angiogenesis because an inhibitory modulator of adenylate cyclase is blocked. However, when endothelial cells are pretreated with pertussis toxin (1 ng/ml) for 2 h they lose the ability to proliferate faster in response to adenosine. Since pertussis toxin interacts with several G proteins other than G_i, adenosine signaling may require the formation of more than one second messenger. For example, pertussis toxin can block the action of G proteins involved in the coupling of surface receptors to the phospholipases A_2 and C. The role of signals (e.g., arachidonate metabolites, inositol phosphate, calcium, and diacylglycerol) generated by these pathways during exposure to adenosine remains undefined.

IV. SUMMARY

The growth of blood vessels occurs under very diverse conditions, e.g., during normal development, exposure to hypoxia, adaptation to increased activity and metabolic demand, wound healing, or in association with many different pathological states (tumors, retinopathy, etc.). It is unlikely that one common stimulus would be responsible for the neovascularization occurring in all these situations. However, available evidence suggests that adenosine may play a role in angiogenesis due to its ability to stimulate endothelial cell proliferation and migration directly.

REFERENCES

1. **Ausprunk, D. H. and Folkman, J.,** Migration and proliferation of endothelial cells in preformed and newly formed blood vessels during tumor angiogenesis, *Microvasc. Res.,* 14, 53, 1977.
2. **Burger, P. C., Chandler, D. B., and Klintworth, G. K.,** Corneal neovascularization as studied by scanning electron microscopy of vascular casts, *Lab. Invest.,* 48, 169, 1983.
3. **Folkman, J. and Cotran, R. S.,** Relation of vascular proliferation to tumor growth, *Int. Rev. Exp. Pathol.,* 16, 207, 1976.
4. **Hudlicka, O.,** Growth of capillaries in skeletal and cardiac muscle, *Circ. Res.,* 50, 451, 1982.
5. **Hudlicka, O. and Tyler, K. R.,** The effect of long-term high-frequency stimulation on capillary density and fibre types in rabbit fast muscles, *J. Physiol. (London),* 353, 435, 1984.
6. **Knighton, D. R., Silver, I. A., and Hunt, T. K.,** Regulation of wound-healing angiogenesis: effect of oxygen gradients and inspired oxygen concentration, *Surgery,* 90, 262, 1981.
7. **Knighton, D. R., Hunt, T. K., Scheuenstuhl, H., Halliday, B. J., Werb, Z., and Banda, M. J.,** Oxygen tension regulates the expression of angiogenesis factor by macrophages, *Science,* 221, 1283, 1983.
8. **Banchero, N.,** Capillary density of skeletal muscle in dogs exposed to simulated altitude, *Proc. Soc. Exp. Biol. Med.,* 148, 435, 1975.
9. **Patz, A.,** Current concepts of the effect of oxygen on the developing retina, *Curr. Eye Res.,* 3, 159, 1984.
10. **Bresnick, G. H., Engerman, R., Davis, M. D., De Venecia, G., and Myers, F. L.,** Patterns of ischemia in diabetic retinopathy, *Trans. Am. Acad. Ophthalmol. Otolaryngol.,* 81, 694, 1976.
11. **Kumar, D., West, D., Shahabuddin,, S., Arnold, F., Haboubi, N., Reid, H., and Carr, T.,** Angiogenesis factor from human myocardial infarct, *Lancet,* 2, 364, 1983.
12. **Ziada, A. M., Hudlicka, O., Tyler, K. R., and Wright, A. J.,** The effect of long-term vasodilatation on capillary growth and performance in rabbit heart and skeletal muscle, *Cardiovasc. Res.,* 18, 724, 1984.
13. **Meininger, C. J., Schelling, M. E., and Granger, H. J.,** Adenosine and hypoxia stimulate proliferation and migration of endothelial cells, *Am. J. Physiol.,* 255 (*Heart Circ. Physiol.* 26), H554, 1988.
14. **Dusseau, J. W., Hutchins, P. M., and Malbasa, D. S.,** Stimulation of angiogenesis of the chick chorioallantoic membrane, *Circ. Res.,* 59, 163, 1986.
15. **Adair, T. H., Montani, J.-P., Strick, D. M., and Guyton, A. C.,** Vascular development in chick embryos: a possible role for adenosine, *Am. J. Physiol.,* 256 (*Heart Circ. Physiol.* 25), H240, 1989.
16. **Langille, B. L., Reidy, M. A., and Kline, R. L.,** Injury and repair of endothelium at sites of flow disturbances near abdominal aortic coarctations in rabbits, *Arteriosclerosis,* 6, 146, 1986.
17. **Hudlicka, O. and Ziada, A. M. A. R.,** The effect of long-term administration of adenosine, xanthine derivative HWA 285 and prazosin on capillarization in the heart and skeletal muscle, in *Adenosine: Receptors and Modulation of Cell Function,* Stefanovich, V., Rudolphi, K., and Schubert, P., Eds., IRL Press, Oxford, 1985, 373.
18. **Tornling, G., Unge, G., Skoog, I., Ljungqvist, A., Carlsson, S., and Adolfsson, J.,** Proliferative activity of myocardial capillary wall cells in dipyridamole treated rats, *Cardiovasc. Res.,* 12, 692, 1978.
19. **Mattfeldt, T. and Mall, G.,** Dipyridamole-induced capillary endothelial cell proliferation in the rat heart—a morphometric investigation, *Cardiovasc. Res.,* 17, 229, 1983.
20. **Tornling, G., Adolfsson, J., Unge, G., and Ljungqvist, A.,** Capillary neoformation in skeletal muscles of dipyridamole treated rats, *Arzneim. Forsch.,* 30, 791, 1980.
21. **Meininger, C. J. and Granger, H. J.,** Mechanisms leading to adenosine-stimulated proliferation of microvascular endothelial cells, *Am. J. Physiol.,* 258 (*Heart Circ. Physiol.* 27), H198, 1990.
22. **Meininger, C. J., Hawker, and Granger, H. J.,** unpublished observations.

V. Actions of Adenosine on Body Systems

Chapter 20

INFLUENCE OF ADENOSINE ON CARDIAC ACTIVITY

M. G. Collis

TABLE OF CONTENTS

I. INTRODUCTION

Over half a century ago Drury and Szent-Gyorgyi[1] published their classic paper on the effects of adenine compounds on the heart. They demonstrated sinus slowing and atrioventricular (AV) conduction block with adenylic acid derived from heart muscle extracts and with adenosine obtained from yeast. These findings have been confirmed many times in subsequent years. It is now known that adenosine can affect atrial contractility, the Purkinje system, and the ventricular myocytes in addition to its effects on the sinoatrial (SA) and AV nodes. This article briefly reviews each of these effects, the receptors and second messengers that are thought to mediate them, and their potential physiological significance. It is important to realize in the latter context that the demonstration of an effect of exogenous adenosine does not prove that local endogenous adenosine levels are sufficient to evoke this effect. It may be that it is only in situations of metabolic imbalance that endogenous adenosine plays a role in the control of cardiac function.

I. ATRIAL FUNCTION

A. EFFECT ON SINOATRIAL NODE

Adenosine slows the heart rate by a direct action on the SA node. This effect has been noted in most species, including man.[2,3] In addition to its direct depressant action on SA node cells, adenosine can produce a shift of the pacemaker from primary pacemaker cells towards secondary pacemakers, which are closer to the cristae terminalis.[4] The electrophysiological effect of the purine on SA nodal cells is characterized by a prolongation of diastolic period as a consequence of a reduced rate of phase 4 depolarization. There is little effect of adenosine on the amplitude or duration (phases 0 to 3) of the action potential.[2] The decrease in rate of firing of the SA node is accompanied by a hyperpolarization. The ionic mechanisms underlying this hyperpolarizing effect have been investigated further in voltage clamp experiments. In these experiments adenosine had no effect on the inward calcium current (I_{Ca}), on the hyperpolarization-activated inward current (I_f) that occurs in phase 4, or on the delayed rectifier potassium current (I_K) which terminates the action potential, but it caused a small time-independent outward current which accounts for its hyperpolarizing action. This time-independent outward current is carried by potassium ions ($I_{K\ ado}$) and has inward-rectifying properties.[5] However, it is distinct from the background-inward rectifier I_{K1}, which is absent in SA nodal cells.

When the sinus rate is enhanced by catecholamines, adenosine has an additional action. Under these circumstances the enhanced I_{Ca} and I_f evoked by β-adrenergic stimulation are attenuated by the purine. This indicates that adenosine can act via two distinct mechanisms on the mammalian sinus node — a direct action to enhance K^+ conductance and an indirect action which opposes the effects of catecholamines. This dual action of adenosine is also seen in other parts of the atrial myocardium (Figure 1).

B. EFFECT ON ATRIAL CONTRACTILITY

Adenosine has a negative inotropic effect on the atrial myocytes, and, as with its action on the SA node, two mechanisms can be observed. A direct negative inotropic effect occurs which is associated with hyperpolarization of the resting membrane potential and a shortening of the action potential duration. This is due to an increase in potassium conductance[6] (Figure 1). In addition to this direct effect, adenosine has an indirect action to antagonize the effects of positive inotropic stimuli that enhance levels of cyclic adenosine 5′-monophosphate (cAMP)[7] (Figure 1). The direct and indirect effects of adenosine on atrial contractility occur over the same concentration range.

Actions of Adenosine on the Heart

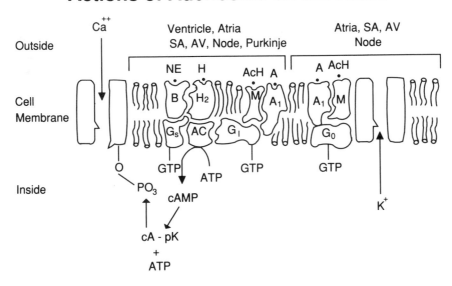

FIGURE 1. Schematic of the effects of adenosine on the heart. In all cardiac tissue, adenosine (A) interacts with an A_1 receptor linked via G_I to adenylate cyclase. This inhibits the effects of catechol-amines (NE) and histamine (H) via β_1 (B) and H_2 receptors to enhance adenylate cyclase activity. The level of cAMP in the cells regulates phosphorylation of the calcium slow channel via cAMP-dependent protein kinase (cA-pK). In atrial and AV nodal tissue, adenosine can act via a cAMP-independent pathway. In these tissues A_1 adenosine receptor activation enhances K^+ conductance via a GTP-dependent regulatory protein, G_O. This causes hyperpolarization and negative inotropic, chron-otropic, and dromotropic effects. The effects of adenosine on cardiac activity are very similar to those of muscarinic (M) receptor activation by acetylcholine (ACh).

C. EFFECT ON AV CONDUCTION

Drury and Szent-Gyorgyi[1] first showed that adenosine and related compounds can block cardiac conduction. There appear, however, to be species differences in the sensitivity of the AV node to this inhibitory effect of adenosine, with the guinea pig and the rat being more sensitive than the dog. Adenosine can cause AV block in man.[8] Adenosine prolongs the atria-His bundle conduction time, but does not alter conduction from the His bundle to the ventricle. This suggests an action in the AV node, and direct recordings have demonstrated that adenosine attenuates the calcium-based action potential of the AV nodal N cells.[9] Adenosine does not alter the conduction of impulses through atrial tissue despite its marked effects on the atrial action potential.[10] It is unclear at present whether the effect of adenosine on the N cells is via a direct effect upon their Ca^{++} conductance or is due to an increase in K^+ conductance, as is seen in other parts of the atrial myocardium. The response does appear to involve a guanosine 5'-triphosphate (GTP)-dependent regulatory protein, since it is blocked by pertussis toxin which produces adenosine 5'-diphosphate (ADP) ribosylation of G_I and G_O.[11]

D. RECEPTORS

The effects of adenosine on atrial rate, force, and AV nodal conduction are blocked by alkylxanthines and enhanced by substances that inhibit the transport of adenosine into the myocardial cell.[12-14] These observations indicate that adenosine acts at a cell-surface receptor which is of the P_1 type. Cell-surface adenosine receptors have been subdivided into A_1 and A_2 receptors, based on the potency or affinity of a series of adenosine analogs (see Chapter 13 of this volume). A further basis for the subclassification of adenosine receptors is by their effect on adenylate cyclase activity; A_1 receptor activation causing inhibition of the enzyme and A_2 receptor activation causing stimulation.[15] Agonist potency studies using isolated cardiac tissue have generally supported the presence of an A_1 receptor. Thus, for the prolongation of AV conduction time in the guinea pig heart, the order of agonist potency is N^6-cyclopentyladenosine > R-N^6 phenylisopropyladenosine (R-PIA) > 5'N-ethylcarbox-amideadenosine (NECA) > 2-chloroadenosine, which is the expected order for an A_1 sub-type.[9] For the direct negative inotropic and chronotropic actions of adenosine analogs, there has been surprisingly little difference detected between the potency of R-PIA and NECA.[12,13] For these effects, however, there is a consistently high stereoselectivity for R-PIA (approx-imately 100-fold), which is indicative of an A_1 receptor. The equivalent potency of NECA and R-PIA, together with the knowledge that most of the direct actions of adenosine on the atrium involve an increase in potassium conductance and reduced calcium fluxes, rather than changes in adenylate cyclase activity, has led Ribeiro[16] to suggest that the receptor that mediates decreases in the rate and force of contraction should be classified as a distinct A_3 subtype. There are, however, considerable difficulties associated with the use of agonist potencies or of second-messenger systems in purine receptor classification, and this further subclassification must be viewed at this stage with caution. Selective antagonists provide a more meaningful basis for receptor classification.[17] Recent studies with the A_1 selective antagonist 1,3-dipropyl-8-cyclopentyl xanthine have supported an A_1 classification for atrial adenosine receptors in the guinea pig, as the xanthine has a high affinity (pA_2 8.4 to 9.2) in this tissue.[18-20] Thus, data with this antagonist do not support the view that guinea pig atrial receptors represent a novel A_3 subtype. The adenosine receptor which mediates a decrease in heart rate in the dog may be of an unusual type. A recent study in conscious, autonomically blocked dogs has shown an order of agonist potency in which NECA > 2-chloroadenosine > R-PIA. A similar order of potency was noted in isolated atria from dogs.[21] This order of agonist potency is more typical of an A_2 receptor than of an A_1 subtype; however, the moderately A_2 selective agonist CV1808 did not slow heart rate. The xanthine adenosine antagonist xanthine amine congener (XAC) was found to be equieffective as an antagonist of the bradycardic and the coronary vasodilator effects of adenosine,[21] and this has been cited as further evidence for a novel subtype, since XAC is A_1 selective in some binding systems. There is, however, controversy concerning the selectivity of this xanthine compound[22,23] and the effects of other antagonists need to be evaluated in the canine heart to confirm the novelty of the adenosine receptor in this tissue.

III. VENTRICULAR FUNCTION

In contrast to the direct effect of adenosine on atrial and AV nodal tissue, the purine has little direct effect on contractility or electrical activity of the ventricular myocytes. This is because ventricular tissue does not contain the K^+ channel which adenosine activates in the atrial myocardium. Adenosine does antagonize the mechanical and electrical effects of catecholamines and of other substances which enhance the activity of adenylate cyclase in guinea pig and rat ventricular tissue.[24,25] There may be significant species differences for this effect, since some authors,[26] but not others,[24] have had difficulty in demonstrating this effect in rabbit heart. The effect can be shown in dog isolated heart preparations.[27] The receptor mediating this effect appears to be of the A_1 subtype based on the relative potencies of adenosine receptor agonists and on the high affinity to the A_1 selective antagonist 1,3-dipropyl-8-cyclopentylxanthine.[20] The ventricular A_1 receptor probably mediates its effect via a guanine nucleotide regulatory protein (G_I or G_O), since pertussis toxin prevents or attenuates the ability of adenosine to inhibit isoprenaline-induced positive inotropic re-sponses.[28]

A number of studies have addressed the effects of adenosine on cAMP levels in ventricular tissue. Since adenosine can only oppose the positive inotropic effects of agents that enhance adenylate cyclase activity, it would seem logical that it should reduce levels of cAMP. A range of effects, however, have been reported in heart membrane preparations which range from an enhancement to a reduction of cAMP levels. One factor which complicates the interpretation of these results is that most cardiac preparations are a mixture of cell types, and adenylate cyclase activity in some of these cells (smooth muscle and endothelial) may be stimulated by adenosine. It has been shown that adenosine analogs only decrease adenylate cyclase activity in cardiac membrane preparations when endothelial cell membranes are removed.[29] An alternative explanation of the ventricular effects of adenosine has been proposed by Heller et al.,[30] who suggest that it inhibits the action of cAMP rather than decreases the concentration of the cyclic nucleotide.

Adenosine and its analogs can also counteract the electrophysiological effects of catecholamines on ventricular tissue. Thus, the prolongation of action potential duration, the enhanced action potential amplitude, and the after potentials and triggered activity caused by β-adrenoceptor stimulants are reduced by adenosine A_1 receptor agonists.[31,32] Similar antiadrenergic effects have also been reported for adenosine on Purkinje tissue. Adenosine can oppose the effects of catecholamines to enhance automaticity and to shorten action potential duration in these specialized conducting tissues.[33] In man, adenosine has been shown to have no direct effect on His-Purkinje cycle length, but to reduce automaticity in the presence of isoprenaline.[34] Although adenosine has no direct effect on action potential parameters in normal Purkinje tissue in the absence of catecholamines, it can oppose the decreased phase 0 dV/dt and the decreased action potential amplitude that occur in fibers subjected to simulated ischemic conditions.[35,36]

The antiadrenergic effect of adenosine and its apparent ability to reverse some of the electrophysiological consequences of ischemia suggest that it might have an antiarrhythmic effect. Studies in the anesthetized rat and the dog have shown an effect of adenosine to reduce the frequency of some arrhythmias resulting from coronary artery occlusions.[35,37] In the dog, adenosine administered into the left ventricle reduced the incidence of ventricular tachycardia, particularly in phase 1b (when catecholamine release occurs). However, intraventricular adenosine increased the incidence of ventricular fibrillation during ischemia. Surprisingly, adenosine delivered into the coronary artery supplying the ischemic zone exacerbated the incidence of ventricular tachycardia in a group of four dogs. On reperfusion, adenosine reduced the incidence of ventricular fibrillation. The mechanisms underlying these pro- and antiarrhythmic effects of adenosine are not clear, but could include alterations in collateral coronary flow, metabolic effects, and changes in myocardial work load (blood pressure was decreased by adenosine), in addition to any cardiac electrophysiological actions of the purine. Adenosine has also been shown to decrease the rate of spontaneous ventricular tachycardia following occlusion of the left anterior descending coronary artery in dogs. This effect appears to be due to an attenuation of the effects of catecholamines, as it was not observed following cardiac denervation.[38] It is interesting to note that adenosine has been shown to inhibit ventricular tachycardia facilitated by exercise or by isoprenaline in patients with structurally normal hearts. This antiarrhythmic effect of adenosine in man appears to be rather selective, as the purine is ineffective against tachycardia due to ventricular re-entry in patients with coronary artery disease, even when this is exacerbated by catecholamines.[34]

IV. PHYSIOLOGICAL/PATHOPHYSIOLOGICAL SIGNIFICANCE

Although adenosine has been shown to reduce sinus rate, atrial contractility, AV conduction velocity, to oppose the effects of catecholamines, and to have both pro- and anti-arrhythmic effects, it is not clear that endogenous levels of the purine are sufficient for these effects to occur *in vivo*. This question can be best addressed by examining the effects of adenosine antagonists. Theophylline can increase cardiac rate, which could indicate a tonic role of adenosine at the SA node, but this is more likely to be a consequence of the ability of theophylline to inhibit phosphodiesterase than to block adenosine receptors since enprofylline, which does not interact with adenosine receptors, exhibits the same effect.[39] Selective and potent adenosine receptor antagonists such as 8-phenyltheophylline and 1,3-dipropyl 8-cyclopentylxanthine, which do not inhibit phosphodiesterases, do not alter cardiac rate in either anesthetized or conscious animals.[39-41] Therefore, there seems little evidence to support the idea that levels of adenosine in the normal unstressed heart are sufficient to influence the activity of the SA node. A similar argument probably applies to the effects of adenosine on AV conduction.

It is when adenosine production is enhanced by ischemia or hypoxia that a role for the endogenous purine is more likely to occur. Sinus bradycardia associated with acute inferior myocardial infarction is one situation which could involve an effect of adenosine on the SA node. Thus, hypoxia- or ischemia-induced bradycardia in dog and guinea pig hearts can be attenuated by adenosine receptor antagonists.[42,43] It is not known whether there are any other clinical situations in which the direct effect of adenosine on the SA node is of relevance. It has been suggested that the purine could be involved in the bradycardia of sick sinus syndrome,[44] but this has not been tested by examining the effects of adenosine antagonists in patients with this disorder.

Studies in isolated guinea pig hearts have also shown that the prolonged AV conduction interval provoked by hypoxia can be ameliorated by a number of adenosine receptor antagonists and by the degradative enzyme adenosine deaminase.[10] These results indicate a role for endogenous adenosine in hypoxia-induced AV conduction block. There is a report that some ischemia-induced AV conduction disturbances in man can be beneficially treated with aminophylline.[45] Data obtained in man with theophylline and its derivatives must be viewed with caution, however, since the phosphodiesterase inhibitory effects of this compound could also cause a decrease in AV conduction time.

Is the atrial and ventricular antiadrenergic effect of adenosine of relevance to the normal control of cardiac inotropic state? It is well known that theophylline is a positive agent and this could be an action mediated via blockade of the negative inotropic effects of endogenous adenosine. When theophylline was compared in the anesthetized dog with the more potent adenosine antagonist 8-phenyltheophylline and with the xanthine enprofylline, which does not block adenosine receptors, the order of potency as positive inotropic agents was enprofylline > theophylline > 8-phenyltheophylline.[39] Thus, the order of potency of these xanthines as adenosine receptor antagonists was the reverse of their order of potency as positive inotropic agents. These results do not support the hypothesis that adenosine normally exerts a negative inotropic influence on the ventricle. In order to investigate whether or not the potent adenosine receptor antagonist 8-phenyltheophylline has a positive inotropic effect in the ischemic myocardium, where adenosine production should be enhanced, its effects have also been examined in dogs subjected to short periods of coronary ligation.[46] Even in the presence of severe transient regional ischemia, a positive inotropic effect of 8-phenyltheophylline could not be detected. Thus it appears that in the dog, a species in which adenosine has been shown to exert an antiadrenergic effect *in vitro,*[27] insufficient adenosine is produced *in vivo* to significantly affect the inotropic state.

The absence of evidence for a role of adenosine in controlling the inotropic state in the dog does not rule out the possibility that there may be considerable species differences in the role of adenosine as a regulator of cardiac contractile function. Adenosine deaminase, an enzyme that inactivates adenosine, has been shown to enhance the positive inotropic response of normoxic and hypoxic rat isolated hearts to isoprenaline.[47,48] Further *in vivo* studies are needed to determine whether these results indicate a true species difference or if they are due to the conditions under which isolated heart preparations are perfused. It is of particular interest that adenosine exerts a pronounced antiadrenergic effect on human isolated myocytes.[50] The potential clinical implications of this can only be evaluated by studies of the effects of adenosine deaminase or of selective adenosine antagonists in man. In a similar fashion, the demonstration that exogenous adenosine can have antiarrhythmic effects under certain circumstances does not prove that the endogenously produced purine ever exerts such an effect. Studies with specific adenosine antagonists in animal models of arrhythmia are required to address this question; however, none have been reported to date.

V. CONCLUSIONS

We have learned a considerable amount about the effects of adenosine on cardiac function since the original observations of Drury and Szent-Gyorgyi.[1] The major issue still to be resolved is the role and significance of these effects in the normal and abnormal control of cardiac function. At the present time it appears that adenosine may play a regulatory role in some species when the heart is metabolically compromised by ischemia or hypoxia. There is little evidence for an involvement of the purine in the control of cardiac excitability in normal physiological states. Further studies utilizing potent and selective adenosine receptor antagonists may reveal new roles for the purine in the control of cardiac function and may lead to the development of useful therapeutic agents.

REFERENCES

1. **Drury, A. N. and Szent-Gyorgyi, A.,** The physiological activity of adenine compounds with special reference to their action upon the mammalian heart, *J. Physiol. (London),* 68, 213, 1929.
2. **Belardinelli, L., West, A., Crampton, R., and Berne, R. M.,** Chronotropic and dromotropic actions of adenosine, in *The Regulatory Function of Adenosine,* Berne, R. M., Rall, T. W., and Rubio, R., Eds., Martinus Nijhoff, The Hague, 1983, 378.
3. **DiMarco, J. P., Sellers, T. D., Berne, R. M., West, G. A., and Belardinelli, L.,** Adenosine: electrophysiologic effects and therapeutic use for terminating paroxysmal supraventricular tachycardia, *Circulation,* 68, 1254, 1983.
4. **West, G. S. and Belardinelli, L.,** Sinus slowing and pacemaker shift caused by adenosine in rabbit SA node, *Pfluegers Arch.,* 403, 66, 1985.
5. **Belardinelli, L., Giles, W. R., and West, A.,** Ionic mechanisms of adenosine actions in pacemaker cells from rabbit heart, *J. Physiol.,* 405, 615, 1988.
6. **Isenberg, G., Cerbia, E., and Klockner, U.,** Ionic channels and adenosine in isolated heart cells, in *Topics and Perspectives in Adenosine Research,* Gerlach, E. and Becker, B. F., Eds., Springer-Verlag, New York, 1987, 323.
7. **Dobson, J. G.,** Adenosine reduces catecholamine contractile responses in oxygenated and hypoxic atria, *Am. J. Physiol.,* 245, H468, 1983.
8. **Honey, R. M., Ritchie, W. T., and Thomson, W. A. R.,** The action of adenosine upon the human heart, *J. Med.,* 23, 485, 1930.
9. **Clemo, H. F. and Belardinelli, L.,** Effect of adenosine on atrioventricular conduction. I. Site and characterization of adenosine action in the guinea pig atrioventricular node, *Circ. Res.,* 59, 427, 1986.
10. **Clemo, H. F. and Belardinelli, L.,** Effect of adenosine on atrioventricular conduction. II. Modulation of atrioventricular node transmission by adenosine in hypoxic isolated guinea pig hearts, *Circ. Res.,* 59, 37, 1986.
11. **Bohm, M., Schmitz, W., Scholz, H., and Wilken, A.,** Pertussis toxin prevents adenosine receptor- and m-cholinoceptor-mediated sinus slowing and AV conduction block in the guinea-pig heart, *Naunyn-Schmeideberg's Arch. Pharmacol.,* 339, 152, 1989.
12. **Collis, M. G.,** Evidence for an A_1 adenosine receptor in the guinea-pig atrium, *Br. J. Pharmacol.,* 78, 207, 1983.
13. **Collis, M. G. and Saville, V. L.,** An investigation of the negative chronotropic effect of adenosine on the guinea-pig atrium, *Br. J. Pharmacol.,* 83, 413, 1984.

14. **Belardinelli, L., Fenton, R. A., West, A., Linden, J., Althaus, J. S., and Berne, R. M.,** Extracellular action of adenosine and the antagonism by aminophylline on the atrioventricular conduction of isolated perfused guinea pig and rat hearts, *Circ. Res.,* 51, 569, 1982.

15. **Van Calker, D., Muller, M., and Hamprecht, B.,** Adenosine regulates via two different types of receptors, the accumulation of cyclic AMP in cultured cells, *J. Neurochem.,* 33, 999, 1979.

16. **Ribeiro, J. A. and Sebastião, A. M.,** Adenosine receptors and calcium: basis for proposing a third (A₃) adenosine receptor, *Prog. Neurobiol.,* 26, 179, 1986.

17. **Collis, M. G.,** Are there two types of adenosine receptors in peripheral tissues?, in *Purines: Pharmacology and Physiology Roles,* Stone, T. W., Ed., Macmillan, New York, 1985, 75.

18. **Collis, M. G., Stoggall, S. M., and Martin, F. M.,** Apparent affintiy of 1,3 dipropyl-8-phenylxanthine for adenosine A₁ and A₂ receptors in isolated tissues from guinea-pigs, *Br. J. Pharmacol.,* 79, 1274, 1989.

19. **Borea, P., Caparrotta, L., De Biasi, M., Fassini, G., Froldi, G., Pandolfo, L., and Ragazzi, E.,** Effects of selective agonists and antagonists on atrial adenosine receptors and their interaction with Bay K8644 and [³H]-nitrendipine, *Br. J. Pharmacol.* 96, 372, 1989.

20. **Von der Leyen, H., Schmitz, W., Scholz, H., Scholz, J., Lohse, M. J., and Schwabe, U.,** Effects of 1,3-dipropyl-8-cyclopentylxanthine (DPCPX), a highly selective adenosine receptor antagonist, on force of contraction in guinea-pig atrial and ventricular cardiac preparations, *Naunyn-Schmiedeberg's Arch. Pharmacol.,* 340, 204, 1989.

21. **Belloni, F. L., Belardinelli, L., Halperin, C., and Hintze, T. H.,** An unusual receptor mediates adenosine-induced S-A nodal bradycardia in dogs, *Am. J. Physiol.,* 256, H1553, 1989.

22. **Collis, M. G., Jacobson, K. A., and Tomkins, D. M.,** Apparent affinity of some 8-phenyl-substituted xanthines at adenosine receptors in guinea-pig aorta and atria, *Br. J. Pharmacol.,* 92, 69, 1987.

23. **Collis, M. G.,** Adenosine receptor sub-types in isolated tissues: antagonist studies, in *Purines in Cellular Signalling: Targets for New Drugs,* Jacobson, K. A., Daly, J. W., and Manganiello, V., Eds., Springer-Verlag, New York, 1990, 48.

24. **Endoh, M. and Yamashita, S.,** Adenosine antagonizes the positive inotropic action mediated via beta but not alpha-adrenoceptors in rabbit papillary muscle, *Eur. J. Pharmacol.,* 65, 445, 1980.

25. **Baumann, G., Schrader, J., and Gerlach, E.,** Inhibitory action of adenosine on histamine- and dopamine-stimulated cardiac contractility and adenylate cyclase in guinea pigs, *Circ. Res.,* 48, 259, 1981.

26. **Hopwood, A. M., Harding, S. E., and Harris, P.,** An anti-adrenergic effect of adenosine on guinea-pig but not rabbit ventricles, *Eur. J. Pharmacol.,* 137, 67, 1987.

27. **Endoh, M. and Kushida, H.,** Inhibitory effects of adenosine on myocardial contractility: comparison with muscarinic receptor stimulation, in *Adenosine and Adenine Nucleotides: Physiology and Pharmacology,* Paton, D. M., Ed., Taylor & Francis, London, 1988, 288.

28. **Wilson, W. W., West, G. A., Hewlett, E. L., and Belardinelli, L.,** Attenuation of the inhibitory effects of adenosine by pertussis toxin in isolated guinea pig hearts, *Fed. Proc. Fed. Am Soc. Exp. Biol.,* 46, 1436, 1987.

29. **Schutz, W., Freissmuth, M., Hausleithner, V., and Tuisl, E.,** Cardiac sarcolemmal purity is essential for the verification of adenylate cyclase inhibition via A₁-adenosine receptors, *Naunyn-Schmeideberg's Arch. Pharmacol.,* 333, 156, 1986.

30. **Heller, T., Kocher, M., Neumann, J., Schmitz, W., Scholz, H., Stemmildt, V., and Stortzel, K.,** Effects of adenosine analogues on force and cAMP in the heart. Influence of adenosine deaminase, *Eur. J. Pharmacol.,* 164, 179, 1989.

31. **Belardinelli, L. and Isenberg, G.,** Actions of adenosine and isoproterenol on isolated mammalian ventricular myocytes, *Circ. Res.,* 53, 287, 1983.

32. **Isenberg, G. and Belardinelli, L.,** Ionic basis for the antagonism between adenosine and isoproterenol on isolated mammalian ventricular myocytes, *Circ. Res.,* 55, 309, 1984.

33. **Rosen, M. R., Danilo, P., and Weiss, R. M.,** Actions of adenosine on normal and abnormal impulse initiation in canine ventricle, *Am. J. Physiol.,* 244, H715, 1983.

34. **Lerman, B. B., Wesley, R. C., DiMarco, J. P., Haines, D. E., and Belardinelli, L.,** Antiadrenergic effects of adenosine on His-Purkinje automaticity. Evidence for accentuated antagonism, *J. Clin. Invest.,* 82, 2127, 1988.

35. **Parratt, J. R., Boachie-Ansah, G., Kane, K. A., and Wainwright, C. L.,** Is adenosine an endogenous antiarrhythmic agent under conditions of myocardial ischemia?, in *Adenosine and Adenine Nucleotide: Physiology and Pharmacology,* Paton, D. M. Ed., Taylor & Francis, London, 1988, 157.

36. **Bailey, J. C. and Rardon, D. P.,** Electrophysiological effects of adenosine and dipyridamole on cardiac purkinje fibres and ventricular myocardium, in *Cardiac Electrophysiology and Pharmacology of Adenosine and ATP: Basic and Clinical Aspects,* Pelleg, A., Michelson, E. L., and Dreifus, L. S., Eds., Alan R. Liss, New York, 1987. 119.

37. **Wainwright, C. L. and Parratt, J. R.,** An antiarrhythmic effect of adenosine during myocardial ischemia and reperfusion, *Eur. J. Pharmacol.,* 145, 183, 1988.

38. **Constantin, L. and Martins, J. B.,** Autonomic control of ventricular tachycardia: effects of adenosine and N⁶-R-1-phenyl-2-propyladenosine, *J. Am. Coll. Cardiol.,* 10, 399, 1987.
39. **Collis, M. G., Keddie, J. R., and Torr, S. R.,** Evidence that the positive inotropic effects of the alkylxanthines are not due to adenosine receptor blockade, *Br. J. Pharmacol.,* 81, 401, 1984.
40. **Bowmer, C. J., Collis, M. G., and Yates, M. S.,** Effect of the adenosine antagonist 8-phenyltheophylline on glycerol-induced acute renal failure in the rat, *Br. J. Pharmacol.,* 88, 205, 1986.
41. **Kellet, R., Bowmer, C. J., Collis, M. G., and Yates, M. S.,** Amelioration of glycerol-induced acute renal failure in the rat with 8-cyclopentyl-1,3-dipropylxanthine, *Br. J. Pharmacol.,* 98, 1066, 1989.
42. **Wesley, R. C., Boykin, M. T., and Boykin, L.,** Role of adenosine as mediator of bradyarrhythmias during hypoxia in isolated guinea pig hearts, *Cardiovasc. Res.,* 20, 752, 1986.
43. **Motomura, S. and Hashimoto, K.,** Reperfusion-induced bradycardia in the isolated, blood-perfused sino-atrial node and papillary muscle preparations of the dog, *Jpn. Heart J.,* 23, 112, 1982.
44. **Watt, A. H.,** Sick sinus syndrome: an adenosine mediated disease, *Lancet,* (Apr.) 6, 786, 1985.
45. **Wesley, R. C., Lerman, B. B., DiMarco, J. P., Berne, R. M., and Belardinelli, L.,** Mechanism of atropine-resistant atrioventricular block during inferior myocardial infarction: possible role of adenosine, *J. Am. Coll. Cardiol.,* 8, 1232, 1986.
46. **Collis, M. G.,** Cardiac and renal actions of adenosine antagonists in vivo, in *Adenosine and Adenine Nucleotides, Physiology and Pharmacology,* Paton, D. M., Ed., Taylor & Francis, London, 1988, 259.
47. **Dobson, J. G., Jr., Ordway, R. W., and Fenton, R. A.,** Endogenous adenosine inhibits catecholamine contractile responses in normoxic hearts, *Am. J. Physiol.,* 251, H455, 1986.
48. **Dobson, J. G.,** Mechanism of adenosine inhibition of catecholamine-induced responses in heart, *Circ. Res.,* 52, 151, 1983.
49. **Sperelakis, N.,** Regulation of calcium slow channels of cardiac and smooth muscles by adenine nucleotides, in *Cardiac Electrophysiology and Pharmacology of Adenosine and ATP,* Pelleg, A., Michelson, E. L., and Dreifus, L. S., Eds., Alan R. Liss, New York, 1987, 135.
50. **Harding, S. E.,** Personal communication.

Chapter 21

ADENOSINE RECEPTOR ACTIVATION AND RENAL FUNCTION

William S. Spielman

TABLE OF CONTENTS

I. INTRODUCTION

Adenosine is capable of regulating a wide range of cellular functions, and this diversity of action is especially apparent in the kidney. Adenosine binding to plasma membrane receptors on a variety of renal cell types results in a vast array of functional responses, including alterations in the rate and distribution of renal blood flow, glomerular filtration, hormone and neurotransmitter release, and tubular reabsorption. Recently, the use of adenosine receptor agonist ligands with relative selectivity for adenosine receptor subtypes have indicated which receptor subtypes mediate each of these renal events. These findings, coupled with the knowledge that kidney cells are capable of releasing endogenous adenosine into the interstitial space, suggest that alterations in the extracellular adenosine concentration may regulate renal function. It is the intention of this brief review to summarize the known renal actions of adenosine and identify the receptor subtypes involved, what signal transduction systems are activated, and the evidence suggesting that endogenous adenosine production acts as a mediator of intrinsic renal regulation.

During the past decade a variety of comprehensive reviews have appeared summarizing the intrarenal role of adenosine, and to which the reader is referred for a more complete historical perspective and overview of the subject.[1-5] Further, many of the chapters in the present volume deal in detail with subjects that are relevant to the general issues of adenosine in the overall regulation of cell function. In particular, the reader is referred to chapters dealing with the formation and metabolism of adenosine by Schrader (Chapter 5), nucleoside transport in cells by Hertz (Chapter 7), and adenosine receptors by Schwabe (Chapter 3) and Green (Chapter 4).

II. PHYSIOLOGICAL ACTIONS OF RENAL ADENOSINE RECEPTOR ACTIVATION

With the recognition of a wide array of renal cellular actions and the continuing development of relatively specific adenosine receptor agonist and antagonist ligands, investigators have undertaken the task of assigning the different renal actions of adenosine to the known adenosine receptor types, as previously identified in other tissues, by comparison of relative agonist and antagonist potencies.

A. HEMODYNAMIC ACTIONS
1. Rate and Distribution of Renal Blood Flow and Glomerular Filtration
The hemodynamic actions of adenosine are varied and demonstrate an interesting interaction with angiotensin.[6,7] The reader is referred to previous reviews for a detailed description.[1,3] Briefly, the infusion of adenosine into the renal artery of dogs results in a prompt fall in blood flow which rapidly wanes, with renal blood flow returning to a value below, at, or above the control level. Despite the return of blood flow toward preinfusion levels during the continued infusion of adenosine, glomerular filtration remains depressed until the infusion is terminated.[8] The vascular response of the kidney is vasoconstriction of the outer cortex accompanied by vasodilation of the deep cortex.[9] Studies directed at the mechanism of the adenosine-induced fall in glomerular filtration rate (GFR) demonstrate a fall in glomerular hydrostatic pressure resulting from a preglomerular vasoconstriction and a more slowly developing postglomerular vasodilation, findings in keeping with the biphasic action on blood flow and the sustained effect on GFR.[8] More recent investigations into the subtype of adenosine receptor responsible for these hemodynamic effects, taking advantage of the nonmetabolized adenosine receptor agonists, convincingly indicate that the vasoconstrictive response is mediated by activation of an A_1 receptor, whereas the renal vasodilation develops from stimulation of the lower affinity A_2 receptor.[10-13]

2. Glomerular Filtration Coefficient

No evidence has been gathered to demonstrate an effect of adenosine on hydraulic conductivity or surface area of the glomerular membrane. Adenosine and adenosine analogs have, however, been reported to stimulate an increase in cAMP production by isolated glomeruli.[14,15] This observation of adenosine-induced cAMP production in glomerular cells, presumably via activation of A_2 receptors, is consistent with the action of other effectors of the glomerular filtration coefficient (K_f) on cAMP, raising the possibility that the decrease in GFR during adenosine infusion is also the result of a decrease in K_f.

3. Tubuloglomerular Feedback

It has long been recognized that an increase in the perfusion rate of fluid in the distal nephron results in a vasoconstrictive response of the afferent arteriole such that the GFR of that nephron is diminished. It has been hypothesized that this phenomenon, termed tubuloglomerular feedback, serves as a mechanism whereby each nephron has the ability to limit large increases in tubular fluid and solute delivery and the consequent alterations in tubular and excretory function.[16] It is proposed that changes in perfusion of the thick ascending limb are detected by a specialized group of epithelial cells, collectively termed the *macula densa*, which in turn signals the afferent arteriole to constrict. Although a detailed summary of the extensive research directed at determining the mediator of this phenomenon is beyond the scope of this discussion, it is sufficient to state that it remains unclear how information is transmitted from the *macula densa* to the adjacent afferent arteriole. An interesting hypothesis relevant to the present discussion is that with an increased delivery of solute to the cells of the *macula densa,* an increased reabsorptive sodium chloride transport ensues, associated with stimulation of ATP hydrolysis and adenosine formation and release from the epithelial cells, resulting in afferent arteriolar vasoconstriction.[1,17] This hypothesis suggests that the *macula densa* might then serve as a ''transport sensor'', and use adenosine as a paracrine signal to reduce delivered solute, and thereby transport, toward normal. The evidence for such a hypothesis has been previously reviewed; however, preliminary evidence was recently presented that appears particularly convincing that the highly specific A_1 antagonist 1,3-dipropyl-8-cyclopentylxanthine (DPCPX) was effective in the inhibition of tubuloglomerular feedback when administered either into the tubule lumen or into the peritubular capillary circulation,[18] suggesting an important role for adenosine in mediating this phenomenon.

B. RENIN SECRETION

Adenosine has long been known to inhibit the release of renin and is the specific topic of a recent review by Churchill and Churchill.[5] Studies both *in vivo*[19] and *in vitro*[5,20-25] have provided evidence that this action is likely a direct effect of adenosine on the renin-secreting cells rather than mediated by adenosine-induced alterations in hemodynamics, tubular reabsorption, or neurotransmitter release. At the same time that adenosine was proposed as the mediator of the vasoconstriction associated with increased perfusion of the *macula densa* (tubuloglomerular feedback), it was also proposed as a mediator of the inhibition of renin release by the *macula densa*.[1] Evidence in support of a role for adenosine in the control of renin release has come principally from studies that have evaluated the action of exogenously administered adenosine, and evidence in support of a role for endogenously produced adenosine in the control of renin release is largely circumstantial. A more direct assessment of the actions of intrarenal adenosine is to elevate endogenous levels of adenosine through pharmacological, physiological, or pathophysiological perturbations while monitoring renal function, and then using adenosine receptor antagonists. Evidence that endogenous adenosine production can suppress renin secretion was obtained in studies in which the administration of maleic acid, a substance which alters ATP formation by the kidney and produces a generalized tubular transport defect resembling the Fanconi syndrome, results in the elevation of renal adenosine production as determined by measurements of adenosine in renal venous blood and urine and results in a marked suppression of renin release.[26] This decrease in renin release was largely reversed by the administration of theophylline.[26] Additional evidence for an inhibitory action of endogenous adenosine on renin release has been obtained in studies in which 1,3-dipropyl-8-(p-sulfophenyl)xanthine (DPSPX), a relative nonselective, but more potent adenosine receptor antagonist than theophylline, administered to sodium-depleted and normal rats, resulted in an increased rate of renin release.[27] Evidence for adenosine mediating *macula densa* control of renin release was reported by Itoh et al.,[28] who demonstrated that in the microdissected glomerulus, renin release was lower when the *macula densa* was left intact as compared to when the *macula densa* segment was removed. Furthermore, this difference was abolished with the administration of theophylline, suggesting that adenosine produced by the *macula densa* segment was responsible for the inhibition of renin release. Finally, preliminary data obtained using a single glomerulus, with an attached, perfused *macula densa* segment, demonstrated that the inhibition of renin release by elevated sodium chloride was inhibited by the A_1 adenosine antagonist, DPCPX.[27] Collectively, these findings are supportive of a role for endogenous adenosine in regulating both basal renin release and the renin response to sodium restriction, and that these effects may be mediated by a *macula densa* mechanism.

Although most theories concerning the role of adenosine in renin release stress its inhibitory action, Churchill and Churchill[5] point out in their review that the recent use of adenosine receptor agonist ligands has revealed both an inhibitory and a stimulatory action of adenosine mediated by activation of the A_1 and A_2 receptor populations, respectively. In general, studies using either isolated perfused rat kidneys or rat renal cortical slices have demonstrated that renin release can either be inhibited or stimulated by activation of adenosine receptors, depending on the dose of agonist. The rank order of potency for the adenosine agonists on the renin response indicates that inhibition of renin release is by activation of the A_1 receptor, whereas stimulation is the result of activation of A_2 receptors. In addition, it has been reported that pretreatment of rats with pertussis toxin blocked the actions of N^6-cyclohexyladenosine (CHA) to inhibit renin release in kidney slices.[30] This observation suggests that the renin-inhibitory effect of adenosine at the A_1 receptor is coupled to a guanine nucleotide binding protein (G protein). Because a number of G proteins are now known to act as pertussis toxin substrates and are inactivated by ADP ribosylation, this observation alone cannot be taken as evidence that the A_1 receptor-mediated inhibition is the result of the suppression of adenylate cyclase activity. As is discussed in detail below, adenosine A_1 receptor activation has recently been shown to stimulate turnover of membrane inositol phospholipids with the resultant stimulation of inositol triphosphate production and mobilization of cytosolic free Ca^{2+} via a pertussis toxin-sensitive G protein in renal epithelial cells. This observation raises the possibility of a similar mechanism existing in juxtaglomerular cells. The hypothesis that adenosine acts to inhibit renin release via elevation of cytosolic Ca^{2+} has been discussed previously,[5,24] and is particularly appealing in light of the vasoconstriction due to activation of A_1 receptors in afferent arteriolar vascular smooth-muscle cells, the cells from which the renin-secreting cells apparently derive.

Recently, it was demonstrated in isolated juxtaglomerular cells that activation of A_1 receptors is associated with an elevation of cyclic guanosine 5'-monophosphate (cGMP), but not with changes in either cytosolic calcium or cAMP, suggesting the involvement of yet another second-messenger system for adenosine.[31] This interesting observation fails to adequately explain why adenosine activation of the A_1 receptor results in vasoconstriction at the afferent arteriole, a result directionally opposite that expected with elevations of cGMP. Furthermore, the previously mentioned pertussis toxin sensitivity of the response is difficult to resolve with elevation of cGMP as there are no reports of G protein coupling to guanylate cyclase.

C. ERYTHROPOIETIN SECRETION

Erythropoietin is secreted from an unknown site in the renal cortex and is important in stimulating the production of red blood cells in the bone marrow. It was recently reported that radioiron incorporation into red cells of exhypoxic polycythemic mice, indicative of erythropoietin production, was inhibited by A_1 receptor activation and stimulated by A_2 receptor activation.[32] In a preliminary report by the same workers, activation of A_1 receptors inhibited, and A_2 receptor activation stimulated, erythropoietin production by a renal carcinoma cell-culture system.[33] These intriguing findings raise the possibility that adenosine, produced from the degradation of ATP due to limited oxygen availability, is involved in the regulation of erythropoietin production.

D. NEUROTRANSMITTER RELEASE

Adenosine acts to inhibit the release of neurotransmitter from postganglionic sympathetic neurons.[34] In the kidney, adenosine acts at a prejunctional A_1 receptor that reduces the release of norepinephrine.[35-37] Curiously, the postjunctional effects of adenosine increase the sensitivity of the kidney to transmitter. The effects of nerve stimulation, for example, in the presence of elevated adenosine levels, are enhanced despite diminished transmitter release.[35] Though it has been suggested that the inhibition of neurotransmitter release by adenosine could be the initiating factor for many of the other observed renal actions of adenosine, it appears unlikely that this is the case, since adenosine-induced inhibition of renin release and changes in hemodynamics and excretion occur in the absence of functional renal nerves.[1,5] While it is generally agreed that neurotransmitter release is inhibited by activation of A_1 adenosine receptors, the postreceptor mechanism remains unclear. Neither increases nor decreases in cAMP appear to account for the presynaptic effects of adenosine, and various hypotheses for the inhibition of transmitter release have been suggested.[38,39] To date, no studies have directly determined the effect of adenosine analogs on postsynaptic second messengers in renal sympathetic nerves.

E. EXCRETORY AND TUBULAR EPITHELIAL EFFECTS OF ADENOSINE

Elevation of intrarenal adenosine by exogenous administration or by pharmacological manipulation of endogenous adenosine has been reported to result in a fall of urine flow and solute excretion.[8,40] Despite the fact that adenosine has been reported to change the fractional excretion rate of sodium, the concomitant changes in GFR and renal blood flow have generally made it difficult to determine in the intact kidney if adenosine has direct actions to effect tubular reabsorption. In a recent study, however, in which adenosine infusions had no effect on either GFR or renal blood flow, it was observed that urinary volume and sodium excretion were increased, leading the authors to suggest a direct tubular effect for adenosine.[41]

A direct action of adenosine and adenosine analogs has been reported in a wide variety of nonrenal epithelia and was recently summarized.[2] In the isolated perfused rabbit collecting tubule, adenosine receptor agonists have been shown to both inhibit and stimulate hydraulic conductivity, depending on the dose and agonist used.[42] Using recently developed immunoselection techniques and the subsequent development of a clonally expanded cell line, we have investigated the presence of adenosine receptors and signaling mechanisms in tubular cells from the rabbit renal cortical collecting tubule (RCCT) and the rabbit thick ascending limb.[43-45]

Cyclic-AMP production was stimulated by micromolar concentrations of each analog with a rank order of potency of NECA > R-PIA > CHA, indicating activation of adenylate cyclase by an A_2 receptor. At nanomolar concentrations, the receptor agonists produced an inhibition of basal cAMP production. Furthermore, stimulation of cAMP in RCCT cells by isoproterenol or vasopressin was inhibited by simultaneous addition of CHA, an A_1 receptor agonist. This inhibitory action of CHA on hormone-stimulated cAMP production was inhibited by the specific A_1 receptor antagonist, DPCPX, or by prior treatment of the cells with pertussis toxin, indicating that activation of A_1 adenosine receptors in RCCT cells regulates adenylate cyclase activity. Similar evidence for both A_1 and A_2 receptors coupled via G proteins to adenylate cyclase was also reported for cultured cells of the thick ascending limb[45] and from a recently established cell line derived from the cortical collecting tubule.[44] The only notable difference of the renal epithelial adenosine receptors and their coupling to adenylate cyclase when compared to adenosine receptors described in brain, fat cells, and platelets is that in the renal epithelial cells studied, both stimulatory (A_2) and inhibitory receptor (A_1) appear to exist in the same cell type, contrary to the current perception that cells may demonstrate one, but not both, receptor subtypes.[46] Although the functional significance of A_1 and A_2 receptor populations on collecting tubule and thick limb cells is not entirely clear, the ability of adenosine to regulate basal and hormone-stimulated cAMP production via activation of two distinct receptor populations makes it a potentially important regulator of hormonally controlled transport.

Several observations in nonrenal tissues suggest that adenosine may alter cell function independently of changes in cAMP production.[35,47] We have recently reported in cultured thick ascending limb cells, cortical collecting tubule cells, and the 28A cell line that adenosine analogs result in the mobilization of cytosolic free calcium and the increased turnover of inositol phosphates.[44,45,48] The adenosine analog-induced increase in cytosolic calcium and inositol phosphate is inhibited by the specific A_1 receptor antagonist, DPCPX, and by prior treatment of the cells with pertussis toxin. While the blockade of the response by a specific A_1 adenosine receptor antagonist and the observation that the G protein is a pertussis toxin substrate are consistent with a phospholipase C (PLC) response coupled to an A_1 receptor, the lack of significantly different potencies among various adenosine receptor agonist raises the possibility of a previously unidentified adenosine receptor subtype with similarity to the A_1 receptor.

III. FUTURE CONSIDERATIONS AND CONCLUDING COMMENTS

The presence of both stimulatory (A_2) and inhibitory (A_1) adenosine receptors capable of regulating several aspects of renal function illustrates the "dual-control" nature of adenosine as a regulator of kidney cell function. This "dual-control" regulation provides for an interesting model with which to examine renal cellular regulation, and provides the impetus for the development of highly selective adenosine agonist and antagonist ligands with which to therapeutically manipulate renal function. It must be kept in mind, however, that the affinity of the endogenous ligand adenosine for the inhibitory A_1 receptor is generally felt to be 100- to 1000-fold higher than for the stimulatory A_2 receptor. Therefore, if endogenous adenosine participates in the control of renal function, its action at the inhibitory A_1 receptor will most likely dominate.

In the nephron, A_1 adenosine receptors linked to the inhibition of adenylate cyclase may also evoke the mobilization of cytosolic calcium via the turnover of inositol phosphates through the activation of PLC coupled with a pertussis toxin-sensitive G protein. Additional molecular information is needed to determine whether or not a single receptor population can effect both the inhibition of adenylate cyclase and the acceleration of inositol polyphosphate production, or whether or not particular subpopulations are linked to different effector systems. Although GTP-binding proteins link receptor occupancy to changes in both inhibition of cyclase and the increase in inositol phosphate production, the identity of the GTP-binding proteins and the mechanisms involved *in vivo* are not clear. Finally, further work is required to establish which of the possible signaling events induced by occupancy of receptors coupled to adenylate cyclase and for PLC are causal in mediating a physiological event; which are permissive; and which are without functional consequence. It also remains to be determined if the coupling of adenosine receptors to both adenylate cyclase and PLC is unique for renal epithelia or represents a more general pattern of adenosine signal transduction.

REFERENCES

1. **Spielman, W. S. and Thompson, C. I.,** A proposed role for adenosine in the regulation of renal hemodynamics and renin release, *Am. J. Physiol.,* 242, F423, 1982.
2. **Spielman, W. S., Arend, L. J., and Forrest, J. N.,** The renal and epithelial actions of adenosine, in *Topics and Perspecives in Adenosine Research,* Gerlach, E. and Becker, B. F., Eds., Springer-Verlag, Berlin, 1987, 249.
3. **Oßwald, H.,** Adenosine and renal function, in *Regulatory Function of Adenosine,* Berne, R. M., Rall, T. W., and Rubio, R., Eds., Martinus Nijhoff, Boston, 183, 399.
4. **Oßwald, H.,** The role of adenosine in the regulation of glomerular filtration rate and renin secretion, *Trends Pharmacol. Sci.,* 5, 94, 1984.
5. **Churchill, P. C. and Churchill, M. C.,** Effects of adenosine on renin secretion, in *ISI Atlas of Science: Pharmacology,* ISI Press, Philadelphia, 1988, 367.
6. **Spielman, W. S. and Oßwald, H.,** Characterization of the post-occlusive response of renal blood flow in the cat, *Am. J. Physiol.,* 235, F286, 1978.
7. **Spielman, W. S. and Oßwald, H.,** Blockade of postocclusive renal vasoconstriction by an angiotensin II antagonist: evidence for an angiotensin-adenosine interaction, *Am. J. Physiol.,* 237, F463, 1979.
8. **Oßwald, H., Spielman, W. S., and Knox, F. G.,** Mechanism of adenosine-mediated decreases in glomerular filtration rate in dogs, *Circ. Res.,* 43, 465, 1978.
9. **Spielman, W. S., Britton, S. L., and Fiksen-Olsen, M. J.,** Effect of adenosine on the distribution of renal blood flow in dogs, *Circ. Res.,* 46, 449, 1980.
10. **Murray, R. D. and Churchill, P. C.,** The effects of adenosine receptor agonists in the isolated, perfused rat kidney, *Am. J. Physiol.,* 247, H343, 1984.
11. **Murray, R. D. and Churchill, P. C.,** The concentration-dependency of the renal vascular and renin secretory responses to adenosine receptor agonists, *J. Pharmacol. Exp. Ther.,* 232, 189, 1985.
12. **Rossi, N. F., Churchill, P. C., and Churchill, M. C.,** Pertussis toxin reverses adenosine receptor-mediated inhibition of renin secretion in rat renal cortical slices, *Life Sci.,* 40, 481, 1987.
13. **Rossi, N. F., Churchill, P. C., and Amore, B.,** Mechanism of adenosine receptor induced renal vasoconstriction in the rat, *Am. J. Physiol.,* 255, H885, 1988.
14. **Abboud, H. and Dousa, T. P.,** Action of adenosine on cyclic $3',5'$-nucleotides in glomeruli, *Am. J. Physiol.,* 244, F633, 1983.
15. **Freissmuth, M., Hausleithner, V., Tuisl, E., Nannoff, C., and Schuetz, W.,** Glomeruli and microvessels of the rabbit kidney contain both A_1 and A_2 adenosine receptors, *Naunyn-Schmiedeberg's Arch. Pharmacol.,* 335, 438, 1987.
16. **Schnermann, J., Briggs, J., Kritz, W., Moore, L. C., and Wright, F. S.,** Control of glomerular vascular resistance by the tubuloglomerular feedback mechanism, in *Renal Pathophysiology,* Leaf, A. and Giebish, G., Eds., Raven Press, New York, 1980.
17. **Oßwald, H., Hermes, H. H., and Nabakowski, G.,** Role of adenosine in signal transmission of tubuloglomerular feedback, *Kidney Int.,* 22 (Suppl.), S136, 1982.
18. **Schnermann, J. and Briggs, J.,** Inhibition of tubuloglomerular feedback-induced vasoconstriction during adenosine 1-receptor blockade, *FASEB J.,* 3 (Abstr.), A541, 1989.
19. **Arend, L. J., Haramati, A., Thompson, C. I., and Spielman, W. S.,** Adenosine-induced decrease in renin release: dissociation from hemodynamic effects, *Am. J. Physiol.,* 247, F447, 1984.
20. **Barchowsky, A., Data, J. L., and Whorton, A. R.,** Inhibition of renin release by analogues of adenosine in rabbit renal cortical slices, *Hypertension (Dallas),* 9, 619, 1987.
21. **Churchill, P. C. and Bidani, A. K.,** Renal effects of selective adenosine receptor agonists, *Am. J. Physiol.,* 252, F299, 1987.
22. **Churchill, P. C., Jacobson, K. A., and Churchill, M. C.,** XAC, a functionalized congener of 1,3-dialkyxanthine, antagonizes A_1 adenosine receptor-mediated inhibition of renin secretion in vitro, *Arch. Int. Pharmacodyn. Ther.,* 290, 293, 1987.

23. **Churchill, P. C., Rossi, N. F., and Churchill, M. C.,** Renin secretory effects of N^6-cyclohexyladenosine: effects of dietary sodium, *Am. J. Physiol.,* 252, F872, 1987.

24. **Churchill, P. C.,** Second messenger in renin secretion, *Am. J. Physiol.,* 249, F175, 1985.

25. **Churchill, P. C., and Churchill, M. C.,** A_1 and A_2 adenosine receptor activation inhibits and stimulates renin secretion of rat renal cortical slices, *J. Pharmacol. Exp. Ther.,* 232, 589, 1985.

26. **Arend, L. J., Thompson, C. I., Brandt, M. A., and Spielman, W. S.,** Elevation of intrarenal adenosine by maleic acid decreases GFR and renin release, *Kidney Int.,* 30, 656, 1986.

27. **Kuan, C. J., Wells, J. N., and Jackson, E. K.,** Endogenous adenosine restrains renin release during sodium restriction, *J. Pharmacol. Exp. Ther.,* 249, 110, 1989.

28. **Itoh, S., Carretero, O. A., and Murray, R. D.,** Possible role of adenosine in the macula densa mechanism of renin release in rabbits, *J. Clin. Invest.,* 76, 1412, 1985.

29. **Weihprecht, H., Lorenz, J. N., Schnermann, J., and Briggs, J. P.,** Effect of adenosine₁ receptor blockade on macula densa mediated renin release from the isolated perfused juxtaglomerular apparatus (JGA), 22nd Annual Meeting, Am. Society Nephrology, 1989, 193A.

30. **Rossi, N. F., Churchill, P. C., Jacobson, K. A., and Leahy, A. E.,** Further characterization of the renovascular effects of N^6-cyclohexyladenosine in the isolated perfused rat kidney, *J. Pharmacol. Exp. Ther.,* 240, 911, 1987.

31. **Kurtz, A., Bruna, R. D., Pfeilschifter, J., and Bauer, C.,** Role of cGMP as second messenger of adenosine in the inhibition of renin release, *Kidney Int.,* 33, 798, 1988.

32. **Ueno, M., Brooking, J., Beckman, B., and Fisher, J. W.,** A_1 and A_2 adenosine receptor regulation of erythropoietin production, *Life Sci.,* 43, 229, 1988.

33. **Ueno, M., Brooking, J., Beckman, B., and Fisher, J. W.,** Adenosine receptor regulation of erythropoietin secretion, *Kidney Int.,* 31 (Abstr.), 290, 1987.

34. **Ginsborg, B. L. and Hirst, G. D. S.,** The effect of adenosine on the release of transmitter from the phrenic nerve of the rat, *J. Physiol. (London),* 224, 629, 1972.

35. **Hedqvist, P. and Fredholm, B. B.,** Effects of adenosine on adrenergic neurotransmission: prejunctional inhibition and postjunctional enhancement, *Naunyn-Schmiedeberg's Arch. Pharmacol.,* 293, 217, 1976.

36. **Hedqvist, P., Fredholm, B. B., and Olundh, S.,** Antagonistic effects of theophylline and adenosine on adrenergic neuroeffector transmission in the rabbit kidney, *Circ. Res.,* 43, 592, 1978.

37. **Snyder, S. H.,** Adenosine as a neuromodulator, *Annu. Rev. Neurosci.,* 8, 103, 1985.

38. **Fredholm, B. B. and Dunwiddie, T. V.,** How does adenosine inhibit transmitter release? *Trends Pharmacol. Sci.,* 9, 130, 1988.

39. **Silinsky, E. M.,** Inhibition of transmitter release by adenosine: are Ca^{2+} currents depressed or are the intracellular effects of Ca^{2+} impaired?, *Trends Pharmacol. Sci.,* 7, 180, 1986.

40. **Arend, L. J., Thompson, C. I., and Spielman, W. S.,** Dipyridamole decreases glomerular filtration in the sodium-depleted dog, *Circ. Res.,* 56, 242, 1985.

41. **Miyamoto, M., Yagil, Y., Larson, T., Robertson, C., and Jamison, R. L.,** Effect of intrarenal adenosine on renal function and medullary blood flow in the rat, *Am. J. Physiol.,* 255, F1230, 1988.

42. **Dillingham, M. A. and Anderson, R. J.,** Purinergic regulation of basal and arginine vasopressin-stimulated hydraulic conductivity in rabbit cortical collecting tubule, *J. Membr. Biol.,* 88, 277, 1985.

43. **Arend, L. J., Sonnenberg, W. K., Smith, W. L., and Spielman, W. S.,** A_1 and A_2 adenosine receptors in rabbit cortical collecting tubule cells: modulation of hormone-stimulated cAMP, *J. Clin. Invest.,* 79, 710, 1987.

44. **Arend, L. J., Gusovsky, F., Handler, J. S., Rhim, J. S., and Spielman, W. S.,** Adenosine-sensitive phosphoinositide turnover in a newly established renal cell line, *Am. J. Physiol.,* 256, F1067, 1989.

45. **Burnatowska-Hledin, M. A. and Spielman, W. S.,** Effects of adenosine on cAMP production and cytosolic Ca^{2+} in cultured rabbit medullary thick limb cells, *Am. J. Physiol.,* 260, C143, 1991.

46. **Jacobson, K. A.,** Chemical approaches to the definition of adenosine receptors, in *Adenosine Receptors,* Cooper, D. M. F. and Londos, C., Eds., Alan R Liss, New York, 1988, 43.

47. **Souness, J. E. and Chagoya de Sanchez, V.,** The stimulation of [1-^{14}C]glucose oxidation in isolated fat cells by N^6-methyladenosine: an effect independent of cAMP, *FEBS Lett.,* 125, 249, 1981.

48. **Arend, L. J., Burnatowska-Hledin, M. A., and Spielman, W. S.,** Adenosine receptor-mediated calcium mobilization in cortical collecting tubule cells, *Am. J. Physiol.,* 255, C581, 1988.

Chapter 22

PURINERGIC REGULATION OF GASTROINTESTINAL MOTILITY AND SECRETION

M. A. Cook

TABLE OF CONTENTS

I. INTRODUCTION

The gastrointestinal (GI) system was among the first biological systems in which regulatory roles for purine nucleotides and nucleosides were implicated. The existence, in the intestine, of inhibitory nerves which are neither adrenergic nor cholinergic was recognized over 25 years ago[1] and the hypothesis that such nerves use ATP as a transmitter was proposed by Burnstock et al.[2] in 1970. The subsequent activity in this area has generated a considerable literature dealing with the actions of purines on both motility and secretion. It is not surprising that this organ system has been the focus of investigative activity in this regard. The presence of the enteric nervous system, with a complexity found nowhere else outside the central nervous system, functionally linked to relatively easily measurable parameters such as contraction or secretion, has ensured attention and has yielded both insights and inconsistencies. Regulatory roles for purines include putative neurotransmitter and cotransmitter functions as well as neuromodulatory and local or paracrine activity. The role of purines as transmitters in the GI tract is controversial and the evidence varies both with the region of the gut examined and with the species. The evidence for neuromodulatory roles is perhaps more convincing. The ability of purines to alter function supports the presence of functional purine receptors on several cell types in the GI tract. The presence of these receptors suggests functional involvement of purines notwithstanding a shortage of information about source and extracellular concentration as well as the stimuli for release at several loci. This brief treatment examines the activity of purines on two of the major functions of the GI system, motility and secretion. For additional analysis the reader is referred to appropriate reviews.[3-9]

II. MOTILITY

The role of purines in the regulation of GI motility may be assessed by examining the evidence for a purine or purines mediating the effects of nonadrenergic, noncholinergic (NANC) nerves in the GI tract, by examining the evidence for the release of purines from such nerves, by examining the actions of exogenous purines on postsynaptic structures, and by examining the presynaptic actions of both nucleotides and nucleosides on neurotransmitter or neuromediator release from enteric nerves. This section examines such evidence for each of the discrete, functionally distinct regions within the GI tract. Wherever possible, functional evidence concerning the role of purines in GI motor function will be emphasized.

A. ESOPHAGUS

The innervation of both the lower esophageal sphincter (LES) and the body of the esophagus includes NANC inhibitory nerves.[10-13] Electrical field stimulation of circular smooth muscle-nerve preparations of the body of the opossum esophagus gives rise to inhibitory junction potentials (IJPs), which are not abolished by atropine, guanethidine, or indomethacin, but are tetrodotoxin sensitive. The nature of the mediator(s) released from these nerves remains unknown. Burnstock[14] and Cohen[15] provided early support for the mediator at the LES being a purine such as ATP, but later studies[16] do not support this conclusion. While both ATP and adenosine produce inhibitory responses of the opossum esophagus on intra-arterial administration, tachyphylaxis to either does not affect vagally mediated sphincter relaxation. Dipyridamole, an adenosine uptake inhibitor, enhances the inhibitory effect of adenosine, but does not alter the nerve-mediated relaxation. It is possible that vasoactive intestinal polypeptide (VIP) or calcitonin gene-related peptide may function as NANC mediators although the possible corelease of ATP with peptide from these nerves has been suggested.[17] Daniel et al.[12] conclude that, at the body of the opossum esophagus, neither a purine nor a VIP serve as the NANC mediator. In this tissue, exogenous purines do not cause significant hyperpolarization, and apamin has no effect on the IJP. However, purines do reversibly inhibit the amplitude of the IJP, and actions of purines, either nucleotides or adenosine, at P_1 receptors may be responsible for prejunctional inhibition of the NANC transmitter release. At the esophagus, then, the evidence for purinergic transmission is not convincing, but it is possible that purines exert a neuromodulatory influence on the release of excitatory mediators.

B. STOMACH

The stomach exhibits NANC relaxation on stimulation of the vagus nerves in the presence of atropine, guanethidine, and adrenalectomy.[18] It is clear that the innervation of the stomach includes fibers which are NANC in nature and give rise to IJPs in the smooth muscle on stimulation,[19,20] and the evidence has been reviewed.[3,4,21,22] These fibers are probably important in mediating descending inhibitory reflexes[21,23] which constitute part of the control of peristalsis. The identity of the mediator released from NANC nerves in the stomach is not known. Some evidence supports such a role for ATP, although many inconsistencies exist. The release of ATP or adenosine from the stomachs of both guinea pig and toad on stimulation of vagal inhibitory nerves has been reported[2] and the inhibitory actions of ATP on the stomachs of several species documented. The nucleotide causes relaxations of gastric musculature in the guinea pig,[2,24] rat,[2,25] rabbit,[26] mouse,[2] cat,[27] and toad.[2] However, desensitization of ATP does not block NANC responses of guinea pig fundic strips,[28] notwithstanding a previous report to the contrary.[29] Furthermore, ATP-induced relaxation of this tissue are not antagonized by arylazido aminopropionyl ATP (ANAPP₃),[30] although this compound is a relatively weak P_{2y} receptor antagonist. Desensitization with α,β-methylene ATP blocks relaxation to vagal stimulation in the cat stomach,[27] but this could not be confirmed in the ferret,[31] a species in which ATP causes contraction of the stomach, or in the rabbit.[26] Vagal relaxation of the stomach in rabbits is antagonized by reactive blue 2, a putative P_{2y} receptor antagonist,[32,33] although the specificity of this compound has not been established.[8] Incubation of guinea pig fundic strips with *Botulinum* toxin[24] does not alter NANC responses or the response to ATP, showing that corelease from cholinergic nerves is unlikely.

It has been suggested that, in the guinea pig stomach, theophylline antagonizes the responses to both inhibitory nerve stimulation and to the purines.[29] However, opposite findings were reported by Baer and Frew,[28] and subsequently, 8-phenyltheophylline was shown to antagonize ATP-mediated relaxation, but not that due to field stimulation.[34] Although theophylline and related methylxanthines may act as functional antagonists, especially at high concentrations,[28,34,35] it is unlikely that they specifically block NANC transmission at the stomach. In experiments using rat fundic strips, De Beurme and Lefebvre[36] showed that in the rat fundus, trypsin and α-chymotrypsin antagonize the relaxations induced by VIP and partially antagonize those due to NANC stimulation (transmural), suggesting that the peptide may function as an inhibitory transmitter.

In the atropinized rabbit *in vivo*, vagal stimulation gives rise to inhibition which is not affected by guanethidine or adrenergic antagonists.[26] Intra-arterial infusion of ATP and adenosine 5'-monophosphate (AMP), but not adenosine, elicits similar inhibitory responses which are not affected by theophylline or desensitization with α,β-methylene ATP. However, both responses were diminished or abolished in the presence of reactive blue 2, which may be consistent with an action at P_{2y} receptors. Prejunctional actions of either ATP or adenosine at P_1 receptors on cholinergic, and possibly other excitatory nerves, resulting in inhibition of transmitter release are also likely.[37] Adenosine infusion reduces the atropine-sensitive gastric contractions induced by vagal stimulation, and similar inhibition is seen *in vitro*, although postjunctional enhancement of field-stimulated contractions also occurs. Gastric motility is clearly modified by purines acting at specific receptors, but the existence of a purinergic innervation is still equivocal.

C. DUODENUM

Few reports of studies on the duodenum as a distinct region of the intestine are extant. Transmural (field) or chemical stimulation of this tissue gives rise to NANC relaxation,[38] as does exogenous ATP.[39,40] Adenosine also causes relaxation, although it is much less potent. None of these responses are affected by 8-phenyltheophylline, suggesting a post-junctional locus of action. Reactive blue 2 inhibits the response to ATP, suggesting an action at P_{2y} receptors, while exposure to the enzyme nucleotide pyrophosphatase (EC 3.6.1.9.) which, among several activities, converts ATP to 5'-AMP, inhibits relaxations to transmural stimulation. This is unusually convincing evidence for the involvement of a purine, possibly ATP, in NANC inhibitory transmission.

D. ILEUM

The innervation of this region of the GI tract has been extensively studied, and the present model of enteric neuronal circuitry derives from such studies, carried out largely in the guinea pig.[23] The existence and coexistence of both classical and nonclassical neuro-mediators in enteric nerves in the myenteric and submucous plexuses are firmly established, suggesting that functional roles for such mediators exist and emphasizing the increment in complexity which the model implies. The evidence for a purine functioning as a neurotrans-mitter at the ileum is contradictory. There is, however, support for prejunctional actions of ATP and adenosine as inhibitors of the release of excitatory mediators. The occurrence of NANC responses, both excitatory and inhibitory, is well established.[2,41-45] It is possible that the mediator responsible for the excitatory effects is substance P (SP).[46] However, the identity of the inhibitory mediator(s) involved is not clear. Exogenous ATP gives rise to both inhibitory and excitatory responses at the ileum,[38,43,47,48] although caution must be exercised in making comparisons between species and between regions of the ileum. It is probable that the pharmacology of the terminal ileum differs from that of the remainder of the ileum,[49] and this clearly requires additional scrutiny in the light of current knowledge.

Desensitization with ATP does not diminish NANC inhibitory responses in guinea pig ileum[46] or in rat distal ileum,[40] and the peristaltic reflex, which includes NANC inhibition of the longitudinal muscle and is elicited by raising intraluminal pressure, is not abolished by exposure to high concentrations of ATP, which abolish the response of longitudinal muscle to the nucleotide.[50] Further, nucleotide pyrophosphatase, which abolishes NANC responses in the duodenum (see above), does not affect such responses in the rat ileum.[40]

The inhibitory responses to both ATP and adenosine at the ileum appear to be due largely to actions at prejunctional (P_1) sites[45,51-58] on excitatory nerves. Most studies on ileum have demonstrated the modulation of cholinergic transmission by the purines, although recent work has revealed similar inhibition of transmission mediated by one or more tachykinins.[59] The site of action is at cell-surface[60] P_1 receptors located on the excitatory nerves, and these have usually been characterized as the A_1 subtype on the basis of the potency of adenosine analogs in producing inhibition.[61] However, there is pharmacological evidence suggesting that the receptors at this site are heterogeneous and include both A_1 and A_2 subtypes.[62] The activity of ATP at these receptors probably involves some conversion of ATP to AMP and adenosine, which are more potent at the P_1 receptor.[63] Adenosine partially inhibits the [^{3}H]-acetylcholine release, evoked by high external potassium concentration, from synaptosomes (isolated terminal varicosities) derived from guinea pig ileum myenteric plexus, while com-pletely inhibiting the electrically evoked release.[64] A similar preparation was used to dem-onstrate a positive correlation between the binding affinity of labeled adenosine analogs at the nerve endings and their efficacy as inhibitors of the electrically stimulated contraction of the ileum.[65] Measurement of the release of the excitatory tachykinins SP and neurokinin A from a perfused preparation of enteric synaptosomes has permitted direct determination of the inhibitory actions of several adenosine analogs on depolarization-induced transmitter output.[66] These experiments have demonstrated the inhibitory activity of several adenosine analogs, including the putative, highly selective A_2 agonist (2-[p-(carboxyethyl)phenylethylamino]-5'-N-ethylcarboxamidoadenosine) (CGS 21680),[67,68] on evoked SP release. Interestingly, the efficacy displayed by the compound at this neural locus (EC_{50} is between 1 and 10 μM) is that expected from interaction at A_1 receptors,[69] as no activity was observed at nanomolar concentrations. The inhibitory effect of CGS 21680 on evoked release of SP-LI from perfused enteric synaptosomes is shown in Figure 1.

The activity of theophylline and related xanthines as antagonists at P_1 receptors provides evidence for activity of endogenous adenosine at these receptors in the ileum. Electrically stimulated contractions of the guinea pig ileum are enhanced by 8-p-sulfophenyltheophylline,[70] and the output of bioassayable acetylcholine is incremented by theophylline.[55] Similarly, theophylline enhances the atropine-insensitive contractile activity of the ileum in response to electrical stimulation.[71] Endogenous adenosine may derive from the breakdown of ATP and use of the firefly luciferin-luciferase method has provided evidence for such release from enteric synaptosomes.[72-74] It seems probable that much of the ATP is released from varicosities of noradrenergic origin.[8]

Electrophysiological studies on the actions of adenosine at the ileum show that the nucleoside causes membrane hyperpolarization, decreases input resistance, and enhances postspike hyperpolarizing potentials.[75] These effects were observed in AH type 2 myenteric neurons and demonstrate the presence of purine receptors on cell somata in addition to those on endings.

The effect of adenosine and other purines on functional measures of motility of the intact ileum in situ has not been extensively studied. The frequency of occurrence of the migrating myoelectric complex (MMC) is increased approximately 50% by both adenosine and N^6-cyclohexyladenosine, while 5'-N-ethylcarboxamidoadenosine (NECA) reduces the frequency by about 50%.[76] Dipyridamole also increased the MMC frequency, suggesting a role for endogenous adenosine.

E. TAENIA CAECI

The occurrence of NANC responses in preparations of guinea pig taenia coli[2,77,78] is well established. Relaxation of the smooth muscle on stimulation of the intramural nerves in the presence of adrenergic and cholinergic blockade is followed by a rebound contraction, the mechanism of which is distinct from the relaxation.[79] The ability of exogenous ATP to generate identical responses provided impetus for the development of the purinergic nerve hypothesis.[2-4] Stimulation of NANC nerves produces a IJPs in the smooth muscle, the amplitudes of which are determined by the extracellular K^+ concentration.[80] The reversal potential is similar to that of K^+, suggesting that the mediator causes an increase in K^+ conductance, a probable mechanism in light of the sensitivity of the NANC response to blockade by apamin,[81] which is known to block a K^+ channel.[82] Apamin also blocks the ATP-mediated hyperpolarization response in the taenia,[81-83] but the appropriateness of apamin as a tool to study NANC transmission is questioned because it is unable to antagonize NANC responses in other tissues[9,12] and it blocks epinephrine effects at taenia.[84] White[8] has commented on the lack of specificity of some other putative P_2 receptor antagonists.

The actions of ATP and several other purine nucleotides on the taenia have suggested that the receptors mediating the response are of the P_{2y} subtype. ATP causes relaxation and subsequent contraction, mimicking the NANC response. The stable isostere α,β-methylene ATP, which is not degraded by ectonucleotidases,[85] similarly gives rise to smooth-muscle hyperpolarization in taenia. ATP is considerably more potent than adenosine in causing relaxation of the taenia, although the range of concentrations required (10^{-6} M to 10^{-4} M) is higher than for several other neurotransmitters. Theophylline does not antagonize the responses to ATP,[86] indicating lack of activity at P_1 receptors.

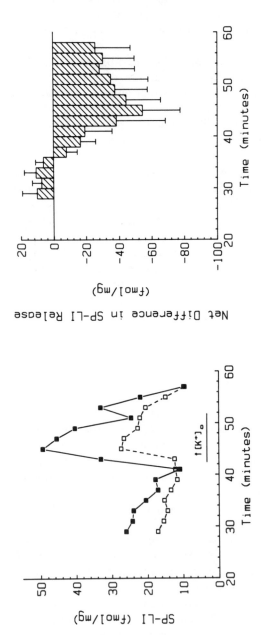

FIGURE 1. Left panel — Perfusion profile showing release of substance P-like immunoreactivity (SP-LI) from enteric synaptosomes in the absence (solid line) and presence (dashed line) of CGS 21680 (10 μM). Synaptosomes were depolarized by exposure to a 50-mM increment in $[K^+]_o$ at the bar. Right panel — Histogram showing the net difference in SP-LI release from synaptosomes in the presence of CGS 21680 (10 μM) from that under control conditions. Data are means ± SEM from three experiments performed as in left panel.

Evidence for release of purines from the taenia provides mixed support for its putative transmitter role. The taenia takes up [³H]-adenosine and converts it to [³H]-ATP, while transmural stimulation causes wholly or partially TTX-sensitive release of tritiated purines.[87,88] Approximately one third of the total released purines come from neural sources and the rest from muscle. Direct measurement of ATP release has been reported,[89,90] although in the latter report the authors could not detect such release unless high-frequency, high-current stimulation was applied. This release was not TTX sensitive, which raises the possibility of a nonneural origin, possibly from muscle. Studies with *Goniopora* toxin, a polypeptide which increases transmitter release on stimulation, reveal enhanced NANC responses, increased amplitude of the IJP and enhanced efflux of tritiated material following preloading with tritiated adenosine.[91] Both the IJP and the ³H efflux show TTX sensitivity. Additional evidence for the involvement of ATP arises from the ability of the enzyme, nucleotide pyrophosphatase, to reduce the responses of taenia to both exogenous ATP and NANC stimulation.[92] As in the duodenum (above), this appears to provide convincing support for a role for purine nucleotides, presumably ATP, in the mediation of NANC responses. The relative lack of antagonistic activity of ANAPP₃ against the inhibitory responses of taenia to stimulation[93] does not provide support for such a role. However, the weak activity of this antagonist at P_{2y} receptors, as well as nonspecific effects at this tissue, may not sustain this conclusion.

Adenosine also exerts an inhibitory effect at the taenia,[86,94,95] and this may be mediated by an A_2 subtype[96] of the P_1 receptor, contrasting with the evidence suggesting probable A_1 mediation at the ileum. The responses to adenosine are antagonized competitively by 8-phenyltheophylline, although insensitivity of adenosine-mediated inhibition at this tissue has also been reported.[86]

F. COLON, RECTUM, ANOCOCCYGEUS, RECTOCOCCYGEUS, AND INTERNAL ANAL SPHINCTER

Stimulation of the pelvic nerves in the cat reveals the existence of biphasic NANC responses at the colon.[97] The excitatory component of the response is blocked by desensitization with large doses of α,β-methylene ATP,[98] while responses to histamine or acetylcholine are not affected. These results are consistent with ATP functioning as a neuromediator and acting at P_{2x} receptors on colonic and rectal smooth muscle. Part of the response may involve cotransmission with Met-enkephalin. ATP causes a concentration-dependent reduction in velocity of propulsion in rabbit colon as well as a delay in the onset of the distension-induced propulsive wave.[99] Direct inhibition of the circular smooth muscle or NANC relaxation on stimulation are not blocked by theophylline, while the delay in onset and the velocity of propulsion are, suggesting that P_1 receptors are involved in the latter responses.

The muscularis mucosae of opossum distal colon responds to ATP with concentration-dependent relaxations,[100] while adenosine 5′-diphosphate (ADP), AMP, and adenosine are without effect. This tissue gives rise to NANC inhibitory responses to electrical stimulation and, while the actual mediator is not known, several candidate mediators, including VIP and the prostaglandins, are known not to be involved. The pharmacology of colonic muscularis mucosae appears distinct from other GI smooth muscles.

The involvement of purines in the regulation of the anococcygeus muscle is unclear. NANC inhibition occurs in this muscle from rat,[101,102] rabbit,[102] mouse,[103,104] and cat.[105] An elevated tone is required in order to observe the response, and agents such as 5-HT, carbachol, and guanethidine have been used. Pretreatment of tissues with 6-hydroxydopamine does not alter these responses. Release of ATP from rat and rabbit anococcygeus occurs on stimulation[102] and this is not affected by the presence of atropine or guanethidine, but is abolished by TTX. Exogenous ATP causes contraction of the resting anococcygeus from the rat[106] and mouse.[103] In preparations with elevated tone, ATP causes relaxation in the rabbit[102] and mouse,[104] but further contracts the muscle in the rat.[106] Burnstock et al.[102] reported that the contractile response to ATP was abolished by treatment with indomethacin, implicating the prostaglandins as mediators. It was subsequently shown that the tone of the tissue is of critical importance and that the magnitude of contractions decreased as the tone diminished, providing a basis for the apparent blockade by the prostaglandin synthesis inhibitor. Adenine nucleosides also cause contraction in the high-tone rat anococcygeus and these responses are not blocked by theophylline. However, prejunctional inhibition of adrenergic transmission by both ATP and purine nucleosides acting at P_1 receptors is theophylline sensitive.[106] In the mouse, field stimulation of the anococcygeus produces a two-component relaxation which is not affected by apamin.[104] The response to ATP in this tissue is also insensitive to apamin.

Evidence for the involvement of ATP in the excitatory response to field stimulation is somewhat clearer than for its role as an inhibitory mediator. The fast excitatory junction potential in rat anococcygeus is mimicked by exogenous ATP and is not blocked by adrenergic or cholinergic antagonists.[107] Evidence for the presence of a purinergic NANC inhibitory innervation of the rectococcygeus muscle in the rabbit includes the detection of ATP release on field stimulation and the inhibitory actions of exogenous ATP on preparations with high tone.[108]

The internal anal sphincter (IAS) may also posses a purinergic NANC innervation. The IJP produced on stimulation of the guinea pig IAS and the relaxation to ATP are both apamin sensitive, and ATP is more potent than adenosine.[109] This tissue also relaxes to VIP, which is, however, sensitive to treatment with α-chymotrypsin, while the response to ATP or to stimulation is not affected. Human IAS also relaxes in response to ATP.[110]

G. GALLBLADDER

The filling function of the gallbladder requires controlled relaxation of the smooth muscle, and there is evidence for the presence of NANC innervation in this organ,[111] although the identity of the mediator(s) is unknown. Adenosine is able to relax longitudinal strips of gallbladder, both in the absence and presence of prior contraction with cholecystokinin.[112] The effects of adenosine on resting tissues are antagonized by theophylline. Relaxation with ATP is only apparent in the presence of indomethacin. In its absence, ATP causes contraction of gallbladder strips. The role of purines in regulation of gallbladder tone is not clear. Although there is no compelling evidence for a transmitter function, the demonstrable presence of purine receptors on gallbladder smooth muscle suggests putative functional roles for the purines. There seems not to be evidence about the presence or otherwise of prejunctional (P_1) receptors on intramural nerves in the gallbladder.

III. SECRETION

The role of purines in the regulation of GI secretory functions is perhaps less distinct than their putative roles in the regulation of motility. No comprehensive hypotheses concerning putative regulatory mechanisms have yet emerged and the volume of information available is limited. This section deals with possible functional roles for purines with respect to gastric secretion, intestinal electrolyte secretion, and pancreatic exocrine secretion. No attempt has been made to assess the influence of changes in vascular caliber within the GI tract, brought about by purines, on secretory function. The reader is referred to Chapter 16, this volume, for relevant information.

A. GASTRIC SECRETION

The ability of theophylline to enhance gastric acid secretion has been known for some time.[113,114] The mechanism underlying this effect is probably antagonism of adenosine receptors, which may be located in the gastric fundus, rather than the inhibition of phosphodiesterase activity, which had been presumed. In anesthetized dogs, the infusion of adenosine in the gastric artery inhibits gastric acid secretion stimulated by either histamine or methacholine.[115] This inhibition is prevented by concentrations of theophylline which are insufficient to alter phosphodiesterase activity. In Shay rats, in which gastric acid secretion is stimulated by the reflex response to pyloric ligation,[116] subcutaneous or intraduodenal administration of R-phenylisopropyladenosine (R-PIA) inhibits both acid output and the volume of gastric secretion, and these responses are reduced by prior subcutaneous administration of theophylline.[117] Similar responses to adenosine or its analogs on the basal acid secretion of conscious rats have been reported. Glavin et al.[118] have shown inhibition of acid secretion by R-PIA, while the stereoisomer S-PIA in similar concentrations was ineffective. The inhibitory response to R-PIA is blocked by 8-phenyltheophylline. The i.p. administration of NECA similarly inhibits basal acid secretion in conscious rats as well as inhibiting the volume of gastric secretion.[119] Interestingly, R-PIA does not inhibit the volume of gastric secretion, while it elevates its pH.

Different results have been reported by Puurunen et al.[120] in urethane-anesthetized rats. Intravenous administration of NECA, R- and S-PIA, or 2-chloroadenosine results in dose-related stimulation of gastric acid secretion which is sensitive to the methylxanthines and which is blocked by atropine or vagotomy. Vagal mediation of the response is implied. Intracerebroventricular administration of adenosine analogs in Shay (pylorus ligated) rats inhibits acid secretion as well as the volume of gastric secretion and pepsin output.[120,121] Theophylline (i.v.) or 8-phenyltheophylline (s.c. or i.p.) had no effect on the inhibition, while depletion of brain monoamines attenuated the inhibitory effect of NECA. An action of the nucleosides to decrease the vagal stimulation of gastric secretion may be a mechanism involved.

The locus of action of purines on gastric secretion is not clear. Given the central actions of the purines, the effects observed following peripheral administration could arise from movement of these compounds across the blood-brain barrier. However, this does not seem likely in view of the reported inability of several adenosine analogs to gain access to the brain in significant, i.e., biologically effective, concentrations.[122] There are clearly multiple target sites available for the interaction of purines with the complex process of control of gastric secretion. One such site may be the parietal cell itself.

Adenosine acts at isolated canine parietal cells to inhibit histamine-stimulated acid secretion as measured by the accumulation of [^{14}C]-aminopyrine, and this response is reduced by theophylline and 8-phenyltheophylline.[123] In the same experimental system, carbachol-stimulated secretion is not altered by the purines. There is also evidence for the release of endogenous adenosine by the canine parietal cell preparation, as the uptake of [^{14}C]-aminopyrine is enhanced by 8-phenyltheophylline and adenosine deaminase and is decreased by dipyridamole.[124] Contrary findings have been reported by Puurunen et al.,[125] using an isolated parietal cell preparation from the rat. In this model, R-PIA has no effect on the [^{14}C]-aminopyrine uptake stimulated by either histamine or carbachol. It is not clear why this apparent species difference occurs. Additional support for a parietal cell site of action of adenosine comes from quantitative cytochemistry performed on histological sections of guinea pig stomach fundus.[126] Microdensitometric measurement of a reaction product in parietal cells, which reflects hydroxyl ion production, allows measurement of the effects of R-PIA, NECA, and adenosine on acid production (which parallels hydroxyl ion formation) to be assessed. The analogs inhibit acid production stimulated by histamine or gastrin, but not that stimulated by carbachol or dibutyryl cyclic AMP (cAMP). It is intriguing that the concentration of adenosine required (10^{-6} M) is not different from that required in other systems, while the secretagogue histamine, which was used at 10^{-14} M, appears to be several orders more potent in this system. The inhibitory effect of the purines is reversed by 1,3-diethylphenylxanthine. These findings may support the presence of specific receptors for adenosine on parietal cells linked to the production and/or secretion of acid.

There is some evidence for the ability of adenosine to inhibit gastrin release from antral mucosal fragments in the rat.[127] This is an additional mechanism by which the purines may modulate gastric acid secretion. A summary of the reported activity of purines on gastric secretion is presented in Table 1.

B. ELECTROLYTE SECRETION

Stimulation of enteric nerves results in secretomotor responses at the intestinal epithelium. The mediators responsible for the secretion of chloride ions, and therefore water, by ileal mucosa include acetylcholine in humans and guinea pigs, but not in rabbits.[128] Possible involvement of purines as mediators in this species was sought and has been excluded because, in Ussing chamber experiments using rabbit ileal mucosal tissue, ANAPP$_3$ does not reduce the increment in chloride secretion (measured as the short-circuit current) produced by field stimulation. It is, however, able to reduce the response to ATP, ADP, AMP, and adenosine by approximately 50%.[128] The weak antagonist potency of ANAPP$_3$ at P$_{2y}$ receptors, as well as the lack of additional evidence, e.g., with theophylline, may not have permitted accurate assessment of the involvement of a purine or purines in the mediation of the response. The evidence for the presence of purine receptors on secretory epithelial cells is supported in studies using mucosal strips of rabbit colon in Ussing chambers.[129] The serosal, but not the luminal, addition of adenosine causes an increase in both short-circuit current and conductance resulting from electrogenic secretion of chloride ions. Active transport of sodium is not altered in the presence of the purine. A cell-surface receptor for adenosine, located on the basolateral membrane, is implicated in the mediation of this response which is partially inhibited by theophylline and is mimicked by AMP and cyclic guanosine 5'-monophosphate.

Experiments using a human colonic epithelial cell line (T$_{84}$) grown in monolayers have shown that adenosine, R-PIA, and NECA are able to induce sustained increases in chloride secretion when added on either apical or basolateral surfaces of the cells. However, addition on the basolateral surface is more potent.[130] Addition of NECA on the basolateral surface produces a greater increase in intracellular cAMP levels than does addition on the apical surface. The existence of a purine receptor on the basolateral surface of these cells seems probable.

C. PANCREATIC EXOCRINE SECRETION

Purines may be able to influence pancreatic exocrine secretion. In the dog, close-arterial administration of adenosine to the vascularly isolated, self-hemoperfused pancreas increases secretin-stimulated output of pancreatic juice and bicarbonate.[131] Adenosine has no effect on the basal secretion rate or bicarbonate or protein content. Pretreatment with theophylline reverses the potentiation. Similar results are obtained with ATP, while inosine is without effect.[132] It is not certain whether or not ATP must be degraded to adenosine for activity. Both ATP and adenosine cause vasodilation and decrease perfusion pressure in the isolated pancreas. However, since the perfusion is carried out at a constant flow rate it seems unlikely that vasodilation is responsible for the response.

ACKNOWLEDGMENT

The author is indebted to M. James for invaluable assistance. The author's research cited in this chapter has been supported by the Medical Research Council of Canada.

TABLE 1
Reported Activity of Purines on Gastric Secretion

Species	Anesthetic	Route	Stimulus[a]	Effects of adenosine or analogs on:			Effects of methylxanthines	Ref.
				H⁺	Volume	Pepsin		
Dog	Pentobarbitol	i.a.	Histamine or methacholine	Inhibition			Blocks inhibition	115
Rat	None[b]	s.c.	Pyloric ligation[b]	Inhibition	Inhibition		Reduces inhibition	117
	Urethane	i.v.	Pyloric ligation[b]	Stimulation			Blocks stimulation	120
	None[b]	i.c.v.	Pyloric ligation[b]	Inhibition	Inhibition	Inhibition	No effect	120
	None[b]	i.c.v.	Pyloric ligation[b]	Inhibition	Inhibition		No effect	121
	Urethane	i.c.v.	Pyloric ligation	No effect				121
	None	s.c.	Basal	Inhibition			Blocks inhibition	118
	None	i.p.	Basal	Inhibition	Inhibition[c]			119

Isolated Parietal Cell Preparation

Species	Anesthetic	Route	Stimulus[a]	H⁺	Volume	Pepsin	Effects of methylxanthines	Ref.
Dog			Histamine	Inhibition[d]			Reduced inhibition	123
			Carbachol	No effect				123
Rat			Histamine or carbachol	No effect[d]				125

Quantitative Cytochemistry of Gastric Fundus

Species	Anesthetic	Route	Stimulus[a]	H⁺	Volume	Pepsin	Effects of methylxanthines	Ref.
Guinea pig			Histamine or gastrin	Inhibition			Blocks inhibition	126
			Carbachol	No effect				126

[a] Indicates means used to elicit acid secretion.
[b] Pyloric ligation stimulates acid secretion by vagal reflexes; ether or chloral hydrate anesthesia was used briefly during surgery or for euthanasia prior to collection of gastric secretion.
[c] NECA inhibited volume response, R-PIA did not.
[d] Measured by [¹⁴C]aminopyrine accumulation technique.

REFERENCES

1. **Burnstock, G., Campbell, G., Bennett, M., and Holman, M. E.,** Innervation of the guinea pig taenia coli: are there intrinsic inhibitory nerves which are distinct from sympathetic nerves?, *Int. J. Neuropharmacol.,* 3, 163, 1964.
2. **Burnstock, G., Campbell, G., Satchell, D., and Smythe, A.,** Evidence that adenosine triphosphate or a related nucleotide is the transmitter substance released by non-adrenergic inhibitory nerves in the gut, *Br. J. Pharmacol.,* 40, 668, 1970.
3. **Burnstock, G.,** Purinergic nerves, *Pharmacol. Rev.,* 24, 509, 1972.
4. **Burnstock, G.,** Purinergic transmission, in *Handbook of Psychopharmacology,* Iverson, L. L., Iverson, S. D., and Snyder, S. H., Eds., Plenum Press, New York, 1975, 131.
5. **Burnstock, G.,** A basis for distinguishing two types of purinergic receptor, in *Cell and Membrane Receptors for Drugs and Hormones: A Multidisciplinary Approach,* Bolis, L. and Straub, R. W., Eds., Raven Press, New York, 1978, 107.
6. **Burnstock, G.,** Neurotransmitters and trophic factors in the autonomic nervous system, *J. Physiol. (London),* 313, 1, 1981.
7. **Burnstock, G.,** The changing face of autonomic neurotransmission, *Acta Physiol. Scand.,* 126, 67, 1986.
8. **White, T. D.,** Role of adenine compounds in autonomic neurotransmission, *Pharmacol. Ther.,* 38, 129, 1988.
9. **Daniel, E. E., Collins, S. M., Fox, J. E. T., and Huizinga, J. D.,** Pharmacology of drugs acting on gastrointestinal motility, in *Handbook of Physiology,* Sect. 6, Vol. 1, Wood, J. D., Ed., American Physiological Society, Bethesda, MD, 1989, chap. 19.
10. **Goyal, R. K. and Rattan, S.,** Nature of the vagal inhibitory innervation of the lower esophageal sphincter, *J. Clin. Invest.,* 55, 1119, 1975.
11. **Tuch, A. and Cohen, S.,** Neurogenic basis of lower esophageal sphincter relaxation, *J. Clin. Invest.,* 52, 14, 1973.
12. **Daniel, E. E., Helmy-Elkholy, A., Jager, L. P., and Kannan, M. S.,** Neither a purine nor VIP is the mediator of inhibitory nerves of Opossum oesophageal smooth muscle, *J. Physiol. (London),* 336, 243, 1983.
13. **Lundgren, O.,** Vagal control of the motor functions of the lower esophageal sphincter and the stomach, *J. Auton. N.S.,* 9, 185, 1983.
14. **Burnstock, G.,** Past and current evidence for the purinergic nerve hypothesis, in *Physiological and Regulatory Functions of Adenosine and Adenine Nucleotides,* Baer, H. P. and Drummond, G. I., Eds., Raven Press, New York, 1979, 3.
15. **Cohen, S.,** Augmentation of the neural inhibitory response of the lower esophageal sphincter, *Proc. Soc. Exp. Biol. Med.,* 145, 1004, 1974.
16. **Rattan, S. and Goyal, R. K.,** Evidence against purinergic inhibitory nerves in the vagal pathway to the opossum lower esophageal sphincter, *Gastroenterology,* 78, 898, 1980.
17. **Goyal, R. and Paterson, W. G.,** Esophageal motility, in *Handbook of Physiology,* Sect. 6, Vol. 1, Wood, J. D., Ed., American Physiological Society, Bethesda, MD, 1989, chap. 22.
18. **Martinson, J. and Muren, A.,** Excitatory and inhibitory effects of vagus stimulation on gastric motility in the cat, *Acta Physiol. Scand.,* 57, 309, 1963.
19. **Beani, L., Bianchi, C., and Crema, A.,** Vagal non-adrenergic inhibition of guinea pig stomach, *J. Physiol. (London),* 217, 259, 1971.
20. **Beck, C. S. and Osa, J.,** Membrane activity in guinea-pig gastric sling muscle: a nerve-dependent phenomenon, *Am. J. Physiol.,* 220, 1397, 1971.
21. **Furness, J. B. and Costa, M.,** The nervous release and the action of substances which affect intestinal muscle through neither adrenoreceptors nor cholinoreceptors, *Philos. Trans. R. Soc. London Ser. B.,* 265, 123, 1973.
22. **Andrews, P. L. R.,** The non-adrenergic non-cholinergic innervation of the stomach, *Arch. Int. Pharmacodyn.,* Suppl. 280, 84, 1986.
23. **Furness, J. B. and Costa, M.,** Influence of the enteric nervous system on motility, in *The Enteric Nervous System,* Furness, J. B. and Costa, M., Eds., Churchill Livingstone, New York, 1987, chap. 7.
24. **Paul, M. L. and Cook, M. A.,** Lack of effect of botulinum toxin on nonadrenergic, noncholinergic inhibitory responses of the guinea pig fundus, *in vitro, Can. J. Physiol. Pharmacol.,* 58, 88, 1980.
25. **Huggins, N. and Cook, M. A.,** Effect of efferent denervation of the vagus nerve on nonadrenergic inhibitory responses obtained from the rat stomach, *in vitro, Can. J. Physiol. Pharmacol.,* 60, 1303, 1982.
26. **Beck, K., Calamai, F., Staderini, G., and Susini, T.,** Gastric motor responses elicited by vagal stimulation and purine compounds in the atropine-treated rabbit, *Br. J. Pharmacol.,* 94, 1157, 1988.
27. **Delbro, D. and Fandriks, L.,** Inhibition of vagally induced non-adrenergic, non-cholinergic gastric relaxation by P_2-purinoceptor desensitization, *Acta Physiol. Scand.,* 120, C18, 1984.

28. **Baer, H. P. and Frew, R.,** Relaxation of guinea-pig fundic strip by adenosine, adenosine triphosphate and electrical stimulation: lack of antagonism by theophylline or ATP treatment, *Br. J. Pharmacol.*, 67, 293, 1979.

29. **Okwuasaba, F. K., Hamilton, J. R., and Cook, M. A.,** Relaxation of guinea-pig fundic strip by adenosine, adenine nucleotides and electrical stimulation: antagonism by theophylline and desensitization to adenosine and its derivatives, *Eur. J. Pharmacol.*, 46, 181, 1977.

30. **Fedan, J. S., Hogaboom, G. K., O'Donnell, J. P., and Westfall, D. P.,** Use of photoaffinity labels as P$_2$-purinoceptor antagonists, in *Methods in Pharmacology, Volume 6, Methods used in Adenosine Research,* Paton, D. M., Ed., Plenum Press, New York, 1985, 279.

31. **Andrews, P. L. R. and Lawes, I. N. C.,** Characteristics of the vagally driven non-adrenergic, non-cholinergic inhibitory innervation of ferret gastric corpus, *J. Physiol. (London)*, 363, 1, 1985.

32. **Kerr, D. J. B. and Krantis, A.,** A new class of ATP antagonists, *Proc. Aust. Physiol. Pharmacol. Soc.*, 10, 156, 1979.

33. **Burnstock, G., Hopwood, A. M., Hoyle, C. H. V., Reilly, W. M., Saville, V. L., Stanley, M. D. A., and Warland, J. J. I.,** Reactive blue-2 selectively antagonises the relaxant responses to ATP and its analogues which are mediated by the P$_{2y}$ purinoceptor, *Br. J. Pharmacol.*, 89, 857P, 1986.

34. **Frew, R. and Lundy, P. M.,** Evidence against ATP being the nonadrenergic, noncholinergic transmitter in guinea-pig stomach, *Eur. J. Pharmacol.*, 81, 333, 1982.

35. **Huzinga, J. D. and Den Hertog, A.,** Inhibition of fundic strips from guinea-pig stomach: the effect of theophylline on responses to adenosine, ATP and intramural nerve stimulation, *Eur. J. Pharmacol.*, 63, 259, 1980.

36. **De Beurme, F. A. and Lefebvre, R. A.,** Influence of α-chymotrypsin and trypsin on the non-adrenergic non-cholinergic relaxation in the rat gastric fundus, *Br. J. Pharmacol.*, 91, 171, 1987.

37. **Gustafsson, L.,** Influence of adenosine on responses to vagal nerve stimulation in the anesthetized rabbit, *Acta Physiol. Scand.*, 111, 263, 1981.

38. **Maggi, C. A., Manzini, S., and Meli, A.,** Evidence that GABA$_A$ receptors mediate relaxation of rat duodenum by activating intramural nonadrenergic, noncholinergic neurons, *J. Auton. Pharmacol.*, 4, 77, 1984.

39. **Manzini, S., Maggi, C. A., and Meli, A.,** Further evidence for involvement of adenosine-5'-triphosphate in non-adrenergic non-cholinergic relaxation of the isolated rat duodenum, *Eur. J. Pharmacol.*, 113, 399, 1985.

40. **Manzini, S., Maggi, C. A., and Meli, A.,** Pharmacological evidence that at least two different non-adrenergic non-cholinergic inhibitory systems are present in the rat small intestine, *Eur. J. Pharmacol.*, 123, 229, 1986.

41. **Kosterlitz, H. W.,** Intrinsic intestinal reflexes, *Am. J. Dig. Dis.*, 12, 245, 1967.

42. **Hirst, G. D. S. and McKirdy, H. C.,** A nervous mechanism for descending inhibition in guinea-pig small intestine, *J. Physiol. (London)*, 238, 129, 1974.

43. **Kazic, T. and Milosavljevic, D.,** Influence of pyridylisatogen tosylate on contractions produced by ATP and by purinergic stimulation in the terminal ileum of the guinea-pig, *J. Pharm. Pharmacol.*, 29, 542, 1977.

44. **Bauer, V. and Kuriyama, H.,** The nature of non-cholinergic, non-adrenergic transmission in longitudinal and circular muscles of the guinea-pig ileum, *J. Physiol. (London)*, 332, 375, 1982.

45. **Watt, A. J.,** Direct and indirect effects of adenosine 5'-triphosphate on guinea-pig ileum, *Br. J. Pharmacol.*, 77, 725, 1982.

46. **Matusak, O. and Bauer, V.,** Effect of desensitization induced by adenosine 5'-triphosphate, substance P, bradykinin, serotonin, γ-aminobutyric acid and endogenous noncholinergic-nonadrenergic transmitter in the guinea-pig ileum, *Eur. J. Pharmacol.*, 126, 199, 1986.

47. **Ferrero, J. D., Cocks, T., and Burnstock, G.,** A comparison between ATP and bradykinin as possible mediators of the responses of smooth muscle to non-adrenergic, non-cholinergic nerves, *Eur. J. Pharmacol.*, 63, 295, 1980.

48. **Bauer, V. and Kuriyama, H.,** Evidence for non-cholinergic, non-adrenergic transmission in the guinea-pig ileum, *J. Physiol. (London)*, 330, 95, 1982.

49. **Munro, A. F.,** Effects of autonomic drugs on the responses of isolated preparations from the guinea-pig intestine to electrical stimulation, *J. Physiol. (London)*, 120, 41, 1953.

50. **Weston, A. H.,** The effect of desensitization to adenosine triphosphate on the peristaltic reflex in guinea-pig ileum, 47, 606, 1973.

51. **Vizi, E. S. and Knoll, J.,** The inhibitory effect of adenosine and related nucleotides on the release of acetylcholine, *Neuroscience*, 1, 391, 1976.

52. **Sawynok, J. and Jhamandas, K. H.,** Inhibition of acetylcholine release from cholinergic nerves by adenosine, adenine nucleotides and morphine: antagonism by theophylline, *J. Pharmacol. Exp. Ther.*, 197, 379, 1976.

53. **Gustafsson, L., Hedqvist, P., Fredholm, B. B., and Lundgren, G.,** Inhibition of acetylcholine release in guinea-pig ileum by adenosine, *Acta Physiol. Scand.,* 104, 469, 1978.

54. **Hayashi, E., Mori, M., Yamada, S., and Kunitomo, M.,** Effects of purine compounds on cholinergic nerves. Specificity of adenosine and related compounds on acetylcholine release in electrically stimulated guinea pig ileum, *Eur. J. Pharmacol.,* 48, 297, 1978.

55. **Cook, M. A., Hamilton, J. T., and Okwuasaba, F. K.,** Coenzyme A is a purine nucleotide modulator of acetylcholine output, *Nature,* 271, 768, 1978.

56. **Dowdle, E. B. and Maske, R.,** The effects of calcium concentrations on the inhibition of cholinergic neurotransmission in the myenteric plexus of guinea-pig ileum by adenine nucleotides, *Br. J. Pharmacol.,* 71, 245, 1980.

57. **Paul, M. L., Miles, D. L., and Cook, M. A.,** The influence of glycosidic conformation and charge distribution on activity of adenine nucleosides as presynaptic inhibitors of acetylcholine release, *J. Pharmacol. Exp. Ther.,* 222, 241, 1982.

58. **Moody, C. J. and Burnstock, G.,** Evidence for the presence of P_1-purinoceptors on cholinergic nerve terminals in the guinea-pig ileum, *Eur. J. Pharmacol.,* 77, 1, 1982.

59. **Christofi, F. L., McDonald, T. J., and Cook, M. A.,** Adenosine receptors are coupled negatively to release of tachykinin(s) from nerve endings, *J. Pharmacol. Exp. Ther.,* 253, 290, 1990.

60. **Okwuasaba, F. K., Hamilton, J. T., and Cook, M. A.,** Evidence for the cell surface locus of presynaptic purine nucleotide receptors in the guinea-pig ileum, *J. Pharmacol. Exp. Ther.,* 207, 779, 1978.

61. **Gustafsson, L. E., Wiklund, N. P., Lundin, J., and Hedqvist, P.,** Characterization of pre- and post-junctional adenosine receptors in guinea-pig ileum, *Acta Physiol. Scand.,* 123, 195, 1985.

62. **Christofi, F. L. and Cook, M. A.,** Possible heterogeneity of adenosine receptors present on myenteric nerve endings, *J. Pharmacol. Exp. Ther.,* 243, 302, 1987.

63. **Moody, C. J., Meghji, P., and Burnstock, G.,** Stimulation of P_1-purinoceptors by ATP depends partly on its conversion to AMP and adenosine and partly on direct action, *Eur. J. Pharmacol.,* 97, 47, 1984.

64. **Shinozuka, K., Maeda, T., and Hayashi, E.,** Effects of adenosine on ^{45}Ca uptake and [^{3}H]acetylcholine release in synaptosomal preparation from guinea-pig ileum myenteric plexus, *Eur. J. Pharmacol.,* 113, 417, 1985.

65. **Christofi, F. L. and Cook, M. A.,** Affinity of various purine nucleosides for adenosine receptors on purified myenteric varicosities compared to their efficacy as presynaptic inhibitors of acetylcholine release, *J. Pharmacol. Exp. Ther.,* 237, 305, 1986.

66. **Broad, R. M. and Cook, M. A.,** Unpublished data.

67. **Jarvis, M. F., Schulz, R., Hutchinson, A. J., Hoi Do, U., Sills, M. A., and Williams, M.,** [^{3}H]CGS 21680, a selective A_2 adenosine receptor agonist directly labels A_2 receptors in rat brain, *J. Pharmacol. Exp. Ther.,* 251, 888, 1989.

68. **Phillis, J. W.,** The selective adenosine A_2 receptor agonist, CGS 21680 is a potent depressant of cerebral cortical neuronal activity, *Brain Res.,* 509, 328, 1990.

69. **Lupica, C. R., Cass, W. A., Zahniser, N. R., and Dunwiddie, T. V.,** Effects of the selective adenosine A_2 receptor agonist CGS 21680 on *in vitro* electrophysiology, cAMP formation and dopamine release in rat hippocampus and striatum, *J. Pharmacol. Exp. Ther.,* 252, 1134, 1990.

70. **Wiklund, N. P. and Gustafsson, L. E.,** On the nature of endogenous purines modulating cholinergic neurotransmission in the guinea-pig ileum, *Acta Physiol. Scand.,* 131, 11, 1987.

71. **Bartho, L., Petho, G., and Ronai, Z.,** Theophylline-sensitive modulation of non-cholinergic excitatory neurotransmission in the guinea-pig ileum, *Br. J. Pharmacol.,* 86, 315, 1985.

72. **White, T. D.,** Release of ATP from isolated myenteric varicosities by nicotinic agonists, *Eur. J. Pharmacol.,* 79, 333, 1982.

73. **White, T. D. and Leslie, R. A.,** Depolarization-induced release of adenosine 5'-triphosphate from isolated varicosities derived from the myenteric plexus of the guinea-pig small intestine, *J. Neurosci.,* 2, 206, 1982.

74. **White, T. D. and Al-Humayyd, M.,** Acetylcholine releases ATP from varicosities isolated from guinea-pig myenteric plexus, *J. Neurochem.,* 40, 1069, 1983.

75. **Palmer, J. M., Wood, J. D., and Zafirov, D. H.,** Purinergic inhibition in the small intestinal myenteric plexus of the guinea-pig, *J. Physiol. (London),* 387, 357, 1987.

76. **Feit, C. and Roche, M.,** Action de l'adénosine sur la motricité intestinale après une ischémie mésentérique expérimentale chez le chien, *Gastroenterol. Clin. Biol.,* 12, 803, 1988.

77. **Burnstock, G., Campbell, G., and Rand, M. J.,** The inhibitory innervation of the taenia of the guinea-pig caecum, *J. Physiol. (London),* 182, 504, 1966.

78. **Cocks, T. and Burnstock, G.,** Effects of neuronal polypeptides on intestinal smooth muscle; a comparison with non-adrenergic, non-cholinergic nerve stimulation and ATP, *Eur. J Pharmacol.,* 54, 251, 1979.

79. **Burnstock, G., Cocks, T., Paddle, B., and Staszewska-Barczak, J.,** Evidence that prostaglandin is responsible for the 'rebound contraction' following stimulation of non-adrenergic, non-cholinergic ('purinergic') inhibitory nerves, *Eur. J. Pharmacol.,* 31, 360, 1975.

80. **Jager, L. P.,** The effects of catecholamines and ATP on the smooth muscle cell membrane of the guinea pig taenia coli, *Eur. J. Pharmacol.,* 25, 372, 1974.

81. **Shuba, M. F. and Vladimirova, I. A.,** Effect of apamin on the electrical responses of smooth muscle to adenosine 5′-triphosphate and to non-adrenergic, non-cholinergic nerve stimulation, *Neuroscience,* 5, 853, 1980.

82. **Maas, J. J., Den Hertog, A., Ras, R., and Van Den Akker, J.,** The action of apamin on guinea-pig taenia caeci, *Eur. J. Pharmacol.,* 67, 265, 1980.

83. **McKenzie, I. and Burnstock, G.,** Evidence against vasoactive intestinal polypeptide being the non-adrenergic, non-cholinergic inhibitory transmitter released from nerves supplying the smooth muscle of the guinea-pig taenia coli, *Eur. J. Pharmacol.,* 67, 255, 1980.

84. **Muller, M. J. and Baer, H. P.,** Apamin, a non-specific antagonist of smooth muscle relaxants, *Naunyn-Schmiedeberg's Arch. Pharmacol.,* 311, 105, 1980.

85. **Welford, L. A., Cusack, N. J., and Hourani, S. M. O.,** ATP analogues and the guinea-pig taenia coli: a comparison of the structure-activity relationship of ectonucleotidases with those of the P$_2$-purinoceptor, *Eur. J. Pharmacol.,* 129, 217, 1986.

86. **Small, R. C. and Weston, A. H.,** Theophylline antagonizes some effects of purines in the intestine but not those of intramural inhibitory nerve stimulation, *Br. J. Pharmacol.,* 67, 301, 1979.

87. **Su, C., Bevan, J. A., and Burnstock, G.,** [^{3}H]Adenosine triphosphate: release during stimulation of enteric nerves, *Science,* 173, 336, 1971.

88. **Rutherford, A. and Burnstock, G.,** Neuronal and non-neuronal components in the overflow of labeled adenyl compounds from guinea-pig taenia coli, *Eur. J. Pharmacol.,* 48, 195, 1978.

89. **Burnstock, G., Cocks, T., Tasakov, L., and Wong, H.,** Direct evidence for ATP release from non-adrenergic, non-cholinergic ('purinergic') nerves in the guinea-pig taenia coli and bladder, *Eur. J. Pharmacol.,* 49, 145, 1978.

90. **White, T. D., Potter, P., Moody, C., and Burnstock, G.,** Tetrodotoxin-resistant release of ATP from guinea-pig taenia coli and vas deferens during electrical field stimulation in the presence of luciferin-luciferase, *Can. J. Physiol. Pharmacol.,* 59, 1094, 1981.

91. **Fujiwara, M., Hong, S-C., and Muramatsu, I.,** Effects of goniopora toxin on non-adrenergic, non-cholinergic response and purine nucleotide release in guinea-pig taenia coli, *J. Physiol. (London),* 326, 515, 1982.

92. **Satchell, D. G.,** Nucleotide pyrophosphatase antagonizes responses to adenosine 5′-triphosphate and non-adrenergic, non-cholinergic inhibitory nerve stimulation in the guinea-pig isolated taenia coli, *Br. J. Pharmacol.,* 74, 319, 1981.

93. **Westfall, D. P., Hogaboom, G. K., Colby, J., O'Donnell, J. P., and Fedan, J. S.,** Direct evidence against a role of ATP as the nonadrenergic, noncholinergic inhibitory transmitter in guinea-pig taenia coli, *Proc. Natl. Acad. Sci. U.S.A.,* 79, 7041, 1982.

94. **Den Hertog, A., Pielkenrood, J., and Van Den Akker, J.,** Responses evoked by electrical stimulation, adenosine triphosphate, adenosine and 4-aminopyridine in taenia caeci of the guinea-pig, *Eur. J. Pharmacol.,* 109, 373, 1985.

95. **Den Hertog, A., Pielkenrood, J., and Van Den Akker, J.,** Effector mechanisms for α,β-methylene ATP and ATP derivatives in guinea-pig taenia caeci, *Eur. J. Pharmacol.,* 110, 95, 1985.

96. **Burnstock, G., Hills, J. M., and Hoyle, C. H. V.,** Evidence that the P$_1$-purinoceptor in the guinea-pig taenia coli is an A$_2$ subtype, *Br. J. Pharmacol.,* 81, 533, 1984.

97. **Fasth, S., Hulten, L., and Nordgren, S.,** Evidence of a dual pelvic nerve influence on large bowel motility in the cat, *J. Physiol. (London),* 298, 159, 1980.

98. **Hedlund, H., Fandriks, L., Delbro, D., and Fasth, S.,** Effect of α,β-methylene ATP on distal colonic and rectal motility—a possible involvement of P$_2$-purinoceptors in pelvic nerve mediated non-adrenergic, non-cholinergic contraction, *Acta Physiol. Scand.,* 127, 425, 1986.

99. **Tonini, M., Onori, L., Lecchini, S., Frigo, G., Perucca, E., and Crema, A.,** Mode of action of ATP on propulsive activity in rabbit colon, *Eur. J. Pharmacol.,* 82, 21, 1982.

100. **Percy, W. H. and Christensen, J.,** Pharmacological characterization of opossum distal colonic muscularis mucosae in vitro, *Am. J. Physiol.,* 250, G98, 1986.

101. **Gillespie, J. S.,** The rat anococcygeus muscle and its response to nerve stimulation and to some drugs, *Br. J. Pharmacol.,* 45, 404, 1972.

102. **Burnstock, G., Cocks, T., and Crowe, A. R.,** Evidence for purinergic innervation of the anococcygeus muscle, *Br. J. Pharmacol.,* 64, 13, 1978.

103. **Gibson, A. and Wedmore, C. V.,** Responses of the isolated anococcygeus of the mouse to drugs and to field stimulation, *J. Auton. Pharmacol.,* 1, 225, 1981.

104. **Gibson, A. and Tucker, J. F.,** The effects of vasoactive intestinal polypeptide and of adenosine 5′-triphosphate on the isolated anococcygeus muscle of the mouse, *Br. J. Pharmacol.,* 77, 97, 1982.

105. **Gillespie, J. S. and McGrath, J. C.,** The response of the cat anococcygeus muscle to nerve or drug stimulation and a comparison with the rat anococcygeus, *Br. J. Pharmacol.,* 50, 109, 1974.

106. **Stone, T. W.,** Purine receptors in the rat anococcygeus muscle, *J. Physiol. (London),* 335, 591, 1983.
107. **Byrne, N. G. and Large, W. A.,** Comparison of the biphasic excitatory junction potential with membrane responses to adenosine triphosphate and noradrenaline in the rat anococcygeus muscle, *Br. J. Pharmacol.,* 83, 751, 1984.
108. **Cocks, T., Crowe, R., and Burnstock, G.,** Non-adrenergic, non-cholinergic (purinergic?) inhibitory innervation of the rabbit rectococcygeus muscle, *Eur. J. Pharmacol.,* 54, 261, 1979.
109. **Lim, S. P. and Muir, T. C.,** Neuroeffector transmission in the guinea-pig internal anal sphincter: an electrical and mechanical study, *Eur. J. Pharmacol.,* 128, 17, 1986.
110. **Burleigh, D. E., D'Mello, A., and Parks, A. G.,** Responses of isolated human internal anal sphincter to drugs and electrical field stimulation, *Gastroenterology,* 77, 484, 1979.
111. **Davison, J., Al-Hassani, M., Crowe, R., and Burnstock, G.,** Non-adrenergic, inhibitory innervation of the guinea-pig gallbladder, *Pfluegers Arch.,* 377, 43, 1978.
112. **Naughton, P., Baer, H. P., Clanachan, A. S., and Scott, G. W.,** Adenosine and ATP effects on isolated guinea pig gallbladder, *Pfluegers Arch.,* 399, 42, 1983.
113. **Krasnow, S. and Grossman, M. I.,** Stimulation of gastric secretion in man by theophylline ethylenediamine, *Proc. Soc. Exp. Biol. Med.,* 71, 335, 1949.
114. **Foster, L. J., Trudeau, W. L., and Goldman, A. L.,** Bronchodilator effects on gastric acid secretion, *J. Am. Med. Assoc.,* 241, 2613, 1979.
115. **Gerber, J. G., Fadul, S., Payne, N. A., and Nies, A. S.,** Adenosine: a modulator of gastric acid secretion *in vivo, J. Pharmacol. Exp. Ther.,* 231, 109, 1984.
116. **Shay, H., Sun, C. H., and Gruenstein, M.,** A quantitative method for measuring spontaneous gastric secretion in the rat, *Gastroenterology,* 26, 906, 1954.
117. **Scarpignato, C., Tramacere, R., Zappia, L., and Del Soldato, P.,** Inhibition of gastric acid secretion by adenosine receptor stimulation in the rat, *Pharmacology,* 34, 264, 1987.
118. **Glavin, G. B., Westerberg, V. S., and Geiger, J. D.,** Modulation of gastric acid secretion by adenosine in conscious rats, *Can. J. Physiol. Pharmacol.,* 65, 1182, 1987.
119. **Westerberg, V. S. and Geiger, J. D.,** Adenosine analogs inhibit gastric secretion, *Eur. J. Pharmacol.,* 160, 275, 1989.
120. **Puurunen, J., Aittakumpu, R., and Tanskanen, T.,** Vagally mediated stimulation of gastric acid secretion by intravenously administered adenosine derivatives in anaesthetized rats, *Acta Pharmacol. Toxicol.,* 58, 265, 1986.
121. **Puurunen, J. and Huttunen, P.,** Central gastric antisecretory action of adenosine in the rat, *Eur. J. Pharmacol.,* 147, 59, 1988.
122. **Brodie, M. S., Lee, K., Fredholm, B. B., Stahle, L., and Dunwiddie, T. V.,** Central versus peripheral mediation of responses to adenosine receptor agonists: evidence against a central mode of action, *Brain Res.,* 415, 323, 1987.
123. **Gerber, J. G., Nies, A. S., and Payne, N. A.,** Adenosine receptors on canine parietal cells modulate gastric acid secretion to histamine, *J. Pharmacol. Exp Ther.,* 233, 623, 1985.
124. **Gerber, J. G. and Payne, N. A.,** Endogenous adenosine modulates gastric acid secretion to histamine in canine parietal cells, *J. Pharmacol. Exp. Ther.,* 244, 190, 1988.
125. **Puurunen, J., Ruoff, H-J., and Schwabe, U.,** Lack of direct effect of adenosine on the parietal cell function in the rat, *Pharmacol. Toxicol.,* 60, 315, 1987.
126. **Heldsinger, A. A., Vinik, A. I., and Fox, I. H.,** Inhibition of guinea-pig oxyntic cell function by adenosine and prostaglandins, *J. Pharmacol. Exp. Ther.,* 237, 351, 1986.
127. **Harty, R. F. and Franklin, P. A.,** Effects of exogenous and endogenous adenosine on gastrin release from rat antral mucosa, *Gastroenterology,* 86, 1107, 1984.
128. **Hubel, K. A.,** Electrical stimulus-secretion coupling in rabbit ileal mucosa, *J. Pharmacol. Exp. Ther.,* 231, 577, 1984.
129. **Grasl, M. and Turnheim, K.,** Stimulation of electrolyte secretion in rabbit colon by adenosine, *J. Physiol. (London),* 346, 93, 1984.
130. **Barrett, K. E., Huott, P. A., Shah, S. S., Dharmsathaphorn, K., and Wasserman, S. I.,** Different effects of apical and basolateral adenosine on colonic epithelial cell line T_{84}, *Am. J. Physiol.,* 256, C197, 1989.
131. **Yamagishi, F., Homma, N., Haruta, K., Iwatsuki, K., and Chiba, S.,** Adenosine potentiates secretin-stimulated pancreatic exocrine secretion in the dog, *Eur. J. Pharmacol.,* 118, 203, 1985.
132. **Yamagishi, F., Homma, N., Haruta, K., Iwatsuki, K., and Chiba, S.,** Effects of three purine-related compounds on pancreatic exocrine secretion in the dog, *Clin. Exp. Pharmacol. Physiol.,* 13, 425, 1986.

Chapter 23

EFFECTS OF ADENOSINE INFUSION ON THE CONSCIOUS MAN

Alf Sollevi

TABLE OF CONTENTS

I. PHARMACOKINETIC CONSIDERATIONS

Adenosine is subjected to extremely rapid elimination in contact with tissues with a plasma half-time in human blood in the order of 5 to 10 s. This rapid elimination, primarily by cellular uptake, causes marked loss of adenosine when administered intravenously. This is schematically illustrated in Figure 1, where a bolus peak concentration and a continuous infusion concentration are shown declining along the circulation, and especially across the lungs. The degree of this elimination is dependent on the site of administration (peripheral or central vein), the transit time of blood (cardiac output), the elimination capacity of blood cells (reduced by low hematocrit and low temperature), the elimination capacity of the endothelium, and finally the distance to target receptors/organs. It is therefore evident that adenosine may cause markedly different interindividual responses to a given dose. For this reason, the site of administration will be given as peripheral vein (PV), central vein (CV), and pulmonary artery (PA) in this overview. The degree of adenosine elimination in the human circulation has not been quantified in detail, but in the dog (with slower elimination than in humans) approximately 70% of elevated adenosine is lost across the lung circulation.[2] It is therefore likely that at least 90% of an intravenous dose is eliminated before reaching its site of action in the arterial circulation. Furthermore, it is relevant in this context to comment on the striking difference in biological effects between administration as an intravenous bolus and a continuous infusion. Whereas bolus infections produce dose-dependent negative chronotropic and dromotropic effects,[3] such effects are rarely seen during infusion even at very high doses.[4] On the contrary, infusion in awake subjects is associated with a positive chronotropic effect.[5] Adenosine infusion is also more associated with systemic vasodilation.[4-7] The transient cardiac electrophysiological effects of a bolus injection may be explained by two factors. First, a bolus injection causes higher peak plasma concentrations than infusions, as illustrated in Figure 1. Second, electrophysiological effects require higher blood adenosine concentrations than vascular effects, due to the diffusion barriers for adenosine across the vessel wall and possibly also due to the requirement of higher adenosine concentration at the receptors for inhibitory chronotropic and dromotropic responses. The transient biological effect of a bolus (seconds)[3] is probably too short for establishing major vasodilatory effects.

II. GENERAL HEMODYNAMIC EFFECTS

The first study in healthy volunteers assessing the effects of intravenous adenosine infusions on heart rate and blood pressure in relation to sympathoadrenal activation was published as late as 1986.[7] With infusions of 80 to 140 μg/kg/min PV, there was a dose-dependent increase in systolic pressure and a decrease in diastolic arterial pressure, and thus an increase in pulse pressure. There was also a dose-dependent increase in heart rate. This response is consistent with a vasodilatory effect, but the mean arterial blood pressure was unaffected.[5,7,8] The maintenance of the mean arterial pressure has later been explained by a compensatory increase in cardiac output (Figure 2), mediated by an increase in both heart rate and stroke volume, thus increasing cardiac output by 100% in healthy volunteers at an infusion rate of 80 μg/kg/min CV.[9] Adenosine is a resistance vessel dilator, while the influence on capacitance vessels seems to be minor or absent.[13]

The vasodilatory effect of adenosine is also associated with autonomic reflex activation, expressed by elevated circulating plasma catecholamines, as well as by increased efferent sympathetic discharge.[7,9,14] This autonomic reflex response is responsible for the maintenance of the mean arterial blood pressure during adenosine-induced systemic vasodilation, while impairment of this autoregulatory function is associated with the hypotensive effects of adenosine. Therefore, in patients with severe autonomic failure, adenosine infusion causes

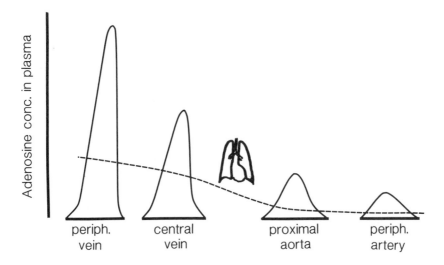

FIGURE 1. A schematic illustration of the rapid decline of adenosine concentration in plasma from peripheral venous to peripheral arterial blood, in conjunction with peripheral intravenous bolus (wave shape) or intravenous infusion (dotted line) of adenosine.

reduced blood pressure without reflex inotropic and chronotropic responses.[5] Similarly, during anesthesia with impaired autonomic reflexes, a dose-dependent hypotension is obtained.[11,12] The fact that an intact autonomic reflex mechanism may prevent a fall in systemic blood pressure during adenosine infusion indicates that adenosine-induced presynaptic inhibition of catecholamine release[15] is absent at doses that can be given in awake subjects, i.e., up to 200 μg/kg/min PV.

When adenosine is given during anesthesia to induce controlled hypotension, there is no tachyphylaxis to the blood pressure lowering effect. This is mainly due to the adenosine-mediated inhibition of renin release, leading to an unchanged plasma renin activity, thereby blocking the renin-angiotensin activation that is of importance for the tachyphylaxis during hypotension.[11,16] A second factor may be an inhibition of the autonomic reflex activation, caused by the basal anesthesia as such or combined with presynaptic inhibitory and ganglion blocking effects of adenosine at high doses (>200 μg/kg/min).

III. REGIONAL HEMODYNAMIC EFFECTS

In the *heart*, the first clinical study examining coronary flow effects of adenosine in humans was conducted in patients during coronary bypass surgery, where graft flow was quantified intraoperatively.[10,17] This vascular bed was very sensitive as evidenced by a doubling of the flow at 30 μg/kg/min PA, while the systemic hemodynamic parameters were essentially unaffected. The vasodilatory effect was dose dependent and was maintained during continous infusion.[17,19] The absolute coronary flow increase was most prominent in healthy subjects,[18] and the flow response was unrelated to myocardial work load.[17,19] Further, the effect was reversible within a few minutes after discontinuation. Coronary vasodilation can be demonstrated even at high and hypotensive doses during anesthesia.[12] When given as a bolus into the coronary arteries of coronary artery disease (CAD) patients, a maximal flow increase of 160% has been demonstrated (similar to papaverine), but 3 out of 12 subjects developed bradyarrhythmias.[20] A *hypertensive* effect during intracoronary adenosine infusion at a high dose (2.2 mg/min) has also been demonstrated, suggesting an afferent reflex activation, since this pressure response was lacking in patients with denervated hearts.[21]

Marked adenosine-induced coronary vasodilation may be associated with intramyocardial

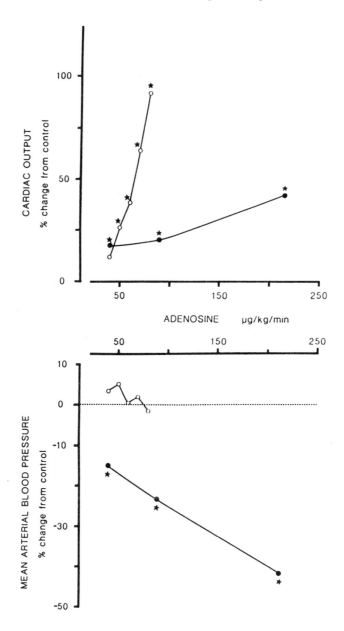

FIGURE 2.   Effect of intravenous adenosine infusion on mean arterial blood pressure and cardiac output in healthy volunteers (O—O),[9] and in anesthetized patients (●—●).[10,12] Results significantly different from control are indicated by a star.

redistribution of flow with subsequent ischemia in subjects with CAD. This has been indicated in some patients during clinical hypotension,[4,12] and in CAD subjects upon dose titration as documented by typical chest pain and ischemic ECG.[22]

In the *brain,* it has been shown repeatedly since the early work of Berne et al.[23] in 1974 that adenosine dilates cerebral vessels when topically applied. There are, however, species differences in the response to intravascular infusion, probably due to the function and location of the blood-brain barrier. With the exception of the dog and the cat, intravascular adenosine dilates cerebral vessels.[13] In humans, i.v. infusion produces a marked flow increase at high doses (200 to 500 µg/kg/min CV), which are hypotensive in anesthetized patients.[24] The fact that adenosine increases cerebral blood flow could further indicate that the cerebral autoregulatory mechanism is impaired. Such dose-dependent impairment of autoregulation has also recently been demonstrated in pigs.[25]

In the *kidney,* i.v. adenosine up to 50 μg/kg/min PV in healthy volunteers lacks vascular effects, while a minor increase in the tubular reabsorption of water causes reduced urine output.[26] The renal effect of higher doses in volunteers is not known. However, in anesthetized subjects there was a dose-dependent (60 to 90 μg/kg/min CV) reduction in glomerular filtration, renal blood flow, and urine output.[27] There was a 60 to 70% reduction in these variables at the highest dose; during controlled hypotension even more pronounced, but transient renal vasoconstriction has been seen.[48] However, in conscious subjects with maintained blood pressure during adenosine infusion, the renovascular effects may be quite different from the response in anesthetized patients.

In the normal *pulmonary circulation,* adenosine dilates vessels at high doses (>200 μg/kg/min CV), thereby increasing the lung shunting, which leads to lowered arterial P_aO_2;[4] at lower doses the influence is minor. Adenosine has, on the other hand, significant dilatory effects at doses below 50 μg/kg/min PV when pulmonary vascular resistance is abnormally elevated.[28]

The flow effects in the *skin, skeletal muscle, subcutaneous fat,* and *splanchnic area* have also been investigated in volunteers.[9] Skin is as sensitive to adenosine infusion as the coronary vessels, but above 50 μg/kg/min CV there is no further increase in flow. Splanchnic flow increases at 60 μg/kg/min (CV), while no effect is seen in fat and muscle at the highest dose of 80 μg/kg/min. Since muscle blood flow is markedly elevated by local infusion,[29] the noted differences in organ flow reactivity may reflect different sensitivity rather than variability in the distance from the infusion site to the tissue studied. Another factor to take into account is the adenosine-induced increase in sympathetic discharge,[14] which can counteract regional dilatation and subsequently cause flattening of the dose-response curve (skin) or abolished flow effects (muscle, fat).

In summary, there seem to be regional differences in sensitivity with respect to vasodilation. The splanchnic area accounts for a quantitatively large proportion of the elevated cardiac output during adenosine infusion.[9]

IV. SUBJECTIVE SYMPTOMS

There is a dose-dependent appearance of symptoms when adenosine is infused at doses above approximately 50 μg/kg/min (PV) (Table 1). Especially, pain and dyspnea reactions limit the individual maximally tolerable dose. With bolus administration, there are similar types of pain/discomfort symptoms, but of a transient duration (60 to 90 sec).[3] Some of the sensations during infusion, such as flush, headache, palpitation, and blocked sinus, can be related to vascular effects. Stimulation of the respiration[5,30-32] and painful reactions[33,34] cannot be explained by vascular effects of adenosine, but rather by stimulation of afferent reflexes.

A. RESPIRATORY STIMULATION

A respiratory stimulating effect of adenosine has been noted already in early works,[35] but only recently has attention been focused on this effect in clinical studies.[5,30-32] The major effect is a deeper respiration after bolus injections[30,34] and the onset is earlier than pain and cardiac electrophysiological effects.[34] During both bolus and infusion administration there are dose-dependent increases in ventilation volume without changes in respiratory rate.[5] Furthermore, respiratory stimulation was noted only when infusion proximal to the origin of the carotid arteries was performed.[5] This response was inhibited by hyperventilation and theophylline treatment, and enhanced by adenosine uptake-inhibiting drugs, whereas inhibition of cholinergic, adrenergic, and opioid receptors had no influence.[36] Since experimental

TABLE 1
Subjective Symptoms During Adenosine
Infusion in Man

Facial flush
Chest pain/discomfort
Radiating pain (neck, throat, limbs, back, and abdomen)
Blocked sinuses of the nose
Headache
Palpitation
Dyspnea, tachypnea
Tingling sensation in the arms
Nausea

Note: Subjective symptoms (dose dependent and individually variable) reported during adenosine infusion to
healthy volunteers and patients. Information is obtained from all available clinical studies.

studies have demonstrated adenosine-mediated chemoexcitatory effects in the carotid bodies associated with regulation of breathing[37] and since the activation can be attenuated by adenosine receptor antagonism,[36,37] it is probable that a carotid body reflex is involved in the respiratory stimulating effect of exogenous adenosine.[5,32]

B. PAIN SYMPTOMS

Systemic adenosine causes a variety of pain reactions in healthy subjects (Table 1), and the dose-dependent retrosternal pain is most frequently present during bolus injection.[33,34] Adenosine also produces pain during regional administration in the human forearm and at topical skin application,[38,39] indicating that it generates afferent algogenic signals unrelated to ischemia. Clinical doses of theophylline reduced the chest pain intensity in healthy subjects.[33] The mechanism for this sensory activation is not known, but autonomic and opioid receptor antagonism do not modify the response.[34] Adenosine-mediated pain may be involved in the pathophysiology of ischemic pain, and Sylvén and co-workers[40] have shown that patients with CAD experience chest pain with adenosine similar to the pain of angina pectoris. Also, the analysis of time to peak intensity of pain after bolus injection could indicate a myocardial origin of the adenosine-mediated response. Others state that the re-semblance of adenosine-induced pain in patients with evidence of silent and painful myo-cardial ischemia indicates that the chest pain is unlikely to be entirely of cardiac origin.[41] It is finally noteworthy that high i.v. doses of adenosine (>60 μg/kg/min PV) in CAD patients may generate pain, both by some direct sensory activation and by induction of intramyocardial flow redistribution leading to ischemia.

V. EFFECT ON THE AIRWAY

Inhalation of nebulized adenosine (0.67 to 6.7 mg/ml) induces dose-dependent and marked bronchoconstriction in allergic and nonallergic asthmatics, while this effect is lacking in normal subjects.[42] This bronchoconstriction is inhibited by adenosine receptor antagonism (theophylline).[43] The maximal effect after a 1-min inhalation is seen within 5 min, while the response lasts for more than 30 min.[42] Since adenosine is most labile in biological tissues, it is unlikely that the long-lasting airway response is a direct effect. The time course of adenosine effects is parallel to that of exogenous histamine, and subjects with nonspecific bronchial hyperreactivity are less influenced by adenosine.[44] This would indicate that the effect involves an activation of luminal mast cells, and the adenosine effect is also antagonized by sodium cromoglycate.[45,46] Systemic adenosine infusion had no effect on asthmatic basal bronchial tone or methacholine-stimulated bronchoconstriction at i.v. doses up to 50 µg/kg/min PV.[47]

VI. METABOLIC EFFECTS

Central venous infusion of adenosine (50 to 80 µg/kg/min) in healthy volunteers is associated with a dose-dependent reduction of arterial glycerol (-50%), without changes in glucose and lactate levels.[9] The splanchnic clearance of these substrates was unaffected, suggesting that the lowered systemic glycerol concentration was mediated by a peripheral antilipolytic effect of adenosine. Therefore, at an adenosine dose where adipose tissue blood flow was unaffected, significant antilipolytic effects were seen, although circulating catecholamines were elevated. This indicates that human fat cells are highly sensitive to the antilipolytic action of adenosine.

REFERENCES

1. **Klabunde, R. E.,** Dipyridamole inhibition of adenosine metabolism in human blood, *Eur. J. Pharmacol.,* 93, 21, 1983.
2. **Sollevi, A., Lagerkranser, M., Andreen, M., and Irestedt, L.,** Relationship between arterial and venous adenosine levels and vasodilatation during ATP- and adenosine-infusion in dogs, *Acta Physiol. Scand.,* 120, 171, 1984.
3. **DiMarco, J. P., Sellers, T. D., Berne, R. M., West, G. A., and Belardinelli, L.,** Adenosine: electrophysiologic effects and therapeutic use for terminating paroxysmal supraventricular tachycardia, *Circulation,* 68, 1254, 1983.
4. **Öwall, A., Gordon, E., Lagerkranser, M., Lindquist, C., Rudehill, A., and Sollevi, A.,** Clinical experience with adenosine for controlled hypotension during cerebral aneurysm surgery, *Anesth. Analg.,* 66, 229, 1987.
5. **Biaggioni, I., Olafsson, B., Robertson, R. M., Hollister, A. S., and Robertson, D.,** Cardiovascular and respiratory effects of adenosine in conscious man, *Circ. Res.,* 61, 779, 1987.
6. **Sollevi, A., Lagerkranser, M., Irsestedt, L., Gordon, E., and Lindquist, C.,** Controlled hypotension with adenosine in cerebral aneurysm surgery, *Anesthesiology,* 61, 400, 1984.
7. **Biaggioni, I., Onrot, J., Hollister, A. S., and Robertson, D.,** Humoral and hemodynamic effects of adenosine infusion in man, *Life Sci.,* 39, 2229, 1986.
8. **Fuller, R. W., Maxwell, D. J., Conradson, T-B. G., Dixon, C. M. S., and Barnes, P. J.,** Circulatory and respiratory effects of infused adenosine in conscious man, *Br. J. Clin. Pharmacol.,* 24, 309, 1987.
9. **Edlund, A., Sollevi, A., and Linde, B.,** Haemodynamic and metabolic effects of infused adenosine in man, *Clin. Sci.,* 79, 131, 1990.
10. **Torssell, L., Sollevi, A., Öhqvist, G., and Ekeström, S.,** Adenosine-induced coronary vasodilation during coronary bypass surgery, *Anesthesiology,* 63, A34, 1985.
11. **Öwall, A., Järnberg, P-O., Brodin, L-Å. and Sollevi, A.,** Effects of adenosine-induced hypotension on myocardial hemodynamics and metabolism in fentanyl anesthetized patients with peripheral vascular disease, *Anesthesiology,* 68, 416, 1988.
12. **Öwall, A., Lagerkranser, M., and Sollevi, A.,** Effects of adenosine-induced hypotension on myocardial hemodynamics and metabolism during cerebral aneurysm surgery, *Anesth. Analg.,* 67, 228, 1988.
13. **Sollevi, A.,** Cardiovascular effects of adenosine in man; possible clinical implications, *Prog. Neurobiol.,* 27, 319, 1986.
14. **Biaggioni, I., Mosqueda-Garcia, R., Robertson, R. M., and Robertson, D.,** Adenosine increases sympathetic nerve traffic in man, *Circulation,* 80 (Suppl.), II-90, 1989.
15. **Fredholm, B. B., Gustafsson, L. E., Hedqvist, P., and Sollevi, A.,** Adenosine in the regulation of neurotransmitter release in the peripheral nervous system, in *Regulatory Function of Adenosine,* Berne, R. M., Rall, T. W., and Rubio, R., Eds., Martinus Nijhoff, The Hague, 1983, 479.
16. **Zäll, S., Edén, E., Winsö, I., Volkmann, R., Sollevi, A., and Ricksten, S-E.,** Controlled hypotension with adenosine or sodium nitroprusside during cerebral aneurysm surgery: effects on renal hemodynamics, excretory function and renin release, *Anesth. Analg.,* 71, 631, 1990.

17. **Torssell, L., Ekeström, S., and Sollevi, A.,** Adenosine-induced increase in graft flow during coronary bypass surgery, *Scand. J. Thorac. Cardiovasc. Surg.,* 23, 235, 1989.

18. **Edlund, A., Sollevi, A., and Conradsson, T.,** Adenosine—a coronary vasodilator in vivo. Effect of theophylline and enprophylline, *Jpn. J. Pharmacol.,* 52 (Suppl. 11), 116P, 1990.

19. **Zäll, S., Milocco, I., and Ricksten, S-E.,** Effects of adenosine on myocardial hemodynamics and metabolism after coronary artery bypass grafting (CABG), *J. Cardiothorac. Anesth.,* 3 (Suppl. 1), 30A, 1989.

20. **Zijlstra, F., Juilliére, Y., Serruys, P. W., and Roelandt, J. R. T. C.,** Value and limitations of intracoronary adenosine for the assessment of coronary flow reserve, *Catheterization Cardiovasc. Diagnosis,* 15, 76, 1988.

21. **Cox, D. A., Vita, J. A., Treasure, C. B., Selwyn, A. P., and Ganz, P.,** Pressor response to intracoronary adenosine, *Circulation,* 78 (Suppl. 2), 697A, 1988.

22. **Edlund, A., Strååt, E., and Henriksson, P.,** Infusion of adenosine provokes myocardial ischemia in patients with ischemic heart disease, *Clin. Physiol.,* 9, 309, 1989.

23. **Berne, R. M., Rubio, R., and Curnish, R. R.,** Release of adenosine from ischemic brain. Effect on cerebral vascular resistance and incorporation into cerebral adenine nucleotides, *Circ. Res.,* 35, 262, 1974.

24. **Sollevi, A., Ericson, K., Eriksson, L., Lindqvist, C., Lagerkranser, M., and Stone-Elander, S.,** Effect of adenosine on human cerebral blood flow as determined by positron emission tomography, *J. Cerebr. Blood Flow Metab.,* 7, 673, 1987.

25. **Stånge, K., Lagerkranser, M., and Sollevi, A.,** Effect of adenosine-induced hypotension on the cerebral autoregulation in the anesthetized pig, *Acta Anesth. Scand.,* 33, 450, 1989.

26. **Nyberg, G., Nilsson, J-E., Sollevi, A., and Öwall, A.,** The renal effects of adenosine infused in healthy volunteers over 1 to 5 h, *Br. J. Clin. Pharmacol.,* 28, 221P, 1989.

27. **Zäll, S. and Ricksten, S-E.,** Effects of Adenosine on Renal Function, Central Hemodynamics and ECG Signs of Myocardial Ischemia, After Coronary Artery Bypass Surgery, 5th Annual Meeting EACTA, Vienna, May 1990.

28. **Morgan, J. M., McCormack, D., Griffiths, M., and Evans, T. W.,** Evidence for preferential pulmonary vasodilatation by adenosine, *Am. Heart Assoc., New Orleans,* 353 (Abstr.), 1989.

29. **Nowak, J., Wennmalm, M., Edlund, A., Wennmalm, Å., and Fitzgerald, G. A.,** Vascular effects of infused adenosine are not mediated by prostacyclin release in humans, *Am. J. Physiol.,* 252, H598, 1987.

30. **Watt, A. H. and Routledge, P. A.,** Adenosine stimulates respiration in man, *Br. J. Clin. Pharmacol.,* 20, 503, 1985.

31. **Maxwell, D. L., Fuller, R. W., Nolop, E. B., Dixon, C. M. S., and Hughes, J. M. B.,** The effects of adenosine on ventilatory responses to hypoxia and hypercapnea in humans, *J. Appl. Physiol.,* 61, 1762, 1986.

32. **Watt, A. H., Reid, P. G., Stephens, M. R., and Routledge, P. A.,** Adenosine-induced respiratory stimulation depends on the site of infusion. Evidence for an action on the carotid body?, *Br. J. Clin. Pharmacol.,* 23, 486, 1987.

33. **Sylvén, C., Beermann, B., Jonzon, B., and Brandt, R.,** Angina pectoris-like pain provoked by intravenous adenosine in healthy volunteers, *Br. Med. J.,* 293, 227, 1986.

34. **Sylvén, C., Jonzon, B., Brandt, R., and Beermann, B.,** Adenosine provoked angina pectoris-like pain—time characteristics, influence of autonomic blockade and naloxone, *Eur. Heart J.,* 8, 738, 1987.

35. **Drury, A. N. and Szent-Gyorgyi, A.,** The physiological activity of adenine compounds with special reference to their action upon the mammalian heart, *J. Physiol. (London),* 68, 213, 1929.

36. **Jonzon, B., Sylvén, C., Beermann, B., and Brandt, R.,** Adenosine receptor mediated stimulation of ventilation in man, *Eur. J. Clin. Invest.,* 19, 65, 1989.

37. **McQueen, D. S. and Ribeiro, J. A.,** Pharmacological characterization of the receptor involved in chemoexcitation induced by adenosine, *Br. J. Pharmacol.,* 88, 615, 1986.

38. **Sylvén, C., Jonzon, B., Fredholm, B. B., and Kaijser, L.,** Adenosine injection into the brachial artery produces ischemia like pain or discomfort in the forearm, *Cardiovasc. Res.,* 22, 674, 1988.

39. **Bleehen, T. and Keele, C. A.,** Observations on the algogenic actions of adenosine compounds on the human blister base preparation, *Pain,* 3, 367, 1977.

40. **Sylvén, C., Beermann, B., Edlund, A., Lewander, R., Jonzon, B., and Mogensen, L.,** Provocation of chest pain in patients with coronary insufficiency using the vasodilator adenosine, *Eur. Heart J.,* 9 (Suppl. N), 6, 1988.

41. **Crea, F., El-Tamimi, H., Veijar, M., Kaski, J. C., Davies, G., and Maseri, A.,** Adenosine-induced chest pain in patients with silent and painful myocardial ischaemia: another clue to the importance of generalized defective perception of painful stimuli as a cause of silent eschaemia, *Eur. Heart J.,* 9 (Suppl. N), 34, 1988.

42. **Cushley, M. J., Tattersfield, A. E., and Holgate, S. T.,** Inhaled adenosine and guanosine on airway resistance in normal and asthmatic subjects, *Br. J. Clin. Pharmacol.,* 15, 161, 1983.

43. **Cushley, M. J., Tattersfield, A. E., and Holgate, S. T.,** Adenosine induced bronchoconstriction in asthma: antagonism by inhaled theophylline, *Am. Rev. Respir. Dis.,* 129, 380, 1984.

44. **Holegate, S. T., Mann, J. S., Church, M. K., and Cushley, M. J.,** Mechanisms and significance of adenosine-induced bronchoconstriction in asthma, *Allergy,* 42, 481, 1987.
45. **Cushley, M. J. and Holgate, S. T.,** Adenosine induced bronchoconstriction in asthma: role of mast cell mediator release, *J. Allergy Clin. Immunol.,* 75, 272, 1985.
46. **Crimi, N., Palermo, F., Vancheri, C., Oliveri, R., Distefano, S. M., Polosa, R., and Mistretta, A.,** Effect of sodium cromoglycate and nifedipine on adenosine-induced bronchoconstriction, *Respiration,* 53, 74, 1988.
47. **Larsson, K. and Sollevi, A.,** Influence of infused adenosine on bronchial tone and bronchial reactivity in asthma, *Chest,* 93, 280, 1988.
48. **Sollevi, A.,** Unpublished observations.

VI. Influence of Adenosine and ATP on the
    Nervous System

Chapter 24

ATP AS A NEUROTRANSMITTER, COTRANSMITTER, AND NEUROMODULATOR

David P. Westfall, Hugh H. Dalziel, and Karyn M. Forsyth

TABLE OF CONTENTS

I. INTRODUCTION

Classically, it was thought that the sole mediators of chemical neurotransmission at any synapse or neuroeffector junction were either acetylcholine (ACh) or norepinephrine (NE). More recently, however, a variety of substances have been shown to have a neurotransmitter function in the central and peripheral nervous systems.[1,2] This chapter will focus on the growing possibility that the ubiquitous adenine nucleotide, adenosine 5'-triphosphate (ATP), may have a role in chemical neurotransmission. The emphasis will be on the cotransmitter role of ATP in the autonomic nervous system, particularly in regard to the sympathetic innervation of the mammalian urogenital and cardiovascular systems. Studies with these systems has provided the strongest and most extensive support for a role for ATP in the neurotransmission process. The evidence that ATP is a transmitter in the central nervous system (CNS) will also be reviewed. Finally, evidence that ATP plays a neuromodulatory role via prejunctional purinoceptors will be discussed.

II. ATP AS A NEUROTRANSMITTER

Until the early 1960s it was generally accepted that the only neurotransmitters in the peripheral nervous system were ACh and NE. Although the work of the Holtons[3,4] was perhaps the original suggestion of a neurotransmitter role for ATP, the first concrete evidence indicating the existence of nerves which utilize a substance other than ACh or NE as a transmitter came from studies of the nerve-mediated hyperpolarizations and resulting relaxations of the guinea pig taenia coli which persisted in the presence of atropine and guanethidine.[5-7] These responses were termed nonadrenergic, noncholinergic (NANC). Subsequent studies, mainly on the mammalian gut, led to the proposal that ATP, or a related nucleotide, was the NANC neurotransmitter in this tissue,[8,9] and these nerves were termed "purinergic".[10] The evidence in favor of the purinergic nerve hypothesis is intriguing, but incomplete. To date it has not been possible to identify ATP as the *sole transmitter* at any mammalian peripheral or central synapse.[11-13] However, there is substantial evidence that ATP is coreleased with other transmitters and thereby contributes to the neurotransmission process as a *cotransmitter*. This evidence is discussed below.

III. ATP AS A COTRANSMITTER

Burnstock[14] was one of the first to suggest that nerve cells may release more than one transmitter. Since then numerous reports have appeared which support the concept of cotransmission,[15] in particular the release of ATP with the classical peripheral neurotransmitters ACh and NE.[16] The following paragraphs will attempt to review briefly some of the evidence which implicates ATP as a cotransmitter with NE and ACh, especially in the urogenital and cardiovascular systems, where the majority of research has been concentrated to date.

A. CRITERIA FOR ATP AS A COTRANSMITTER

Generally, there are five criteria which should be satisfied in order for a substance to be considered as a transmitter or cotransmitter. The putative transmitter must be shown first to be synthesized and stored in the nerve, second to be released upon stimulation of the nerve, and third to mimic the action of nerve stimulation upon the effector cells when applied exogenously. The fourth criterion is that drugs which antagonize the postjunctional response to the neurally released transmitter must also block the action of the exogenously applied putative transmitter. The fifth criterion is that there must be a mechanism for inactivating the putative transmitter at the neuroeffector junction.

In the case of ATP as a cotransmitter in sympathetic nerves there is ample evidence that these criteria are fulfilled. For example, Lagercrantz[17] has reviewed the evidence for the synthesis and storage of ATP in the varicosities of sympathetic nerves. There is also extensive documentation that sympathetic nerve stimulation promotes the release of endogenous ATP[18-21] or [^{3}H]-ATP after preincubation of tissues with [^{3}H]-adenosine.[22,23] The criterion of mimicry has been fulfilled in that exogenous ATP produces responses in numerous effectors that reproduce the nonadrenergic component of sympathetic nerve stimulation.[23-28] Antagonism of the postjunctional response to sympathetic nerve stimulation and exogenous ATP has been demonstrated for the vas deferens and certain blood vessels using the selective photoaffinity ATP analog, arylazido aminopropionyl-ATP (ANAPP$_3$), or the purinoceptor-desensitizing agent α,β-methylene-ATP.[24,29-39] Neurally released ATP would be expected to be inactivated rapidly at the neuroeffector junction by ectonucleotidases.[40] Furthermore, once adenosine is formed extracellularly from ATP the adenosine may be conserved by neuronal uptake as is suggested by the observation that [^{3}H]-adenosine is taken up into sympathetic nerves.[22,23] The criteria that ATP is a cotransmitter with ACh in certain cholinergic nerves are also well fulfilled[13,41] (*vide infra*).

B. ATP AS A COTRANSMITTER WITH NOREPINEPHRINE

Indications for the costorage of ATP with NE came from investigations which suggested that ATP may be stored and released with catecholamines from vesicles in cat perfused adrenal glands.[42,43] Furthermore, evidence that ATP and NE may be costored in bovine splenic nerve vesicles[17,44,45] and the sympathetic vesicles of the rat vas deferens[46] was provided.

Following the suggestion of the involvement of ATP as the NANC neurotransmitter in the gut[10,47] and Burnstock's[14] proposal of cotransmission in 1976, further indications that ATP may be costored and coreleased with NE in sympathetic nerves became evident. Work on the rat isolated portal vein and aorta showed that after preincubation with [^{3}H]-adenosine, subsequent electrical stimulation of the perivascular nerves resulted in the release of [^{3}H]-purines. This release was sensitive to guanethidine or tetrodotoxin, but not to phenoxybenzamine, indirectly indicating that it was due to stimulation of the perivascular nerves, and not secondary to smooth-muscle contraction.[22] In the nictitating membrane of the reserpinized cat, a residual neurogenic contraction persisted in the presence of an α-antagonist, and since it was closely mimicked by exogenous ATP, a purinergic involvement in the neurotransmission process in this tissue was suggested.[48] Subsequent to these reports, interest in the possibility of cotransmission of ATP with NE has grown and a vast literature has accumulated which provides strong evidence in favor of this process. The majority of the initial work was conducted on the vas deferens of various species and has provided the impetus for a reconsideration of the process of sympathetic neurotransmission. A review of the evidence which supports the theory that ATP is a cotransmitter in sympathetic nerves in the vas deferens and the vasculature follows below.

1. Vas Deferens

The vas deferens is supplied with a dense adrenergic innervation[49] and the NE content has been shown to be high.[50] Stimulation of the sympathetic nerves to the vas deferens of the guinea pig and rat results in a biphasic mechanical response which consists of an initial rapid twitch followed by a maintained contracture.[51]

Despite the classical evidence which suggested that NE was the sole transmitter in sympathetic nerves in the vas deferens,[49,50,52,53] there is growing evidence which indicates that part of the mechanical response, in particular the initial twitch component, may be mediated by another transmitter.[51] Pharmacological analysis of the sympathetic neurogenic responses of the vas deferens and reconsideration of some of the earlier studies have revealed several features which are incompatible with a purely noradrenergic neurotransmission process. For example, resistance of the electrical and part of the mechanical response to α-antagonism and pretreatment with reserpine has been frequently observed.[25,26,29,30,51-58] Furthermore, the adrenergic and nonadrenergic components of the response can be mimicked by exogenous NE and ATP, respectively.[23-26] The nonadrenergic component and the response to exogenous P_2 purinoceptor agonists were blocked by ANAPP$_3$ or α,β-methylene-ATP.[25,26,29,30,56,58,59] Potentiation of only part of the response by cocaine suggests a nonadrenergic involvement.[25,55,59]

As a result of these and other studies, cotransmission of ATP with NE has become the favored explanation for the complicated mechanisms of neurotransmision in the vas deferens. More recently, direct evidence for a sympathetic neurogenic origin of electrically stimulated ATP release and its involvement in the generation of the first phase of the contraction of the guinea pig vas deferens has been presented,[18,19] and has validated previous, less direct studies which implicated ATP as the factor responsible for the nonadrenergic component of the neurogenic response.

When considered collectively, the literature indicates that ATP is released as a functional cotransmitter with NE in the vas deferens, and can be summarized as follows. The rodent vas deferens responds to stimulation of the postganglionic sympathetic nerves with trains of pulses with a biphasic mechanical response. In general, the first phase is probably mediated mainly by ATP and the second mainly by NE, acting on postjunctional P_{2x} purinoceptors and $α_1$ adrenoceptors, respectively. Evidence suggests that these two transmitters are released from the same nerve varicosity as cotransmitters. The result of stimulation of P_{2x} purinoceptors in this tissue by ATP is the production of excitatory junction potentials which can summate to fire action potentials which propagate from one smooth-muscle cell to another. The resulting calcium influx leads to the generation of a major component of the transient first phase of the contraction. The activation of the $α_1$ adrenoceptors by NE results mainly in the second maintained phase of the contraction by a mechanism which is independent of membrane potential change and probably involves the release of intracellular calcium by a phosphoinositide pathway.[60]

Generally, schematic representations of sympathetic cotransmission depict ATP and NE as stored in the same vesicle within the sympathetic varicosity. However, recent evidence suggests that certain prejunctionally acting drugs may differentially affect the release of the two transmitters.[61-64] Therefore, the possibility exists that ATP and NE originate in separate vesicles or different pools of vesicles containing varying proportions of the two cotransmitters.

2. Blood Vessels

Subsequent to work with the gut and vas deferens, which implicated a neurotransmitter role for ATP, this possibility was considered for periarterial sympathetic nerves.[65-67] Indirect evidence that ATP may be released from vascular nerves was initially provided by Su,[22] who demonstrated the neurogenic release of [^{3}H]-adenosine from the rabbit portal vein and aorta in response to perivascular nerve stimulation. These results, together with the observation that the neurogenic electrical and mechanical responses of various vascular preparations to sympathetic nerve stimulation were resistant to α-antagonism, led to the formulation of the hypothesis that ATP may be released as a cotransmitter with NE from perivascular sympathetic nerves, and be responsible for the α-antagonist-resistant component of the neurogenic response. Investigation of the possibility of an involvement of ATP in the α-antagonist-resistant responses in blood vessels has followed a similar approach as those studies in the vas deferens. However, the extent of involvement of ATP as a cotransmitter in blood vessels is subject to both species and regional variation. Table 1 lists examples of blood vessels where there is reasonable evidence for an involvement of ATP as a cotransmitter.

TABLE 1
Examples of Blood Vessels for Which There Is
Evidence That ATP and NE Are Cotransmitters

Blood vessel	Ref.
Rat tail artery	68—71
Rabbit mesenteric artery	31,35,36,72—75
Dog mesenteric artery	37—39,76—78
Guinea pig mesenteric artery	74,79,80
Rabbit ear artery	27,32,36,68,70,80,81
Rabbit saphenous artery	33,34,68
Guinea pig saphenous artery	82
Dog cerebral arteries	83—85
Rabbit hepatic artery	86

It can be seen that there is good evidence for the involvement of ATP in the sympathetic nerve-mediated responses of many isolated vascular preparations. The extent to which ATP contributes to neurotransmission in blood vessels *in vitro* is variable, ranging from an apparently minor contribution (e.g., rat tail artery) through a substantial contribution (e.g., rabbit saphenous and mesenteric arteries), to an almost exclusive contribution (e.g., rat small intrapulmonary artery).[87] Recently, it has been proposed that a purinergic component of the sympathetic pressor response of the pithed rat can be demonstrated using α,β-methylene-ATP;[88] however, the selectivity of this ATP analog under *in vivo* conditions has been questioned[89-91] and emphasizes the need for development of selective purinoceptor antagonists.

C. ATP AS A COTRANSMITTER WITH ACETYLCHOLINE

The corelease of ATP with ACh has been suggested to occur from cholinergic nerves of both the somatic and autonomic nervous systems. For example, the neurogenic release of ATP from the somatic motoneurons of the skeletal neuromuscular junction of the rat diaphragm has been observed directly using the firefly luciferin-luciferase assay, and a common vesicular origin for ATP and ACh was suggested.[92] Furthermore, ATP and its metabolites have been shown to have pre- and postjunctional modulatory effects at the skeletal neuromuscular junction,[93-96] suggesting a physiological role.

Stimulation of the parasympathetic innervation of the bladder of many species results in a contraction which is only partially blocked by atropine, and it has been suggested that the residual response is due to the action of a noncholinergic transmitter,[97] possibly ATP since the residual contraction could be blocked by ANAPP₃ or α,β-methylene-ATP. This is true for the bladder of the cat,[98] guinea pig,[99-103] rat,[99] ferret and marmoset,[102] and rabbit.[104] It was shown that ATP and its analogs closely mimicked the atropine-resistant contraction[97-99,101,105,106] and that repeated exposure to ATP produced tachyphylaxis to its own responses and attenuation of the neurogenic contraction.[99] This evidence suggested an involvement of ATP in the neurogenic contractile response of the bladder. The origin of the ATP (purinergic, adrenergic, or cholinergic nerve) is unclear, although it has been suggested that it is released from cholinergic nerves in the bladder, since the residual NANC response was abolished by botulinum toxin type A, which acts on cholinergic nerves.[107]

IV. ATP AS A TRANSMITTER IN THE CENTRAL NERVOUS SYSTEM

Although the evidence is less extensive than that obtained with the peripheral nervous system, there are some intriguing results suggesting that ATP may function as a transmitter or cotransmitter in the CNS, particularly in regard to sensory modalities. Application of exogenous ATP has been shown to produce a variety of excitatory effects in the CNS,[13] and there are a number of studies showing the electrically induced release of ATP from brain.[13,108-111] Although it has not been established whether or not release of ATP can occur exclusive of other transmitters, it does seem that ATP can be released from central cholinergic nerves. Richardson and Brown[112] have shown, for example, that depolarization with K^+ can cause the release of ATP from affinity purified cholinergic nerve terminals isolated from the caudate nucleus of the rat.

Perhaps the strongest argument for a central transmitter function of ATP comes from studies in the spinal cord. Exogenous ATP produces responses in neurons in the dorsal horn of the spinal cord that mimic the actions of certain primary afferent nerves, especially those associated with nociception and mechanoreception.[113-116] White and colleagues[117] subsequently showed that stimulants such as K^+ and veratridine cause the release of ATP from synaptosomes prepared from spinal cord. Interestingly, the release of ATP from synaptosomes prepared from dorsal horn tissue exceeded that from synaptosomes prepared from ventral horn tissue. These studies lend credence to the early suggestion of the Holtons[3,4] that ATP could be a transmitter released at central terminals of sensory afferent fibers.

V. ATP AS A NEUROMODULATOR

The evidence is substantial that adenine nucleosides and nucleotides, such as adenosine and ATP, can reduce the nerve stimulation-induced release of neurotransmitters from both central and peripheral nerves.[13,96,118-120] The general concept has been that this action is mediated by P_1 purinoceptors. Because ATP is a poor P_1 purinoceptor agonist, while adenosine is a good P_1 purinoceptor agonist, the action of the nucleotide is suggested to occur indirectly, that is, ATP is metabolized by ectonucleotidases to adenosine, which then acts via P_1 purinoceptors. Although some of the effects of ATP could be due to formation of adenosine, there is a growing body of evidence indicating that nucleotides can act per se to inhibit transmitter release without first being degraded to adenosine.[118,122-128] The knowledge that nucleotides can act per se has led to a reexamination of the nature of the purinoceptor that mediates an inhibition of transmitter release. Shinozuka et al.[123,125] have suggested, based on studies with a number of purinoceptor agonists and antagonists in the rat tail artery, that the prejunctional receptor is different from either P_1 or P_2 purinoceptors. As a result, they proposed a third purinoceptor, the P_3 purinoceptor.

Evidence for a unique prejunctional purinoceptor is not restricted to sympathetic nerves. Wiklund, Gustafsson, and Lundin[118] have reported results concerning the regulation of ACh release from parasympathetic nerves by adenine nucleotides and nucleosides that are similar in many respect to those obtained concerning the regulation of NE release in blood vessels. Ribeiro and Sebastião[96,126,127] have suggested that the prejunctional purinoceptors on somatic motor nerve terminals may be unique as well.

Regardless of the specific details of the receptor subtype, it does seem that, at least in autonomic nerves, ATP can act per se as a modulator of neurotransmitter release. This neuromodulatory function of ATP may prove to be just as important as a neurotransmitter function. Indeed, the potential for a neuromodulatory role of ATP, especially in blood vessels, maybe even greater than originally recognized. Not only is there release of ATP from sympathetic nerves in blood vessels, but there is an endothelium-dependent release of large amounts of ATP in response to α-adrenoceptor stimulation.[128]

REFERENCES

1. **Burnstock, G.,** Recent concepts of chemical communication between excitable cells, in *Dale's Principle and Communication Between Neurones,* Osborne, N. N., Ed., Pergamon Press, Oxford, 1983, 7.

2. **Burnstock, G.,** Mechanism of interaction of peptide and non-peptide vascular neurotransmitter systems, *J. Cardiovasc. Pharmacol.,* 10 (Suppl. 12), S74, 1987.

3. **Holton, F. A. and Holton, P.,** The capillary dilator substances in dry powders of spinal roots: a possible role of adenosine triphosphate in chemical transmission from nerve endings, *J. Physiol.,* 126, 124, 1954.

4. **Holton, P.,** The liberation of adenosine triphosphate on antidromic stimulation of sensory nerves, *J. Physiol.,* 145, 494, 1959.

5. **Burnstock, G., Campbell, G., Bennett, M., and Holman, M. E.,** The effects of drugs on the transmission of inhibition from autonomic nerves to the smooth muscle of the guinea-pig taenia coli, *Biochem. Pharmacol.,* 12 (Suppl.), 134, 1963.

6. **Burnstock, G., Campbell, G., Bennett, M., and Holman, M. E.,** Inhibition of the smooth muscle of the taenia coli, *Nature,* 200, 581, 1963.

7. **Burnstock, G., Campbell, G., Bennett, M., and Holman, M. E.,** Innervation of the guinea-pig taenia coli: are there intrinsic inhibitory nerves which are distinct from sympathetic nerves?, *Int. J. Neuropharmacol.,* 3, 163, 1964.

8. **Burnstock, G., Campbell, G., Satchell, D., and Smythe, A.,** Evidence that adenosine triphosphate or a related nucleotide is the transmitter substance released by non-adrenergic inhibitory nerves in the gut, *Br. J. Pharmacol.,* 40, 668, 1970.

9. **Su, C., Bevan, J. A., and Burnstock, G.,** Release of [^{3}H]-ATP during stimulation of enteric nerves, *Science,* 173, 336, 1971.

10. **Burnstock, G.,** Purinergic nerves, *Pharmacol. Rev.,* 24, 509, 1972.

11. **Westfall, D. P., Hogaboom, G. K., Colby, J., O'Donnell, J. P., and Fedan, J. S.,** Direct evidence against a role of ATP as the non-adrenergic, non-cholinergic inhibitory neurotransmitter in guinea-pig taenia coli, *Proc. Natl. Acad. Sci. U.S.A.,* 79, 7041, 1982.

12. **Campbell, G. and Gibbins, I. L.,** Nonadrenergic, noncholinergic transmission in the autonomic nervous system: purinergic nerves, in *Trends in Autonomic Pharmacology,* Kalsner, S., Ed., Urban and Schwarzenberg, Baltimore, 1979, 103.

13. **Phillis, J. W. and Wu, P. H.,** The role of adenosine and its nucleotides in central synaptic transmission, *Prog. Neurobiol.,* 16, 187, 1981.

14. **Burnstock, G.,** Do some nerve cells release more than one transmitter?, *Neuroscience,* 1, 239, 1976.

15. **Cuello, A. C., Ed.,** *Cotransmission,* Macmillan, London, 1982.

16. **Westfall, D. P., Sedaa, K. O., Shinozuka, K., Bjur, R. A., and Buxton, I. L. O.,** ATP as a cotransmitter, in *Annals of The New York Academy of Science,* Dubyak, G. R. and Fedan, J. S., Eds., New York Acad. Sci., New York, 1990, 300.

17. **Lagercrantz, H.,** On the composition of large dense-cored vesicles in sympathetic nerves, *Neuroscience,* 1, 81, 1976.

18. **Kirkpatrick, K. and Burnstock, G.,** Sympathetic nerve-mediated release of ATP from the guinea-pig vas deferens is unaffected by reserpine, *Eur. J. Pharmacol.,* 138, 207, 1987.

19. **Lew, M. J. and White, T. D.,** Release of endogenous ATP during sympathetic nerve stimulation, *Br. J. Pharmacol.,* 92, 349, 1987.

20. **Levitt, B. L., Head, R. J., and Westfall, D. P.,** High pressure liquid chromatographic-fluorometric detection of adenosine and adenine nucleotides: application to endogenous content and electrically-induced release of adenyl purines in the guinea-pig vas deferens, *Anal. Biochem.,* 137, 93, 1984.

21. **Westfall, D. P., Sedaa, K. O., and Bjur, R. A.,** Release of endogenous ATP from rat caudal artery, *Blood Vessels,* 24, 125, 1987.

22. **Su, C.,** Neurogenic release of purine compounds in blood vessels, *J. Pharmacol. Exp. Ther.,* 204, 351, 1975.

23. **Westfall, D. P., Stitzel, R. E., and Rowe, J. N.,** The postjunctional effects and neural release of purine compounds in the guinea-pig vas deferens, *Eur. J. Pharmacol.,* 50, 27, 1978.

24. **Fedan, J. S., Hogaboom, G. K., Westfall, D. P., and O'Donnell, J. P.,** Comparison of contractions of the smooth muscle of the guinea-pig vas deferens induced by ATP and related nucleotides, *Eur. J. Pharmacol.,* 81, 193, 1982.

25. **Sneddon, P. and Westfall, D. P.,** Pharmacological evidence that ATP and noradrenaline are cotransmitters in the guinea-pig vas deferens, *J. Physiol.,* 347, 561, 1984.

26. **Sneddon, P., Westfall, D. P., Colby, J., and Fedan, J. S.,** A pharmacological investigation of the biphasic nature of the contractile response of the rabbit and rat vas deferens to field stimulation, *Life Sci.,* 35, 1903, 1984.

27. **Suzuki, H.,** Electrical responses of smooth muscle cells of the rabbit ear artery to adenosine triphosphate, *J. Physiol.,* 359, 401, 1985.

28. **Kennedy, C. and Burnstock, G.,** Evidence for two types of P_2-purinoceptor in longitudinal muscle of the rabbit portal vein, *Eur. J. Pharmacol.*, 111, 49, 1985.

29. **Meldrum, L. A. and Burnstock, G.,** Evidence that ATP acts as a cotransmitter with noradrenaline in sympathetic nerves supplying the guinea-pig vas deferens, *Eur. J. Pharmacol.* 92, 161, 1983.

30. **Sneddon, P. and Burnstock, G.,** Inhibition of excitatory junction potentials in guinea-pig vas deferens by α,β-methylene ATP: further evidence for ATP and noradrenaline as cotransmitters, *Eur. J. Pharmacol.*, 100, 85, 1984.

31. **Kugelgen, I. V. and Starke, K.,** Noradrenaline and adenosine triphosphate as cotransmitters of neurogenic vasoconstriction in rabbit mesenteric artery, *J. Physiol.*, 367, 435, 1985.

32. **Kennedy, C., Saville, V. L., and Burnstock, G.,** The contributions of noradrenaline and ATP to the responses of the rabbit central ear artery to sympathetic nerve stimulation depend on the parameters of stimulation, *Eur. J. Pharmacol.*, 122, 291, 1986.

33. **Burnstock, G. and Warland, J. J. I.,** A pharmacological study of the rabbit saphenous artery *in vitro*: a vessel with a large purinergic contractile response to sympathetic nerve stimulation, *Br. J. Pharmacol.*, 90, 111, 1987.

34. **Warland, J. J. I. and Burnstock, G.,** Effects of reserpine and 6-hydroxydopamine on the adrenergic and purinergic components of sympathetic nerve responses of the rabbit saphenous artery, *Br. J. Pharmacol.*, 92, 871, 1987.

35. **Ramme, D., Regenold, J. T., Starke, K., Busse, R., and Illes, P.,** Identification of the neuroeffector transmitter in jejunal branches of the rabbit mesenteric artery, *Naunyn Schmeideberg's Arch. Pharmacol.*, 336, 267, 1987.

36. **Muir, T. C. and Wardle, K. A.,** The electrical and mechanical basis of cotransmission in some vascular and non-vascular smooth muscles, *J. Autonom. Pharmacol.*, 8, 203, 1988.

37. **Muramatsu, I.,** Evidence for sympathetic purinergic transmission in the mesenteric artery of the dog, *Br. J. Pharmacol.*, 87, 478, 1986.

38. **Muramatsu, I.,** The effect of reserpine on sympathetic purinergic neurotransmission in the isolated mesenteric artery of the dog: a pharmacological study, *Br. J. Pharmacol.*, 91, 467, 1987.

39. **Muramatsu, I., Ohmura, T., and Oshita, M.,** Comparison between sympathetic adrenergic and purinergic neurotransmission in the dog mesenteric artery, *J. Physiol.*, 411, 227, 1989.

40. **Pearson, J. D.,** Ectonucleotidases: measurement of activity and use of inhibitors, in *Methods in Pharmacology, Vol. 6*, Paton, D. M., Ed., Plenum Press, New York, 1985, 83.

41. **White, T. D.,** Role of adenine compounds in autonomic neurotransmission, *Pharmacol. Ther.*, 38, 129, 1988.

42. **Douglas, W. W. and Poisner, A. M.,** On the relation between ATP splitting and secretion in the adrenal chromaffin cell: extrusion of ATP (unhydrolysed) during the release of catecholamines, *J. Physiol.*, 183, 249, 1966.

43. **Stevens, P., Robinson, R. L., Van Dyke, K., and Stitzel, R.,** Studies on the synthesis and release of adenosine triphosphate-^{3}H in the isolated perfused cat adrenal gland, *J. Pharmacol. Exp. Ther.*, 181, 463, 1972.

44. **Lagercrantz, H. and Stjarne, L.,** Evidence that a major proportion of the noradrenaline is stored without ATP in sympathetic large dense core vesicles, *Nature*, 249, 843, 1974.

45. **Lagercrantz, H., Fried, G., and Dahlin, I.,** An attempt to estimate the in vivo concentrations of noradrenaline and ATP in sympathetic large dense core nerve vesicles, *Acta Physiol. Scand.*, 94, 136, 1975.

46. **Fried, G., Lagercrantz, H., and Hokfelt, T.,** Improved isolation of small noradrenergic vesicles from rat seminal ducts following castration. A density gradient centrifugation and morphological study, *Neuroscience*, 3, 1271, 1978.

47. **Burnstock, G.,** Neural nomenclature, *Nature*, 229, 282, 1971.

48. **Langer, S. Z. and Pinto, J. E. B.,** Possible involvement of a transmitter different from norepinephrine in the residual responses of the cat nictitating membrane after pretreatment with reserpine, *J. Pharmacol. Exp. Ther.*, 196, 697, 1976.

49. **Falck, B.,** Observations on the possibilities of the cellular localization of monoamines by a fluorescence method, *Acta Physiol. Scand.*, 56 (Suppl. 197), 1, 1962.

50. **Sjostrand, N. O.,** Effect of reserpine and hypogastric denervation on the noradrenaline content of the vas deferens of the guinea-pig, *Acta Physiol. Scand.*, 56, 376, 1962.

51. **Swedin, G.,** Biphasic mechanical response of the isolated vas deferens to nerve stimulation, *Acta Physiol. Scand.*, 81, 574, 1971.

52. **Hukovic, S.,** Responses of the isolated sympathetic nerve-ductus deferens preparation of the guinea-pig, *Br. J. Pharmacol.*, 16, 188, 1961.

53. **Birmingham, A. T. and Wilson, A. B.,** Preganglionic and postganglionic stimulation of the guinea-pig isolated vas deferens preparation, *Br. J. Pharmacol.*, 21, 569, 1963.

54. **Ambache, N. and Zar, M. A.,** Evidence against adrenergic motor transmission in the guinea-pig vas deferens, *J. Physiol.*, 216, 359, 1971.

55. **McGrath, J. C.,** Adrenergic and non-adrenergic components in the contractile response of the vas deferens to a single indirect stimulus, *J. Physiol.,* 283, 23, 1978.

56. **Fedan, J. S., Hogaboom, G. K., O'Donnell, J. P., Colby, J., and Westfall, D. P.,** Contribution by purines to the neurogenic response of the vas deferens of the guinea-pig, *Eur. J. Pharmacol.,* 69, 41, 1981.

57. **Stjarne, L. and Astrand, P.,** Relative pre- and post-junctional roles of noradrenaline and adenosine 5'-triphosphate as neurotransmitters of the sympathetic nerves of the guinea-pig and mouse vas deferens, *Neuroscience,* 14, 929, 1985.

58. **Allcorn, R. J., Cunnane, T. C., and Kirkpatrick, K.,** Actions of α,β-methylene ATP and 6-hydroxy-dopamine on sympathetic neurotransmission in the vas deferens of the giunea-pig, rat and mouse: support for cotransmission, *Br. J. Pharmacol.,* 89, 647, 1986.

59. **Sneddon, P., Westfall, D. P., and Fedan, J. S.,** Cotransmitters in the motor nerves of the guinea-pig vas deferens: electrophysiological evidence, *Science,* 218, 693, 1982.

60. **Khoyi, M. A., Westfall, D. P., Buxton, I. L. O., Akhtar-Khavari, F., Rezaei, E., Salaices, M., and Sanchez-Garcia, P.,** Norepinephrine and potassium-induced calcium translocation in rat vas deferens, *J. Pharmacol. Exp. Ther.,* 246, 917, 1988.

61. **Forsyth, K. M. and Pollock, D.,** Clonidine and morphine increase [^{3}H]-noradrenaline overflow in mouse vas deferens, *Br. J. Pharmacol.,* 93, 35, 1988.

62. **Trachte, G. T.,** Angiotensin effects on vas deferens adrenergic and purinergic neurotransmission, *Eur. J. Pharmacol.,* 146, 261, 1988.

63. **Trachte, G. T., Binder, S. B., and Peach, M. J.,** Indirect evidence for separate vesicular neuronal origins of norepinephrine and ATP in the rabbit vas deferens, *Eur. J. Pharmacol.,* 164, 425, 1989.

64. **Ellis, J. L. and Burnstock, G.,** Angiotensin neuromodulation of adrenergic and purinergic co-transmission in the guinea-pig vas deferens, *Br. J. Pharmacol.,* 97, 1157, 1989.

65. **Burnstock, G.,** Dual control of blood pressure by purines, in *Pharmacology,* Rand, M. J. and Raper, C., Eds., Elsevier Science Publishers (Biomedical Division), New York, 1987, 245.

66. **Burnstock, G.,** Present status for purinergic neurotransmission—implications for vascular control, in *Neuronal Messengers in Vascular Function,* Nobin, A., Owman, C., and Arneklo-Nobin, B., Eds., Elsevier Science Publishers (Biomedical Division), New York, 1987, 327.

67. **Burnstock, G.,** Sympathetic, purinergic transmission in small blood vessels, *Trends Pharmacol. Sci.,* 9, 116, 1988.

68. **Holman, M. E. and Suprenant, A. M.,** An electrophysiological analysis of the effects of noradrenaline and α-adrenoceptor antagonists on neuromuscular transmission in mammalian muscular arteries, *Br. J. Pharmacol.,* 71, 651, 1980.

69. **Cheung, D. W.,** Two components in the cellular response of rat tail arteries to nerve stimulation, *J. Physiol.,* 328, 461, 1982.

70. **Suzuki, H. and Kou, K.,** Electrical components contributing to the nerve-mediated contractions of the rabbit ear artery, *Jpn. J. Physiol.,* 33, 743, 1983.

71. **Sneddon, P. and Burnstock, G.,** ATP as a cotransmitter in rat tail artery, *Eur. J. Pharmacol.,* 106, 149, 1984.

72. **Kuriyama, H. and Makita, Y.,** Modulation of noradrenergic transmission in the guinea-pig mesenteric artery: an electrophysiological study, *J. Physiol.,* 335, 609, 1983.

73. **Mishima, S., Miyahara, H., and Suzuki, H.,** Transmitter release modulated by α-adrenoceptor antagonists in the rabbit mesenteric artery: a comparison between noradrenaline outflow and electrical activity, *Br. J. Pharmacol.,* 83, 537, 1984.

74. **Ishikawa, S.,** Actions of ATP and α,β-methylene ATP on neuromuscular transmission and smooth muscle membrane of the rabbit and guinea-pig mesenteric arteries, *Br. J. Pharmacol.,* 86, 777, 1985.

75. **Angus, J. A., Broughton, A., and Mulvany, M. J.,** Role of α-adrenoceptors in constrictor responses of rat, guinea-pig and rabbit small arteries to neural activation, *J. Physiol.,* 403, 495, 1988.

76. **Muramatsu, I., Kigoshi, S., and Oshita, M.,** Nonadrenergic nature of prazosin-resistant, sympathetic contraction in the dog mesenteric artery, *J. Pharmacol. Exp. Ther.,* 229, 532, 1984.

77. **Machaly, M., Dalziel, H. H., and Sneddon, P.,** Evidence for ATP as a cotransmitter in dog mesenteric artery, *Eur. J. Pharmacol.,* 147, 83, 1988.

78. **Omote, S., Kigoshi, S., and Muramatsu, I.,** Selective inhibition by nifedipine of the purinergic component of neurogenic vasoconstriction in the dog mesenteric artery, *Eur. J. Pharmacol.,* 160, 239, 1989.

79. **Hirst, G. D. S. and Neild, T. O.,** Evidence for two populations of excitatory receptors for noradrenaline on arterial smooth muscle, *Nature,* 283, 767, 1980.

80. **Suzuki, H., Mishima, S., and Miyahara, H.,** Effects of reserpine treatment on electrical responses induced by perivascular nerve stimulation in the rabbit ear artery, *Biomed. Res.,* 5, 259, 1984.

81. **Saville, V. L. and Burnstock, G.,** Use of reserpine and 6-hydroxydopamine supports evidence for purinergic cotransmission in the rabbit ear artery, *Eur. J. Pharmacol.,* 155, 271, 1988.

82. **Cheung, D. W. and Fujioka, M.,** Inhibition of the excitatory junction potential in the guinea-pig saphenous artery by ANAPP$_3$, *Br. J. Pharmacol.,* 98, 3, 1986.

83. **Muramatsu, I., Fujiwara, M., and Shibata, S.,** Reactivity of isolated canine cerebral arteries to adenine nucleotides and adenosine, *Pharmacology,* 21, 198, 1980.

84. **Muramatsu, I., Fujiwara, M., Miura, A., and Sakakibara, Y.,** Possible involvement of adenine nucleotides in sympathetic neuroeffector mechanisms of dog basilar artery, *J. Pharmacol. Exp. Ther.,* 216, 401, 1981.

85. **Muramatsu, I. and Kigoshi, S.,** Purinergic and non-purinergic innervation in the cerebral arteries of the dog, *Br. J. Pharamcol.,* 92, 901, 1987.

86. **Brizzolara, A. L. and Burnstock, G.,** Evidence for noradrenergic-purinergic cotransmission in the hepatic artery of the rabbit, *Br. J. Pharmacol.,* 99, 835, 1990.

87. **Inoue, T. and Kannan, M. S.,** Nonadrenergic and noncholinergic excitatory neurotransmission in rat intrapulmonary artery, *Am. J. Physiol.,* 254, H1142, 1988.

88. **Bulloch, J. A. and McGrath, J. S.,** Blockade of vasopressor and vas deferens responses by α,β-methylene ATP in the pithed rat, *Br. J. Pharmacol.,* 94, 103, 1988.

89. **Taylor, E. M. and Parsons, M. E.,** Adrenergic and purinergic neurotransmission in arterial resistance vessels of the cat intestinal circulation, *Eur. J. Pharmacol.,* 164, 23, 1989.

90. **Schlicker, E., Urbanek, E., and Gothert, M.,** ATP, α,β-methylene ATP and suramin as tools for characterisation of vascular P_{2x}-receptors in the pithed rat, *J. Autonom. Pharmacol.,* 9, 357, 1989.

91. **Dalziel, H. H., Gray, G. A., Drummond, R. M., Furman, B. L., and Sneddon, P.,** Investigation of the selectivity of α,β-methylene ATP in inhibiting vascular responses of the rat in vitro and in vivo, *Br. J. Pharmacol.,* 99, 820, 1990.

92. **Silinsky, E. M.,** On the association between transmitter secretion and the release of adenine nucleotides from mammalian motor nerve terminals, *J. Physiol.,* 247, 145, 1975.

93. **Silinsky, E. M.,** Evidence for specific adenosine receptors at cholinergic nerve endings, *Br. J. Pharmacol.,* 71, 191, 1980.

94. **Akasu, T., Hirai, K., and Koketsu, K.,** Increase of acetylcholine receptor sensitivity by adenosine triphosphate; a novel action of ATP on Ach-sensitivity, *Br. J. Pharmacol.,* 74, 505, 1981.

95. **Ribeiro, J. A. and Sebastião, A. M.,** On the role inactivation and origin of endogenous adenosine at the frog neuromuscular junction, *J. Physiol.,* 384, 571, 1987.

96. **Sebastião, A. M. and Ribeiro, J. A.,** On the adenosine receptor and adenosine inactivation at the rat diaphragm neuromuscular junction, *Br. J. Pharmacol.,* 94, 109, 1988.

97. **Ambache, N. and Zar, M. A.,** Non-cholinergic transmission by post-ganglionic motor neurones in the mammalian bladder, *J. Physiol.,* 210, 761, 1970.

98. **Theobald, R. J.,** The effect of arylazido-aminopropionyl ATP on atropine-resistant contractions of the cat urinary bladder, *Life Sci.,* 32, 2479, 1983.

99. **Burnstock, G., Dumsday, B., and Smythe, A.,** Atropine-resistant excitation of the urinary bladder: the possibility of transmission via nerves releasing a purine nucleotide, *Br. J. Pharmacol.,* 44, 451, 1972.

100. **Westfall, D. P., Fedan, J. S., Colby, J., Hogaboom, G. K., and O'Donnell, J. P.,** Evidence for a contribution by purines to the neurogenic response of the guinea-pig urinary bladder, *Eur. J. Pharmacol.,* 87, 415, 1983.

101. **Kasakov, L. and Burnstock, G.,** The use of the slowly degradable analog, α,β-methylene ATP to produce desensitisation of the P_2-purinoceptor: effect on non-adrenergic, non-cholinergic responses of the guinea-pig urinary bladder, *Eur. J. Pharmacol.,* 86, 291, 1982.

102. **Moss, H. E. and Burnstock, G.,** A comparative study of electrical field stimulation of the guinea-pig, ferret and marmoset urinary bladder, *Eur. J. Pharmacol.,* 114, 311, 1985.

103. **Fujii, D.,** Evidence for adenosine triphosphate as an excitatory transmitter in guinea-pig, rabbit and pig urinary bladder, *J. Physiol.,* 404, 39, 1988.

104. **Dean, D. M. and Downie, J. W.,** Contribution of adrenergic and purinergic neurotransmission to contraction in rabbit detrussor, *J. Pharmacol. Exp. Ther.,* 207, 431, 1978.

105. **Brown, C., Burnstock, G., and Cocks, T.,** Effects of adenosine 5'-triphosphate (ATP) and β,γ-methylene ATP on the rat urinary bladder, *Br. J. Pharmacol.,* 65, 97, 1979.

106. **MacKenzie, I. and Burnstock, G.,** Neuropeptide action on the guinea-pig bladder; a comparison with the effects of field stimulation and ATP, *Eur. J. Pharmacol.* 105, 85, 1984.

107. **MacKenzie, I., Burnstock, G., and Dolly, J. O.,** The effects of purified botulinum toxin type A on cholinergic, adrenergic and non-adrenergic, atropine-resistant autonomic neuromuscular transmission, *Neuroscience,* 7, 997, 1982.

108. **White, T. D.,** Direct detection of depolarization-induced release of ATP from a synaptosomal preparation, *Nature,* 267, 67, 1977.

109. **White, T. D.,** Release of ATP from a synaptosomal preparation by elevated extracellular K^+ and by veratridine, *J. Neurochem.,* 30, 329, 1978.

110. **Potter, P. and White, T. D.,** Release of adenosine 5'-triphosphate from synaptosomes from different regions of the rat brain, *Neuroscience,* 5, 1351, 1980.

111. **MacDonald, W. F. and White, T. D.,** Nature of extrasynaptosomal accumulation of endogenous adenosine evoked by K$^+$ and veratridine, *J. Neurochem.,* 45, 791, 1985.

112. **Richardson, P. J. and Brown, S. J.,** ATP release from affinity-purified rat cholinergic nerve terminals, *J. Neurochem.,* 48, 622, 1987.

113. **Jahr, C. E. and Jessel, J. M.,** ATP excites a subpopulation of rat dorsal horn neurones, *Nature,* 304, 730, 1983.

114. **Salt, T. E. and Hill, R. G.,** Excitation of single sensory neurones in the rat caudal trigeminal nucleus by iontophoretically applied adenosine 5'-triphosphate, *Neurosci. Lett.,* 35, 53, 1983.

115. **Fyffe, R. W. and Perl, E. R.,** Is ATP a central synaptic mediator for certain primary afferent fibers from mammalian skin?, *Proc. Natl. Acad. Sci. U.S.A.,* 81, 6890, 1984.

116. **Salter, M. W. and Henry, J. L.,** Effect of adenosine 5'-monophosphate on functionally identified units in the cat spinal dorsal horn. Evidence for a differential effect of adenosine 5'-triphosphate on nociceptive and non-nociceptive units, *Neuroscience,* 3, 815, 1985.

117. **White, T. D., Downie, J. W., and Leslie, R. A.,** Characteristics of K$^+$- and veratridine-induced release of ATP from synaptosomes prepared from dorsal and ventral spinal cord, *Brain Res.,* 334, 372, 1985.

118. **Wiklund, N. P., Gustafsson, L. E., and Lundin, J.,** Pre- and postjunctional modulation of cholinergic neuroeffector transmission by adenine nucleotides. Experiments with agonist and antagonist, *Acta Physiol. Scand.,* 125, 681, 1985.

119. **Paton, D. M.,** Presynaptic neuromodulation mediated by purinergic receptors, in *Purinergic Receptors,* Burnstock, G., Ed., Chapman and Hall, London, 1981, 199.

120. **Su, C.,** Purinergic neurotransmission and neuromodulation, *Pharm. Rev.,* 23, 397, 1983.

121. **Lukacsko, P. and Blumberg, A.,** Modulation of the vasoconstrictor response to adrenergic stimulation by nucleosides and nucleotides, *J. Pharmacol. Exp. Ther.,* 222, 344, 1982.

122. **Husted, S. E. and Nedergaard, O. A.,** Dual inhibitory action of ATP on adrenergic neuroeffector transmission in rabbit pulmonary artery, *Acta Pharmacol. Toxicol.,* 57, 204, 1985.

123. **Shinozuka, K., Bjur, R. A., and Westfall, D. P.,** Characterization of prejunctional purinoceptors on adrenergic nerves of the rat caudal artery, *Naunyn-Schmeideberg's Arch. Pharmacol.,* 338, 221, 1988.

124. **Kugelgen, I. V., Schoffel, E., and Starke, K.,** Inhibition by nucleotides acting at presynaptic P$_2$-receptors of sympathetic neuroeffector transmission in the mouse isolated vas deferens, *Naunyn-Schmeideberg's Arch. Pharmacol.,* 340, 522, 1989.

125. **Shinozuka, K., Bjur, R. A. and Westfall, D. P.,** Effects of α,β-methylene ATP on the prejunctional purinoceptors of the rat caudal artery, *J. Pharmacol. Exp. Ther.,* 254, 900, 1990.

126. **Ribeiro, J. A. and Sebastião, A. M.,** On the type of receptor involved in the inhibitory action of adenosine at the neuromuscular junction, *Br. J. Pharmacol.,* 84, 911, 1985.

127. **Ribeiro, J. A. and Sebastião, A. M.,** Adenosine receptors and calcium: basis for proposing a third (A$_3$) adenosine receptor, *Prog. Neurobiol.,* 26, 179, 1986.

128. **Sedaa, K. O., Bjur, R. A., Shinozuka, K., and Westfall, D. P.,** Nerve and drug-induced release of adenine nucleosides and nucleotides from rabbit aorta, *J. Pharmacol. Exp. Ther.,* 252, 1060, 1990.

Chapter 25

LOCALIZATION AND ACTIONS OF ADENOSINE RECEPTORS IN THE MAMMALIAN RETINA

Christine Blazynski and Maria-Thereza R. Perez

TABLE OF CONTENTS

I. INTRODUCTION

The vertebrate retina provides an excellent model system with which to study information processing. The tissue is easily isolated as an intact piece of central nervous system (CNS), 100 to 500 μm in thickness, containing all the elements required for the collection, processing, and transmission of complicated visual information, such as movement and contrast, to higher visual centers. The morphology and extensive interconnections established by retinal cells are known in great detail, and the presence of neuroactive agents found in other parts of the CNS has been confirmed (for review see Massey and Redburn).[1] Moreover, many transmitter systems have been localized to particular classes of retinal cells, providing a neurochemical framework for mapping the pathways of information flow through this tissue. Because of the well-defined anatomy, neurochemistry, and synaptic physiology, many investigators utilize the retina to study interactions that occur between transmitter systems to process visual information (see Daw et al.[2] for review). We have begun to take this approach to elucidate the role of adenosine in retinal information processing.

A. RETINAL MORPHOLOGY AND CIRCUITRY

Retinal neurons are linearly arranged so that, when viewing the retina in transverse sections, one can readily distinguish the layers formed by cell bodies and their synapses (Figure 1). The first-order neurons, photoreceptors, ideally organized for the detection of light and transduction of this signal into electrical impulses, consist of the specialized outer portions (outer segments and inner segments), cell bodies (outer nuclear layer), and synaptic terminals (outer plexiform layer [OPL]). Two types of photoreceptors, rods and cones, detect light of relatively dim or bright intensity, respectively, and synapse with processes from two types of neurons in the OPL, the horizontal and the bipolar cells, with their soma located in the inner nuclear layer (INL). Horizontal cells make very extensive lateral connections throughout the OPL, while bipolar cells transmit signals from the OPL to the inner retina. Bipolars that connect with rods terminate deep in the inner plexiform layer (IPL) with synapses onto amacrine cells, while cone bipolars synapse onto amacrine and ganglion cells in one of five sublaminae within the IPL. Amacrine cells synapse with bipolar, ganglion and other amacrine cells, providing extensive interconnections. Amacrine cells are found both in the INL (conventionally placed) and in the ganglion cell layer (GCL; displaced amacrines). Ganglion cells, the retinal output neurons, send complex information from the retina to the brain in the form of frequency-dependent spikes. The retina also possesses a major glial cell type, the Müller cell, with processes extending radially from the inner to the outer portions of the retina.

Within the retina are two tonically active circuits, designated ON and OFF. Photoreceptors drive both circuits; these first-order neurons maximally release their transmitter in the dark, and light diminishes the release. Second-order bipolar cells either are depolarized (OFF) or hyperpolarized (ON) in the dark, and light reverses the membrane potentials of the cells. These cells synapse with the output cells of the retina, the ganglion cells. In addition to this direct pathway for information flow are two major levels of lateral processing. In the OPL, horizontal cells connect with one another and feed back onto photoreceptors, while in the IPL, amacrine cell processes convey information from, and to, many cell types. Convergence of the ON and OFF pathways can be demonstrated both in the INL as well as in the GCL. However, information from both circuits ultimately converges at the ganglion cells, resulting in cells that can discriminate directional movement and contrasts.

B. RETINAL TRANSMITTERS

Across species, it has been demonstrated that, in general, particular classes of cells are labeled for distinctive neurotransmitter markers,[1,3] as summarized in Figure 1. It should be noted that some differences are observed between species. For example, interplexiform cells have been labeled for tyrosine hydroxylase-like immunoreactivity (a dopamine marker) in a number of animals such as rat[4,5] and mouse,[6,7] while in the rabbit this has not been demonstrated. And while some species, such as cat and rabbit, specifically accumulate [³H]-serotonin into amacrine cells,[8,9] other mammalian species as demonstrated for rat (Long-Evans) by Churchill et al.[10] and Redburn and Mitchell[11] accumulate [³H]-serotonin within photoreceptor terminals. Electrophysiological studies have begun to elucidate the role of these classical transmitters in retinal information processing (reviewed in Daw et al).[2]

As discussed by Massey and Redburn,[1] immunoreactivity to a number of peptides has been detected in retinas of many species. For almost all reports, immunoreactivity is localized to amacrine cell bodies and their terminals. As demonstrated by Lam and colleagues,[12-15] neuropeptides have been colocalized with each other and with classical neurotransmitters.[12,13,15,19] The functional significance of neuromodulation by peptides in information processing is not as well known.

Recently, evidence has been accumulating which suggests that the purine nucleoside, adenosine, may play a role in retinal processing. The majority of this data has been obtained in mammalian retinas, primarily the rabbit. This evidence is presented in the following discussion.

II. BIOCHEMICAL, PHARMACOLOGIC, AND PHYSIOLOGIC EVIDENCE FOR ADENOSINE RECEPTORS

Two types of extracellular adenosine receptors have been identified in plasma membranes, A_1 and A_2, originally based on their coupling to adenylate cyclase via either inhibitory or stimulatory G proteins, respectively.[16,17] Ligand binding studies have more recently been employed to distinguish receptor subtypes. The A_1 receptor has a high affinity for N^6-substituted derivatives such as N^6-phenylisopropyladenosine (PIA) and N^6-cyclohexyladenosine (CHA). At the low-affinity A_2 receptor, N-ethylcarboxamidoadenosine (NECA) is more potent than R-PIA and CHA. Methylxanthines are potent antagonists at both types of receptors.

Both A_1 and A_2 receptors have been detected and characterized in mammalian retinal homogenates by both biochemical and pharmacologic techniques.[18-20] As illustrated in Figure 2, the mixed A_1-A_2 agonist, NECA, elicits a concentration-dependent inhibition and stimulation of adenylate cyclase in homogenates of rabbit retina, providing direct evidence for the presence of both classes of adenosine receptors. Using incubated rabbit retinas, Blazynski and colleagues[18] have shown that 10 to 100 μM adenosine elicited a 10- to 14-fold increase in cAMP levels; this modulation was antagonized with 3-isobutyl-1-methylxanthine (IBMX). Schorderet[20] has also reported modulation of cAMP levels in incubated retinas by micromolar concentrations of adenosine, R-PIA, and 2-chloroadenosine (CADO).

A_2 receptor-mediated increases in cAMP have been detected in cultures of human and porcine retinal pigmented epithelium (RPE),[21] a layer of epithelium located distal to photoreceptors which is involved in the maintenance of the vitamin A cycle. In intact RPE cells, 100 nM elicited an increase in cAMP, while 1 μM agonist was needed to detect a significant cAMP increase in membrane preparations. Higher concentrations of the A_1-selective agonists R-PIA or N^6-cyclopentyladenosine (CPA) were required to elicit a similar increase in cAMP levels. 8-Cyclopentyltheophylline and 8-sulfophenyltheophylline antagonized the cAMP increases elicited with all of the adenosine analogs.

To date there are no published reports of potencies of various adenosine agonists or antagonists at displacing specific binding of A_1-selective radioligands to mammalian retinal adenosine receptors. However, as discussed later, specific binding sites for [¹²⁵I]-PIA, [³H]-PIA, and [³H]-CHA have been localized in a number of mammalian retinas[22,23] and in one

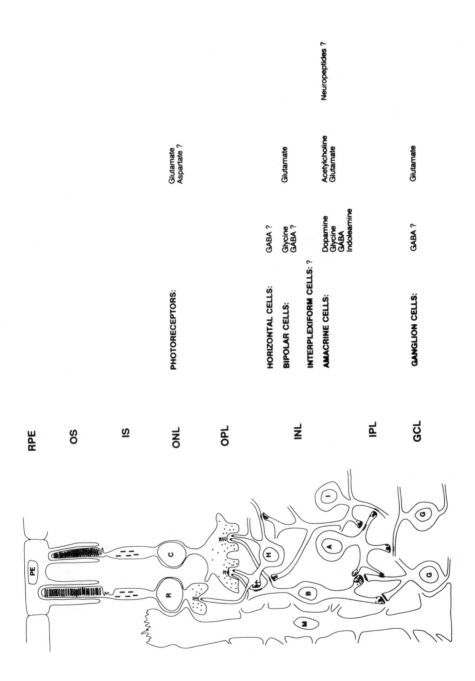

PHOTORECEPTORS: Glutamate
                    Aspartate ?

HORIZONTAL CELLS: GABA ?

BIPOLAR CELLS: Glycine   Glutamate
                    GABA ?

INTERPLEXIFORM CELLS: ?

AMACRINE CELLS: Dopamine   Acetylcholine   Neuropeptides ?
                    Glycine   Glutamate
                    GABA
                    Indoleamine

GANGLION CELLS: GABA ?   Glutamate

RPE
OS
IS
ONL
OPL
INL
IPL
GCL

FIGURE 1. Anatomy of the retina and identification of transmitters localized to cell populations. The rabbit retina, as observed in all vertebrate retinas, has a characteristic layered arrangement of cell bodies and processes. The photoreceptors rods (R) and cones (C), comprise the outer portion of the neural retina and can be differentiated into the following parts or layers: the outer segments (OS) containing the rhodopsin-laden stacks of membranes, the inner segments (IS) containing many mitochondria, the outer nuclear layer (ONL) consisting of the cell bodies of the photoreceptors, and their terminals, situated in the outer plexiform layer (OPL). The inner retina consists of the cell bodies and processes of second- and third-order neurons (bipolar cells, B; horizontal cells, H; amacrine cells, A; and interplexiform cells, I). The second-order cells have their cell bodies situated in the inner nuclear layer (INL), and these cells receive input from the OPL and transmit information to the innerplexiform layer (IPL). Finally, the ganglion cells (G), whose axons converge to form the optic nerve, are situated in the ganglion cell layer (GCL). The retinal glial cell, the Müller cell (M), extends throughout the retina. The variety of transmitters that have been identified in specific subpopulations of retinal cell types is listed (reviewed in Massey and Redburn,[1] Ehinger and Dowling,[2] and Daw et al.[3]).

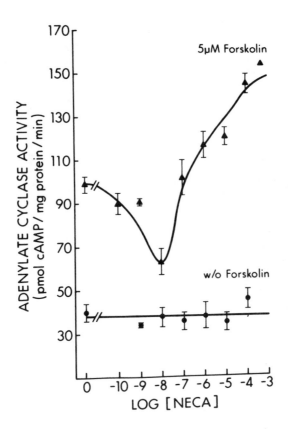

FIGURE 2.   Modulation of adenylate cyclase activity in rabbit retinal homogenates by the mixed A_1-A_2 agonist NECA. Forskolin-activated (submaximal, 10 μM) adenylate cyclase activity was measured in the presence of the indicated concentrations of NECA, using rolipram as a phosphodiesterase inhibitor. Biphasic effects on activity were elicited by NECA in a concentration-dependent manner. The addition of 10 μM IBMX completely antagonized both inhibition and stimulation elicited by NECA (data not shown). In the absence of forskolin, modulation of enzyme activity was not observed.

of these studies it was stated that the relative potencies of adenosine analogs in displacing the binding of [^{125}I]-PIA was consistent with the characteristics of A_1 receptor binding reported for brain.[22] Recently, similar studies have been done using membranes prepared from bovine retinas (Blazynski and Woods, in preparation). Both high- and low-affinity binding sites were detected (K_d of 0.14 and 30 nM, respectively), and the rank order of potency of displacement of [^{3}H]-PIA binding is CPA = R-PIA > CHA > NECA.

It has also been reported that adenosine elicits an increase in glycogen hydrolysis;[24] it appears that this response is mediated via agonist binding to extracellular adenosine receptors, located possibly on glial cells. Identification of the receptor subtype is unclear, since the rank order of potency for this effect was reported as CADO = adenosine > S-PIA > R-PIA > NECA > CHA.

The effects of adenosine on physiologic responses to light have been assessed in an arterially perfused cat eye.[25,26] Micromolar concentrations of adenosine induced striking changes in the response characteristics of an extracellular recording ascribed to the retinal pigmented epithelium (reflected in an increase in the standing potential and depression of the light peak), in the b-wave of the electroretinogram (originating in the inner retina), and a depression of all components of the optic nerve response. These observations clearly indicate that the effects involve activation of adenosine receptors located in different retinal regions. In addition, adenosine was found to promote a slight increase in the perfusion flow rate, reflecting a vasodilatory action of adenosine. However, it appears that vasodilation is only partially responsible for the electrophysiological effects described above.

III. LOCALIZATION OF PURINERGIC MARKERS

Perhaps the best evidence supporting the hypothesis that adenosine plays a role in retinal information flow is provided by morphologic studies. Much of this data, determined in mammalian retinas, has been summarized in Table 1. There is extremely good agreement between the localizations of endogenous stores of adenosine, detected by immunohisto-chemistry,[22,27] and accumulated stores of [^{3}H]-adenosine (or [^{3}H]-PIA or [^{3}H]-CHA).[19,26-29] These two markers, as well as immunoreactivity to adenosine deaminase,[31] involved in the breakdown of adenosine, are all localized within the inner retina and label the majority of cell bodies in the GCL and a large number of cells residing in the proximal half of the INL (closest to the IPL). Glial uptake of adenosine has been reported.[28] Virtually all of the specific binding of A_1-selective radioligands occurs in the inner retina, particularly in the IPL, thereby strongly suggesting that adenosine exerts its effect via A_1 receptors in this region where a great deal of processing occurs. As discussed below, the localization results are quite consistent with the idea that adenosine is involved in regulating the release of neurotransmitters.

Specific binding sites for the mixed A_1-A_2 agonist, [^{3}H]-NECA, are also present in the mammalian retina, predominantly over the outer retina.[23] The relevance of these putative A_2 receptors to retinal physiology has not yet been elucidated. Interestingly, it has recently been reported that the retinal pigmented epithelium also contains these receptors.[21] Additional experiments have demonstrated that binding in the inner retina represents A_1 binding sites.[23]

IV. ADENOSINE-NEUROTRANSMITTER INTERACTIONS

A. ADENOSINE RELEASE
1. K$^+$- and Light-Evoked Release

As subpopulations of retinal cells accumulate exogenous adenosine, endogenous adenosine may well be released from these cells. Indeed, both light- and K$^+$-evoked[35] release of endogenous and [^{3}H]-adenosine derivatives have been reported by Perez and colleagues.[35] Whereas endogenous retinal purines are detected in the form of phosphorylated intermediates, the superfusate from the rabbit retina contains hypoxanthine and inosine (80%), with the remainder comprised of approximately equal amounts of adenosine, xanthine, ATP and ADP.

The K$^+$-evoked release of purines is rapid in onset, with maximal release observed within the period of stimulation, and basal levels of release observed shortly after cessation of stimulus.[35] At K$^+$ concentrations between 8.6 to 23.6 mM, purine release was dramatically reduced in Ca^{++}-free medium.[35,36] At higher concentrations of K$^+$, only a partial reduction in release was observed when Ca^{++} was excluded. The proportions of purines and bases in the superfusate are identical after stimulation-induced release, suggesting that stimulation results in an increased rate of the basal release processes. Increases in the absolute levels of adenosine could be affected by including in the superfusate the adenosine deaminase inhibitor, erythro-9-(2-hydroxy-3-nonyl)adenine, (EHNA) indicating that some of the inosine and hypoxanthine are formed from released adenosine.[35] Little or no adenosine was formed from ATP breakdown via ectoenzymes, as demonstrated using the 5'-nucleotidase inhibitor, AOPCP (α,β-methylene AD).[34]

There are also indications that part of the release may occur by facilitated diffusion. The addition of the purine nucleoside transport inhibitor dipyridamole produced a small,

TABLE 1
Localization of Purinergic Markers

Marker	OS—IS	ONL—OPL	INL	IPL	GCL	Ref.
			Retinal layers			
[³H] Adenosine uptake	None	None	Many cell bodies in proximal (vitreal) half of layer labeled	Very sparse label	Many large cell bodies labeled	19,26—29
Adenosine-like immunoreactivity	None except sparse labeling in monkey and human tissue	None except sparse labeling in monkey and human tissue	Many cell bodies in proximal half labeled	Occasional to dense labeling of processes	Many cell bodies of both ganglion cells and displaced amacrine cells labeled	22,27,30
Adenosine deaminase-like immunoreactivity	None	None	Cell bodies located near the IPL border labeled	Three discrete sublamina labeled	Cell bodies of amacrine cells labeled	31
5'-Nucleotidase	Heaviest reaction product	Positive	Positive	Positive	Positive	32,33
A₁ receptor binding	Sparse	Sparse	High density	Highest density in all species with the exception of human	High density; human, highest density	22,23
[³H]NECA binding sites	Highest density of neural retina	High density	Very low density; displaceable with 10 nM CHA or NEM pretreatment	Very low density; displaceable with 10 nM CHA or NEM pretreatment	Very low density	23

Note: OS, outer segments; IS, inner segments; ONL, outer nuclear layer; OPL, outer plexiform layer; INL, inner nuclear layer; IPL, inner plexiform layer; GCL, ganglion cell layer.

but significant reduction of the K^+-evoked release of adenosine derivatives. This indicates that the transporter, presumably located on the plasma membrane, which is responsible for the removal of adenosine from the extracellular space, operates bidirectionally. Whether purine release occurs by a Ca^{++}-dependent or carrier-mediated mechanism may depend on the pools which are affected by stimuli and/or the type and intensity of stimulus.

Light stimulation also evoked the release of purines from the rabbit retina.[35] The time course of this response to light differed markedly from that evoked by K^+ stimulation. Light induced a slow increase in the release, which became apparent at the end of the stimulation period. It is not yet possible to resolve whether this represents merely a delayed response or an increase in release with light offset. For the most part, light-stimulated release of a number of transmitters from cells localized to the INL in the retina has been measured. However, a light-evoked decrease in the release of a putative photoreceptor transmitter, aspartate, has been reported.[41] Excitatory amino acids have been demonstrated to increase the release of purines and their derivatives from the rabbit retina[37] (see below). Thus a light-evoked decrease in extracellular aspartate may ultimately result in a decrease in purine release, with reversal occurring with the cessation of light. It is notable that flashing light was required to elicit an increase in purine release,[35] suggesting that this release occurs from ON and OFF processes located in the IPL.

2. Tonic- and Neurotransmitter-Evoked Release

The stimulation of specific pathways involved in synaptic communication in the retina has also resulted in the release of purines. Brief pulses of both excitatory (glutamate, aspartate, acetylcholine) and inhibitory (γ-aminobutyric acid [GABA], glycine, dopamine, serotonin) transmitter agonists induced a rapid increase in the release of endogenous and ³H-adenosine-derived radioactivity from the superfused rabbit retina.[37] The effect observed with the agonists is likely to reflect specific activation of the corresponding receptors, since an increase of the release was always seen within the stimulation period and specific antagonists effectively blocked these responses. These results indicate that purinergic activity in the retina is under the influence of these various neurotransmitter systems.

Perez and Ehinger[37] have also provided for the regulation of a tonic release of purines by certain transmitter systems using antagonists for excitatory and inhibitory transmitters. Excitatory amino acid antagonists reduced the rate of release of radioactivity, indicating that a basal glutamate/aspartate input modulates tonic adenosine release. This, in turn, is regulated by inhibitory input, as evidenced by the observation that exposure to inhibitory neurotransmitter antagonists (bicuculline, picrotoxin, strychnine, haloperidol) resulted in increased overflow of radioactivity derived from ³H-adenosine. It was also found that the release of purines evoked by GABA receptor antagonists reflects, at least in part, an indirect effect via the blockade of GABAergic input onto glutamate/aspartate neurons feeding forward to the purine-containing cell.

B. COLOCALIZATION STUDIES

Markers for purinergic cells have been colocalized with markers for both GABA[29,30] and acetylcholine (ACh).[30] Adenosine uptake autoradiography, combined with GABA immunohistochemistry, has demonstrated in rabbit retina that a subpopulation of cells in the INL (positioned where amacrine cells are believed to reside) and GCL contain both markers.[29] Likewise, adenosine-like immunoreactivity has been colocalized in displaced, cholinergic cells of the rabbit.[30] These cells have also been shown to be GABAergic.[31,32] Adenosine-like immunoreactivity has been localized to not only the displaced cholinergic amacrine cells, but also to ganglion cells. Adenosine uptake has also been localized in dopaminergic amacrine cells of the rabbit. These localization results are all consistent with the corelease of adenosine with these classical transmitters.

C. REGULATION OF TRANSMITTER RELEASE

Perez and Ehinger[40] have recently demonstrated that, following loading of the retina with [^{3}H]-choline chloride, presumed to be incorporated into ACh, the depolarization-mediated overflow of radioactivity is inhibited by both R-PIA and NECA. The effect is receptor mediated, as it was completely blocked in the presence of 8-phenyltheophylline. At higher concentrations of the antagonist, both spontaneous and evoked release of ACh were increased. These results indicate that the ACh release from retinal cells is controlled, even under basal conditions, by an adenosinergic tone.

The dose-response curves for both compounds were similar with maximal inhibition (50 to 60%) observed with 1 μM PIA. The stereoisomer, S-PIA, had no effect on release. These results are consistent with both R-PIA and NECA acting at A_1 receptors which have been localized to the IPL.[23] Above 1 μM R-PIA, the inhibition of ACh release was significantly smaller than that elicited with 1 μM. This result may reflect an effect of adenosine agonists on the release of inhibitory transmitters such that the depolarization-induced release of ACh is then increased. It seems unlikely that A_2 receptors are activated, due to their localization over the outer retina.[23]

V. CONCLUSIONS

It is obvious that many, if not of all, of the criteria for establishing a role for adenosine as a neurotransmitter or modulator in the mammalian retina have been met. The stimulation-evoked release of purines has been measured, a physiologic response (modulation of ACh release) established, and localization of receptors and their characterization reported. However, we strongly feel that adenosine plays an even larger role in retinal function than that which has thus far been elucidated. The distinct distributions of A_1 and A_2 receptors within the retina is suggestive of distinct mechanisms by which adenosine can influence cellular activity. As stated earlier, adenosine exerts marked effects in three areas of the posterior eye,[25,26] including the RPE, the inner retina and the optic nerve. These observations, taken together with the demonstration of A_2 receptors in the outer retina on both photoreceptors[23] and pigmented epithelium,[21,23] also suggest that adenosine may be involved in processes other than those of a neuromodulator acting at conventional synapses. Much remains to be elucidated concerning the roles of adenosine in the retina, and hence, the CNS.

ACKNOWLEDGMENTS

The authors acknowledge support from the National Eye Institute (EY02294, C.B.) and the Crafoord Foundation, Crown Princess Margareta's Committee for the Blind and Faculty of Medicine at the University of Lund, Helfrid and Lorentz Nilsson Foundation, RP Foundation, Swedish Medical Research Council (14X-02321) and Swedish Society for Medical Research (M.T.R.P.).

REFERENCES

1. **Massey, S. C. and Redburn, D. A.,** Transmitter circuits in the vertebrate retina, *Prog. Neurobiol.,* 28, 55, 1987.
2. **Daw, N. W., Brunken, W. J., and Parkinson, D.,** The function of synaptic transmitters in the retina, *Annu. Rev. Neurosci.,* 12, 205, 1989.
3. **Ehinger, B. and Dowling, J.,** Retinal neurocircuitry and transmission, in *Handbook of Chemical Neuroanatomy,* Vol. 5, Bjorklund, A., Hokfelt, T., and Swanson, L. W., Eds., Elsevier, New York, 1987, 389.

4. **Nguyen-Legros, J., Berger, B., Vigny, A., and Alvarez, C.**, Tyrosine hydroxylase-like immunoreactive interplexiform cells in the rat retina, *Neurosci. Lett.*, 27, 255, 1981.

5. **Nguyen-Legros, J., Vigny, A., and Gay, M.**, Post-natal development of TH-like immunoreactivity in the rat retina, *Exp. Eye Res.*, 37, 23, 1983.

6. **Versaux, Botteri, C., Nguyen-Legros, J., Vigny, A., and Raoux, N.**, Morphology, density and distribution of tyrosine hydroxylase-like immunoreactive cells in the retina of mice, *Brain Res.*, 301, 192, 1984.

7. **Wulle, I. and Schnitzer, J.**, Distribution and morphology of tyrosine hydroxylase-immunoreactive neurons in the developing mouse retina, *Dev. Brain Res.*, 48, 59, 1989.

8. **Mitchell, C. K. and Redburn, D. A.**, Analysis of pre- and post-synaptic factors of the serotonin system in rabbit retina, *J. Cell Biol.*, 100, 64, 1985.

9. **Osborne, N. N.**, Evidence for serotonin being a neurotransmitter in the retina, in *Biology of Serotonergic Transmission*, Osborne, N. N., Ed., John Wiley & Sons, Chichester, England, 1982, 401.

10. **Churchill, L., Blocker, Y., and Redburn, D. A.**, Specific serotonin uptake in photoreceptor cell terminals as evidenced by loss of specific uptake sites in retinal dystrophic rats, *Invest. Ophthalmol. Vis. Sci.*, Suppl. 25, 85, 1984.

11. **Redburn, D. A. and Mitchell, C. K.**, A light sensitive serotonin system in rat retinal photoreceptors, *Invest. Ophthalmol. Vis. Sci.*, Suppl. 25, 85, 1984.

12. **Li H.-B., Watt, C. B., and Lam, D. K.**, The coexistence of two neuroactive peptides in a subpopulation of retinal amacrine cells, *Brain Res.*, 345, 176, 1985.

13. **Li, H.-B., Watt, C. B., and Lam, D. K.**, The presence of neurotensin in enkaphalinergic and glycinergic amacrine cells in the chicken retina, *Invest. Ophthalmol. Vis. Sci.*, Suppl. 26, 278, 1985.

14. **Watt, C. B., Su, Y. Y. T., and Lam, D. M.-K.**, Interactions between enkephalin and GABA in avian retina, *Nature*, 311, 761, 1984.

15. **Lam, D. M.-K., Hai-Biao, L., Su, T. Y.-Y., and Watt, C. B.**, The signature hypothesis: co-localizations of neuroactive substances as anatomical probes for circuitry analysis, *Vision Res.*, 25, 1353, 1985.

16. **Van Calker, D., Muller, M., and Hamprecht, B.**, Adenosine regulates via two different types of receptors, the accumulation of cyclic AMP in cultured brain cells, *J. Neurochem.*, 33, 999, 1979.

17. **Londos, C., Wolff, J., and Cooper, D. M. F.**, Adenosine as a regulator of adenylate cyclase, in *Purine Receptors, Receptors and Recognition*, Ser. B, Vol. 12, Burnstock, G., Ed., Chapman and Hall, 1981, 289.

18. **Blazynski, C., Kinscherf, D. A., Geary, K. M., and Ferrendelli, J. A.**, Adenosine-mediated regulation of cyclic AMP levels in isolated incubated retinas, *Brain Res.*, 366, 224, 1986.

19. **Blazynski, C.**, Adenosine A_1 receptor-mediated inhibition of adenylate cyclase in rabbit retina, *J. Neurosci.*, 7, 2522, 1987.

20. **Schorderet, M.**, Receptors coupled to adenylate cyclase in isolated rabbit retina, *Neurochem. Int.*, 14, 387, 1989.

21. **Friedman, Z., Hackett, S. F., Linden, J., and Campochiaro, P. A.**, Human retinal pigment epithelial cells in culture possess A_2-adenosine receptors, *Brain Res.*, 492, 29, 1989.

22. **Brass, K. M., Zerbin, M. A., and Snyder, S. H.**, Endogenous adenosine and adenosine receptors localized to ganglion cells of the retina, *Proc. Natl. Acad. Sci. U.S.A.*, 84, 3906, 1986.

23. **Blazynski, C.**, Discrete distributions of adenosine receptors in mammalian retina, *J. Neurochem.*, 54, 648, 1990.

24. **Osborne, N. N.**, [^{3}H]Glycogen hydrolysis elicited by adenosine in rabbit retina: involvement of A_2 receptors, *Neurochem. Int.*, 14, 419, 1989.

25. **Niemeyer, G. and Fruh, B.**, Adenosine and cyclohexyladenosine inhibit the cat's optic nerve action potential, *Experientia*, 45, A8, 1989.

26. **Blazynski, C., Cohen, A. I., Fruh, B., and Niemeyer, G.**, Adenosine: autoradiographic localization and electrophysiologic effects in the cat retina, *Invest. Ophthalmol. Vis. Sci.*, 30, 2533, 1989.

27. **Blazynski, C., Mosinger, J. L., and Cohen, A. I.**, Comparison of adenosine uptake and endogenous adenosine-containing cells in mammalian retina, *Visual Neurosci.*, 2, 109, 1989.

28. **Ehinger, B. and Perez, M. T. R.**, Autoradiography of nucleoside uptake into the retina, *Neurochem. Int.*, 6, 369, 1984.

29. **Perez, M. T. R. and Bruun, A.**, Colocalization of [^{3}H] adenosine accumulation and GABA immunoreactivity in the chicken and rabbit retina, *Histochemistry*, 87, 413, 1987.

30. **Blazynski, C.**, Displaced cholinergic, GABAergic amacrine cells in the rabbit retina also contain adenosine, *Visual Neurosci.*, 3, 425, 1989.

31. **Senba, E., Daddona, P. E., and Nagy, J. I.**, Immunohistochemical localization of adenosine deaminase in the retina of the rat, *Brain Res. Bull.*, 17, 209, 1986.

32. **Scott, T. G.**, The distribution of 5'-nucleotidase in the brain of the mouse, *J. Comp. Neurol.*, 29, 97, 1967.

33. **Kreutzberg, G. W. and Hussain, S. T.,** Cytochemical localization of 5'-nucleotidase activity in retinal photoreceptor cells, *Neuroscience,* 11, 857, 1984.

34. **Perez, M. T. R., Ehinger, B. E., Lindstrom, K., and Fredholm, B. B.,** Release of endogenous and radioactive purines from the rabbit retina, *Brain Res.,* 398, 106, 1986.

35. **Perez, M. T. R., Arner, K., and Ehinger, B.,** Stimulation-evoked release of purines from the rabbit retina, *Neurochem. Int.,* 13, 307, 1988.

36. **Perez, M. T. R. and Ehinger, B.,** Adenosine uptake and release in the rabbit retina, in *Retinal Signal Systems, Degenerations and Transplants,* Agardh, E. and Ehinger, B., Eds., Elsevier, New York, 1986, 163.

37. **Perez, M. T. R. and Ehinger, B.,** Multiple neurotransmitter systems influence the release of adenosine derivatives from the rabbit retina, *Neurochem. Int.,* 15, 411, 1989.

38. **Brecha, N., Johnson, D., Peichl, L., and Wassle, H.,** Cholinergic amacrine cells of the rabbit retina contain glutamate decarboxylase and γ-aminobutyrate immunoreactivity, *Proc. Natl. Acad. Sci. U.S.A.,* 85, 6187, 1988.

39. **Vaney, D. I. and Young, H. M.,** GABA-like immunoreactivity in cholinergic amacrine cells of the rabbit retina, *Brain Res.,* 438, 369, 1988.

40. **Perez, M. T. R. and Ehinger, B.,** Adenosine inhibits evoked acetylcholine release from the rabbit retina, *J. Neurochem.,* 52, S157C, 1989.

41. **Neal, M. J., Collins, G. G., and Massey, S. C.,** Inhibition of aspartate release from the retina of the anesthetized rabbit by stimulation with light flashes, *Neurosci. Lett.,* 14, 241, 1979.

42. **Perez, M. T. R.,** Unpublished observation.

Chapter 26

PURINERGIC EFFECTS IN AUTONOMIC GANGLIA

E. M. Silinsky

TABLE OF CONTENTS

I. INTRODUCTION

Synapses in autonomic ganglia are frequently envisaged as rapid throughput systems in which information is transmitted from the central nervous system to the target effector with considerable celerity, but without significant plasticity or modulation. This generalization, although useful for didactic purposes, is not totally accurate.[1,2] The results to be summarized in this review suggest that purines modulate presynaptic or postsynaptic function at all of the autonomic synapses thus far studied. Moreover, at some synapses, adenosine derivatives may be primary neurotransmitter substances. In this chapter I will consider the actions of purines at ganglia in the three divisions of the autonomic nervous system: (1) the sympathetic division, (2) the parasympathetic division, and (3) the enteric division.

II. ACTIONS OF PURINES AT SYMPATHETIC GANGLIA

A. OVERVIEW

The sympathetic division of the autonomic nervous system prepares the organism for vigorous physical activity, i.e., flight or fight. The sympathetic chain ganglia or paravertebral ganglia are located in paired rows on opposite sides of the spinal cord and are logically arrayed to allow for the near synchronous activation of their effector components. Rapid synaptic communication in the sympathetic division, as in the parasympathetic and enteric divisions, is mediated by the release of acetylcholine (ACh) from preganglionic nerve endings onto nicotinic receptors in the membrane of the postganglionic neuron. This behavior of ACh in sympathetic ganglia is modulated by purines. The activity of these sympathetic neuronal circuits is paralleled and reinforced hormonally by the release of catecholamines from the adrenal medullary cells, which embryologically and functionally are modified sympathetic ganglion cells. Catecholamines are secreted when ACh released from splanchnic nerve endings activates nicotinic receptors on the chromaffin cell. Purines exert both stimulatory and inhibitory effects on adrenal medullary secretion. The stimulatory effects of nicotinic activation of sympathetic chain ganglia and adrenal medullary secretion are superimposed upon the effects of a series of bureaucratic ganglia, the prevertebral ganglia (the celiac ganglion, superior mesenteric ganglion, inferior mesenteric ganglion, and the pelvic-hypogastric plexus). The function of prevertebral ganglia (much as is the function of corporate bureaucrats) is to impede the activity of other divisions of the autonomic nervous system (e.g., they impede digestion, excretion, and sexual arousal, which are predominantly parasympathetic functions). Our recent data suggests that purines also modulate synaptic function in prevertebral ganglia. The details of purine action alluded to above will now be considered in light of these arbitrary but hopefully useful subdivisions of the sympathetic division of the autonomic nervous system.

B. SYMPATHETIC CHAIN GANGLIA

The most studied of these ganglia is the mammalian superior cervical ganglion (considered as a chain ganglion for convenience, as it appears anatomically as an extension of the sympathetic chain in the cranial direction). In the superior cervical ganglion, adenosine and its congeners inhibit the release of ACh presynaptically[4,5] and block Ca currents postsynaptically[4] (see below). The presynaptic effect is extremely common at fast excitatory synapses; indeed, adenosine derivatives inhibit the release of most excitatory neurotransmitters.[3,6,7] Whilst the prejunctional effects of ATP at most synapses are due to the hydrolysis product, adenosine, there are a number of loci including sympathetic chain ganglia where ATP directly inhibits ACh release. In frog lumbar sympathetic chain ganglia, ATP inhibits ACh release[8] and under some conditions this presynaptic effect may reduce the nicotinic excitatory postsynaptic potential (esp) below threshold for action potential generation (Figure

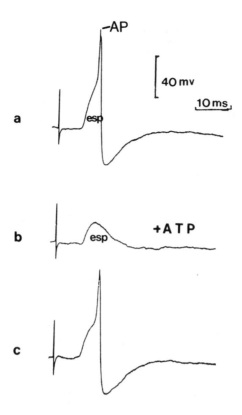

FIGURE 1. Antagonism of cholinergic synaptic transmission in frog sympathetic chain ganglia by ATP. Note that ATP (500 μM in this experiment, but similar results were observed at lower concentrations) reduced the fast nicotinic excitatory synaptic potential (esp) below threshold for action potential (AP) generation. (a) Control Ringer; (b) ATP Ringer; (c) return to control Ringer.

1). Neither adenosine nor 2-chloroadenosine significantly inhibited ACh release from these synapses at concentrations two orders of magnitude greater than that necessary to produce maximal inhibition of ACh release from motor nerve endings of the same species.[9] Figure 2 shows that α,β-methylene ATP was without effect (suggesting that a P_{2x} purinoceptor receptor was not involved in inhibition at preganglionic nerve endings) but theophylline, an adenosine receptor antagonist, blocked the effect of ATP. In bullfrog sympathetic ganglia, the potency order of purines was studied in more detail with ATP > ADP > AMP > adenosine for the inhibition of ACh release (this effect has been attributed to a depolarization of the nerve ending).[10] Because of such results, it has been suggested that a new purinoceptor subtype (P_3)[11] may be necessary to explain some of these effects of ATP (for a review of purine receptor nomenclature, see Silinsky[3] and Chapter 2 of this volume).

These results are similar to those found at some neuro-effector junctions. In guinea pig ileum, ATP, in a theophylline-sensitive manner, exerted a direct action to inhibit ACh release.[12] This effect of ATP was not due to degradation to adenosine and also was not mimicked by α,β-methylene-ATP. In rat caudal artery and vas deferens,[13] ATP in a theophylline-sensitive manner inhibits norepinephrine release directly and not through hydrolysis of the nucleotide to adenosine.

The results at nerve endings in frog sympathetic ganglia, guinea pig ileum, rat caudal artery, and rat vas deferens suggest that ATP may act directly, and not by degradation to adenosine, to inhibit ACh release via a receptor sensitive to theophylline derivatives. Whilst such results may be due to a P_3 receptor it is also possible to consider an alternative nomenclature based upon the traditional concept of defining receptors by selective antago-

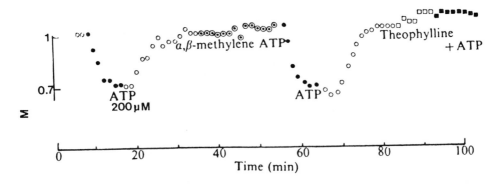

FIGURE 2.   Effects of ATP (200 μM), α,β-methylene-ATP (200 μM), and theophylline (2 mM) on evoked ACh release (M) in frog sympathetic chain ganglia. Evoked ACh release is expressed relative to the control level, which is taken as 1. Ringer solution contained 1 mM Ca^{2+} and 17 mM Mg^{2+}. (Modified from Silinsky, E. M. and Ginsborg, B. L., *Nature*, 305, 327, 1983. With permission.)

nists. Specifically, as P_1 receptors are those which are antagonized by theophylline derivatives, it has been suggested that P_1 receptors be divided into a P_{1X} subtype (the conventional adenosine receptor) and a P_{1Y} subtype (the theophylline-sensitive sites at which ATP is more potent than the other naturally occurring adenosine derivatives).[14] If indeed such a nomenclature is acceptable to the reader, then frog sympathetic ganglia appear to provide the clearest example of a P_{1Y} site.

As alluded to above, postjunctional effects of adenosine derivatives may be observed on sympathetic neurons, but these effects are either modest, transient, or require changes in the normal ionic composition of the bathing fluid to be observed.[4,10] For example, in rat superior cervical ganglia, after the normal somatic action potential is blocked by Na and K channel antagonists, direct electrical stimulation reveals a long-duration Ca action potential which is decreased in duration by adenosine.[4] ATP, but not adenosine, also exerts postjunctional effects in bullfrog sympathetic ganglia;[10] ATP and other nucleotides depolarize these cells by a decrease in potassium conductance. The effect of ATP in bullfrog is likely to be due to block of a specific potassium current, the muscarine-sensitive M current. Thus, upon hyperpolarizing the membrane from -30 mV, a slow inward relaxation and conductance decrease occurs due to the closing of M channels. ATP (30 to 100 μM) decreases the size of this slow relaxation, suggesting that fewer M channels are active in the presence of ATP. Based upon results in bullfrog ganglia and frog neuromuscular junctions, ATP has been suggested to increase the sensitivity of nicotinic receptors to ACh.[10] This effect cannot be of major physiological significance, however, as no effects on the amplitude of spontaneous quantal events at the neuromuscular junction or in the ganglia of the frog have been reported.

C. ADRENAL MEDULLARY CELLS

Until recently, there has been no convincing evidence for the control of catecholamine secretion from the adrenal medulla by purines, possibly because adenosine derivatives are secreted in extremely high concentrations in conjunction with the exocytotic release of catecholamines. When care is taken to reduce extracellular purines, adenosine derivatives produce both excitatory and inhibitory effects on catecholamine secretion.[15-17] Pretreatment with ATP prior to the addition of the secretogogue (nicotinic agonists or K) produces inhibitory effects on catecholamine release whilst simultaneous addition of secretogogue and ATP causes stimulation of secretion. The stimulatory effect of ATP occurs at micromolar concentrations and is associated with complex cellular mechanisms; ATP increases cytosolic Ca, activates phospholipase C, and increases cyclic AMP (cAMP).[17] With respect to the nucleoside, available data are in accordance with the hypothesis that adenosine, via A_2 receptors, stimulates adenyl cyclase and enhances catecholamine secretion. Adenosine also inhibits catecholamine secretion.[15] As the inhibitory effect of adenosine occurs when release is evoked by the calcium ionophore, ionomycin, it appears that membrane ionic channels may not be the target site for this inhibition of catecholamine secretion.[14]

These effects of ATP and adenosine are complex and much remains to be elucidated as to cellular mechanisms of purine action in the adrenal medulla.

D. PREVERTEBRAL GANGLIA

At the time of writing, this reviewer is not aware of any published studies on effects of adenosine derivatives at these sites. Preliminary data suggest that ATP may play a role in rapid synaptic excitation in celiac ganglia. Briefly, ATP (P$_2$) receptor activation depolarizes celiac neurons via a conductance increase to cations; ATP is more potent than ACh in this regard. Single channel studies on outside-outside patches suggest that the cationic channel is an integral part of the P$_2$ receptor.[27]

In summary, the sympathetic division of the autonomic nervous system appears to be a target site for modulation by adenosine derivatives. While postjunctional effects may be observed in prevertebral ganglia, only the presynaptic action is capable of producing physiological modulation of sympathetic chain ganglionic transmission (e.g., see Figure 1). Even this presynaptic effect is modest, however, serving as a fine-tuning device for synaptic function rather than generally producing a global repression of transmitter release.[4] The strength of the fast excitatory synaptic input is befitting a subdivision that acts to produce rapid forceful preparation for vigorous motor activity.

III. ACTIONS OF PURINES AT PARASYMPATHETIC GANGLIA

A. INTRODUCTION

The parasympathetic division of the autonomic nervous system subserves a localized healing function in contrast to the explosive dispensing of energy reserves produced by sympathetic activation. Thus slow local control of ganglia (which are often embedded in the target effector tissue) becomes more significant in this system as rapid, synchronous activation of the parasympathetic division may not be of physiological utility. (To males that have to micturate during sexual arousal, synchronous activation of these two parasympathetic responses can prove quite dysphoric.)

B. ADENOSINE DERIVATIVES AS NEUROTRANSMITTERS

It is in the parasympathetic vesical ganglia in the wall of the cat urinary bladder that the strongest evidence is provided for purinergic transmission.[10,18] Staining with quinacrine, a marker generally used to detect purinergic nerves, suggests that purinergic nerves are present in vesical ganglia.[10] In addition, slow hyperpolarizing postsynaptic potentials evoked by preganglionic nerve stimulation are mimicked by adenosine (Figure 3). These potentials are blocked by caffeine and by adenosine deaminase and are increased by a blocker of adenosine uptake (dipyridamole; Figure 3). The hyperpolarizing effect of adenosine is associated with a conductance increase. The hyperpolarization is decreased in amplitude as the membrane is hyperpolarized beyond the resting potential (near -60 mV) and becomes depolarizing in polarity at approximately -94 mV. The response is blocked in elevated extracellular K (10 mM), increased in low extracellular K (0.5 mM), and does not appear to involve Cl ions. When taken together, these results suggests that adenosine is a likely candidate for an inhibitory transmitter in cat vesical ganglia, acting via P$_1$ purinoceptors to increase a potassium conductance and generate an inhibitory postsynaptic potential.[18] In support of the physiological role of adenosine, very high levels of adenosine deaminase staining were found in the cell bodies of the preganglionic axons in some species;[19] this enzyme would provide a mechanism for terminating the action of adenosine after uptake into the neuronal cytoplasm.

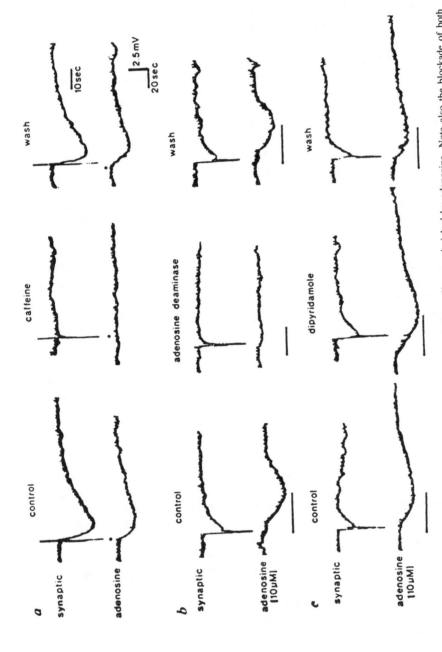

FIGURE 3.   Slow inhibitory postsynaptic potentials in cat vesical ganglia are mimicked by adenosine. Note also the blockade of both synaptic potential and adenosine response by the methylxanthine caffeine (1 m*M* [a], middle trace) and by adenosine deaminase (0.25 IU ml^{-1} [b], middle trace). Both responses are increased in the presence of the adenosine uptake blocker dipyridamole (1 µ*M* [c], middle traces). (From Akasu, T., Shinnick-Gallagher, P., and Gallagher, J. P., *Nature*, 311, 62, 1984. Copyright Macmillan Magazines Limited. With permission.)

Purines also elicit depolarizing responses in cat vesical ganglia,[10] but these are mediated by an ATP receptor. The ATP-induced depolarization was increased in amplitude by membrane hyperpolarization and associated with conductance increase. This response appears to be another example of fast excitatory transmission mediated via a P_2 receptor as a consequence of a nonselective conductance increase to cations (see above and Silinsky[3] for review). ATP and adenosine may be excitatory transmitters in guinea pig intracardiac neurons.[28] The effects of purines on these neurons are highly complex, however.

IV. ACTIONS OF PURINES IN THE ENTERIC NERVOUS SYSTEM

A. OVERVIEW

In the seminal papers on the involuntary nervous system, Langley[20] defined a third division of the autonomic nervous system, namely the enteric division.[20,21] Enteric nerves are contained in two neuronal plexus sandwiched within the walls of the intestine, the myenteric or Auerbach's plexus (between the inner circular and outer longitudinal smoothmuscle layers) and the submucous or Meissner's plexus (between the circular muscle and the mucosa). This division has been termed the ''forgotten nervous system''[20] because it is frequently omitted in classroom accounts of the autonomic nervous system.

B. MYENTERIC PLEXUS

Neurons originating in this plexus send axons to target cells in the circular smoothmuscle layers and to other neurons both in the myenteric and submucous plexus.[20] The main function of these neurons is to control and modulate the contractile activity of intestinal smooth muscle. Earlier studies have demonstrated that excitatory nicotinic transmission in this plexus is modulated by adenosine acting on the cell soma to produce membrane hyperpolarization.[22] This effect is mediated by an increase in potassium conductance, possibly by decreases in cAMP concentrations. More recent studies have found that this K conductance is far more sensitive to ATP than adenosine,[23] with the published experimental records consistent with adenosine acting as a partial agonist. The receptor responsible for this effect is blocked by methylxanthines, and indeed this may be another example of the hypothetical P_{1Y} site. The hyperpolarizing effect of adenosine derivatives appears to be due to an increase in the resting or leak Ca-activated K conductance.[23] A slow synaptic excitation is also produced by ATP in this plexus, with adenosine being less efficacious than ATP. This depolarizing effect has been suggested to be produced by an inhibition of the resting Ca-activated K current via a P_2 receptor.[23] The appropriate pharmacology has yet to be performed to confirm this suggestion, however. As a result of these studies, it has been speculated that ATP might rival substance P as a transmitter in this neuronal plexus.[23] We have also observed rapid excitatory effects of ATP associated with conductance increases in these neurons.[29]

C. SUBMUCOUS PLEXUS

The main function of these neurons is in the regulation of secretory activity and blood flow in the gastrointestinal tract[21] and perhaps in controlling the timing between descending inhibition prior to peristalsis and peristalsis proper.[24] A slow, nonreversing excitatory synaptic potential has been observed in this plexus which thus far is mimicked only by ATP.[25] This potential is due to a conductance decrease to potassium, and it has been suggested that ATP may be the neurotransmitter responsible for this potential.[25] The effect of adenosine was not studied, however, and recent results suggest that an A_2 adenosine receptor may be responsible for this effect of ATP.[30] With regard to presynaptic modulation, A_1 adenosine receptors mediate presynaptic inhibition of ACh release in the submucous plexus.[30] It is of interest that considerable A_2, but not A_1, agonist binding was found in the enteric nervous system.[26]

V. CONCLUSIONS

Each of the three divisions of the autonomic nervous system reveals effects of adenosine derivatives as parts of their physiological profiles. In the sympathetic chain ganglia, the effect is largely modulatory, but in the sympathetic prevertebral ganglia and in the parasympathetic and enteric divisions, adenosine and ATP may serve as primary neurotransmitter substances.

ACKNOWLEDGMENT

This review was supported by a grant from the National Institutes of Health (NS 12782).

REFERENCES

1. **Blackman, G.,** Ganglionic transmission, in *The Peripheral Nervous System,* Hubbard, J. I. Eds., Plenum Press, New York, 1975.
2. **Skok, V. I.,** *Physiology of Autonomic Ganglia,* Igaku Shoin, Medical Publishers, New York, 1973.
3. **Silinsky, E. M.,** Adenosine derivatives and neuronal function, *Semin. Neurosci.,* 1, 155, 1989.
4. **Henon, B. K. and McAfee, D. A.,** The ionic basis of adenosine receptor actions on post-ganglionic neurons in the rat, *J. Physiol.,* 336, 601, 1983.
5. **Alkadhi, K. A., Brown, T. R., and Sabouni, M. H.,** Inhibitory effect of adenosine on transmission in sympathetic ganglia, *Naunyn-Schmiedeberg's Arch. Pharmacol.,* 328, 16, 1984.
6. **Phillis, J. W. and Wu, P.,** Adenosine and adenosine triphosphate as neurotransmitter/neuromodulator in the brain: the evidence is mounting, *Trends Autonom. Pharm.,* 2, 237, 1982.
7. **Ribeiro, J. A. and Sebastião, A. M.,** Adenosine receptors and calcium: basis for proposing a third (A_3) adenosine receptor, *Progr. Neurobiol.,* 26, 279, 1986.
8. **Silinsky, E. M. and Ginsborg, B. L.,** Inhibition of acetylcholine release from preganglionic frog nerves by ATP but not adenosine, *Nature,* 305, 327, 1983.
9. **Silinsky, E. M.,** On the mechanism by which adenosine receptor activation inhibits the release of acetylcholine from motor nerve endings, *J. Physiol.,* 346, 243, 1984.
10. **Akasu, T., Shinnick-Gallagher, P., and Gallagher, P.,** Actions of purines in autonomic ganglia, in *Purines: Pharmacological and Physiological Roles,* Stone, T. W., Ed., Macmillan, London, 1985, 57.
11. **Stone, T.,** Some unresolved problems, in *Purines: Pharmacological and Physiological Roles,* Stone, T. W., Ed., Macmillan, London, 1985, 245.
12. **Wiklund, N. P., Gustafsson, L. E., and Lundin, J.,** Pre- and postjunctional modulation of cholinergic neuroeffector transmission by adenine nucleotides. Experiments with agonist and antagonist, *Acta Physiol. Scand.,* 125, 681, 1985.
13. **Westfall, D. P., Sedaa, K. O., Shinozuka, R. A., Bjur, R. A., and Buxton, I. L. O.,** ATP as a cotransmitter, *Ann. N.Y. Acad. Sci. (Biological Actions of Extracellular ATP),* 603, 300, 1990.
14. **Silinsky, E. M., Hunt, J. M., Solsona, C. S., and Hirsh, J. K.,** Prejunctional actions of adenosine and ATP receptors, *Ann. N.Y. Acad. Sci. (Biological Actions of Extracellular ATP),* 603, 324, 1990.
15. **Wakade, A. R.,** Studies on secretion of catecholamines evoked by acetylcholine or transmural stimulation of the rat adrenal gland, *J. Physiol.,* 313, 463, 1981.
16. **Chern, Y-J., Kim, K-T., Slakey, L., and Westhead, E. W.,** Adenosine receptors activate adenylate cyclase and enhance secretion from bovine adrenal chromaffin cells in the presence of forskolin, *J. Neurochem.,* 50, 1484, 1988.
17. **Kim, K. T., Diverse-Pierluissi, M., Kopell, W. N., and Westhead, E. W.,** ATP effects on secretion and second messenger production in bovine chromaffin cells, *Ann. N.Y. Acad. Sci. (Biological Actions of Extracellular ATP),* 603, 435, 1990.
18. **Akasu, T., Shinnick-Gallagher, P., and Gallagher, J. P.,** Adenosine mediates a slow hyperpolarizing synaptic potentials in autonomic neurons, *Nature,* 311, 62, 1984.
19. **Senba, E., Daddona, P. E., and Nagy, J. L.,** A subpopulation of preganglionic parasympathetic neurons in the rat contain adenosine deaminase, *Neuroscience,* 20, 487, 1987.
20. **North, R. A.,** Pharmacology of the forgotten nervous system, *Trends Pharm. Sci.,* 1, 439, 1980.

21. **Surprenant, A. M.,** Transmitter mechanisms in the enteric nervous system: an electrophysiological vantage point, *Trends Autonom. Pharm.,* 3, 17, 1983.

22. **Palmer, J. M., Wood, J. D., and Zafirov, D. H.,** Purinergic inhibition in the small intestinal myenteric plexus of the guinea-pig, *J. Physiol.,* 387, 357, 1987.

23. **Katayama, Y. and Morita, K.,** Adenosine 5'-triphosphate modulates membrane potassium conductance in guinea-pig myenteric neurons, *J. Physiol.,* 408, 373, 1989.

24. **Hirst, G. D. S.,** Mechanisms of peristalsis, *Br. Med. Bull.,* 35, 263, 1979.

25. **Mihara, S., Katayama, Y., and Nishi, S.,** Slow postsynaptic potentials in neurons of submucous plexus of guinea-pig caecum and their mimicry by noradrenalin and various peptides, *Neuroscience,* 16, 1057, 1985.

26. **Buckley, N. and Burnstock, G.,** Autoradiographic demonstration of peripheral adenosine binding sites using [^{3}H] NECA, *Brain Res.,* 260 374, 1983.

27. **Silinsky, E. M., Gerzanich, V., Vanner, S., and North, R. A.,** in preparation.

28. **Allen, T. G. J. and Burnstock, G.,** The actions of adenosine 5'-triphosphate on guinea-pig intracardiac neurons in culture, *Br. J. Pharmacol.,* 100, 269, 1990.

29. **Gerzanich, V. and Silinsky, E. M.,** unpublished.

30. **Barajas-Lopez, C., Surprenant, A. M., and North, R. A.,** Adenosine A_1 and A_2 receptors mediate presynaptic inhibition and postsynaptic excitation in guinea-pig submucosal neurons, in preparation.

Chapter 27

ADENOSINE AS A NEUROACTIVE COMPOUND IN THE CENTRAL NERVOUS SYSTEM

Trevor W. Stone

TABLE OF CONTENTS

I. ADENOSINE AS A NEUROACTIVE COMPOUND IN THE CENTRAL NERVOUS SYSTEM

Adenosine fulfills some of the classical criteria demanded of a neurotransmitter. It is, for example, released from slices or synaptosomes prepared from brain tissue[1-3] and can be detected also at the exposed surface of the neocortex *in vivo*.[4,5] Indeed, the calcium-dependent release of adenosine and its metabolites, inosine and hypoxanthine, were among the earliest findings to focus attention on adenosine as a neuroactive compound, but the impact of this has been diminished by the realization that most tissues can release adenosine in response to metabolic demand. The amount of release may even be specifically related to discrepancies between that metabolic demand and the available nutrient supply. Adenosine has thus been described as a "retaliatory metabolite",[6] a description which refers to the tendency of adenosine to inhibit cell function and the release of activating neurotransmitters, and to induce local vasodilatation.

Simply fulfilling the neurotransmitter criteria is not therefore sufficient in the case of adenosine to argue for a neurotransmitter function — there is an additional need to demonstrate some specificity of localization, release, and action. It is for this reason that emphasis has been increasingly placed upon the presence and localization of synthesizing or metabolizing enzymes (such as adenosine deaminase, discussed elsewhere in this volume), transporters, and receptors.

A. THE TRANSPORTER LOCALIZATION IN THE CENTRAL NERVOUS SYSTEM

The nucleoside transporter has been labeled with nitrobenzylthioinosine and shown to exist at high density in the nucleus tractus solitarius, superior colliculus, thalamus, nucleus accumbens, substantia nigra, and striatum.[7,8] There are also regions of high density within the spinal cord, especially the substantia gelatinosa, which decline following section of dorsal roots.[8] This implies an association with primary afferent fibers rather than interneurons and provides the kind of evidence needed to postulate a specific neuronal function for adenosine rather than the more general metabolic-related activity. It is still difficult, however, to propose a specific neurotransmitter function in view of evidence that the uptake transporter sites can be detected well before the completion of synapse formation.[9]

B. RECEPTORS

Adenosine receptors have been localized in the central nervous system (CNS) by employing analogs such as N^6-cyclohexyladenosine (CHA). In the rat spinal cord, CHA binding is not decreased by lesions of descending pathways to primary afferents.[10] In rat brain, CHA binding is highest in cerebellar cortex and superficial layers of the superior colliculus, although some regions of cerebral cortex were also heavily labeled in autoradiographic studies. Some binding also occurs in the striatum, medial hypothalamus, and hippocampus.[11]

In a later study by Goodman and Snyder,[12] CHA binding was localized to the cerebellar molecular layer, regions of the hippocampal formation, thalamic nuclei including the geniculate bodies, and also parts of the neocortex, striatum, and accumbens. A common finding[11,12] is the relative paucity of CHA binding in the hypothalamus, an area rich in ADA activity. It is indeed tempting to speculate that this relationship may reflect a protective role of ADA for hypothalamic neurons outside the blood-brain barrier.

There is a significant reduction of adenosine receptor density in the cerebellum of mutant mice lacking granule cells.[13] Similarly, unilateral enucleation produced a depletion of adenosine receptors in the contralateral superior colliculus, suggesting that CHA binding occurred predominantly on retinocollicular projections. In both cases the neuronal pathway involved is an excitatory pathway, and these results prompted Goodman et al.[13] to speculate on a relatively selective distribution of adenosine receptors on excitatory neuronal terminals. However, a study by Shaw et al.[14] has failed to detect any changes in superior colliculus binding of CHA or 5′-N-ethylcarboxamidoadenosine (NECA) following enucleation, decortication, or geniculate lesions in adult cats. This may indicate major species differences in the role of adenosine as a neuromodulator.

In the striatum, Goodman et al.[13] demonstrated that cortical ablation produced no change of CHA binding, and Lloyd and Stone[15] subsequently demonstrated a failure of nigrostriatal lesions to alter significantly striatal CHA binding. Even kainate lesions of striatal cell bodies produced a depression of CHA binding density by only 28%, although Geiger[16] has more recently found a decrease of around 50%.

Lesions of the fornix caused no change of CHA binding density in the rat hippocampus and thalamus.[13] Murray and Cheney[17] reported high binding of CHA in the rat hippocampus and cerebellum and low binding in the hypothalamic region. Lesions of cholinergic, adrenergic, or tryptaminergic afferents to the hippocampus produced no change in CHA binding, although intrahippocampal injections of kainate reduced binding density by around 30%, a value remarkably similar to that reported by Lloyd and Stone[15] in the striatum. Binding in the CA1 region was greater than in the CA3 region and the dentate gyrus.

These findings correlate well with biochemical findings by Wojcik and Neff[18] in which a comparison of mouse mutant strains led to the conclusion that mutants lacking granule cells would not sustain an inhibitory effect of N^6-phenylisopropyladenosine (PIA) on adenylate cyclase activity. This correlates with the reduction of receptor density in such mutants.[13] The striatum yielded results which are more difficult to interpret since a depression of A_1 receptor density (reflected in cyclase inhibition) was diminished significantly by decortication (20% decrease) or intrastriatal injections of kainate (40% depression), but not by 6-hydroxydopamine.[19,20] The kainate result and the absence of any effect of 6-hydroxydopamine treatment would be consistent with an earlier interpretation[15] that CHA binding sites do not exist on dopaminergic afferents and occur on only a fraction of striatal interneurons.

The distribution of A_1 binding in the human brain generally parallels that seen in rats, with highest levels in the hippocampus, neocortex, and striatum. There is appreciably less binding in the cerebellum, however, with radiolabeled CHA being confined to the Purkinje cell layer in human tissue, although the significance of this remains unknown.[21]

A_2 sites localized by NECA binding in the presence of R-PIA appear to be concentrated in regions such as the striatum, accumbens, geniculate nuclei, olfactory tubercle, and amygdala, with lesser concentrations in the hippocampus, cerebellum, and superior colliculus.[22] A similar distribution was noted for NECA binding in the presence of the relatively selective A_1 antagonist 1,3-dipropyl-8-cyclopentylxanthine.

Unfortunately, the correlation between receptor localization and tissue sensitivity is not very satisfactory. While in some regions such as hippocampus there is a correlation between the potency of compounds acting at A_1 receptors in depressing orthodromically evoked synaptic potentials and displacing CHA binding from hippocampal membranes,[23,24] this degree of correlation is not obtained in many other regions. It has already been noted, for example, that lesions of dopaminergic afferent pathways to either the striatum or the hippocampus do not produce significant falls of CHA binding or adenosine-stimulated cyclase in those regions, yet Michaelis et al.[25] reported a clear and potent effect of adenosine in suppressing the depolarization-evoked release of dopamine from synaptosomal preparations from rat striatum.

Similarly, the high density of both adenosine deaminase and CHA binding localized to superficial layers of the superior colliculus does not correlate with the negligible activity of adenosine in inhibiting orthodromically induced synaptic potentials in this area following stimulation of visual input.[26] It remains to be shown whether or not this is due to an artifact of the original localization of CHA binding to superior colliculus terminals, since Shaw et al.[14] were unable to demonstrate retinocollicular terminal localization of CHA binding.

II. EFFECTS ON CENTRAL NERVOUS SYSTEM NEURONS

A. NEURONAL FIRING

Phillis et al.[27] observed that a large range of purines applied by microiontophoresis would depress the firing rate of neurons in the cerebral cortex of methoxyflurane-anesthetized rats. Among the most potent compounds were the adenine derivatives. Subsequent work revealed that the effects of adenosine could be potentiated by uptake inhibitors such as dipyridamole and hexobendine. The responses were also enhanced by papaverine and hydroxynitrobenzyl thioinosine as well as by inhibition of adenosine deaminase (by deoxycoformycin or erythro-9-[2-hydroxy-3-nonyl]adenine) which themselves had potent effects on cell firing rate. ATP and β,γ-mATP were able to depress cells in a similar manner to adenosine, though α,β-mATP was not able to mimic this action, implying the need for metabolism of nucleotides to adenosine.[28,29] This depressant activity of purines extends to many regions of CNS besides the cortex, and Kostopoulos and Phillis[30] showed that neurons in several regions of the rat CNS responded in like manner to adenosine and related compounds.

The methylxanthines block depressant action of purines. Stone and Perkins[29] used stable analogs and dinucleotides of adenine to reveal weak depressant actions which could be blocked by xanthines, supporting the view that there is only a P_1 receptor mediating the depression of cell firing and not a receptor for ATP in the neocortex.

This conclusion has been further supported by use of the synthetic compounds L-AMP-PCP (the L-isomer of β,γ-mATP) and 2-methylthioadenosine-5'-(β,γ-difluoromethylene)-triphosphonate. These compounds are extremely stable[31,32] and the very small amounts of L-adenosine produced by any hydrolysis are not active at adenosine receptors; these compounds have no effect on neuronal excitability or synaptically evoked potentials in the CA1 region of the rat hippocampus *in vitro*, implying an absence of P_2 receptors from cell bodies and terminals in this region.[33]

The nature of the P_1 receptor mediating the depressant responses to adenosine has not been made entirely clear. In two simultaneous reports,[34,35] a range of neurons in the cerebral cortex were tested using analogs of adenosine. The conclusion was drawn that the depressant effects were mediated by a receptor with properties of the A_2 type. This was based partly on the apparently greater potency of NECA and also on the fact that the responses to PIA were relatively weak, especially in view of their long-lasting nature. However, it is difficult to obtain accurate information on the relative potency of compounds using microiontophoresis because of the absence of information on the local drug concentrations and their access to different components of the cell surface.[36] In addition, PIA is more lipid soluble than NECA. Its rate of diffusion through the brain tissue may be limited, thus altering the profile of receptor activation at different regions of the cell surface. Dunwiddie et al.[37] concluded that PIA was in fact more active than NECA in brain slices.

The involvement of a presynaptic inhibition of transmitter release would be consistent with the role of A_1 receptors in view of the evidence for an involvement of A_1 receptors in the inhibition of transmitter release from peripheral neurons. It is interesting to note in passing that electrophysiological studies of afferent nerve terminal excitability[38] as well as studies of synaptic terminal membrane potential using dye[39] indicate that adenosine does not change the membrane potential of nerve terminals.

B. MEMBRANE PROPERTIES

One of the first intracellular studies of adenosine was performed on the rat CA1 hippocampal region *in vitro* by Siggins and Schubert.[40] At concentrations of 1 to 5 μ*M* adenosine-depressed ipsps and epsps with no effect on membrane potential or resistance. At 10 to 20 μ*M*, hyperpolarization was seen with a fall of membrane resistance in about half of the cells studied. Adenosine can act therefore directly upon postsynaptic cells, providing a clear dissociation between the presynaptic and postsynaptic actions of adenosine. Many of these observations were confirmed by Segal,[41] also using CA1 pyramids in rat hippocampal slices: adenosine promoted hyperpolarization with a fall of input resistance and produced a large diminution of synaptic potentials, though these could be separated by localized applications of adenosine: applied iontophoretically onto the dendrites, a decrease of synaptic potentials was seen without postsynaptic hyperpolarization, whereas injection into the soma region produced both effects. The occurrence of hyperpolarization with increased conductance tempted Segal[41] to suggest the involvement of an increased potassium conductance in the adenosine response.

Similar results were obtained more recently by Trussell and Jackson,[42] who found that adenosine evoked a hyperpolarization of cultured mouse striatal neurons with an outward current showing a degree of voltage dependence. These authors suggested that these effects could be mediated by a primary increase of potassium current.

A hyperpolarization of CA1 neurons accompanied by reduced input resistance has been documented,[43,44] and further supporting evidence for a postsynaptic inhibitory action of adenosine was adduced by DiCori and Henry,[45] who reported a depression of glutamate-evoked excitation of CA1 pyramids by adenosine 5′-monophosphate (and ATP), probably after metabolism to adenosine.

III. ADENOSINE AS A NEUROMODULATOR

A. POSTSYNAPTIC EFFECTS
1. Interaction with Noradrenaline
One of the first effects of adenosine to be demonstrated on central nervous tissue was that of modifying the activity of adenylate cyclase. Sattin and Rall[46] reported that adenosine would activate adenylate cyclase in homogenates of guineapig neocortex and, in addition, that it would potentiate the stimulatory effect of noradrenaline. Although it is now widely recognized that cyclase is by no means invariably linked to adenosine receptor activation, it is interesting that a postjunctional enhancement of noradrenaline has been noted in other tissues, especially vas deferens and some vascular muscle. By applying adenosine and noradrenaline together, it has also been possible to demonstrate a mutual potentiation of the depression of neuronal firing in the CNS.[47] There may also be as yet unexplored interactions between adenosine and noradrenaline comparable with the ability of adenosine to modify the time course of desensitization and resensitization in peripheral tissues.[48]

2. Interaction with Acetylcholine
Both ATP and adenosine appear to modulate tissue sensitivity to acetylcholine. The potentiation of acetylcholine by ATP at nicotinic receptors on skeletal muscle was first described by Ewald,[49] but the nature of the interaction, an increased postjunctional membrane conductance, was only revealed by the later voltage clamp studies of Akasu et al.[50] Since then it has become clear that adenosine in particular can also enhance or suppress the sensitivity of tissues to muscarinic agonists, the nature of the interaction depending on the tissue being considered.[51]

A similar modulation of transmitter sensitivity probably occurs in the CNS. The application of acetylcholine or other muscarinic agonists such as carbachol causes excitation of pyramidal neurons in the hippocampus, probably by suppressing the M current potassium channels. When adenosine is now coadministered, either by microiontophoresis to the same individual neurons or by addition into the superfusion medium of a hippocampal slice, the muscarinic excitation can be profoundly suppressed.[52] The potency and specificity of this action can be demonstrated by a comparison with other excitatory agents such as the amino acid *N*-methyl-D-aspartate (NMDA) and the glutamate analogs kainic and quisqualic acids.[52]

3. Amino Acids

Despite the apparent selectivity of the interaction with hippocampal muscarinic receptors, it has become apparent that there may be modulatory actions of adenosine at amino acid receptors elsewhere in the CNS. In the neocortex, for example, it has recently been observed that adenosine can increase the mass depolarization of neurons by NMDA.[53] While the significance of this is still unclear, particularly its relationship to the important neurophysiological properties of NMDA receptors, it has proved to be a useful phenomenon for studying chronic changes of adenosine sensitivity. The acute *in vitro* potentiation of NMDA responses, for example, can be prevented by theophylline,[53] indicating mediation of the interaction through conventional extracellular P_1 (xanthine-sensitive, or nucleoside) receptors.

Several groups have previously found that the chronic treatment of animals with theophylline leads to an upregulation in the number of adenosine P_1 receptors, examined by the measurement of binding site density. In the neocortical functional system, however, there was actually a decrease in the sensitivity to adenosine to the extent that virtually no enhancement of NMDA responses could be detected after 2 weeks of theophylline (daily) treatments.[53] This observation emphasizes the possible dissociation between estimates of receptor number in binding studies and the actual functional efficacy of neuroactive compounds in general.

4. Calcium Channels

It will be clear from other sections of this volume that adenosine can modify calcium fluxes across cell membranes as well as the availability of intracellular calcium to excitation and secretion processes. In addition, several reports have now appeared indicating some form of relationship between adenosine receptors and binding sites for the dihydropyridine class of calcium channel activators and blockers. Thus, dihydropyridines can reduce selectively the binding of purine ligands to CNS membranes[54-56] while enhancing their effects on neuronal firing,[57] possibly by an interaction with the uptake transporter.[58]

In order to examine further the possible existence of such a relationship between adenosine and dihydropyridines in the CNS, their combined effects have been examined on hippocampal slices. The dihydropyridine calcium antagonist nifedipine, as well as the channel agonist BAYK 8644, were both able to potentiate the inhibitory effect of adenosine on evoked population potentials in the CA1 region of the hippocampus.[59] In contrast, however, the same channel ligands did not potentiate the effects of 2-chloroadenosine; indeed, both produced a significant reduction of response size.[59] Since adenosine, but not 2-chloroadenosine, is a substrate for the uptake transporter, it is likely that the enhancement of adenosine responses is a consequence of uptake inhibition, a suggestion which would be consistent with the earlier finding of some uptake inhibition by dihydropyridines.

The reduction of responses to the 2-chloro analog, however, seem more likely to reflect a direct interaction at closely related receptor sites, but a problem which needs attention in any explanation is that both the channel agonist BAYK 8644 and antagonist nifedipine share the same activity. Other channel blockers such as nitrendipine and nimodipine do not show activity in this system.[59] Clearly, there are many interesting possible explanations which demand further work. One such possibility is that a specific association exists between a low-affinity binding site for 2′-substituted dihydropypridines (BAYK 8644 and nifedipine) and the adenosine receptor, but that no such relationship obtains for the high-affinity binding site which has been related to calcium channel function and to which all the tested dihydropyridines are known to bind. Since speculation continues that there may exist endogenous ligands for dihydropypridine receptors, it is an exciting possibility that one such ligand may have a modulatory interaction with adenosine receptors.

B. PRESYNAPTIC EFFECTS

A general discussion of the ability of adenosine to suppress transmitter release is to be found elsewhere in this volume. It is important to note, however, that such an inhibition of release has been widely studied in the CNS.

Adenosine and a number of metabolically stable analogs have been shown to inhibit the release of glutamate, for example, from hippocampal neurons[60] and of acetylcholine from neocortical cells.[61] In both of these cases, however, it has also been suggested that adenosine receptors can have a biphasic modulatory effect on release, the A_1 receptors being responsible for the inhibitory action and an A_2 site causing an increase of release. Both of these results have been obtained from studies in which the release of the relevant transmitter was being monitored directly, and it is perhaps curious that no equivalent biphasic action has been demonstrated electrophysiologically in the CNS. This may of course reflect a different localization of the A_1 and A_2 receptors on discrete neuronal or terminal populations, only some of which can be studied electrophysiologically in easily identified pathways.

The effects of adenosine on glutamate and acetylcholine release would be consistent with the earlier suggestions from autoradiographic work that receptors appeared to exist primarily on the terminals of excitatory neurons.[12,13] Certainly, the suppression of inhibitory transmitter release is far weaker, if it exists at all: adenosine only inhibits the release of γ-aminobutyric acid, for example, at concentrations in excess of 1 mM.[62] Similarly, the release of noradrenaline is only decreased by around 25% at high concentrations of adenosine analogs.[63]

The mechanism of these presynaptic effects has been examined by several groups. The inhibition of glutamate release from neurons in the hippocampus can be blocked by pertussis toxin,[64] an agent which selectively inactivates the G protein coupling receptor activation to the inhibition of adenylate cyclase. However, the additional discovery that other agents modifying the adenylate cyclase system did not alter the inhibition of glutamate release led the authors to propose the involvement of a novel G protein which did not necessarily couple to cyclase.

Recent work has also revealed interesting ionic conditions for presynaptic inhibition by purines. The depression of glutamate release by adenosine from stratum radiatum axons of the hippocampus, for example, studied electrophysiologically, requires the presence of magnesium ions.[65] This is a concentration-dependent phenomenon. It is not yet clear whether this magnesium requirement reflects a role in receptor activation or a coupling to a second-messenger system or ion channel, but the phenomenon may be an important one to consider when attempts are made to explore epileptiform activity in the CNS by increasing neuronal excitability in magnesium-free media.

It should be clear that adenosine can act at several distinct sites, and in a variety of ways, to modify neuronal function or transmitter sensitivity and release. Although the evidence for a conventional neurotransmitter function remains poor, therefore, the arguments are strong that adenosine may have a major physiological role to play as a modulator of CNS activity.[66,67]

REFERENCES

1. **McIlwain, H.,** Regulatory significance of the release and action of adenine derivatives in cerebral systems, *Biochem. Soc. Symp.,* 36, 69, 1972.
2. **Shimizu, H., Creveling, C. R., and Daly, J.,** Stimulated formation of cyclic AMP in cerebral cortex: synergism between electrical activity and biogenic amines, *Proc. Natl. Acad. Sci. U.S.A.,* 65, 1033, 1970.
3. **Stone, T. W., Newby, A. C., and Lloyd, H. G. E.,** Adenosine release, in *Adenosine Receptors,* Williams, M., Ed., Humana Press, New York, 1990, chap. 6.
4. **Jhamandas, K. J. and Dumbrille, A.,** Regional release of (3H)-adenosine derivatives from rat brain *in vivo*: effect of excitatory amino acids, opiate agonists and BZD, *Can. J. Physiol. Pharmacol.,* 58, 1262, 1980.
5. **Perkins, M. N. and Stone, T. W.,** *In vivo* release of (3H)-purines by quinolinic acid and related compounds, *Br. J. Pharmacol.,* 80, 263, 1983.
6. **Newby, A. C.,** Adenosine and the concept of retaliatory metabolites, *Trends Biochem. Sci.,* 9, 42, 1984.
7. **Bisserbe, J.-C., Patel, J., and Marangos, P. J.,** Autoradiographic localization of adenosine uptake sites in rat brain using (^{3}H) NBTI, *J. Neurosci.,* 5, 544, 1985.
8. **Geiger, J. D. and Nagy, J. I.,** Localization of (^{3}H) NBTI binding sites in rat spinal cord and primary afferent neurones, *Brain Res.,* 347, 321, 1985.
9. **Morgan, P. F., Montgomery, P., and Marangos, P. J.,** Ontogenetic profile of the adenosine uptake sites in rat forebrain, *J. Neurochem.,* 49, 852, 1987.
10. **Geiger, J. D., LaBella, F. S., and Nagy, J. I.,** Characterisation and localisation of adenosine receptors in rat spinal cord, *J. Neurosci.,* 4, 2303, 1984.
11. **Lewis, M. E., Patel, J., Edley, S. M., and Marangos, P. J.,** Autoradiographic visualization of rat brain adenosine receptors using N^6-cyclohexyl-(^{3}H)-adenosine, *Eur. J. Pharmacol.,* 73, 109, 1981.
12. **Goodman, R. R. and Snyder, S. H.,** Autoradiographic localization of adenosine receptors in rat brain using (^{3}H)cyclohexyladenosine, *J. Neurosci.,* 2, 1230, 1982.
13. **Goodman, R. R., Kuhar, M. J., Bester, L., and Snyder, S. H.,** Adenosine receptors: autoradiographic evidence for their location on axon terminals of excitatory neurons, *Science,* 220, 967, 1983.
14. **Shaw, C., Hall, S. E., and Cynader, M.,** Characterisation, distribution and ontogenesis of adenosine binding sites in cat visual cortex, *J. Neurosci.,* 6, 3218, 1986.
15. **Lloyd, H. G. E. and Stone, T. W.,** Cyclohexyladenosine binding in rat striatum, *Brain Res.,* 334, 385, 1985.
16. **Geiger, J. D.,** Localization of (^{3}H) CHA and (^{3}H) NBTI binding sites in rat striatum and superior colliculus, *Brain Res.,* 363, 404, 1986.
17. **Murray, T. F. and Cheney, D. L.,** Neuronal location of N^6-cyclohexyl (^{3}H)-adenosine binding sites in rat and guinea pig brain, *Neuropharmacol.,* 21, 575, 1982.
18. **Wojcik, W. J. and Neff, N. H.,** Adenosine A1 receptors are associated with cerebellar granule cells, *J. Neurochem.,* 41, 759, 1983.
19. **Wojcik, W. J. and Neff, N. H.,** Differential location of adenosine A1 and A2 receptors in striatum, *Neurosci. Lett.,* 41, 55, 1983.
20. **Wojcik, W. J. and Neff, N. H.,** Location of adenosine release and adenosine A2 receptors to rat striatal neurones, *Life Sci.,* 33, 755, 1983.
21. **Fastbom, J., Pazos, A., Probst, A., and Palacios, J. M.,** Adenosine A1 receptors in human brain: characterisation and autoradiographic visualisation, *Neurosci. Lett.,* 65, 127, 1986.
22. **Lee, K. S. and Reddington, M.,** Autoradiographic evidence for multiple CNS binding sites for adenosine derivatives, *Neurosci.,* 19, 535, 1986.
23. **Reddington, M., Lee, K. S., and Schubert, P.,** An A1 adenosine receptor, characterised by (^{3}H)cyclohexyladenosine binding, mediates the depression of evoked potentials in a rat hippocampal slice preparation, *Neurosci. Lett.,* 28, 275, 1982.
24. **Lee, K. S., Schubert, P., Reddington, M., and Kreutzberg, G. W.,** Adenosine receptor density and the depression of evoked neuronal activity in the rat hippocampus in vitro, *Neurosci. Lett.,* 37, 81, 1983.
25. **Michaelis, M. L., Michaelis, E. K., and Myers, S. L.,** Adenosine modulation of synaptosomal dopamine release, *Life Sci.,* 24, 2083, 1979.
26. **Okada, Y. and Saito, M.,** Inhibitory action of adenosine, 5HT and GABA, on the PSP of slices from olfactory cortex and superior colliculus in correlation to the level of cyclic AMP, *Brain Res.,* 160, 368, 1979.
27. **Phillis, J. W., Kostopoulos, G. K., and Limacher, J. J.,** Depression of corticospinal cells by various purines and pyrimidines, *Can. J. Physiol. Pharmacol.,* 52, 1226, 1974.
28. **Phillis, J. W. and Edstrom, J. P.,** Effects of adenosine analogs on rat cerebral cortical neurons, *Life Sci.,* 19, 1041, 1976.
29. **Stone, T. W. and Perkins, M. N.,** Adenine dinucleotide effects on rat cortical neurones, *Brain Res.,* 229, 241, 1981.

30. **Kostopoulos, G. K. and Phillis, J. W.,** Purinergic depression of neurones in different areas of the rat brain, *Exp. Neurol.,* 55, 719, 1977.
31. **Cusack, N. J., Hourani, S. M. O., Loizou, G. D., and Welford, L. A.,** Pharmacological effects of isopolar phosphonate analogues of ATP on P2-purinoceptors in guinea-pig taenia coli and urinary bladder, *Br. J. Pharmacol.,* 90, 791, 1987.
32. **Hourani, S. M. O., Welford, L. A., and Cusack, N. J.,** L-AMP-PCP, an ATP receptor agonist in guinea-pig bladder, is inactive on taeni coli, *Eur. J. Pharmacol.,* 108, 197, 1985.
33. **Stone, T. W. and Cusack, N. J.,** Absence of P2-purinoceptors in hippocampal pathways, *Br. J. Pharmacol.,* 97, 631, 1989.
34. **Phillis, J. W.,** Evidence for an A2 like adenosine receptor on cerebral cortical neurones, *J. Pharm. Pharmacol.,* 34, 453, 1982.
35. **Stone, T. W.,** Purine receptors involved in the depression of neuronal firing in cerebral cortex, *Brain Res.,* 248, 367, 1982.
36. **Stone, T. W.,** *Microiontophoresis and Pressure Ejection,* Vol. 8, IBRO Handbook Series: Methods in the Neurosciences, John Wiley & Sons, Chichester, England, 1985.
37. **Dunwiddie, T. V., Basile, A. S., and Palmer, M. R.,** Electrophysiological responses to adenosine analogs in rat hippocampus and cerebellum: evidence for mediation by adenosine receptors of the A1 subtype, *Life Sci.,* 34, 37, 1984.
38. **Stone, T. W.,** Adenosine and related compounds do not affect nerve terminal excitability in rat CNS, *Brain Res.,* 182, 198, 1980.
39. **Creveling, C. R., McNeal, E. T., McCulloh, D. H., and Daly, J. W.,** Membrane potentials in cell free preparations from guinea-pig cerebral cortex: effect of depolarizing agents and cyclic nucleotides, *J. Neurochem.,* 35, 922, 1980.
40. **Siggins, G. R. and Schubert, P.,** Adenosine depression of hippocampal neurones in vitro: an intracellular study of dose-dependent actions on synaptic and membrane potentials, *Neurosci. Lett.,* 23, 55, 1981.
41. **Segal, M.,** Intracellular analysis of a postsynaptic action of adenosine in the rat hippocampus, *Eur. J. Pharmacol.,* 79, 193, 1982.
42. **Trussell, L. O. and Jackson, M. B.,** Adenosine-activated potassium conductance in cultured striatal neurons, *Proc. Natl. Acad. Sci. U.S.A.,* 82, 4857, 1985.
43. **Greene, R. W. and Haas, H. L.,** Adenosine actions on CA1 pyramidal neurones in rat hippocampal slices, *J. Physiol.,* 366, 110, 1985.
44. **Haas, H. L. and Greene, R. W.,** Adenosine enhances afterhyperpolarization and accommodation in hippocampal pyramidal cells, *Pfluegers Arch.,* 402, 244, 1984.
45. **DiCori, S. and Henry, J. L.,** Effects of ATP and AMP on hippocampal neurones of the rat in vitro, *Brain Res. Bull.,* 13, 199, 1984.
46. **Sattin, A. and Rall, T. W.,** The effect of adenosine and adenine nucleotides on the cyclic AMP content of guinea-pig cerebral cortex slices, *Mol. Pharmacol.,* 6, 13, 1970.
47. **Stone, T. W. and Taylor, D. A.,** An electrophysiological demonstration of a synergistic interaction between norepinephrine and adenosine in the cerebral cortex, *Brain Res.,* 147, 396, 1978.
48. **Long, C. J. and Stone, T. W.,** Effects of adenosine on adrenoceptor sensitivity in rat vas deferens, *Eur. J. Pharmacol.,* 132, 11, 1986.
49. **Ewald, D. A.,** Potentiation of postjunctional cholinergic sensitivity of rat diaphragm muscle by high-energy phosphate adenine nucleotides, *J. Membr. Biol.,* 29, 47, 1976.
50. **Akasu, T., Hirai, K., and Koketsu, K.,** Increase of acetylcholine receptor sensitivity by ATP: a novel action of ATP on acetylcholine sensitivity, *Br. J. Pharmacol.,* 74, 505, 1981.
51. **Gustafsson, L. E., Wiklund, N. P., and Cederqvist, B.,** Apparent enhancement of cholinergic transmission in rabbit bronchi via adenosine A2 receptors, *Eur. J. Pharmacol.,* 120, 179, 1986.
52. **Brooks, P. A. and Stone, T. W.,** Purine modulation of cholinomimetic responses in the rat hippocampal slice, *Brain Res.,* 458, 106, 1988.
53. **Mally, J., Connick, J. H., and Stone, T. W.,** Theophylline down-regulates adenosine receptor function, *Brain Res.,* 509, 141, 1990.
54. **Morgan, P. F., Tamborska, E., Patel, J., and Marangos, P. J.,** Interactions between calcium channel compounds and adenosine systems in brain of rat, *Neuropharmacology,* 26, 1693, 1987.
55. **Hu, P. S., Lindgren, E., Jacobson, K. A., and Fredholm, B. B.,** Interaction of dihydropyridine calcium channel agonists and antagonists with adenosine receptors, *Pharmacol. Toxicol.,* 61, 121, 1987.
56. **Cheung, W., Shi, M. M., Young, J. D., and Lee, G. M.,** Inhibition of radioligand binding to A1 adenosine receptors by BayK 8644 and nifedipine, *Biochem. Pharmacol.,* 36, 2183, 1987.
57. **Phillis, J. W., Swanson, H., and Barraco, R. A.,** Interactions between adenosine and nifedipine in the rat cerebral cortex, *Neurochem. Int.,* 6, 693, 1984.
58. **Marangos, P. J., Finkel, M. S., Verma, A., Maturi, M. F., Patel, J., and Patterson, R. F.,** Adenosine uptake sites in dog heart and brain; interactions with calcium antagonists, *Life Sci.,* 35, 1109, 1984.

59. **Bartrup, J. T. and Stone, T. W.,** Dihydropyridines alter adenosine sensitivity in the rat hippocampal slice, *Br. J. Pharmacol.,* 101, 97, 1990.
60. **Dolphin, A. C. and Archer, E. R.,** An adenosine agonist inhibits and a cyclic AMP analogue enhances the release of glutamate but not GABA from slices of rat dentate gyrus, *Neurosci. Lett.,* 43, 49, 1983.
61. **Spignoli, G., Pedata, F., and Pepeu, G.,** A1 and A2 adenosine receptors modulate acetylcholine release from brain slices, *Eur. J. Pharmacol.,* 97, 341, 1984.
62. **Hollins, C. and Stone, T. W.,** Adenosine inhibition of GABA release from slices of rat cerebral cortex, *Br. J. Pharmacol.,* 69, 107, 1980.
63. **Jonzon, B. and Fredholm, B. B.,** Adenosine receptor mediated inhibition of noradrenaline release from slices of the rat hippocampus, *Life Sci.,* 35, 1971, 1984.
64. **Fredholm, B. B., Proctor, W., Van der Ploeg, I., and Dunwiddie, T. V.,** In vivo pertussis toxin treatment attenuates some, but not all, adenosine A1 effects in slices of the rat hippocampus, *Eur. J. Pharmacol.,* 172, 249, 1989.
65. **Bartrup, J. T. and Stone, T. W.,** Interactions of adenosine and magnesium on rat hippocampal slices, *Brain Res.,* 463, 374, 1988.
66. **Stone, T. W.,** Purine receptors and their pharmacological potential, *Adv. Drug Res.,* 18, 291, 1989.
67. **Stone, T. W. and Simmonds, H. A.,** *Purines: Basic and Clinical Aspects,* Kluwer Publishers, Dordrecht, The Netherlands, 1991.

Chapter 28

BEHAVIORAL ACTIONS OF ADENOSINE AND RELATED SUBSTANCES

Robin A. Barraco

TABLE OF CONTENTS

I. INTRODUCTION

A. NEURAL MECHANISMS OF BEHAVIOR

Although behavior is observed as a continuous stream of effector events, many lines of evidence indicate that it is discontinuously encoded in the nervous system.[1] The ultimate questions of neurobehavioral research are (1) how the neural networks which generate behavioral patterns are interrelated and controlled and (2) what ultimately determines and elicits a sequence of behavior patterns, particularly in response to drug administration. The investigation of behavior, using neurochemical and neuropharmacological approaches, must be conducted, in the long run, at multiple levels of analyses since the behavioral repertoire available to an animal is expressed with a hierarchy of complexity and varying degrees of central nervous system (CNS) involvement.

In the meantime, it has become apparent, especially over the past 2 decades, that the integrative capacity of the nervous system must accrue from the enormous variability of the intercellular transmission process between neurons. It is also evident that the chemical vocabulary by which central neurons communicate is extraordinarily varied and complex. This conundrum of intercellular signaling now includes a multitude of coexisting transmitters, modulators, and synaptic regulators,[2,3] many of which may affect a throng of second-messenger systems and ion channels via activation of an increasing array of membrane receptor subtypes.[4] It follows therefore that brain function and therefore behavior cannot be understood simply in terms of the "hardwired" circuitry of the component structures in the nervous system, nor can it be revealed through an ever guileless vigil of synaptic microevents and effector-coupling responses in isolated tissue preparations from these same structures. Nevertheless, the elegant findings from anatomical tracing studies and the numerous revelations from *in vitro* studies about synaptic microevents involved in signal transduction have subsequently provided crucial morphological correlates and an overall neuropharmacological context to pursue investigations of behavior. However, behavior is a property of even the simplest neuronal networks and it ensues inherently from the integration of the many complex, discontinuous cellular events encoded within neuronal ensembles. Thus, understanding the integrative mechanisms of neuronal functioning can only be achieved, in the final analysis, through assessments with intact preparations, and by employing *in vivo* investigations wherein the integrative response, the behavior itself, can be assayed directly. Indeed, a formidable challenge facing contemporary neurobehavioral research is to develop novel *in vivo* paradigms by which neuronal responses at the cellular level can be ultimately examined as integrated and intact phenomenal events at the behavioral level.

B. ADENOSINE AS A HOMEOSTATIC NEUROMODULATOR

Much evidence has emerged over the past decade that adenosine plays a physiological role in both the vertebrate[5] and invertebrate nervous systems.[6] Whether adenosine acts as a neuromodulator,[5] a retaliatory metabolite,[7] a homeostatic regulatory substance,[3] or a neurotransmitter[8] is unclear, since the criteria generally employed in evaluating neuroactive substances as modulators, transmitters, or regulatory factors have become increasingly indefinite.[9] Nevertheless, the physiological and pharmacological actions of adenosine receptor agonists and antagonists are well documented.[8-11] At the cellular level, numerous studies have shown that adenosine and its analogs inhibit the firing of central neurons,[8] depress the release of both excitatory and inhibitory neurotransmitters in brain tissue,[10] and modulate ion channels.[11] These neuroactive actions of adenosine appear to be mediated via methylxanthine (MX)-sensitive extracellular receptors and may be accomplished by modulating adenylate cyclase (AC) systems,[12] inositol phospholipid metabolism,[13] or by modulating calcium homeostasis in central tissues via direct action on the calcium channel.[11] From a behavioral perspective, much work has revealed a wide range of behavioral responses following the administration of adenosine and its analogs,[5,14,15] and many of these behavioral effects can be mimicked by drugs which alter endogenous levels of brain adenosine, including adenosine uptake inhibitors and adenosine deaminase inhibitors.[16]

In view of the wide array of adenosine receptors on neuronal surfaces[5] and the numerous, well-characterized, receptor-mediated actions of adenosine on cellular mechanisms in nervous tissue[17] and in light of the fact that adenosine and related substances exert pronounced effects on neural substrates mediating a variety of behaviors,[17,18] it follows that adenosine may serve as a useful probe to not only explore the physiological roles of transmitter substances in synaptic function, including receptor subtypes and related second-messenger systems, but to also gain insight about how neuronal events are ultimately integrated at the cellular level to generate behavioral events at the systems level. Indeed, there are some intriguing parallels for the actions of adenosine at the cellular and systems levels. For example, there is much evidence indicating the existence of purinergic tone in the CNS[8] whereby adenosine regulates the proportional release of neurotransmitter substances in order to maintain homeostasis in the synaptic microenvironment. Likewise, under certain pathophysiologic conditions, alterations in adenosine levels may serve to stabilize cellular energy levels, thereby maintaining overall tissue homeostasis.[5] Thus, in regard to the formulation of investigative strategies by neurobehavioral scientists, adenosine appears to be a powerful analytical vehicle wherein multiple findings on the actions of adenosine at the molecular, cellular, and tissue levels can conceivably be extrapolated to thereby explain adenosine-mediated response events at the behavioral level.

II. BEHAVIORAL ROLE OF ADENOSINE IN THE NERVOUS SYSTEM

A. ADENOSINE, AROUSAL LEVELS, AND PSYCHOACTIVE DRUGS

In contrast to the plethora of studies using binding methodology and *in vitro* preparations to investigate the electrophysiological and biochemical properties of adenosine and related drugs at the cellular level,[10] there has been relatively much less *in vivo* work examining the neuromodulatory role of adenosine at the behavioral level. What appears to be a predominantly consistent finding from the majority of behavioral studies is that adenosine exerts pronounced effects on arousal levels (see below).[8,14,17,19] Moreover, associated with the emergent notions that adenosine depressed neuronal activity at the cellular level[8,20] and that adenosine markedly affected arousal levels at the behavioral level,[8,9,14,21] it was further proposed that the effects on CNS arousal of a number of centrally acting drugs may be exerted, at least in part, via purine-linked mechanisms.[8,20,22,23] For example, with the recognition that commonly used central stimulants, such as the MXs, caffeine, and theophylline, were potent antagonists of the depressant effects of adenosine on neuronal firing[8,20] and displaced adenosine analogs in binding studies,[9,21] it became apparent that the central stimulant action of these compounds might be mediated via the blockade of central adenosine receptors. It was subsequently proposed that the sedative properties of a number of psychoactive drugs, used therapeutically as anxiolytic, hypnotic, and anticonvulsant agents, may be mediated in part via central purine-linked mechanisms.[8,22,23]

The ensuing findings from a variety of behavioral studies were readily forthcoming in providing direct *in vivo* support for this notion. Indeed, the observations from the *in vivo* behavioral studies confirmed many of the *in vitro* and electrophysiological findings at the cellular level relating to the interactions of certain psychoactive drugs with central adenosinergic systems.[8,22,23] Thus, the proposal that central adenosinergic systems may play a role in some of the CNS mechanisms of psychoactive drug action was subsequently supported by findings from numerous behavioral studies. These included studies examining the behavioral interactions on locomotor activity of adenosine and diazepam[24] and related benzodiazepine agonists and antagonists,[25] meprobamate,[26] and phenothiazines.[27] Similarly, behavioral interactions were demonstrated between adenosine and diazepam and/or barbiturates in mediating muscle relaxation[28] or anticonflict behaviors,[29] while other studies have shown interactions with alcohol in mediating motor incoordination and ethanol-induced hypnosis.[30]

Furthermore, there have been other studies reporting the behavioral interactions of adenosine with anticonvulsant agents, such as carbamazepine and phenytoin, in seizure behaviors and epilepsy;[31-33] with chlorpromazine and chlordiazepoxide in operant responding paradigms;[34-36] with carbamazepine in clonidine-induced aggressive behaviors;[37] with morphine in nociception, analgesia, and central mechanisms of opiate tolerance and dependency;[38] with imipramine and related tricyclics on the efficacy of antidepressive therapy in affective disorders;[39] and interactions of adenosinergic mechanisms with the proconvulsant γ-aminobutyric acid (GABA) antagonists, bicuculline and picrotoxin, have been shown for both cortical- and amygdaloid-kindled seizures,[40] while seizures induced by MXs were shown to be modified by the benzodiazepine receptor antagonist RO 15-1788.[41]

In fact, in view of reported discrepancies in the GABA hypothesis for the actions of benzodiazepines and in light of the aforementioned, well-established behavioral interactions between benzodiazepines and adenosine, it has been proposed that the cellular interactions of benzodiazepines and adenosine systems on central neurons may explain some of the anomalies that exist between the benzodiazepine-GABA hypothesis and actual findings from behavioral studies.[25,42] Moreover, at the clinical level the adenosine receptor antagonist aminophylline has been shown to be a potent and effective agent in reversing diazepam-induced anesthesia and EEG changes in humans.[43] Finally, highly selective interactions between adenosine receptor agonists and antagonists with carbamazepine have been once again recently demonstrated in behavioral studies examining drug-induced activity syndromes and spontaneous behaviors.[44]

In most of the earlier work examining the behavioral actions of adenosine and its analogs, only a relatively limited repertoire of adenosine analogs were available for use as adenosine receptor agonists and antagonists. As a result, until the past few years, the principal adenosine analogs used as agonists in the vast majority of behavioral studies included N^6-phenyliso-propyladenosine (PIA) — R and S isomers; 5'-N-ethylcarboxamidoadenosine (NECA); N^6-cyclohexyladenosine (CHA); and 2-chloroadenosine (CADO). Although many of the earlier behavioral studies in the adenosine field made attempts to ascertain whether or not a given adenosine receptor subtype actually mediated a specific behavioral response, the small repertoire of selective adenosine agonists available at the time placed severe limits on these seminal efforts to characterize the behavioral actions of adenosine receptor subtypes through traditional pharmacological methods such as establishing a rank potency order for agonists. Consequently, in the following section an overview of the behavioral actions of adenosine will be made without reference to specific adenosine receptor subtypes. This topic will be discussed in more detail in a subsequent section which deals with the recent behavioral studies using selective agonists and antagonists.

B. BEHAVIORAL ACTIONS OF ADENOSINE AND ITS ANALOGS: AN OVERVIEW

In general terms, with either parenteral or central routes of administration, adenosine, its analogs, and related agents potentiating the actions of endogenous adenosine have been shown to generally exert pronounced effects on arousal levels and to thereby elicit a wide variety of behavioral responses,[8,9,14,18,19] many of which appear to be specifically mediated by adenosine receptors since antagonism by MXs has been demonstrated.[21]

The numerous behavioral responses elicited by adenosine include sedation, hypoactivity, and depression of spontaneous and/or evoked locomotor activity;[21,44-55] inhibition of opiate withdrawal behaviors[38,56] and inhibition of apomorphine-induced rotation behavior;[57] anticonvulsant actions;[33,51,58-62] motor incoordination and ataxia;[30,46,49] antinocisponsive effects and analgesia;[38,48,49,63-66] muscle relaxation,[28] hypothermia,[48,67-69] and hypnogenic activity;[39,70-72] decrements in schedule-controlled operant behaviors[34,35,73-75] and xanthine-induced alterations of self-stimulation behaviors;[76] inhibition of aggressive behaviors,[37,77,78] alterations of food intake,[79-83] other ingestive behaviors,[29,84] and taste perception;[85] effects on some central mechanisms mediating regulation of cardiorespiratory behaviors;[15,18,86-89] effects on cognitive behaviors, including affectual states and memory;[90-92] central modulation of gastric secretions, stress mechanisms, and stress-induced ulcer formation;[93-99] finally, adenosine and its analogs exhibit some anxiolytic actions[29] and also demonstrate similar properties with antipsychotic agents in a variety of behavioral assays.[100]

Furthermore, many of the behavioral effects of adenosine analogs can be mimicked by drugs which manipulate endogenous levels of brain adenosine. For instance, adenosine uptake inhibitors can potentiate the sedative effects of adenosine in mice,[16,52] mimic the hypoactivity produced by adenosine and its analogs in mice,[16,101] and potentiate the antinocisponsive effects of adenosine in mice.[64] Likewise, adenosine deaminase inhibitors produced sedative and hypnotic effects after intravenous injection in humans,[102,103] intraperitoneal (i.p.) injection in rats,[104] or central injections in mice.[16] In one study, an adenosine deaminase inhibitor was shown to produce a profound decrease in spontaneous motor activity in mice and rats following parenteral injections.[103] Interestingly, in this latter study, the cortical EEG and righting reflex remained normal, while locomotor activity was severely depressed.[103] Further, apomorphine-induced rotation behavior was inhibited by erythro-9-(2-hydroxy-3-nonyl)adenine (EHNA), an adenosine deaminase inhibitor, which mimicked the effects of PIA in this paradigm.[57] It is also noteworthy that certain drugs, such as the phosphodiesterase-inhibitor rolipram and forskolin, are known to elevate endogenous levels of cyclic AMP, mimicking many of the behavioral effects of adenosine analogs — particularly hypothermia, hypoactivity, and anticonvulsant activity.[105-108] Similarly, direct central application of analogs of cAMP produce responses which resemble many of the behavioral responses elicited by centrally administered adenosine analogs, including hypoactivity, anticonvulsant activity, hypotension, and depression of respiration.[15,89,108,109]

Moreover, in view of the consensus that the central stimulant effects of xanthines likely involve the antagonism of central adenosine receptors,[110,111] it is not surprising therefore that many of the behavioral effects of adenosine are opposite to those of the xanthines. For example, the locomotor effects of adenosine and its analogs are opposite to those elicited by xanthines such as caffeine or theophylline, both of which cause increases in behavioral activity,[50,101,112] enhancement of mood and memory,[92] anxiogenic and stress activity,[95] and decreases in nociceptive thresholds.[38,48,64] Xanthines also exhibit proconvulsant activity on kindled seizures,[60-62] a potentiation of apomorphine-induced rotation behavior,[57] increases in blood pressure and respiration following central administration,[89,113] and hyperthermia.[69] Also, the relative locomotor stimulant potencies of xanthines in rodents parallel their affinities for binding sites labeled with [³H]CHA, and behaviorally depressant actions of R-PIA are transformed by xanthines into stimulant actions.[21,110,111] 3-Isobutyl-1-methylxanthine (IBMX), which has about the same affinity as caffeine for adenosine receptors, does not display locomotor stimulation at any dose examined.[16,50,101] Moreover, some xanthines show paradoxical and biphasic effects.[16,50,101] At lower doses, caffeine depresses locomotor activity, while it stimulates activity at higher doses.[50,101] In contrast, IBMX was a locomotor depressant at all doses examined,[16,101] whereas no depression is observed with any dose of theophylline.[50] The fact that the behavioral effects of xanthines vary considerably with different derivatives suggests that the stimulant and depressant effects of xanthines on behavior may be mediated by different mechanisms.[16,44,101]

In Table 1 are summarized the major behavioral actions of adenosine from numerous

TABLE 1
Behavioral Actions of Adenosine

Depression of locomotor activity
Decreased schedule-controlled operant responding
Antinociception/analgesia
Sedative/hypnotic
Anticonvulsant
Hypothermic
Respiratory depression (central)
Hypotension
Suppression of food intake
Inhibition of aggressive behaviors
Inhibition of stress-induced ulcer formation
Cognitive activities: affective states, memory, neuroleptic action

studies employing both central and parenteral routes of administration. Since there are other chapters in this monograph reviewing in detail the behavioral studies related to the hypnogenic, anticonvulsant, and antinocisponsive/analgesic actions of adenosine, the remaining portion of this chapter will mainly deal with those areas wherein the behavioral actions of adenosine have been more extensively examined. The following sections therefore will include a relatively thorough review of the studies examining the behavioral actions of adenosine on locomotor and related stereotypic behaviors. Additionally, there will be a more brief consideration of the behavioral actions of adenosine on food intake, operant responding behaviors, and behavioral tolerance.

C. CENTRAL VS. PERIPHERAL ACTIONS OF ADENOSINE

As described above, when administered by parenteral injection, adenosine and its analogs have been shown to have a variety of behavioral actions, including inhibition of spontaneous motor activity, motor incoordination, analgesia, hypnotic activity, changes in respiration, anticonvulsant activity, and alterations in food intake (Table 1). Adenosine also influences numerous peripheral systems. Its cardiovascular actions result in profound falls in blood pressure; it causes bronchial constriction, hypothermia, affects renal blood flow and renin release, and depresses acetylcholine and norepinephrine release at autonomic nervous system terminals.[5,19]

A problem arising in many of the behavioral studies to date has been that, following peripheral administration, it has been difficult to determine whether the behavioral responses are of central origin or secondary to the decreases in blood pressure and/or body temperature. In fact, in a recent study[69] it was clearly shown that the hypothermia or hyperthermia elicited by peripheral administration of adenosine agonists or antagonists, respectively, is not due to specific interactions with thermosensitive pathways in the brain. Instead, these behavioral responses appear to be nonspecific effects, likely mediated by activation of a variety of other nonspecific mechanisms.[69] Moreover, other studies have indicated that, following parenteral administration, frequently used adenosine analogs such as R-PIA and NECA may not cross the intact blood-brain barrier in the hippocampus in pharmacologically active amounts.[114] These results suggest that many of the reported behavioral actions that have been ascribed to peripherally administered adenosine analogs may have been mediated indirectly via actions at peripheral sites, or in regions of the brain where the blood-brain barrier is more permeable, such as the area postrema.

The dangers inherent in overlooking the peripheral effects of adenosine in any assessment of its purported central actions are exemplified in a report that maintenance of body temperature of mice given i.p. CADO or NECA can abolish the anticonvulsant actions of these compounds.[115] The hypotensive and hypothermic actions of adenosine are likely to be caused primarily by peripheral actions, with inhibition of central compensatory mechanisms as an additional factor.[116,117] Moreover, it is also known that parenterally administered adenosine and its analogs can selectively affect cerebral blood flow in regions like the striatum, and this may be a confounding factor in behavioral studies wherein adenosine analogs are administered behaviorally.[118,119] On the other hand, there have been a few recent studies which have directly measured, or indirectly estimated, levels of adenosine agonists and antagonists in the brain following parenteral administration. These findings will be discussed in a subsequent section considering the more recent behavioral work using selective analogs for adenosine receptor subtypes. It should be emphasized, however, that the confounding peripherally mediated actions of adenosine can be minimized in behavioral studies either by using an extensive series of selective agonists and antagonists or by comparing the behavioral actions of adenosinergic drugs following both central and peripheral administration (see below).

With these considerations in mind, the subsequent account of the behavioral actions of adenosine on locomotor activity will initially focus on the earlier results obtained with direct parenchymal injections or intracerebroventricular (i.c.v.) administration of adenosine and its analogs into the brain. With this route of administration it has been possible to dissociate the locomotor and hypotensive effects of adenosine.[54,55,88] Finally, in a subsequent section, the more recent studies examining the effects of selective adenosine receptor agonists and antagonists on locomotor activity, regardless of the route of administration, will be considered separately.

D. EFFECTS ON LOCOMOTOR ACTIVITY
1. Behavioral Studies: an Overview

A decrease in spontaneous locomotor activity (SLA) following i.c.v. injections of adenosine and ATP was noted in early experiments on cats[45] and rats.[46] In an extensive series of experiments on i.c.v. administration of adenosine analogs in mice, NECA was found to be the most potent depressant of locomotor activity, with a number of N^6-substituted analogs also displaying considerable activity.[14,16,53,56] In the i.c.v. studies, NECA exhibited approximately 100-fold greater potency than R-PIA in depressing SLA.[53] Very similar findings were reported for the depression of SLA following direct bilateral injections of NECA and R-PIA into the striatum of mice, wherein NECA was found to be approximately 100-fold more potent than R-PIA.[55] However, in view of the paucity of studies examining the effects of i.c.v.-administered adenosine analogs on SLA in rats, it is especially notable that NECA is only five- to tenfold more potent than R-PIA in depressing SLA following i.c.v. injections in rats.[54] Depression of locomotor activity by NECA and R-PIA was antagonized by caffeine.[14,16,53,55] Further, there were no effects whatsoever on cardiovascular parameters following bilateral striatal (b.l.s.) injections.[55] In contrast, there is some hypotension elicited by adenosine analogs following i.c.v. injections in rats[87] and mice.[88] Nevertheless, the depressant effects of NECA and R-PIA on SLA following i.c.v. administration can be readily dissociated from their hypotensive actions in both rats and mice.[54,55] In fact, the threshold dosages of adenosine analogs required to elicit significant depression of SLA are relatively similar following administration via either the b.l.s. or the i.c.v. route.[53-55,88] However, at the higher dosages of NECA and R-PIA, it is likely that the hypotension evoked via the i.c.v. route contributes somewhat to the observed behavioral sedation. Further, the data from these blood pressure studies suggests that the contributing effect of hypotension to behavioral sedation elicited by i.c.v. adenosine analogs was potentially a more confounding factor for R-PIA, particularly in mice, than was the case with NECA.[53-55,88] In any case, the depression of SLA elicited by both NECA and R-PIA can be antagonized by caffeine following either i.c.v. administration in mice[53] and rats[54] or direct b.l.s. administration in mice.[55]

In SLA studies, i.c.v. injections of adenosine itself were considerably less active than analogs of adenosine,[16,101] but this was undoubtedly due to tissue uptake and metabolism. The action of adenosine appears to be terminated by its removal from the synaptic cleft either by active neuronal uptake, coupled to the intracellular enzyme adenosine kinase, or by deamination to inosine by the enzyme adenosine deaminase. As a result, evidence for a neuromodulatory role of endogenously released adenosine in the brain was further sought in experiments with inhibitors of adenosine transport and adenosine deaminase. Three potent transport inhibitors, dipyridamole, dilazep, and papaverine inhibited locomotor activity when injected i.c.v. The effects of dilazep and papaverine, but not of dipyridamole, were antagonized by caffeine.[16,101] EHNA was used as an inhibitor of adenosine deaminase. This compound also depressed locomotor activity and its actions were antagonized by caffeine.[16]

A paradoxic effect was observed when caffeine administration preceded an i.c.v. injection of adenosine, in that low doses of adenosine now stimulated locomotor activity.[16,101] In contrast, when mice were given NECA after caffeine, there were no dose combinations which produced stimulation.[16,53,101] The observations with adenosine find a parallel in an earlier study reporting that after pretreatment with caffeine, peripherally administered R-PIA in doses which previously depressed activity, now caused a pronounced locomotor stimulation.[21] Low doses (0.005 to 0.07 mg/kg) of R-PIA can stimulate locomotor activity, even in the absence of caffeine.[50] This suggests the presence of a heterogeneous population of receptors mediating the behavioral effects of adenosine, some of which produce sedation while others cause stimulation. Caffeine may compete more effectively with adenosine and R-PIA at the receptor subtype responsible for the depressant actions of adenosine, thus uncovering effects mediated by the stimulant receptor.

Somewhat unexpected findings were also observed with IBMX. This substance depressed locomotor activity in mice when administered i.c.v.,[101] even though it is an adenosine antagonist.[110,111] Indeed, IBMX was able to antagonize the locomotor depressant actions of adenosine itself.[101] Its failure to stimulate locomotor activity may be related to its phosphodiesterase (PDE)-inhibiting properties, as a variety of PDE inhibitors can cause behavioral depression in mice.[106,108] Support for this possibility comes from experiments with forskolin, which stimulates the enzyme adenylate cyclase and enhances cAMP levels in the brain.[106] Forskolin depresses locomotor activity when injected i.c.v.[105] or i.p.[108]

Another type of motor activity affected by adenosine is the rotational behavior associated with modulation of striatal dopaminergic function. Local applications of NECA into the striatum induced ipsilateral rotation of rats following systemic apomorphine injection.[120] R-PIA was less effective than NECA, whereas 2′,5-dideoxyadenosine antagonized the response to NECA, as did theophylline. Rotation behavior induced by apomorphine in rats with unilateral lesions of the nigrostriatal pathway was inhibited by R-PIA, by the PDE inhibitor rolipram, and by EHNA.[57] By contrast, theophylline and 8-phenyltheophylline caused a potentiation of rotation behavior, presumably by antagonizing endogenous adenosine.[57]

2. Adenosine Receptors in the Nervous System: Interactions with Other Neurotransmitter Systems

It is likely that adenosine and its analogs exert receptor-mediated actions on locomotor activity by modulating endogenous transmitter function in structures such as the striatum. Much of the work on adenosine relates to its modulating actions on AC, wherein two extracellular adenosine receptor subtypes have been postulated. Those designated A_1 receptors are associated with decreases in cAMP, while A_2 receptors are associated with increased cAMP levels.[8,12,121] Additionally, there is also evidence that adenosine receptors may be directly coupled to ion channels.[11] In addition to biochemical criteria, adenosine receptor subtypes are often distinguished by their relative affinities for specific agonists.[5,10,11,122,123] In the earlier behavioral studies, NECA was considered to be a specific A_2 agonist, since

NECA was reported to be relatively selective for the A_2 receptor in binding studies.[9,123] Similarly, NECA was reported to be A_2 selective in some biochemical assays, since it only stimulated AC in forskolin-pulsed hippocampal slices, whereas R-PIA produced decreases (via A_1) and increases (via A_2) in this preparation.[124] It was subsequently established, however, that NECA actually exerts mixed A_1-A_2 agonist actions on adenosine receptors.[122,123] However, in contrast to the N^6-substituted analogs of adenosine, such as PIA and CHA, NECA does not appear to be a substrate for the nucleoside transporter.[125] Hence any prolonged behavioral actions observed with NECA would be expected to be predominantly mediated via extracellular receptors. In view of the lack of understanding about the role of adenosine receptor subtypes in mediating actions of adenosine at the behavioral level, the strategy adopted for many behavioral studies therefore was to examine the behavioral physiology of adenosine by using NECA, an analog with mixed agonist properties, which thereby more closely resembles the receptor-mediated behavioral actions of adenosine, the naturally occurring endogenous nucleoside.[14-16,24,53,59]

In any event, it has been proposed that A_1 receptors on presynaptic terminals mediate inhibition of neurotransmitter release in CNS tissues,[10,126] but this principally may only hold true for the hippocampal slice[127] since other findings do not support such an exclusive assignation for the A_1 receptor.[6,11,128-133] Moreover, there is evidence that the presynaptic inhibition of neurotransmitter release by adenosine is not necessarily mediated by decreased levels of cAMP.[126,127] Although the functional role of A_2 receptors has been less apparent, recent work from ligand-binding assays has suggested the existence of A_2 receptor subtypes.[122,123,128] Additionally, there is evidence that A_2 receptors are localized postsynaptically on intrinsic neurons[129-133] and may exist in multiple-affinity states.[133,134] Other studies have shown that A_1 receptors may also be located at postsynaptic sites in some tissues.[135] In some instances, there is also evidence that adenosine receptors are functionally linked to other neurotransmitter substances[127,135-142] wherein second-messenger systems may be shared through common G protein-linked receptors.[143-145]

Information processing and neural integration in the brain evidently involve communication among neurons through release of chemical messengers at synapses, many of which may coexist in individual neurons. Moreover, there is considerable evidence that coexisting neurotransmitters/neuromodulators in central neurons may be coreleased and thereby act simultaneously on pre- and postsynaptic receptors. Many of these pre- and postsynaptic multiple-transmitter receptor interactions in the striatum have been shown to mediate locomotor activity and related stereotypic behaviors.[44,108,131,132,135,136] Specifically, dopaminergic receptor interactions in the striatum appear to principally mediate locomotor behaviors, with other catecholamines serving a secondary or modulating role.[132] It is notable therefore that adenosine modulates the pre- and postsynaptic actions of catecholamines, particularly DA in striatal tissues, and that adenosine receptors are linked with DA receptors wherein they share a common effector-coupling system.[128,132,135,146-148] A large body of evidence indicates that membrane receptors coexisting on neurons and mediating similar functional responses may share common signal-transduction systems. Specifically, G proteins may serve as common binding sites on membranes coupling membrane receptors to receptor-regulated second-messenger systems and/or specific ion channels.[144]

For example, receptors that activate AC, such as DA (D_1), adenosine (A_2), and β-adrenergic receptors, apparently do so by activation of a specific G protein (i.e., G_s). Likewise, a separate G protein (G_i) is likely involved in the inhibition of AC by specific receptors including GABA$_B$, muscarinic (M_2), adenosine (A_1), dopaminergic (D_2), and adrenergic (α_2) receptors.[144] Moreover, these G proteins form complexes with guanine nucleotides, and it has been shown for some G protein-coupled receptors that guanine nucleotides can modulate neurotransmitter affinity for a receptor subtype including the adenosine A_1, dopaminergic (D_2), β-adrenergic, and α_2-adrenergic receptors.[143] Thus, a single transmitter may activate several members of the G protein family by means of different receptor subtypes and thereby influence multiple-effector systems. These interactions may explain the diverse effects of a neurotransmitter such as norepinephrine or a neuromodulator such as adenosine which can either stimulate or inhibit AC or stimulate phosphoinositide turnover.[4,13,145]

Since adenosine receptors are widely arrayed on neuronal surfaces and adenosine receptor activation occurs at both pre- and postsynaptic loci,[5] the pharmacologic effects of adenosine may be indirect, inhibiting the release of a variety of conventional neurotransmitters from presynaptic terminals, likely via A_1 receptors, or direct, by affecting postsynaptic processes involving A_1 and A_2 receptors.[131] The presynaptic inhibitory effect of adenosine receptors on neurotransmitter release has been predominantly demonstrated for acetylcholine,[127] noradrenaline,[137,139,140] DA,[148], serotonin,[142] and glutamate,[138] but not for GABA.[5,10,11,26] It is therefore likely that adenosine more potently inhibits excitatory neurotransmission. While there is good evidence that cAMP is associated with facilitation of monoaminergic neurotransmitter release,[149] it is not clear whether or not reductions in cAMP are associated with decreased neurotransmitter release.[126] Nevertheless, presynaptic inhibition of neurotransmitter release is likely mediated by adenosine receptors via G protein-linked mechanisms.

Much recent work has clearly shown that adenosine receptors are functionally linked to receptor systems involved in catecholaminergic neurotransmission. In a number of studies in CNS tissues, it has been shown that there is a functional coupling between presynaptic adenosine A_1 receptors and adrenergic α_2-autoreceptors which mediates the presynaptic release of noradrenaline.[139,140,147] Evidence from a recent study suggests that a common pool of G proteins couple A_1 receptors and serotonin (5-HT$_{1A}$) receptors to the same effector mechanisms in the hippocampus.[142] Additional interactions have been reported, for instance, between three presynaptic, release-inhibiting receptor systems — adrenergic α_2, opioid kappa, and adenosine A_1 — whereby activation of one receptor system attenuates the inhibition exerted by the two remaining systems.[150] As a point of caution with regard to the receptor-mediated actions of adenosine, it is important to emphasize that catecholamines can exert mixed agonist actions on adrenergic receptors in brain tissue, and it has been shown that activation of α-receptors by norepinephrine can inhibit or augment cAMP accumulation mediated by β-receptor activation.[147] Moreover, 8-phenyltheophylline, an adenosine receptor antagonist, significantly inhibited the α-receptor-mediated augmentation of the cAMP response, further suggesting a complex functional interaction between adrenergic receptors and adenosine receptors in nervous tissue.[147]

In a corresponding manner, a remarkably similar account has emerged for the functional interaction between adenosine A_2 receptors and dopaminergic receptors in the striatum.[136] For example, high-affinity A_2 receptors (i.e., A_{2a})[122] are specifically distributed in the rat brain in regions such as the striatum, also demonstrating discrete distribution of the DA D_2 receptor, suggesting that modulation of dopaminergic activity is an important function of the A_2 receptor.[129,131,132] A number of observations have also indicated a functional interaction between adenosine A_2 receptors and dopaminergic D_1 receptors in the striatum.[129,135,136] It has been shown, for example, that DA autoreceptors and adenosine A_2 receptors interact antagonistically in the striatum to control dopaminergic transmission.[151]

In a related study examining dopaminergic mechanisms mediating locomotor activity in 6-hydroxydopamine (6-OHDA)-lesion rats, a pronounced interaction between DA agonists and postsynaptic A_2 receptors was shown.[129] A number of other observations also indicate an interaction between adenosine A_2 receptors and dopaminergic systems. Administration of adenosine receptor agonists in rats leads to a desensitization of both adenosine A_2 and DA D_1 receptors in the neostriatum.[152] Also, manipulation of the dopaminergic pathway by treatment of rats with 6-OHDA or reserpine leads to a sensitization of A_2 receptors.[135] The observation that maximally stimulating concentrations of A_2 receptor agonists and D_1 receptor agonists on AC are additive in their effects suggests that these receptors are associated with different pools of AC catalytic subunit.[153]

In fact, it has been proposed that endogenous adenosine may exert a tonic inhibition of the dopaminergic system via postsynaptic mechanisms mediated by the A_2 receptors.[132] Furthermore, in rat synaptosomes, a presynaptic interaction was shown between adenosine A_2 receptors and DA autoreceptors. Indeed, the results indicated that presynaptic DA (D_2) autoreceptors and presynaptic adenosine A_2 receptors interact antagonistically to regulate tyrosine hydroxylase activity, presumably by exerting opposite inputs on a presynaptic AC system.[151] These observations suggest an interaction between A_2 receptors and dopaminergic systems in the striatum, and it is likely that the A_2 receptor in most of these studies is associated with neurons that are postsynaptic to the dopaminergic system.[132,136]

Moreover, there also appear to be interactions between dopaminergic systems and adenosine A_1 receptors in the striatum, since it has been shown that A_1 receptors inhibit the presynaptic release of DA.[148] In this study, the inhibitory effects of parenterally administered adenosine agonists on striatal DA release (measured as 3-methoxytyramine) was examined. Although NECA, CHA, and R-PIA were equipotent in inhibiting DA release, the observation that the relatively selective A_1 antagonist, 8-cyclopentyl-1,3-dipropylxanthine, CPDX, was effective in blocking the inhibitory actions of CHA, led the authors to conclude that stimulation of A_1 receptors reduced striatal DA release.[148] In fact, there may also be selective interactions in the striatum between adenosine A_1 receptors and adrenergic receptors, since acute administration of R-PIA, a selective A_1 agonist, significantly decreased central catecholaminergic neurotransmission in all brain regions examined, but the noradrenergic levels in the striatum were the most affected.[146] Taken together, these studies support the notion that central dopaminergic (and adrenergic) systems may interact with both A_1 and A_2 adenosine receptors in the striatum and that these interactions at the receptor level may mediate the locomotor effects of adenosine analogs at the behavioral level.

3. Dopaminergic/Adenosinergic Interactions in the Striatum and the Locomotor Depressant Actions of Adenosine

It has been proposed that studying the properties of interneuronal interactions in the striatum mediating spontaneous or evoked locomotor activity may help reveal the more general principles of brain functioning which govern neural pattern generation involved in a wide range of other behavioral responses.[154] In view of the aforementioned, well-characterized functional interactions between both adenosine A_1 and A_2 receptors and striatal neurotransmitter systems, it follows that adenosinergic drugs may be used as powerful probes to investigate the integrative mechanisms of synaptic function in the striatum, including the organizational principles of striatal interneuronal networks, which mediate spontaneous and evoked locomotor behaviors. Thus, by contrast with the extensive *in vitro* work on hippocampal slices wherein only the role of A_1 receptors has been mostly examined and wherein a specific and assayable behavioral response for the hippocampal A_1 receptor has not as yet been established at the *in vivo* or behavioral levels, *in vitro* work on the striatum, in comparison, may provide observations at the cellular level which can then conceivably be extrapolated to understand the role of adenosine receptors at the behavioral level.

In any case, the striatal dopaminergic neurotransmitter system appears to be the most important and primary system which mediates locomotor activity.[155] In general, manipulations which augment dopaminergic function stimulate locomotor activity, whereas those that reduce dopaminergic function depress locomotor activity. For instance, elevation of central DA levels by peripheral administration of L-dopa enhances SLA particularly when it is administered with a decarboxylase inhibitor, which prevents the conversion of L-dopa to DA in brain capillaries.[155] In addition, a number of reports suggest that amphetamine-induced locomotor activity is mediated by stimulating DA release, while noradrenaline (NA) plays only a secondary or negligible role in the response.[155] Conversely, marked inhibition of SLA has been observed following agents which specifically inhibit DA release without inhibiting DA or noradrenaline synthesis.[155] Moreover, i.p. administration of DA receptor antagonists, including haloperidol, fluphenazine, and sulpiride have been shown to be potent inhibitors of locomotion.[156,157]

Normal striatal DA receptor function appears to involve a complex interaction between D_1 receptors, which are positively linked to stimulation of AC activity, and D_2 receptors, which are negatively linked to AC.[158,159] Complete expression of DA-mediated physiological events and motor behaviors apparently requires simultaneous stimulation of both D_1 and D_2 receptor subtypes.[158] D_2 receptors appear to be the site of action of neuroleptic drugs; however, specific behaviors mediated by the D_1 receptor have also been described.[157]

Numerous lines of evidence from a variety of behavioral, pharmacologic, and neuro-chemical studies have indicated a crucial interaction between striatal adenosine and DA receptors in the mediation of locomotor behaviors. For example, a functional linkage between DA and adenosine receptors modulating striatal AC has been demonstrated.[153] It has also been shown that there is a differential location of cyclase-linked A_1 and A_2 receptors in the striatum wherein A_2 receptors were associated with intrinsic striatal neurons, whereas A_1 receptors appeared more diffuse, being associated with corticostriatal nerve terminals in addition to intrinsic neurons.[160] The potency rank order for inhibiting cyclase via the A_1 receptor was R-PIA > CADO > NECA, whereas the potency order for stimulating cyclase via the A_2 receptor was NECA > CADO > R-PIA. It is notable that the EC_{50} for the NECA-induced stimulation of cyclase in the striatum was in the nanomolar range, suggesting that there is a high-affinity A_2 receptor linked to striatal AC.[160] Similar findings were reported in a study examining the topographical distribution of adenosine-sensitive AC systems in rat striatum wherein adenosine itself was more potent than R-PIA, as was CADO, in stimulating cyclase.[161] Moreover, the distribution of the stimulatory adenosine receptor resembled the distribution of DA-sensitive AC and DA content in the striatum; these enzymes were predominantly located on striatal interneurons or on cell bodies of striatal neurons projecting out of this structure.[161]

In other studies, the adenosinergic modulation of striatal DA transmitter function has been examined directly. For example, in synaptosomal preparations of guinea pig striatum, CADO was found to increase DA synthesis.[151] This stimulatory effect of CADO on DA synthesis was markedly depressed by the D_2 agonist quinpirole, as well as DA itself. Since D_2-DA agonists inhibit AC, whereas CADO has been shown to activate striatal AC, the combined results suggested that DA autoreceptors and adenosine A_2 receptors interact antagonistically in controlling striatal DA synthesis and release.[151]

Adenosinergic modulation of striatal neurotransmitter release has also been examined *in vivo*, wherein adenosine was reported to modulate not only DA release from intrinsic striatal neurons, but it also appeared to inhibit glutamate release from corticostriatal terminals.[162] Similarly, CADO, administered i.c.v., was shown to selectively inhibit striatal cholinergic transmitter function via a MX-sensitive mechanism. The data also indicated that the depressant effects of adenosine agonists on the striatum, in contrast to the hippocampus, were mediated by A_2 receptors associated with intrinsic striatal neurons.[163] In this context, it is notable that cholinergic-dopaminergic-adenosinergic interactions on rat striatal AC activity were recently examined.[164] Thus, whereas NECA, DA, and forskolin stimulated AC activity and these effects were augmented substantially with the addition of purified protein kinase C (C-kinase) in the presence of guanosine 5'-triphosphate (GTP); the muscarinic agonist oxotremorine inhibited striatal AC activity and this effect was abolished by C-kinase. These results indicate that DA, adenosine, and acetylcholine may regulate striatal AC via common G protein-linked mechanisms.[164]

Furthermore, the cellular and biochemical effects of adenosine on striatal DA function appear, in fact, to be relevant to the behavioral actions of parenterally administered adenosine analogs on locomotor activity. It has been shown, for example, that i.p. administration of the adenosine analogs, NECA, R-PIA, and CHA, at behaviorally active doses which depress SLA, produced an *in vivo* dose-dependent decrease in DA synthesis and DA release from rat striatal DA nerve terminals, and these effects on striatal DA function were blocked by MXs at the i.p. dosages which also block agonist-induced depression of SLA.[165] In another recent study, the effects of selective A_1 and A_2 adenosine agonists on striatal DA release and metabolism were investigated *in vivo* using the technique of microdialysis.[166] The adenosine analogs were administered locally via microdialysis, and DA metabolites were subsequently measured from the striatal dialysates. CADO at concentrations of 1 to 10 μM and R-PIA and NECA at concentrations of 10 μM decreased DA and levels of DA metabolites. The inhibitory action of CADO was blocked by the selective A_1 antagonist, 8-cyclopentyl-1,3-dimethylxanthine (CTP), suggesting mediation by A_1 receptors. CADO, at higher concentrations, however, increased DA release, and this effect was not inhibited by CTP. This stimulatory action on extracellular DA concentration was also seen with 10 μM 5'-N-cyclopropyl-carboxamidoadenosine, (CPCA), a relatively specific A_2 agonist.[166] These results suggest that striatal DA release can be both inhibited and enhanced by adenosine, presumably via A_1 or A_2 receptors, respectively.

From a behavioral perspective, numerous studies have examined the interactions of striatal adenosine and DA systems on locomotor activity. The increase in locomotor activity following MX administration[50,57] is manifest as contralateral rotational behavior in rats unilaterally lesioned in the nigrostriatal DA pathway, an effect similar to that observed following administration of DA agonists.[167] The xanthines can also potentiate the effects of DA agonists as well as the ipsilateral rotating elicited by amphetamine. These effects can be antagonized by haloperidol, a mixed DA receptor antagonist, suggesting that DA is mediating the effects of xanthines. Since MXs do not interact directly with DA receptors,[167] the behavioral actions of MXs are likely mediated via antagonism of striatal adenosine receptors, which thereby affects striatal DA levels and DA receptor function.

Other studies suggest that the behavioral responses produced by the MXs may be related to striatal DA transmitter function. In fact, MXs even produced contralateral rotation, a characteristic DA-mediated response, following direct administration into the denervated striatum of 6-OHDA-lesion rats.[168] Moreover, when NECA was administered into the left caudate nucleus along with parenterally administered apomorphine, rats rotated to the left. Interestingly, the combination of NECA and apomorphine was substantially more potent than the combination of R-PIA and apomorphine, in eliciting rotation, suggesting the involvement of the A_2 receptor in this behavior.[120] Indeed, the fact that adenosine receptor agonists can affect locomotor activity and apomorphine-induced stereotypic behaviors in a manner similar to the classical DA antipsychotics[169] has led to attempts to develop a novel class of adenosine agonists which may have therapeutic potential as nondopaminergic, antipsychotic agents.[100]

On the other hand, the specific role of adenosine receptor subtypes in DA-mediated behaviors will require extensive further work, since the role of DA receptor subtypes in stereotypic behaviors is not well understood. As mentioned previously, it is thought that D_1 receptors (positively linked to cyclase) and D_2 receptors (not linked or negatively linked to cyclase) must be activated simultaneously for the full expression of various dopaminergic responses in normal rats.[157-159] Other studies have shown a complex interaction of striatal DA receptor subtypes in mediating stereotyped behaviors. For example, the potentiation of apomorphine-induced stereotypic behaviors by acute administration of DA-depleting agents appears to be mediated by both the stimulation of D_1 receptors and the blockade of a postsynaptic DA receptor subtype. Nevertheless, there was a good correlation for the time course of cAMP accumulation in the striatum with the time-course for potentiation of apomorphine-induced stereotyped behavior, suggesting that stimulation of striatal AC must first occur for the activation of stereotypic behaviors.[170]

4. Persistent Arguments Related to Diffusion of Adenosine Analogs, Potency Order, and Receptor Subtypes

At this point, it may be appropriate to consider the notion put forth by some investigators that lipophilic adenosine analogs (i.e., PIA and other A_1 agonists) have a slower onset of action since they equilibrate slowly with neural tissue, and this may explain the lower potency of these substances in some test systems compared to NECA or other A_2 agonists.[10,171-173] At first thought, it would seem that this type of argument is unessential since it would seem to render as fatuous the ranking of agonist potency orders in behavioral studies. The argument has been principally forthcoming from electrophysiological observations of the depressant effects of adenosine analogs in the hippocampal slice preparation wherein A_1 agonists, such as R-PIA, apparently require a considerably longer period to equilibrate in the tissues and exert maximal depressant effects.[171] However, the problem with this argument may arise from the slice preparation itself, wherein usually only a unitary response is assayed, i.e., depression of field excitatory postsynaptic potential (EPSP), which is purportedly mediated by the A_1 receptor.[171] Thus, although A_2 receptors have also been identified in the hippo-campal regions,[131,133,136] it has not been possible to ascertain their functional role(s) as yet using the assay of field EPSP depression in the hippocampal slice.[10,171]

Nevertheless, since functionally distinct roles for A_1 and A_2 receptors have been estab-lished for many peripheral tissues[5] and in view of the fact that A_1 and A_2 receptors have been identified with a heterogeneous and specific distribution in the CNS,[5,122,123,129-131] it follows therefore that the A_2 receptor would by now appear to have been associated with some functional activity in regions of the nervous system other than the striatum. Surprisingly, up until the aforementioned recent work on the role of A_2 receptors and DA transmitter functions in the striatum, there had not been much effort directed toward investigating the functional role of A_2 receptors.

Be that as it may, it seems that the argument about the limited diffusion of R-PIA due to its lipophilicity is usually made as an effort to explain the inactivity or low potency of R-PIA.[171] For example, such an argument was recently used to explain the lower antino-ciceptive potency of R-PIA in a hot-plate latency test following direct intrathecal (i.t.) injections of R-PIA or NECA.[172] However, whereas R-PIA had a similar potency as NECA in the tail-flick latency test, it was significantly less effective in the hot-plate test. Remark-ably, even though R-PIA had the identical peak onset of action as did NECA (i.e., within 15 min) in the tail-flick test, the argument was put forth that R-PIA was less potent in the hot-plate test due to its poorer penetration into the lumbar spinal cord. The data warrant a more straightforward explanation, namely, that R-PIA did not exhibit any peak antinocis-ponsive action in the hot-plate test during the entire 2-h test period, in contrast to NECA, which exhibited a peak antinocisponsive action in the hot-plate test 15 min after i.t. admin-istration, and this was the same time of onset of peak antinociception for both NECA and R-PIA in the tail-flick test.[172] Thus, it is difficult to accept the diffusion argument for only the hot-plate test and not the tail-flick test in this study.

In another recent study examining the protective effects of a series of adenosine analogs against generalized seizures following direct injection into the rat prepiriform cortex (PPC), the same diffusion argument appeared once again.[173] In the seizure study, NECA was again the most potent analog, followed by CHA, CPA and then R-PIA. Nonetheless, from an analysis of the diffusion of tritiated drugs from the PPC injection site, the argument was made that the higher potency of NECA (the ED_{50} dose was fivefold lower than for R-PIA) was due to its greater diffusion in the PPC compared to R-PIA. Similar diffusion data were not presented for CHA, CPA (which had an ED_{50} dose fourfold higher than NECA), CADO, or the other analogs tested.[173] Thus, it seems that there is a persistent notion that the potency of NECA is due to its better diffusion.

The question to be asked is whether or not this "diffusion/onset of action" argument creates an unnecessary paradox for establishing agonist potency orders of adenosine analogs for behavioral assays. Indeed, the argument of restricted access has also been used to explain the lack of *in vivo* effects of R-PIA, NECA, and CADO on hippocampal-evoked field EPSPs following i.p. or i.c.v. administration.[114] Only dorsal i.c.v. administration of NECA appeared to be effective in depressing hippocampal EPSPs. Brain levels of these analogs were extremely low or negligible following i.p. injections of 2 mg/kg of NECA, 9.1 mg/kg of R-PIA, and 10 mg/kg of CADO.[114] In contrast to these findings, other studies have shown that R-PIA readily gains entry into the brain in appreciable levels following i.p. administration of locomotor depressant doses[50] and at i.p. dosages showing potent antinocisponsiveness.[174] Further, CHA, R-PIA, and NECA were shown to decrease striatal DA synthesis following i.p. administration of behaviorally active doses, further demonstrating that systemically administered adenosine analogs gain access to the brain.[165]

In view of the fact that the diffusion characteristics for adenosine analogs have not been systematically examined, the argument concerning the "limited diffusion and/or restricted access" of R-PIA remains as only a hypothetical notion since it is not supported by findings from any other studies. In fact, in a recent study examining the effects of R-PIA and NECA on SLA, the time course of the hypoactivity syndrome was determined following i.p. injections of varying doses of each analog. The data clearly show that R-PIA and NECA showed very similar time courses at all dosages following i.p. or subcutaneous injections. Further, it was clearly shown that NECA was substantially more potent in depressing SLA.[44] Indeed, in a recent study the agonist rank order potency was examined for a series of selective A_1 and A_2 receptor agonists on motor coordination and in depressing spontaneous and amphetamine-evoked locomotor activity following both i.p. and i.c.v. administration.[156] Both the time course for depression of SLA and the maximal potency of each agonist on SLA were compared between the i.p. and the i.c.v. routes. All analogs were more potent in depressing SLA following i.c.v. injection than i.p. injections. Potency after i.c.v. administration was correlated with potency during the first hour and i.p. potency was correlated with the second hour. The analogs also showed similar time courses following i.c.v. and i.p. administration, and the rank order of potency was similar whether by i.c.v. or i.p. administration. CHA was slightly less effective via the i.p. route. NECA was consistently the most potent analog tested for SLA and the other behaviors examined.[156]

Similar findings regarding time course and onset of action were reported in blood pressure studies with adenosine analogs, wherein the onset of hypotension and the time to the maximum hypotensive response following i.c.v. injections were virtually identical for NECA and R-PIA in studies with both rats and mice.[54,87-89] In fact, the onset for hypotension was even more rapid for R-PIA in the i.c.v. studies with mice.[88] Nevertheless, NECA was still more potent that R-PIA in eliciting hypotension following i.c.v. injections.[88] Finally, in a study comparing the blood pressure responses of NECA and R-PIA following direct injections into the fourth ventricle, once again the time course for analog-induced hypotension was virtually the same for NECA and R-PIA. Nevertheless, NECA was still more potent in eliciting hypotension than was R-PIA.[89]

Taken together, these more recent studies examining the onset of action and the time course for behavioral responses elicited by NECA and R-PIA, whether on locomotor depression following parenteral administration or on blood pressure responses following intraventricular injections, clearly show that R-PIA readily penetrates into brain tissue from the brain ventricles or from the periphery. It follows that the greater behavioral potency of NECA relative to R-PIA in these studies cannot be dismissed, in view of the similar patterns for onset and time course, with arguments that the lower potency of R-PIA is due to a slower onset of action or due to restricted access to the site of action.

Finally, it is worth considering the possibility that the limited diffusion in tissues reported for R-PIA[171-173] may be due to the fact that both R-PIA and CHA, unlike NECA, are potent inhibitors of the adenosine transporter,[125] wherein as substrates for the transporter they even show higher affinities for transport sites in some tissues than adenosine itself.[175] In contrast to NECA, these same analogs have also been shown to be transported into cells and to accumulate intracellularly.[176] Finally, PIA and CHA may not be good agonists for behavioral studies since they are phosphorylated *in vivo* into active derivatives by adenosine kinase, whereas this is not the case for NECA.[177] Thus, it is conceivable that the restricted diffusion and longer equilibration periods associated with some A_1 agonists, like PIA, are due to the transport and accumulation of these analogs intracellularly, whereby they may even exert intracellular actions which are not mediated by either A_1 or A_2 receptors.

5. Recent Studies on Adenosine-Induced Locomotor Depression Using Selective Ligands for A_1 and A_2 Receptors

In more recent behavioral work on the role of adenosine receptors in mediating locomotor depression and related stereotypic behavioral patterns, the use of selective ligands for adenosine receptors, accompanied by increasingly more work on the role of striatal DA function in stereotypic behaviors, has allowed for more detailed examination of the adenosine/DA receptor interactions mediating adenosine-induced locomotor depressions. Additionally, the recent development of inbred strains of mice, which differ in their behavioral sensitivity to MXs, holds much promise for studies on threshold striatal adenosine receptors in mediating locomotor activity.

In this regard, for example, two strains of mouse, CBA/J and SWR/J, have been identified which markedly differ in their sensitivities to caffeine: CBA/J are more sensitive to the locomotor stimulating, hyperthermic, and lethal effects of caffeine as compared to SWR/J mice.[178] While caffeine sensitivity in these mice has been shown to be genetically determined with different genetic components controlling sensitivity to different MXs, the exact mechanisms underlying the behavioral sensitivity is unknown. However, a recent quantitative autoradiography study examined the possibility that the behavioral sensitivity to caffeine in the CBA/J mouse strain was mediated by differences in regional adenosine receptor affinity or density.[178] The results indicated that this behavioral sensitivity may be related to A_1 and A_2 receptors. More specifically, the densities of A_1 receptors in the hippocampus and cerebellum of CBA/J mice, as well as A_2 receptors in the striatum, were greater compared to SWR/J mice. In contrast, no differences were found in striatal DA D_1 or D_2 receptors between the mouse strains. Moreover, in the striatum, wherein equal densities of A_1 and A_2 receptors are found, the only significant differences in the density of A_2 receptors were seen in the striatum of CBA/J mice. In fact, a selective localization of high-affinity A_2 receptors (i.e., A_{2a}) was seen in the striatum of both strains of mice. These results suggest that the differential sensitivity ot MXs between these mouse strains may involve a genetically mediated difference in regional adenosine receptor densities.[178]

In the meantime, in a recent and quite extensive study, the correlation of the behavioral effects of various MX analogs with their activity as adenosine receptor antagonists and as PDE inhibitors was examined.[179] The data showed that caffeine analogs, such as 1,3,7-tripropargylxanthine (TPPX), which stimulated SLA behavior, showed high correlations with their potency as adenosine receptor antagonists, and these MX analogs also showed weak activity as inhibitors of various species of PDE. In contrast, MX analogs such as IBMX, which were potent inhibitors of PDE, were behavioral depressants regardless of whether or not they showed a high potency as adenosine receptor antagonists.

The most potent MX stimulant was TPPX, which also showed the highest affinity for the A_2 receptor of all the MX analogs examined. On the other hand, 1,3,7-tripropylxanthine (TPX) and IBMX, which exhibit similar antagonist potencies as does TPPX on adenosine receptors, but with higher selectivity for the A_1 receptor, produce behavioral depression. Whereas TPPX also reverses the behavioral depression elicited by CHA, TPX does not.[179]

The lack of antagonism by TPX of CHA-induced depression of SLA, in addition to the paradoxical effects of other A_1-selective ligands, such as IBMX, on locomotor activity argue against a functional role for A_1 receptors in mediating adenosine-induced locomotor depression. For example, in contrast to TPX, IBMX at low dosages reverses the depressant effects of the A_1 agonist, CHA,[21] and adenosine itself.[101] Caffeine in combination with IBMX also elicits a marked stimulation of SLA with a magnitude much greater than that for caffeine alone;[101] in addition, depressant doses of R-PIA[21] and adenosine[101] can be converted to stimulant doses by caffeine. In fact, low doses of R-PIA can stimulate locomotor activity even in the absence of caffeine.[50]

Furthermore, one of the A_2-selective MX analogs examined in the Choi et al.[179] study, 3,7-dimethyl-1-proparglyxanthine (DMPX), was a very potent behavioral stimulant and it was reported to selectively reverse NECA-induced behavioral depression compared to CHA-induced depression.[179] DMPX also showed a high selectivity for the A_2 receptor. In contrast, the lack of behavioral activity of enprofylline, a relatively selective A_1 antagonist, compared to the stimulant properties of caffeine is notable, since both xanthines are nearly equipotent at brain adenosine A_1 receptors. However, enprofylline is manyfold less potent at the A_2 receptor than caffeine.[179]

Taken together, these recent studies with MX analogs, along with earlier behavioral studies, indicate that adenosine-mediated activation of the A_2 receptor, which has been shown likely to occur in the striatum,[55] mediates the behavioral depression of adenosine analogs. In view of the fact that rolipram, a potent inhibitor of PDE with little activity at adenosine receptors, potently depresses SLA[179] and in view of the finding that elevating brain levels of cAMP, either with forskolin[105] or by other means,[108] similarly depresses SLA, then it is conceivable that stimulation of striatal AC via adenosine A_2/DA D_1 receptor-linked interactions may elicit locomotor depressant actions. Indeed, it has been shown that there is a unique high-affinity A_2 adenosine receptor positively coupled to AC and selectively located in the striatum (i.e., A_{2a}) in contrast to the ubiquitous, lower-affinity A_2 receptor (i.e., A_{2b}) found in other regions.[160,164,180]

Indeed, selective interactions between D_1 receptor-mediated stimulation of striatal AC and adenosine receptors has been directly demonstrated.[135,160,164] For example, R-PIA was shown to exert dose-dependent biphasic actions on DA-stimulated cyclase in the striatum, presumably mediated via both A_1 and A_2 receptors. Further, the data indicated that the antagonistic interactions between adenosine A_1 and DA D_1 receptors via a common pool of cyclase were shown to be mediated via postsynaptic mechanisms.[135] Interestingly, the PIA-mediated inhibition of cyclase was enhanced and its stimulatory actions attenuated by administration of a D_1 receptor antagonist.[135] Finally, a possible explanation for the anomalous effect of some MXs on locomotor activity was recently forthcoming, whereby it was shown that both caffeine and theophylline, at brain concentrations associated with behavioral stimulation,[21] inhibited forskolin-stimulated cAMP accumulation in rat cerebral cortical slices, whereas theobromine and IBMX, both of which lack behavioral stimulant effects, did not inhibit forskolin-stimulated accumulation of cAMP.[181] These results support the notion that the behavioral stimulant actions of MXs are likely mediated by blockade of striatal A_2 receptors and that the locomotor depressant actions of adenosine analogs are mediated via activation of striatal A_2 receptors.

Fortunately, recent behavioral studies with selective ligands for adenosine receptor subtypes have shed some light on the adenosine receptor-mediated mechanisms involved in locomotor depression. For example, Carney et al.[182] examined the mechanisms of NECA-induced locomotor depression with theobromine, a highly selective A_1 receptor antagonist which is devoid of behavioral stimulant activity. Moreover, theobromine does not block A_2 receptor-mediated increases in striatal cAMP, whereas caffeine, a nonspecific, mixed antagonist at A_1 and A_2 receptors, blocks NECA-induced increases in striatal slice activity.[110,180,182] Indeed, theobromine failed to block NECA-mediated decreases in locomotor activity, but it effectively antagonized CHA-induced decreases in activity. In contrast, the hypothermic effects of CHA and NECA were similarly blocked by theobromine, suggesting that hypothermia may be mediated via activation of A_1 receptors. These findings strongly suggest that the stimulant effects of MXs may be selectively related to effects on NECA-sensitive systems.[182]

In a recent study, Seale et al.[183] examined the effects on adenosine agonists and antagonists on SLA of the MX-sensitive inbred mouse strains, SWR and CBA. The behavioral stimulant actions of both caffeine and theophylline were reduced in SWR, which have been shown to have a selectively decreased density of striatal A_2 receptors.[178] The locomotor depressant actions of CHA, R-PIA, and NECA were also examined. SWR were shown to be significantly more sensitive to NECA-induced depression of SLA, whereas CHA and R-PIA were equally potent in the two strains of mice. Although it is likely that both A_1 and A_2 receptors influence locomotor activity, the observations from this study are consistent with a major role for A_2 receptor-mediated effects in the SWR strain, since NECA is much more potent at A_2 receptors than CHA and R-PIA. In contrast to the selective actions of NECA on SLA, there was a significant reduction in potency for the hypothermia elicited by both NECA and CHA in the CBA strain. Moreover, the difference in susceptibility to NECA-induced depression of locomotor activity in these studies was pharmacologically specific and not due to pharmacokinetic differences, since the brain levels of NECA and theophylline did not differ between the two strains. The authors suggest that there may be a direct association between inherent behavioral responsiveness to the MXs caffeine and theophylline, and the adenosine receptor agonist NECA.[183] Indeed, similar conclusions were drawn by Buckholtz and Middaugh,[184] wherein they found strain-related differences in time-course potency for caffeine-stimulated motor activity in two inbred mouse strains, but they did not find differences between the strains for the binding of CHA in the brain nor for the depression of SLA by R-PIA. The authors conclude that a strain difference in A_2 receptors is the likely mechanism for the strain-related responses in the behavioral reaction to caffeine.[184]

In a related series of studies, Durcan and Morgan[185,186] systematically examined the behavioral (locomotor activity) actions of a series of adenosine analogs in mice. They found NECA to be substantially more potent than CHA and the highly selective A_1 agonist CPA. These behavioral effects of NECA were apparently centrally mediated, since they were antagonized by parenteral administration of theophylline, but not by peripheral administration of the adenosine receptor antagonist 8-(p-sulfophenyl)theophylline, which does not readily cross the blood-brain barrier.[186] In a second study, Durcan and Morgan[185] again examined the behavioral actions of an even more extensive series of adenosine analogs. They found that the behavioral potency of these adenosine analogs correlated extremely well with their reported A_2 receptor affinity, but very poorly with regard to their reported A_1 receptor affinity. These results probably provided the most compelling evidence up to that time that A_2 receptors were involved in the hypomobility induced by adenosine analogs.[185]

In another recent series of studies, the behavioral actions of a series of selective adenosine agonists and antagonists were examined.[44,156] In addition, the behavioral interactions of these adenosine analogs with the antipsychotic and anticonvulsant drug, carbamazepine[44] (CBZ), and with DA antagonists[156] were also investigated. In the study with selective adenosine receptor antagonists,[44] chronic CBZ treatment produced a pattern of interaction in behavioral and biochemical assays identical to that of PD 115,199, a selective antagonist for A_2 receptors. By contrast, PD 116,948, a selective antagonist for A_1 receptors, antagonized the hypoactivity syndrome elicited by CBZ. Overall, the authors conclude that adenosine-induced hypoactivity is probably mediated by A_2 receptors and that chronic CBZ treatment down-regulates A_2 receptors and this may be the specific mechanism whereby CBZ exerts its therapeutic actions.[44]

Related findings were reported for studies with adenosine analogs and DA antagonists.[156] The possibility that some of the therapeutic actions of antipsychotic drugs may be mediated via A_2 receptors was supported by the findings that A_2 agonists (NECA, CV-1808, CADO) attenuate amphetamine-stimulated locomotion without producing ataxia, a property of DA antagonist antipsychotics, whereas CPA, CHA, and R-PIA reduce the effects of amphetamine only at ataxic doses.[156] However, in other tests adenosine agonists reduced apomorphine-induced climbing only at doses that caused ataxia, which differs from the behavioral profile of DA antagonists. Moreover, in the study by Heffner et al.,[156] the agonist potency order for depression of SLA of a series of adenosine analogs with selectivity for A_1 and A_2 receptors were compared following i.p. and i.c.v. administration. The rank order of potency for locomotor depression was similar after i.c.v. and i.p. administration with NECA $\gg$ CHA, R-PIA > CPA > CV-1808, CADO > S-PIA. Potency after i.c.v. administration was significantly correlated with potency during the first hour and the second hour after i.p. administration. The absolute and relative potencies of NECA, R-PIA, CADO, and S-PIA after i.p. and especially i.c.v. administration, wherein NECA was 100-fold more potent than R-PIA, were very similar to previous reports on the locomotor depressant effects of adenosine analogs following i.c.v. administration.[53]

Finally, in another recent study examining the locomotor depressant actions of novel adenosine analogs following i.p. administration, it was shown that activation of A_2 receptors by a highly selective A_2 agonist (APEC), an amine derivative of the A_2 agonist, CGS 21680, elicits very potent depressant effects on SLA.[187] The agonist potency order in depressing locomotor activity was NECA $\gg$ APEC > CHA. Moreover, the locomotor depressant effects of APEC were not reversed by A_1-selective antagonists.[187] Thus, these results again draw attention to the major role played by A_2 receptors in the locomotor depressant effects of adenosine analogs. Nonetheless, it is likely that A_1 receptors also play a role in mediating the locomotor depressant actions of adenosine analogs, although this possibility has not yet been established by research findings.

E. ADENOSINE AND FOOD INTAKE

Adenosine, and to a lesser extent inosine, produced a significant suppression of food intake following subcutaneous administration.[79,80] Intracerebroventricular administration of adenosine and adenine, but not inosine, suppressed food and water intake in rats.[81]

MX adenosine antagonists can enhance taste,[85] including in humans.[188] Taste potentiation is thought to be due to local modulation of endogenously released adenosine at the level of the taste receptors themselves or the nerve endings of the chorda tympani associated with taste cells of the anterior tongue. Human subjects taking aminophylline or theophylline can have an increased sensitivity to both taste and odors that is reversed when the drugs are discontinued.[188] Increased taste and smell sensitivity with the MXs may be one of the factors that leads the elderly to consume greater amounts of tea and coffee than younger individuals. Caffeine excreted in the saliva would enhance taste sensation, and could thus influence appetite. Caffeine has been reported to increase food consumption by rats,[82,85] but whether this reflects central or peripheral sites of action is unknown.

On the other hand, in a more recent study it was reported that R-PIA stimulated feeding in rats following i.c.v. injections at a time in the day when rats normally eat very little food or none at all.[83] NECA and CADO had no effect on feeding. Moreover, the PIA-elicited increase in feeding was not antagonized by caffeine, whether administered in large doses i.c.v. or i.p. The failure of CADO and NECA to stimulate feeding and the inability of caffeine to block PIA-induced feeding suggest that R-PIA-induced feeding is not mediated via A_1 or A_2 receptors. In contrast, the opioid antagonist naloxone blocked R-PIA-induced eating.[83] These results suggest that a specific behavior response elicited by R-PIA is mediated by an unknown mechanism not involving adenosine receptors. Interestingly, it has also been reported that the stimulatory action of R-PIA on cAMP accumulation in hippocampal slices was also insensitive to theophylline.[44] Therefore, the possibility cannot be ignored that R-PIA may exert actions on feeding behavior and cAMP accumulation in hippocampal slices which are not mediated by adenosine receptors.

F. ADENOSINE, OPERANT RESPONSES, AND BEHAVIORAL TOLERANCE
1. Schedule-Controlled Operant Behaviors
Peripherally administered analogs of adenosine decrease schedule-controlled behavior of rats and squirrel monkeys.[35,36,73,74,189,190] When the behavioral effects of a series of metabolically stable analogs of adenosine were studied in squirrel monkeys responding under a fixed interval schedule of stimulus-shock termination, all of the drugs depressed the responding rate. NECA and CADO were as potent as, or more effective than, R-PIA, CHA, and CPA, suggesting an action mediated by A_2 receptors.[190-192] Overall, these studies on adenosine receptors in operant responding paradigms with rats and with both New World and Old World monkeys suggest that the behavioral suppressant effects of adenosine agonists are primarily mediated by activation of A_2 recognition sites.[190-192] Moreover, these studies have also clearly shown that the psychomotor stimulant effects of MXs are linked to their antagonistic actions at these same sites.[192]

2. Behavioral Tolerance and Adenosine
Tolerance develops to the effects of administered R-PIA on response rate of animals responding on a fixed-ratio schedule.[34] Tolerance to R-PIA conferred tolerance to CHA, but no cross-tolerance was observed for the rate-suppressing effects of 4,5,6,7-tetrahydroisoxazolo [5,4-c] pyridin-3-ol (THIP; a GABA mimetic), diazepam, pentobarbital, ketamine, clonidine, amphetamine, and caffeine. The lack of cross tolerance to THIP, diazepam, ketamine, and pentobarbital is interesting in that these agents are effective sedative-hypnotics, producing similar gross behavioral signs as R-PIA. Cross-tolerance studies appear to offer exciting possibilities for the evaluation of the involvement of adenosine receptors in the central action of other sedative-hypnotic agents.

The development of tolerance to caffeine has been evaluated in rodent locomotor paradigms. After a 2-week exposure of rats to caffeine via their drinking water, the stimulatory actions of caffeine on mesencephalic reticular neurons was abolished and locomotor stimulation was attenuated.[193] This was accompanied by an increase in the number of CHA binding sites without any change in receptor affinity. Similar results were observed in mice after chronic administration of caffeine. The dose-response curves of R-PIA for both analgesia and locomotor depression were shifted to the left and the B_{max} values for R-PIA and diethylphenylxanthine were increased without any alterations in affinity.[38,194] By contrast, in mice given R-PIA chronically for 14 d, the dose-response curves for both analgesia and locomotor depression were shifted to the right. The dose-response curve for the locomotor effects of caffeine was shifted to the left, and caffeine exhibited greater antagonist activity against the analgesic action of R-PIA. Interestingly, following R-PIA administration there was no change in the brain K_D or B_{max} values of either labeled R-PIA or DPX.

III. CONCLUDING REMARKS

Much research effort has been invested over the past 2 decades in examining the physiologic role of adenosine in brain function and behavior. On the other hand, the seminal observations on the behavioral effects of adenosine have been known for over half a century, namely that adenosine produces sedation and generally decreases levels of arousal. The specific behavioral actions of adenosine — its hypnogenic, locomotor depressant, anticonvulsant, antinocisposive, hypotensive, and sedative actions — have subsequently been systematically examined by numerous investigators. Indeed, a large body of meritorious work has gradually emerged which provides documentation for the potent actions of adenosine and related drugs on a variety of behavioral responses. However, it was only until very recently, primarily due to the development of selective agonists and antagonists for adenosine receptors, that much of the research on the behavioral actions of adenosine has undergone more rapid development.

In particular, extensive work has been recently conducted on the locomotor depressant actions of adenosine analogs. Simultaneously, related studies were being carried out on striatal DA function and its role in locomotor behaviors. As a result, especially in the past few years, there has been much research work examining the interactions of DA and adenosine receptor systems in the striatum and the role these receptor interactions play in the mediation of stereotypic locomotor behaviors. Thus, in view of the functional coupling of striatal adenosinergic and dopaminergic systems at both cellular and behavioral levels, research work on the striatal actions of adenosine appears to hold much promise for neurobehavioral investigation, since it may ultimately be possible to extrapolate *in vitro* findings from studies on the cellular actions of striatal adenosine receptors to related observations at the behavioral level wherein parallel *in vivo* studies are employed to characterize the actions of adenosine receptor ligands on locomotor behaviors mediated by the striatum.

ACKNOWLEDGMENT

The author gratefully acknowledges the meritorious work and praiseworthy efforts of L. McCraw in the preparation and typing of the manuscript.

REFERENCES

1. **Bentley, D. and Konishi, M.,** Neural control of behavior, *Annu. Rev. Neurosci.,* 1, 35, 1978.
2. **Bartfai, T., Iverfeldt, K., and Fisone, G.,** Regulation of the release of coexisting neurotransmitters, *Annu. Rev. Pharmacol. Toxicol.,* 28, 285, 1988.
3. **Schmitt, F. O.,** Molecular regulators of brain function, *Neurosci.,* 13, 991, 1984.
4. **Bloom, F. E.,** Neurotransmitters: past, present and future directions, *FASEB J.,* 2, 32, 1988.
5. **Williams, M.,** Adenosine: the prototypic neuromodulator, *Neurochem. Int.,* 14, 249, 1989.
6. **Barraco, R. A. and Stefano, G. B.,** Pharmacological evidence for the modulation of monoamine release of adenosine in the invertebrate nervous system, *J. Neurochem.,* 54, 2002, 1990.
7. **Newby, A. C.,** Adenosine and the concept of "retaliatory metabolites", *Trends Biochem.,* 9, 42, 1984.
8. **Phillis, J. W. and Wu, P. H.,** The role of adenosine and its nucleotides in central synaptic neurotransmission, *Prog. Neurobiol.,* 16, 187, 1981.
9. **Snyder, S. H.,** Adenosine as a neuromodulator, *Annu. Rev. Neurosci.,* 8, 103, 1985.
10. **Dunwiddie, T. V.,** The physiological role of adenosine in the central nervous system, *Int. Rev. Neurobiol.,* 27, 63, 1985.
11. **Ribeiro, J. A. and Sebastião, A. M.,** Adenosine receptors and calcium: basis for proposing a third (A_3) adenosine receptor, *Prog. Neurobiol.,* 26, 179, 1986.
12. **Phillis, J. W. and Barraco, R. A.,** Adenosine, adenylate cyclase and transmitter release, in *Advances in Cyclic Nucleotide and Protein Phosphorylation Research,* Vol. 19, Greengard, P. and Robison, G. A., Eds., Raven Press, New York, 1985, 243.
13. **Rubio, R., Bencherif, M., and Berne, R. M.,** Inositol phospholipid metabolism during and following synaptic activation: role of adenosine, *J. Neurochem.,* 52, 797, 1989.
14. **Barraco, R. A.,** Behavioral actions of adenosine analogs, in *Purines: Pharmacology and Physiological Role,* Stone, T. W., Ed., Macmillan, New York, 1985, 27.
15. **Barraco, R. A., Janusz, C. A., Schoener, E. P., and Simpson, L. L.,** Cardiorespiratory function is altered by picomole injections of 5'-*N*-ethylcarboxamidoadenosine into the nucleus tractus solitarius of rats, *Brain Res.,* 507, 234, 1990.

16. **Phillis, J. W., Barraco, R. A., DeLong, R. E., and Washington, D. O.,** Behavioral characterization of centrally administered adenosine analogs, *Pharm. Biochem. Behav.,* 24, 263, 1986.

17. **Phillis, J. W.,** Brain adenosine and purinergic modulation of central nervous system excitability, in *The Brain as an Endocrine Organ,* Foa, P. P., Ed., Springer-Verlag, New York, 1988, 210.

18. **Barraco, R. A.,** Adenosinergic modulation of brainstem mechanisms involved in cardiovascular control, in *Adenosine and Adenine Nucleotides: Physiology and Pharmacology,* Paton, D., Ed., Taylor & Francis, Basingstoke, England, 1988, 233.

19. **Phillis, J. W.,** Behavioral and other actions of adenosine in the central nervous system, in *Neurotransmitters and Cortical Function. From Molecules to Mind,* Avoli, M., Rader, T. A., Dykes, R. W., and Gloor, P., Eds., Plenum Press, New York, 1988, 403.

20. **Phillis, J. W. and Kostopoulos, G. K.,** Adenosine as a putative transmitter in the cerebral cortex. Studies with potentiators and antagonists, *Life Sci.,* 17, 1085, 1975.

21. **Snyder, S. H., Katims, J. J., Annau, Z., Bruns, R. F., and Daly, J. W.,** Adenosine receptors and the behavioral actions of methylxanthines, *Proc. Natl. Acad. Sci. U.S.A.,* 78, 3260, 1981.

22. **Phillis, J. W. and Wu, P. H.,** Adenosine mediates sedative action of various centrally active drugs, *Med. Hypotheses,* 9, 361, 1982.

23. **Wu, P. H. and Phillis, J. W.,** Uptake by central nervous tissues as a mechanism for the regulation of extracellular adenosine concentrations, *Neurochem. Int.,* 6, 613, 1984.

24. **Barraco, R. A., Phillis, J. W., and DeLong, R. E.,** Behavioral interaction of adenosine and diazepam in mice, *Brain Res.,* 323, 159, 1984.

25. **Phillis, J. W. and O'Regan, M. H.,** The role of adenosine in the central actions of benzodiazepines, *Prog. Neuropsychopharmacol. Biol. Psychiatry,* 12, 389, 1988.

26. **DeLong, R. E., Phillis, J. W., and Barraco, R. A.,** A possible role of endogenous adenosine in the sedative action of meprobamate, *Eur. J. Pharmacol.,* 118, 359, 1985.

27. **DeLong, R. E., Barraco, R. A., and Phillis, J. W.,** Behavioral interaction of adenosine and trifluoperazine in mice, *Neurosci. Lett.,* 53, 101, 1985.

28. **Turski, L., Schwarz, M., Turski, W. A., Ikonomidou, C., and Sontag, K. H.,** Effect of aminophylline on muscle relaxant action of diazepam and phenobarbitone in genetically spastic rats, further evidence for a purinergic mechanism in the action of diazepam, *Eur. J. Pharmacol.,* 103, 99, 1984.

29. **Commissaris, R. L., McCloskey, T. C., Damian, G. M., Brown, B. D., Barraco, R. A., and Altman, H. J.,** Antagonism of the anti-conflict effects of phenobarbital, but not diazepam by the A-1 adenosine agonist L-PIA, *Psychopharmacologia,* 102, 283, 1990.

30. **Dar, M. S.,** CNS effects and behavioral interactions with ethanol of centrally administered dilazep and its metabolites in mice, *Eur. J. Pharmacol.,* 164, 303, 1989.

31. **Phillis, J. W.,** Interactions of the anticonvulsants diphenylhydantoin and carbamazepine with adenosine on cerebral cortical neurons, *Epilepsia,* 25, 765, 1984.

32. **Popoli, P., Benedetti, M., and de Carolis, A. S.,** Anticonvulsant activity of carbamazepine and N^6-L-phenylisopropyladenosine in rabbits. Relationship to adenosine receptors in central nervous system, *Pharm. Biochem. Behav.,* 29, 533, 1988.

33. **Chin, J. H.,** Adenosine receptors in brain: neuromodulation and role in epilepsy, *Ann. Neurol.,* 26, 695, 1989.

34. **Spencer, D. G., Caldwell, P., and Emmett-Oglesby, M. W.,** Tolerance to N^6-(L-phenylisopropyl) adenosine: contribution of behavioral mechanisms and cross-tolerance profile, *Neuropharmacology,* 23, 671, 1984.

35. **Coffin, V. L. and Spealman, R. D.,** Modulation of the behavioral effects of chlordiazepoxide by methylxanthines and analogs of adenosine in squirrel monkeys, *J. Pharmacol. Exp. Ther.,* 235, 724, 1985.

36. **Holloway, F. A., Modrow, H. E., and Michaelis, R. C.,** Methylxanthine discrimination in the rat: possible benzodiazepine and adenosine mechanisms, *Pharm. Biochem. Behav.,* 22, 815, 1985.

37. **Fujiwara, Y., Takeda, T., Kazahaya, Y., Otsuki, S., and Sandyk, R.,** Inhibitory effects of carbamazepine on clonidine-induced aggressive behavior in mice, *Int. J. Neurosci.,* 42, 77, 1988.

38. **Ahlijanian, M. K. and Takemori, A. E.,** Effects of (-)-N^6-(R-phenylisopropyl)-adenosine (PIA) and caffeine on nociception and morphine-induced analgesia, tolerance and dependence in mice, *Eur. J. Pharmacol.,* 112, 171, 1985.

39. **Sattin, A.,** Adenosine receptors: some behavioral implications for depression and sleep, *Monogr. Neural Sci.,* 10, 1982, 1984.

40. **Mingo, N. S. and Burnham, W. M.,** Secondary generalization in non-kindled rats following acute administration of GABA-complex and adenosine antagonists, *Electroencephalogr. Clin. Neurophysiol.,* 75, 444, 1990.

41. **Albertson, S. M., Bowyer, J. F., and Paule, M. G.,** Modification of the anticonvulsant efficacy of diazepam by RO 15-1788 in the kindled amygdaloid seizure model, *Life Sci.,* 31, 1597, 1982.

42. **Phillis, J. W. and O'Regan, M. H.,** Benzodiazepine interaction with adenosine systems explains some anomalies in GABA hypothesis, *TIPS,* 9, 153, 1988.

43. **Marrousu, F., Marchi, A., De Martino, M. R., Saba, G., and Gessa, G. L.,** Aminophylline antagonizes diazepam-induced anesthesia and EEG changes in humans, *Psychopharmacologia,* 85, 69, 1985.

44. **Elphick, M., Taghavi, Z., Powell, T., and Godfrey, P. P.,** Chronic carbamazepine down-regulates adenosine A$_2$ receptors: studies with putative selective adenosine antagonists PD 115,199 and PD 116,948, *Psychopharmacologia,* 100, 522, 1990.

45. **Feldberg, W. and Sherwood, S. L.,** Injections of drugs into the lateral ventricle of the cat, *J. Physiol. (London),* 123, 148, 1954.

46. **Buday, P. V., Carr, C. J., and Miya, T. S.,** A pharmacologic study of some nucleosides and nucleotides, *J. Pharm. Pharmacol.,* 13, 290, 1961.

47. **Marley, E. and Nistico, G.,** Effect of catecholamines and adenosine derivatives given into the brain of fowls, *Br. J. Pharmacol. Chemother.,* 46, 619, 1972.

48. **Vapaatalo, H., Onken, D., Neuvonen, P. J., and Westerman, E.,** Stereospecificity in some central and circulatory effects of phenylisopropyladenosine (PIA), *Arzneim. Forsch. (Drug Res.),* 25, 407, 1975.

49. **Crawley, J. N., Patel, J., and Marangos, P. J.,** Behavioral characterization of two long-lasting adenosine analogs: sedative properties and interaction with diazepam, *Life Sci.,* 29, 2623, 1981.

50. **Katims, J. J., Annau, Z., and Snyder, S. H.,** Interactions in the behavioral effect of methylxanthines and adenosine derivatives, *J. Pharmacol. Exp. Ther.,* 227, 167, 1983.

51. **Dunwiddie, T. W. and Worth, T.,** Sedative and anticonvulsive effects of adenosine analogs in mouse and rat, *J. Pharmacol. Exp. Ther.,* 220, 70, 1982.

52. **Crawley, J. N., Patel, J., and Marangos, P. J.,** Adenosine uptake inhibitors potentiate the sedative effects of adenosine, *Neurosci. Lett.,* 36, 169, 1983.

53. **Barraco, R. A., Coffin, V. L., Altman, H. J., and Phillis, J. W.,** Central effects of adenosine analogs on locomotor activity in mice and antagonism of caffeine, *Brain Res.,* 272, 392, 1983.

54. **Barraco, R. A., Aggarwal, A. K., Phillis, J. W., Moron, M. A., and Wu, P. H.,** Dissociation of the locomotor and hypotensive effects of adenosine analogs in the rat, *Neurosci. Lett.,* 48, 139, 1984.

55. **Barraco, R. A. and Bryant, S.,** Depression of locomotor activity following bilateral injections of adenosine analogs into the striatum of mice, *Med. Sci. Res.,* 15, 421, 1987.

56. **Collier, H. O. J. and Tucker, J. F.,** Novel form of drug-dependence on adenosine in guinea pig ileum, *Nature,* 302, 618, 1983.

57. **Fredholm, B. B., Herrera-Marschitz, M., Jonzon, B., Lindstrom, K., and Ungerstedt, U.,** On the mechanism by which methylxanthine enhances apomorphine-induced rotation behavior in the rat, *Pharmacol. Biochem. Behav.,* 19, 535, 1983.

58. **Maitre, M., Ciesielski, L., Lehman, A., Kempf, E., and Mandel, P.,** Protective effects of adenosine and nicotinamide against audiogenic seizure, *Biochem. Pharmacol.,* 23, 2807, 1974.

59. **Barraco, R. A., Swanson, T. H., Phillis, J. W., and Berman, R. F.,** Anticonvulsant effects of adenosine analogs on amygdaloid-kindled seizures in rats, *Neurosci. Lett.,* 46, 317, 1984.

60. **Albertson, T. E., Stark, L. J., Joy, R. M., and Bowyer, J. F.,** Aminophylline and kindled seizures, *Exp. Neurol.,* 81, 703, 1983.

61. **Dragunow, M. and Goddard, G. V.,** Adenosine modulation of amygdala kindling, *Exp. Neurol.,* 84, 654, 1984.

62. **Berman, R. F., Jarvis, M. F., and Lupica, C. R.,** Adenosine involvement in kindled seizures, in *Kindling 4,* Wada, J. A., Ed., Raven Press, New York, in press.

63. **Post, C.,** Antinociceptive effects in mice after intrathecal injection of 5'-N-ethylcarboxamide adenosine, *Neurosci. Lett.,* 51, 325, 1984.

64. **Yarbrough, G. G. and McGuffin-Clineschmidt, J. C.,** *In vivo* behavioral assessment of central nervous system purinergic receptors, *Eur. J. Pharmacol.,* 76, 137, 1981.

65. **Holmgren, M., Hednar, T., Nordberg, G., and Mellstrand, T.,** Antinociceptive effects in the rat of an adenosine analog, N^6-phenylisopropyladenosine, *J. Pharm. Pharmacol.,* 35, 679, 1983.

66. **Haulica, I., Nemtu, C., Petrescu, G. H., Frasin, M., Slatineanu, S., and Nacu, C.,** The influence of adenosine upon thermoalgesic sensitivity, *Physiologie (Buc),* 21, 167, 1984.

67. **Dascombe, M. J. and Milton, S. A.,** The effects of cyclic adenosine 3',5'-monophosphate and other adenine nucleotides on body temperature, *J. Physiol. (London),* 250, 143, 1975.

68. **Wager-Srdar, S. A., Oken, M. M., Morley, J. E., and Levine, A. S.,** Thermoregulatory effects of purines and caffeine, *Life Sci.,* 33, 2431, 1983.

69. **Wang, X. L., Lee, T. F., and Wang, L. C. H.,** Do adenosine antagonists improve cold tolerance by reducing hypothalamic adenosine activity in rats, *Brain Res. Bull.,* 24, 389, 1990.

70. **Haulica, I., Abadei, L., Branisteanu, D., and Topoliceanu, F.,** Preliminary data on the possible hypnogenic role of adenosine, *J. Neurochem.,* 21, 1019, 1973.

71. **Virus, R. M., Djuricic-Nedelson, M., Radulovacki, M., and Green, R. D.,** The effects of adenosine and 2'-deoxycoformycin in sleep and wakefulness in rats, *Neuropharmacology,* 22, 1401, 1983.

72. **Radulovacki, M., Virus, R. M., Djuricic-Nedelson, M., and Green, R. D.,** Adenosine analogs and sleep in rats, *J. Pharmacol. Exp. Ther.,* 228, 268, 1984.

73. **Coffin, V. L. and Carney, J. M.,** Behavioral pharmacology of adenosine analogs, in *Physiology and Pharmacology of Adenosine Derivatives,* Daly, J. W., Kuroda, Y., Phillis, J. W., Shimizu, H., and Ui, M., Eds., Raven Press, New York, 1983, 267.

74. **Coffin, V. L. and Spealman, R. D.,** Behavioral and cardiovascular effects of analogs of adenosine in cynomolgus monkeys, *J. Pharmacol. Exp. Ther.,* 241, 76, 1987.

75. **Logan, L. and Carney, J. M.,** Antagonism of the behavioral effects of L-phenylisopropyladenosine (L-PIA) by caffeine and its metabolites, *Pharmacol. Biochem. Behav.,* 21, 375, 1984.

76. **Valdes, J. J., McGuire, P. S., and Annau, Z.,** Xanthines alter behavior maintained by intracranial electrical stimulation and an operant schedule, *Psychopharmacologia,* 76, 325, 1982.

77. **Ushijima, I., Katsuragi, T., and Furukawa, T.,** Involvement of adenosine receptor activities in aggressive responses produced by clonidine in mice, *Psychopharmacologia,* 83, 335, 1984.

78. **Palmour, R. M., Lipowski, C. J., Simon, C. K., and Ervin, F. R.,** Adenosine analogs inhibit fighting in isolated male mice, *Life Sci.,* 44, 1293, 1989.

79. **Capogrossi, M. C., Francendese, A., and Digirolamo, M.,** Suppression of food intake by adenosine and inosine, *Am. J. Clin. Nutr.,* 32, 1762, 1979.

80. **Levine, A. S. and Morley, J. E.,** Purinergic regulation of food intake, *Science,* 217, 77, 1982.

81. **Levine, A. S. and Morley, J. E.,** Effect of intraventricular adenosine on food intake in rats, *Pharmacol. Biochem. Behav.,* 19, 23, 1983.

82. **Wager-Srdar, S., Levine, A. S., and Morley, J. E.,** Food intake: opioid/purine interactions, *Pharmacol. Biochem. Behav.,* 21, 33, 1984.

83. **Levine, A. S., Grace, M., Krahn, D. D., and Billington, C. J.,** The adenosine agonist N^6-R-phenylisopropyladenosine (R-PIA) stimulates feeding in rats, *Brain Res.,* 477, 280, 1989.

84. **Cooper, S. J.,** Caffeine-induced hypodipsia in water-deprived rats: relationships with benzodiazepine mechanisms, *Pharm. Biochem. Behav.,* 17, 481, 1982.

85. **Diaz, C.,** Methyl xanthines enhance taste: evidence for modulation of taste by adenosine receptor, *Pharm. Biochem. Behav.,* 22, 195, 1985.

86. **Hedner, T., Hedner, J., Wessberg, P., and Jonason, J.,** Regulation of breathing in the rat: indication of a role of central adenosine mechanisms, *Neurosci. Lett.,* 33, 147, 1982.

87. **Barraco, R. A., Phillis, J. W., Campbell, W. R., Marcantonio, D. R., and Salah, R. S.,** The effects of central injections of adenosine analogs on blood pressure and heart rate in the rat, *Neuropharmacology,* 25, 675, 1986.

88. **Barraco, R. A., Stair, R. E., Campbell, W. R., and Shehin, S. E.,** Central effects of adenosine analogs on blood pressure and heart rate in the mouse, *Can. J. Cardiol.,* 3, 205, 1987.

89. **Barraco, R. A., Campbell, W. R., Schoener, E. P., Shehin, S. E., and Parizon, M.,** Cardiovascular effects of microinjections of adenosine analogs into the fourth ventricle of rats, *Brain Res.,* 424, 17, 1987.

90. **Kuroda, Y.,** "Tracing circuit" model for the memory process in human brain: roles of ATP and adenosine derivatives for dynamic change of synaptic connections, *Neurochem. Int.,* 14, 309, 1989.

91. **Winsky, L. and Harvey, J. A.,** Retardation of associative learning in the rabbit by an adenosine analog as measured by classical conditioning of the nictitating membrane response, *J. Neurosci.,* 6, 2684, 1986.

92. **Loke, W. H.,** Effects of caffeine on mood and memory, *Physiol. Behav.,* 44, 367, 1988.

93. **Ushijima, I., Mizuki, Y., and Yamada, M.,** Development of stress-induced gastric lesion involves central adenosine A_1-receptor stimulation, *Brain Res.,* 339, 351, 1985.

94. **Geiger, J. D. and Glavin, G. B.,** Adenosine receptor activation in brain reduces stress-induced ulcer formation, *Eur. J. Pharmacol.,* 115, 185, 1985.

95. **Boulenger, J. P., Marangos, P. J., Zander, K. J., and Hanson, J.,** Stress and caffeine: effects on central adenosine receptors, *Clin. Neuropharmacol.,* 9, 79, 1986.

96. **Westerberg, V. S. and Geiger, J. D.,** Central effects of adenosine analogs on stress-induced gastric ulcer formation, *Life Sci.,* 41, 2201, 1987.

97. **Glavin, G. B., Westerberg, V. S., and Geiger, J. D.,** Modulation of gastric acid secretion by adenosine in conscious rats, *Can. J. Physiol. Pharmacol.,* 65, 1182, 1987.

98. **Puuranen, J. and Huttunen, P.,** Central gastric antisecretory action of adenosine in the rat, *Eur. J. Pharmacol.,* 147, 59, 1988.

99. **Westerberg, V. S. and Geiger, J. D.,** Adenosine analogs inhibit gastric acid secretion, *Eur. J. Pharmacol.,* 160, 275, 1989.

100. **Bristol, J. A., Bridges, A. J., Bruns, R. F., Downs, D. A., Heffner, J. G., Moos, W. H., Ortwine, D. F., Szotek, D. L., and Trivedi, B. K.,** The search for purine and ribose-substituted adenosine analogs with potential clinical application, in *Adenosine and Adenine Nucleotides: Physiology and Pharmacology,* Paton, D., Ed., Taylor & Francis, Basingstoke, England, 1988, 17.

101. **Coffin, V. L., Taylor, J. A., Phillis, J. W., Altman, H. J., and Barraco, R. A.,** Behavioral interaction of adenosine and methylxanthines on central purinergic systems, *Neurosci. Lett.,* 47, 91, 1984.

102. **Major, P. P., Agarwal, R. P., and Kufe, D. W.,** Deoxycoformycin: neurological toxicity, *Cancer Chemother. Pharmacol.,* 5, 193, 1981.

103. **Mendelson, W. B., Kuruvilla, A., Watlington, T., Goehl, K., Paul, S. M., and Skolnick, P.,** Sedative and electroencephalographic actions of erythro-9-(2-hydroxy-3-nonyl)-adenine (EHNA): relationship to inhibition of brain adenosine deaminase, *Psychopharmacologia,* 79, 126, 1983.

104. **Radulovacki, M., Virus, R. M., Djuricic-Nedelson, M., and Green, R. D.,** Hypnotic effects of deoxycoformycin in rats, *Brain Res.,* 271, 392, 1983.

105. **Barraco, R. A., Phillis, J. W., and Altman, H. J.,** Depressant effect of forskolin on spontaneous locomotor activity in mice, *Gen. Pharm.,* 16, 521, 1984.

106. **Wachtel, H.,** Species differences in behavioral effects of rolipram and other adenosine cyclic 3',5'-monophosphate phosphodiesterase inhibitors, *J. Neural Transm.,* 56, 139, 1983.

107. **Sano, M., Seto-Ohshima, A., and Mizutani, A.,** Forskolin suppresses seizures induced by pentylenetetrazol in mice, *Experientia,* 40, 1270, 1984.

108. **Wachtel, H. and Loschmann, P. A.,** Effects of forskolin and cyclic nucleotides in animal models predictive of antidepressant activity: interactions with rolipram, *Psychopharmacologia,* 90, 430, 1986.

109. **Barraco, R. A., Schoener, E. P., Polsek, P. M., Simpson, L. L., Janusz, C. J., and Parizon, M.,** Respiratory effects of cyclic AMP following microinjections into the nucleus tractus solitarius of rats, *Neuropharmacology,* 27, 1285, 1988.

110. **Daly, J. W., Bruns, R. F., and Snyder, S. H.,** Adenosine receptors in the central nervous system: relationship to the central actions of methylxanthines, *Life Sci.,* 28, 2083, 1981.

111. **Bruns, R. F., Daly, J. W., and Snyder, S. H.,** Adenosine receptors in brain membranes: binding of N^6-cyclohexyl[^{3}H]adenosine and 1,3-diethyl-8-[^{3}H]phenylxanthine, *Proc. Natl. Acad. Sci. U.S.A.,* 77, 5547, 1980.

112. **Thithapanda, A., Maling, H. M., and Gillette, J. R.,** Effects of caffeine and theophylline on activity of rats in relation to brain xanthine concentrations, *Proc. Soc. Exp. Biol. Med.,* 139, 582, 1972.

113. **Eldridge, F. L. and Millhorn, D. E.,** Role of adenosine in the regulation of breathing, in *Topics and Perspectives in Adenosine Research,* Gerlach, E. and Becker, B. V., Eds., Springer-Verlag, Berlin, 1987, 586.

114. **Brodie, M. S., Lee, K., Fredholm, B. B., Stahle, L., and Dunwiddie, T. V.,** Central versus peripheral mediation of responses to adenosine receptor agonists: evidence against a central mode of action, *Brain Res.,* 415, 323, 1987.

115. **Bowker, H. M. and Chapman, A. G.,** Adenosine analogs. The temperature-dependence of the anticonvulsant effect and inhibition of ^{3}H-aspartate release, *Biochem. Pharmacol.,* 35, 2949, 1986.

116. **Jonzon, B., Bergquist, A., Li, Y.-O., et al.,** Effects of adenosine and two stable adenosine analogs on blood pressure, heart rate, and colonic temperature in the rat, *Acta Physiol. Scand.,* 126, 491, 1986.

117. **Evoniuk, G., Jacobson, K. A., Shamin, M. T., Daly, J. W., and Wurtman, R. J.,** A1- and A2-selective adenosine antagonists: *in vivo* characterization of cardiovascular effects, *J. Pharmacol. Exp. Ther.,* 242, 882, 1987.

118. **Puiroud, S., Pinard, E., Miller, M., and Seylaz, J.,** Systemically administered adenosine increases caudate blood flow in rabbits, *Neurosci. Lett.,* 80, 224, 1987.

119. **Nehlig, A., Vasconcelos, A., Dumont, I., and Boyet, S.,** Effects of caffeine, L-phenylisopropyladenosine and their combination on local cerebral blood flow in the rat, *Eur. J. Pharmacol.,* 179, 271, 1990.

120. **Green, R. D., Proudfit, H. K., and Yeung, S. M. H.,** Modulation of striatal dopaminergic function by local injection of 5'-N-ethylcarboxamide adenosine, *Science,* 218, 58, 1982.

121. **Londos, C. D., Cooper, M. F., and Wolff, J.,** Subclasses of external adenosine receptors, *Proc. Natl. Acad. Sci. U.S.A.,* 77, 2551, 1980.

122. **Bruns, R. F., Lu, G. H., and Pugsley, T. A.,** Characterization of the A_2 adenosine receptor labeled by [^{3}H] NECA in rat striatal membranes, *Mol. Pharmacol.,* 29, 331, 1986.

123. **Bruns, R. F.,** Adenosine receptor binding assays, in *Receptor Biochemistry and Methodology, Vol. 11, Adenosine Receptors,* Cooper, D. M. F. and Londos, C., Eds., Alan R. Liss, New York, 1988, 43.

124. **Fredholm, B. B., Jonzon, B., and Lindstrom, K.,** Adenosine receptor mediated increases and decreases in cyclic AMP in hippocampal slices treated with forskolin, *Acta Physiol. Scand.,* 117, 461, 1983.

125. **Geiger, J. D., Johnston, M. E., and Yago, V.,** Pharmacological characterization of rapidly accumulated adenosine by dissociated brain cells from adult rat, *J. Neurochem.,* 51, 283, 1988.

126. **Fredholm, B. B.,** Presynaptic adenosine receptors, *ISI Atlas of Science: Pharmacology,* 1988, 257.

127. **Duner-Engstrom, J. and Fredholm, B. B.,** Evidence that prejunctional adenosine receptors regulating acetylcholine release from rat hippocampal slices are linked to an N-ethylmaleimide-sensitive G-protein, but not to adenylate cyclase or dihydropyridine-sensitive Ca^{2+}-channels, *Acta Physiol. Scand.,* 134, 119, 1988.

128. **Florio, C., Traversa, U., Vertua, R., and Puppini, P.,** 5'-N-Ethylcarboxamidoadenosine binds to two different adenosine receptors in membranes from the cerebral cortex of the rat, *Neuropharmacology,* 27, 85, 1988.

129. **Criswell, H., Mueller, R. A., and Breese, G. R.,** Assessment of purine-dopamine interactions in 6-hydroxydopamine-lesioned rats: evidence for pre- and postsynaptic influences by adenosine, *J. Pharmacol. Exp. Ther.,* 244, 493, 1988.

130. **Choca, J. I., Green, R. D., and Proudfit, H. K.,** Adenosine A_1 and A_2 receptors of the substantia gelatinosa are located predominantly on intrinsic neurons: an autoradiography study, *J. Pharmacol. Exp. Ther.,* 247, 757, 1988.

131. **Erfurth, A., Reddington, M., and Schmauss, M.,** Studies on binding sites for adenosine receptor ligands in rat brain: an approach to the specification of adenosinergic functions, *Pharmacopsychiatry,* 21, 326, 1988.

132. **Popoli, P., Caporali, M. G., and de Carolis, A. S.,** The role of the purinergic system in the control of stereotype: relationship to D-1/D-2 dopamine receptor activity, *Pharmacol. Biochem. Behav.,* 32, 203, 1989.

133. **Reddington, M., Erfurth, A., and Lee, K. S.,** Heterogeneity of binding sites for N-ethylcarboxamido [^{3}H] adenosine in rat brain: effects of N-ethylmaleimide, *Brain Res.,* 399, 232, 1986.

134. **Stone, G. A., Jarvis, M. F., Sills, M. S., Weeks, B., Snowhill, E. W., and Williams, J.,** Species differences in high-affinity adenosine A_2 binding sites in striatal membranes from mammalian brain, *Drug Dev. Res.,* 15, 31, 1988.

135. **Abbracchio, M. P., Colombo, F., DiLuca, M., Zaratin, P., and Cattabeni, F.,** Adenosine modulates the dopaminergic function of the nigro-striatal system by interacting with striatal dopamine dependent adenylate cyclase, *Pharmacol. Res. Commun.,* 19, 275, 1987.

136. **Alexander, S. P. and Reddington, M.,** The cellular localization of adenosine receptors in rat neostriatum, *Neuroscience,* 28, 645, 1989.

137. **Allgaier, C., Hertting, G., and Kugelgen, O. V.,** The adenosine receptor-mediated inhibition of noradrenaline release possibly involves a N-protein and is increased by α_2-autoreceptor blockade, *Br. J. Pharmacol.,* 90, 403, 1987.

138. **Fredholm, B. B., Fastbom, J., and Lindgren, E.,** Effects of N-ethylmaleimide and forskolin on glutamate release from rat hippocampal slices. Evidence that prejunctional adenosine receptors are linked to N-proteins, but not to adenylate cyclase, *Acta Physiol. Scand.,* 127, 381, 1986.

139. **Fredholm, B. B. and Lindgren, E.,** Effects of N-ethylmaleimide and forskolin on noradrenaline release from rat hippocampal slices. Evidence that prejunction adenosine and α-receptors are linked to N-proteins but not to adenylate cyclase, *Acta Physiol. Scand.,* 130, 95, 1987.

140. **Fredholm, B. B. and Lindgren, E.,** Protein kinase C activation increases noradrenaline release from the rat hippocampus and modifies the inhibitory effect of α_2-adrenoceptor and adenosine A_1-receptor agonists, *Naunyn-Schmiedeberg's Arch. Pharmacol.,* 337, 477, 1988.

141. **Fredholm, B. B., Proctor, W., Van der Ploeg, I., and Dunwiddie, T. V.,** In vivo pertussis toxin treatment attenuates some, but not all, adenosine A_1 effects in slices of the rat hippocampus, *Eur. J. Pharmacol.,* 172, 249, 1989.

142. **Zgombick, J. M., Beck, S. G., Mahle, C. D., Craddock, Royal, B., and Maayani, S.,** Pertussis toxin-sensitive guanine nucleotide-binding protein(s) couple adenosine A_1 and 5-hydroxytryptamine receptors to the same effector systems in rat hippocampus: biochemical and electrophysiological studies, *Mol. Pharmacol.,* 35, 484, 1989.

143. **Ramkumar, V. and Stiles, G. L.,** Reciprocal modulation of agonist and antagonist binding to A_1 adenosine receptors by guanine nucleotides is mediated via a pertussis toxin-sensitive G protein, *J. Pharmacol. Exp. Ther.,* 246, 1194, 1988.

144. **Worley, P. F., Baraban, J. M., and Snyder, S. H.,** Beyond receptors: multiple second-messenger systems in brain, *Ann. Neurol.,* 21, 217, 1987.

145. **Limbird, L. E.,** Receptors linked to inhibition of adenylate cyclase: additional signaling mechanisms, *Fed. Proc. Fed. Am. Soc. Exp. Biol.,* 2, 2686, 1988.

146. **Birch, B. D., Louie, G. L., Vickery, R. G., Gaba, D. M., and Maze, M.,** L-Phenylisopropyladenosine (L-PIA) diminishes halothane anesthetic requirements and decreases noradrenergic neurotransmission in rats, *Life Sci.,* 42, 1355, 1988.

147. **Pilc, A. and Enna, S. J.,** Activation of alpha-2 adrenergic receptors augments neurotransitter-stimulated cyclic AMP accumulation in rat brain cerebral cortical slices, *J. Pharmacol. Exp. Ther.,* 237, 725, 1986.

148. **Wood, P. L., Kim, H. S., Boyar, W. C., and Hutchinson, A.,** Inhibition of nigrostriatal release of dopamine in the rat by adenosine receptor agonists: A_1 receptor mediation, *Neuropharmacology,* 28, 21, 1989.

149. **Schoffelmeer, A. N. M., Wardeh, G., and Mulder, A. H.,** Cyclic AMP facilitates the electrically evoked release of radiolabled noradrenaline, dopamine and 5-hydroxytryptamine from rat brain slices, *Naunyn-Schmiedeberg's Arch. Pharmacol.,* 330, 74, 1985.

150. **Limberger, N., Spath, L., and Starke, K.,** Presynaptic α_2-adrenoceptor, opioid K-receptor and adenosine A_1-receptor interactions on noradrenaline release in rabbit brain cortex, *Naunyn-Schmiedeberg's Arch. Pharmacol.,* 338, 53, 1988.

151. **Onali, P., Olianas, M. C., and Bunse, B.,** Evidence that adenosine A_2 and dopamine autoreceptors antagonistically regulate tyrosine hydroxylase activity in rat striatal synaptosomes, *Brain Res.*, 456, 302, 1988.

152. **Porter, N. M., Radulovacki, M., and Green, R. D.,** Desensitization of adenosine and dopamine receptors in rat brain after treatment with adenosine analogs, *J. Pharmacol. Exp. Ther.*, 244, 218, 1988.

153. **Premont, J., Perez, M., and Bockaert, J.,** Adenosine-sensitive adenylate cyclase in rat striatal homogenates and its relationship to dopamine- and Ca^{2+}-sensitive adenylate cyclases, *Mol. Pharmacol.*, 13, 662, 1977.

154. **Stein, P. S. G.,** Motor systems, with specific reference to the control of locomotion, *Annu. Rev. Neurosci.*, 1, 61, 1978.

155. **Fishman, R. H. B., Feigenbaum, J. J., Yanai, J., and Klawans, H. L.,** The relative importance of dopamine and norepinephrine in mediating locomotor activity, *Prog. Neurobiol.*, 20, 55, 1983.

156. **Heffner, T. G., Wiley, J. N., Williams, A. E., Bruns, R. F., Coughenour, L. L., and Downs, D. A.,** Comparison of the behavioral effects of adenosine agonists and dopamine antagonists in mice, *Psychopharmacologia*, 98, 31, 1989.

157. **McGonigle, P., Boyson, S. J., Reuter, S., and Molinoff, P. B.,** Effects of chronic treatment with selective and non-selective antagonists on the subtypes of dopamine receptors, *Synapse*, 3, 74, 1989.

158. **Broaddus, W. C. and Bennett, J. P., Jr.,** Postnatal development of striatal dopamine function. I. An examination of D_1 and D_2 receptors, adenylate cyclase regulation and presynaptic dopamine markers, *Dev. Brain Res.*, 52, 265, 1990.

159. **Broaddus, W. C. and Bennett, J. P., Jr.,** Postnatal development of striatal dopamine function. II. Effects of neonatal 6-hydroxydopamine treatments on D_1 and D_2 receptors, adenylate cyclase activity and presynaptic dopamine function, *Dev. Brain Res.*, 52, 273, 1990.

160. **Wojcik, W. J. and Neff, N. H.,** Differential location of adenosine A_1 and A_2 receptors in striatum, *Neurosci. Lett.*, 41, 55, 1983.

161. **Premont, J., Perez, M., Blanc, G., Tassin, J., Thierry, A., Herve, D., and Bockaert, J.,** Adenosine-sensitive adenylate cyclase in rat brain homogenates: kinetic characteristics, specificity, topographical, subcellular and cellular distribution, *Mol. Pharmacol.*, 16, 790, 1979.

162. **O'Neill, R. D.,** Adenosine modulation of striatal neurotransmitter release monitored in vivo using voltammetry, *Neurosci. Lett.*, 63, 11, 1986.

163. **Forloni, G., Fisone, G., Consolo, S., and Ladinsky, H.,** Qualitative differences in the effects of adenosine analogs on the cholinergic systems of rat striatum and hippocampus, *Naunyn-Schmiedeberg's Arch. Pharmacol.*, 334, 86, 1986.

164. **Arand-Srivastava, M. B. and Srivastava, A. K.,** Modulation of adenylate cyclase activity by Ca^{2+}, phospholipid-dependent protein kinase in rat brain striatum, *Mol. Cell. Pharmacol.*, 92, 91, 1990.

165. **Myers, S. and Pugsley, T. A.,** Decrease in rat striatal dopamine synthesis and metabolism in vivo by metabolically stable adenosine receptor agonists, *Brain Res.*, 375, 193, 1986.

166. **Zetterstrom, T. and Fillenz, M.,** Adenosine agonists can both inhibit and enhance in vivo striatal dopamine release, *Eur. J. Pharmacol.*, 180, 137, 1990.

167. **Williams, M. and Jarvis, M. F.,** Adenosine antagonists as potential therapeutic agents, *Pharmacol. Biochem. Behav.*, 29, 443, 1988.

168. **Ballasim, M., Herrera-Marschitz, M., Casas, M., and Ungerstedt, U.,** Striatal adenosine levels measured in vivo by microdialysis in rats with unilateral dopamine denervation, *Neurosci. Lett.*, 83, 338, 1987.

169. **Jarvis, M. F. and Williams, M.,** Adenosine and dopamine function in the CNS, *TIPS*, 8, 330, 1987.

170. **Vasse, M. and Protais, P.,** Potentiation of apomorphine-induced stereotyped behavior by acute treatment with dopamine depleting agents: a potential role for an increased stimulation of D_1 dopamine receptors, *Neuropharmacology*, 28, 931, 1989.

171. **Dunwiddie, T. V. and Fredholm, B. B.,** Adenosine receptors mediating inhibitory electrophysiological responses in rat hippocampus are different from receptors mediating cyclic AMP accumulation, *Naunyn-Schmiedeberg's Arch. Pharmacol.*, 326, 294, 1984.

172. **Fastbom, J., Post, C., and Fredholm, B. B.,** Antinociceptive effects and spinal distribution of two adenosine receptor agonists after intrathecal administration, *Pharmacol. Toxicol.*, 66, 69, 1990.

173. **Franklin, P. H., Zhang, G., Tripp, E. D., and Murray, T. F.,** Adenosine A_1 receptor activation mediates suppression of (-)-bicuculline methiodide-induced seizures in rat prepiriform cortex, *J. Pharmacol. Exp. Ther.*, 251, 1229, 1989.

174. **Ahlijanian, M. K. and Takemori, A. E.,** Cross-tolerance studies between caffeine and (-)-N^6-(phenylisopropyl)-adenosine (PIA) in mice, *Life Sci.*, 38, 577, 1986.

175. **Clanachan, A. S., Heaton, T. P., and Parkinson, F. E.,** Drug interactions with nucleoside transport systems, in *Topics and Perspectives in Adenosine Research*, Gerlach, E. and Beckers, B. F., Eds., Springer-Verlag, Berlin, 1987, 118.

176. **Perez, M. T. R. and Ehinger, B.**, Adenosine uptake and release in the rabbit retina, in *Retinal Signal Systems, Degenerations and Transplants*, Agardh, E. and Ehinger, B., Eds., Elsevier Science, New York, 1986, 163.

177. **Lin, B. B., Hurley, M. C., and Fox, I. H.**, Regulation of adenosine kinase by adenosine analogs, *Mol. Pharmacol.*, 34, 501, 1988.

178. **Jarvis, M. F. and Williams, M.**, Differences in adenosine A-1 and A-2 receptor density revealed by autoradiography in methylxanthine-sensitive and insensitive mice, *Pharm. Biochem. Behav.*, 30, 707, 1988.

179. **Choi, O. H., Shamin, M. T., Padgett, W. L., and Daly, J. W.**, Caffeine and theophylline analogs: correlation of behavioral effects with activity as adenosine receptor antagonists and as phosphodiesterase inhibitors, *Life Sci.*, 43, 387, 1988.

180. **Daly, J. W., Butts-Lamb, P., and Padgett, W.**, Subclasses of adenosine receptors in the central nervous system: interaction with caffeine and related methylxanthines, *Cell. Mol. Neurobiol.*, 3, 69, 1983.

181. **Mante, S. and Minneman, K. P.**, Caffeine inhibits forskolin-stimulated cyclic AMP accumulation in rat brain, *Eur. J. Pharmacol.*, 175, 203, 1990.

182. **Carney, J. M., Cao, W., Logan, L., Rennert, O. M., and Seale, T. W.**, Differential antagonism of the behavioral depressant and hypothermic effects of 5'-(N-ethylcarboxamide) adenosine by theobromine, *Pharm. Biochem. Behav.*, 25, 769, 1986.

183. **Seale, T. W., Abla, K. A., Cao, W., Parker, K. M., Rennert, O. M., and Carney, J. M.**, Inherent hyporesponsiveness to methylxanthine-induced behavioral changes associated with supersensitivity to 5'-N-ethylcarboxamideoadenosine (NECA), *Pharm. Biochem. Behav.*, 25, 1271, 1986.

184. **Buckholtz, N. S. and Middaugh, L. D.**, Effects of caffeine and L-phenylisopropyladenosine on locomotor activity of mice, *Pharm. Biochem. Behav.*, 28, 179, 1987.

185. **Durcan, M. J. and Morgan, P. F.**, Evidence for adenosine A_2 receptor involvement in the hypomobility effects of adenosine analogs in mice, *Eur. J. Pharmacol.*, 168, 285, 1989.

186. **Durcan, M. J. and Morgan, P. F.**, NECA-induced hypomotility in mice: evidence for a predominantly central site of action, *Pharm. Biochem. Behav.*, 32, 487, 1989.

187. **Nikodijevic, O., Daly, J. W., and Jacobson, K. A.**, Characterization of the locomotor depression produced by an A_2-selective adenosine agonist, *FEBS Lett.*, 261, 67, 1990.

188. **Schiffman, S. S., Gill, J. M., and Diaz, C.**, Methylxanthines enhance taste: evidence for modulation of taste by adenosine receptor, *Pharmacol. Biochem. Behav.*, 22, 195, 1985.

189. **Coffin, V. L. and Carney, J. M.**, Effects of selected analogs of adenosine on schedule-controlled behavior in rats, *Neuropharmacology*, 25, 1141, 1986.

190. **Spealman, R. D. and Coffin, V. L.**, Behavioral effects of adenosine analogs in squirrel monkeys: relation to adenosine A_2 receptors, *Psychopharmacologia*, 90, 419, 1986.

191. **Spealman, R. D. and Coffin, V. L.**, Discriminative-stimulus effects of adenosine analogs: mediation by adenosine A_2 receptors, *J. Pharmacol. Exp. Ther.*, 246, 610, 1988.

192. **Spealman, R. D.**, Psychomotor stimulant effects of methylxanthines in squirrel monkeys: relation to adenosine antagonism, *Psychopharmacologia*, 95, 19, 1988.

193. **Chou, D. T., Khan, S., Forde, J., et al.**, Caffeine tolerance: behavioral, electrophysiological, and neurochemical evidence, *Life Sci.*, 36, 2347, 1985.

194. **Ahlijanian, M. K. and Takemori, A. E.**, Cross-tolerance studies between caffeine and (-)-N^6-(phenyl-isopropyl)-adenosine (PIA) in mice, *Life Sci.*, 38, 577, 1986.

Chapter 29

ADENOSINE AND EPILEPTIC SEIZURES

Michael Dragunow

TABLE OF CONTENTS

I. INTRODUCTION

Adenosine has powerful inhibitory neuromodulatory actions in the nervous system that are achieved by a number of effects, the most important of which are its depression of excitatory synaptic transmission and its direct postsynaptic hyperpolarizing actions. Because epileptic seizures are associated with excessive activity of neurons, it seems logical to assume that an inhibitory substance such as adenosine might play some role in seizure disorders. In the 16 years since the initial observation by Maitre et al.[1] that adenosine has anticonvulsant effects on audiogenic seizures in mice, a great deal has been discovered about the role of adenosine in epilepsy.

II. ANTICONVULSANT EFFECTS OF ADENOSINE AND ADENOSINE ANALOGS

A. *IN VITRO* SEIZURE MODELS

The results of a large number of studies have shown that adenosine and adenosine analogs potently suppress epileptiform activity in hippocampal and neocortical slices of rat brain in a theophylline-reversible manner. This epileptiform activity (interictal spikes, pre-seizure-like multiple population spikes, epileptiform afterdischarges) in hippocampal slices has been induced by a variety of means including penicillin in combination with elevated potassium,[2] low calcium and high magnesium mediums,[3,4] and bicuculline.[5] More recently, a neocortical slice model where epileptiform activity was induced by removing magnesium ions has been investigated.[6] Adenosine and its analogs potently suppress epileptiform activity in all these varied models. Also, soluflazine, an adenosine uptake inhibitor, blocks epileptiform activity in guinea pig hippocampal slices induced by low magnesium.[7]

B. *IN VIVO* SEIZURE MODELS

Adenosine and its analogs, injected centrally or systemically, have powerful anticonvulsant effects on a number of seizure models including audiogenic seizures in mice,[1] strychnine, pentylenetetrazol-, methionine sulfoximine-, kynurenic-, pilocarpine and lithium-, pilocarpine-, bicuculline-, penicillin-, and caffeine-induced seizures in mice and rats,[8-15] and on kindled seizures in rats.[18-24] The adenosine uptake blocker, papaverine, has anticonvulsant effects on pentylenetetrazol- and strychnine-induced seizures[25] and on amygdala-[21,26] and hippocampal-kindled[27] seizures (although it has proconvulsant effects on theophylline-induced seizures in rats).[28]

A controversy which has recently arisen in the area is whether the anticonvulsant effects of adenosine are primarily due to block of seizure initiation or seizure spread or both. In the chemoconvulsant models adenosine agonists have been shown to raise the threshold (dose of drug) for seizure onset.[12] However, because only the behavioral seizure is measured it is not possible to determine whether this anticonvulsant effect is due to block of seizure initiation or spread. The question is perhaps best addressed in the kindling model where seizure initiation (afterdischarge threshold) can be dissociated from seizure spread (motor seizure intensity, duration, and threshold, and latency to motor seizure). In this seizure model, adenosine analogs and uptake blockers greatly reduce the duration and intensity of the motor seizure and also increase the latency to motor seizure.[18,19,21,22,24,27]

These results suggest that adenosine has powerful anticonvulsant effects on seizure spread and duration. Similarly, the adenosine agonist L-phenylisopropyladenosine (L-PIA) has been found to prevent the spread of epileptic activity induced by penicillin from one cerebral hemisphere to the other.[13] However, adenosine analogs and uptake blockers also prevent seizure initiation. Papaverine produces a large elevation of afterdischarge threshold in the hippocampus[27] and L-PIA and the adenosine agonist N-ethylcarboxamidoadenosine (NECA) raise the seizure threshold in amygdala, hippocampus, and striatum.[24] Furthermore, although L-PIA prevented the initiation of pilocarpine-induced seizures in rats, it did not stop seizures once they had started.[11]

Thus, there is evidence to suggest that adenosine analogs administered exogenously have powerful anticonvulsant effects on both seizure initiation and spread. However, whether or not endogenous adenosine similarly inhibits both seizure initiation and spread can only be resolved by investigating the effects of adenosine antagonists on initiation vs. spread in the kindling model (see Section III.C).

C. RECEPTOR SUBTYPE(S) MEDIATING ANTICONVULSANT EFFECTS

Adenosine receptors have been classified according to the scheme: A_1 linked to adenylate cyclase, and activation of the receptor inhibits cyclic AMP accumulation; A_2 linked to adenylate cyclase, and activation of the receptor elevates levels of cAMP.[29,30] The existence of A_3 receptors has also been proposed,[31,32] although whether or not these A_3 receptors actually exist as a distinct class of adenosine receptor will have to await further studies. Nevertheless, the putative A_3 receptor site has been proposed to mediate the anticonvulsant effects of adenosine on the basis that it is enriched in brain membranes compared with other organs.[33]

The results of a number of varied studies suggest that the A_1 receptor probably mediates the anticonvulsant effects of adenosine in most brain regions. Firstly, recent autoradiographic studies of rat brain using A_2-selective radioactive ligands have shown that A_2 adenosine receptors are highly concentrated in the basal ganglia (caudate-putamen, nucleus accumbens, and olfactory tubercle),[34] but are not found in the neocortex or hippocampus, whereas A_1 receptors are found in many brain regions including the cortex, the hippocampus, the thalamus, the striatum, and the cerebellum.[35] Thus, in *in vitro* studies of epileptiform activity in neocortical and hippocampal slices, the anticonvulsant effects of adenosine are likely to be mediated by A_1 receptors. Indeed, the rank order of potency of adenosine agonists in inhibiting epileptiform activity and the block by A_1-selective adenosine antagonists of these effects also suggest A_1 receptor mediation.[36,37] Also, the inhibitory neurophysiological effects of adenosine and its inhibitory effects on neurotransmitter release appear to be mediated by A_1 receptors,[36,38] and these actions are likely to be involved in its anticonvulsant effects. Furthermore, pertussis toxin, which uncouples adenosine A_1 (and other) receptors from G_i proteins (ADP-ribosylates G_i proteins) produces seizures after intrahippocampal injection.[39] Finally, recent studies have shown a temporal relationship between upregulation of A_1 adenosine receptors and decreased sensitivity to bicuculline-induced seizures following chronic theophylline administration to rats.[40,41]

Thus, there is good evidence to suggest that adenosine produces its *in vitro* and *in vivo* anticonvulsant effects by activating A_1 receptors that are coupled to an inhibitory G protein. However, although the A_1 receptor has been defined on the basis of coupling in an inhibitory manner to adenylate cyclase, the role of cAMP inhibition in mediating these effects is unclear.[42] The basal ganglia are important sites for seizure generation, indicating that a role for A_2 receptors in some of the anticonvulsant effects of adenosine cannot be excluded. Indeed, the anticonvulsant effects of NECA on caudate-kindled seizures appear to be mediated by activation of A_2 receptors in this brain region.[24] A complete understanding of the receptor subtype involved in the anticonvulsant effects of adenosine will have to await studies using selective adenosine antagonists.

D. SITE OF ANTICONVULSANT ACTION

Presumably, the site of anticonvulsant action of adenosine is determined by the distribution of anticonvulsant adenosine receptors, because the anticonvulsant effects of adenosine are mediated by extracellular receptors. Bearing in mind that A_2 receptors in the striatum might still play some anticonvulsant role, it seems most likely that the site of anticonvulsant effect of adenosine will therefore be determined by the distribution of A_1 receptors in the brain. These receptors are widely distributed in the brain in the limbic system, the neocortex, the thalamus, the cerebellum, and the spinal cord,[35] suggesting that there are many potential sites of action and that adenosine may work throughout the brain to dampen down seizures. This distribution of A_1 receptors also follows the distribution of excitatory amino acid receptors.

The work of Szot et al.[41] and of Sanders and Murray,[40] showing that upregulation of A_1 receptors following chronic theophylline leads to reduced sensitivity to convulsants suggests that if we can identify the brain regions in which the adenosine A_1 receptor upregulation occurs, then these regions might be central to the anticonvulsant effects of adenosine. Also, recent studies have shown that chronic treatment with adenosine antagonists not only increases the number of A_1 receptors, but also causes a shift of the A_1 receptors to a high-affinity state via increased coupling of A_1 receptors to G proteins and an increase in the α_1-subunit of the G_i protein.[42-44] Guanosine 5'-triphosphate (GTP) reduces agonist binding to A_1 receptors by shifting them from a high- to a low-affinity state, but the effect of GTP on agonist binding is reduced in rats treated chronically with theophylline, suggesting that theophylline treatment increases the number of receptors in the high-affinity state (i.e., coupled to G proteins). Autoradiographic studies[42,45] have shown that A_1 receptor upregulation, measured with [^{3}H]CHA, occurs following chronic caffeine in the striatum, the limbic system, the motor and sensory cortex, the thalamus, and the cerebellum, and following theophylline in the frontoparietal cortex, the caudate-putamen, the nucleus accumbens, and in the central grey of the midbrain. Thus, these brain regions may be important sites for mediating the anticonvulsant effects of adenosine.

Recently, Franklin et al.[17] have shown that the adenosine agonist 2-chloroadenosine (CADO) injected into the rat prepiriform cortex blocks seizures induced by bicuculline and that this anticonvulsant effect of CADO is blocked by an adenosine antagonist. This result suggests that perhaps the prepiriform cortex, which is an important site of seizure generation,[46] is involved in the anticonvulsant effects of adenosine. If this hypothesis is true, then adenosine antagonists injected into the prepiriform cortex ought to block the anticonvulsant effects of adenosine. Also, the work of Rosen and Berman[24] suggests that the amygdala, the hippocampus, and the caudate are all sites of anticonvulsant action for adenosine. Other important sites of seizure generalization, such as the substantia nigra, are unlikely to be involved in the anticonvulsant effects of adenosine because this region has a low concentration of adenosine receptors.

E. MECHANISM OF ANTICONVULSANT ACTION

Adenosine exerts potent inhibitory actions both pre- and postsynaptically on neurons, and it is likely that its anticonvulsant properties result from these inhibitory effects on nerve cells. Adenosine depresses excitatory postsynaptic potentials and hyperpolarizes neurons, and may account for the late afterhyperpolarization that follows depolarization of hippocampal pyramidal cells.

Adenosine A_1 receptors are located on axon terminals of excitatory neurons[35,47] and also postsynaptically. Presynaptically, adenosine depresses excitatory postsynaptic potentials and inhibits release of neurotransmitters such as glutamate, aspartate, and acetylcholine.[48,49] The inhibitory effects of adenosine on neurotransmitter release seems to require magnesium ions[50] and to be due to activation of an A_1 receptor linked to a G_i protein, since pertussis toxin, which inactivates G_i proteins via ADP ribosylation, blocks the inhibitory effects of adenosine on glutamate release from cultured cerebellar granule cells.[38] More recently, pertussis toxin has been shown to block the adenosine-induced suppression of synaptic responses in the hippocampal slice.[39] Similarly, N-ethylmaleimide, which also inactivates G proteins linked to the A_1 receptor, blocks adenosine depression of acetylcholine, glutamate, and noradrenaline release from hippocampal slices.[51,52] However, although an inhibitory G protein is clearly involved in these actions of adenosine, adenylate cyclase or dihydropyridine-sensitive calcium channels would appear not to be involved.[51,52]

The potent inhibitory effects of exogenous adenosine and analogs on the release of acetylcholine and the excitatory amino acids may play a major role in their anticonvulsant effects because excitatory amino acids play important roles in seizure initiation and spread. Thus, adenosine may terminate the action of excitatory amino acids by preventing their release and thereby inhibiting the initiation and spread of seizures. Endogenous adenosine also inhibits aspartate and glutamate release because the A_1-selective adenosine antagonist 8-phenyltheophylline increases the release of aspartate and glutamate.[48] Furthermore, a number of studies suggest that the release of excitatory amino acids can lead to adenosine release and that this effect is mediated by both N-methyl-D-aspartate (NMDA) and non-NMDA glutamate receptors.[53] Thus, perhaps adenosine release occurring during seizure activity results from glutamate receptor activation and leads to inhibition of further glutamate release (i.e., a negative feedback system). Certainly, ischemia-induced elevations of brain adenosine levels seem to be due to excitatory amino acid release because they are blocked by NMDA receptor antagonists.[54] The ability of adenosine agonists to inhibit the neurotoxic effects of glutamate agonists such as kainic acid and quinolinic acid[55,56] further supports an interaction between adenosine and excitatory amino acid systems.

The interaction of adenosine with ion channels might also be involved in its anticonvulsant effects. Simoes et al.[57] have shown that CHA inhibits veratridine-stimulated sodium uptake by rat brain synaptosomes, suggesting that adenosine might inhibit voltage-dependent sodium channels. If levels of adenosine either basally or after seizures are sufficient to exert this effect in vivo, then it is likely that it would contribute to the anticonvulsant effects of adenosine.

Adenosine can inhibit calcium fluxes in a number of systems,[37] and this action may account for its presynaptic effects on transmitter release, its postsynaptic hyperpolarizing effects, and its anticonvulsant effects because calcium is involved in seizure generation. Furthermore, there is some evidence to suggest that an NMDA-evoked synaptic calcium influx is controlled by endogenous adenosine in hippocampal slices[37] and adenosine can directly depress glutamate responses in the hippocampus.[58] Thus, adenosine may produce its anticonvulsant effects not only by inhibiting the release of excitatory amino acid neurotransmitters, but also by inhibiting the direct calcium-dependent, postsynaptic actions of these neurotransmitters.

Also, adenosine turns on a 4-aminopyridine-sensitive potassium channel (the A current) in hippocampal pyramidal cells, and this action has been proposed to account for its anticonvulsant effects.[59] Indeed, the postsynaptic actions of adenosine appear to be due mainly to the activation of a G_i protein-coupled potassium channel.[60]

Recent data which might have bearing on the mechanism of action of adenosine in seizures suggest that the inhibitory effects of adenosine in the hippocampus are blocked by translocation of protein kinase C,[61] either directly by the addition of phorbol esters or indirectly by NMDA agonists[62] or muscarinic agonists.[63] It has been suggested that this results because activation of protein kinase C by muscarinic agonists leads to inactivation of the G protein coupling the adenosine receptor to potassium channels. Thus, during seizure activity when NMDA and muscarinic receptors might be expected to be activated, the postsynaptic inhibitory effects of adenosine might be reduced.

III. PROCONVULSANT AND CONVULSANT EFFECTS OF ADENOSINE ANTAGONISTS

A. *IN VITRO* SEIZURE MODELS

Adenosine antagonists have "proconvulsant" and "convulsant" effects on *in vitro* hippocampal and neocortical seizure models. Not only do they reverse the "anticonvulsant" effects of adenosine agonists, but they promote epileptiform activity induced by other convulsants (e.g., penicillin, bicuculline, kainic acid) in these models and can themselves generate burst discharges when administered alone.[2,6,37,64-67] Furthermore, adenosine deaminase is "proconvulsant" in these models, and the adenosine requirement for magnesium ions may account for the proconvulsant effects of low magnesium mediums. In area CA3 of guinea pig hippocampus, the A_1-selective adenosine antagonist 1,3-dipropyl-8-cyclopentylxanthine (DPCPX) initiated irreversible, self-sustained epileptiform activity.[65] These results indicate that endogenous adenosine exerts an antiepileptiform effect on hippocampal and neocortical slices and that in area CA3, block of A_1 receptors can lead to epileptiform activity.

B. *IN VIVO* SEIZURE MODELS

High doses of nonspecific adenosine antagonists such as caffeine and theophylline are convulsant in animals and humans. However, at these high doses (>200 mg/kg i.p. in rats) these drugs have a number of neurochemical actions besides adenosine antagonism so that it seems very unlikely that adenosine antagonism alone accounts for these seizures (see Dragunow[68] for a discussion of the data). However, at lower doses caffeine and theophylline produce many of their pharmacological effects by blocking the action of endogenous adenosine.[8] At these lower doses (10 to 50 mg/kg i.p.) these drugs produce proconvulsant effects in a number of *in vivo* seizure models, including chemically induced seizures (e.g., pilocarpine, kainic acid, pentylenetetrazol), electroshock seizures, and kindled seizures (see Dragunow[68-70] for a review of these data). Furthermore, the specific adenosine antagonist 8-cyclopentyltheophylline also has proconvulsant effects, indicating that the proconvulsant effects of caffeine and theophylline are due to adenosine antagonism.[71] These *in vivo* and *in vitro* results suggest that endogenous adenosine acts as a natural anticonvulsant in the nervous system as previously suggested.[69]

C. SEIZURE INITIATION VS. SPREAD?

Adenosine agonists can inhibit the initiation and spread of seizures (see Section II.B). However, does endogenous adenosine similarly inhibit both seizure initiation and spread? The results of *in vitro* seizure models, where adenosine antagonists can generate epileptiform activity alone and promote epileptiform activity induced by other convulsants, suggest that endogenous adenosine prevents initiation of these seizure-like discharges. However, *in vivo* the situation is not so clear. We have used electrically induced seizures where seizure initiation and spread can be clearly dissociated to investigate the role of endogenous adenosine in seizure initiation and spread. We and others found that although adenosine antagonists can promote the spread of seizures, they do not lower the threshold for seizure initiation.[22,72,73] These results have led us to speculate that adenosine is brought into play as an endogenous anticonvulsant by the seizure itself and that its main function is to prevent seizure spread and to terminate seizures.[22] The observation that brain adenosine levels are rapidly and greatly elevated after seizure onset supports this hypothesis (see later).

D. ADENOSINE ANTAGONISTS PROLONG SEIZURES, SUGGESTING THAT ENDOGENOUS ADENOSINE TERMINATES SEIZURES

As mentioned above, one of the most potent and consistent effects of adenosine antagonists on kindled and electrically induced seizures is their ability to prolong seizure duration and enhance seizure spread.[18,21,22,72-78] In the presence of a subconvulsive dose of caffeine, theophylline, or 8-cyclopentyltheophylline, partial seizures that would normally last only a few seconds are converted to fully generalized seizures that can last many minutes.[22,71,73] Caffeine and theophylline also prolong the duration of seizures in humans.[79,80] Conversely, adenosine agonists convert generalized seizures to partial seizures.[19] These results strongly support our hypothesis that endogenous adenosine inhibits seizure spread and terminates seizure discharges in the brain.

IV. ADENOSINE AND STATUS EPILEPTICUS

Seizures in both animals and humans will usually arrest spontaneously and abruptly (in 1 or so minutes) and the brain will remain seizure free for some time thereafter. We believe that this spontaneous arrest of brief seizures is brought about by endogenous adenosine, although the postseizure-free period may involve other mechanisms (e.g., opioids, prostaglandins, and perhaps adenosine metabolites such as inosine; see Dragunow[70] for a review). Sometimes seizures do not spontaneously terminate and a protracted and recurrent seizure state called status epilepticus (SE) ensues. A number of years ago we suggested that SE might be partly caused by a loss of anticonvulsant adenosine mechanisms.[69] The results of a number of recent studies support this hypothesis. Turski et al.[16] found that CADO prevented and aminophylline facilitated pilocarpine-induced SE in rats, and Morrisett et al.[11] found that L-PIA blocked SE in rats induced by a combination of pilocarpine and lithium. Albertson[81] found that subconvulsive doses of aminophylline produced SE in kindled rats given repeated stimulation. We have found that the specific adenosine antagonist 8-cyclopentyltheophylline produces SE in nonkindled rats stimulated recurrently at 1-min intervals in the hippocampus (Dragunow, in preparation). Treiman and Handforth[82] have found similar results using a 5-min recurrent stimulation paradigm with CADO and aminophylline. Terminal SE also developed in rat pups given hourly amygdala stimulation in the presence of aminophylline.[78] In another model of SE, Eldridge et al.[83] found that theophylline and 8-cyclopentyltheophylline greatly prolonged the ictal phase of recycling seizures induced by cortical application of penicillin in cats, whereas the adenosine uptake blocker dipyridamole lengthened the interictal phase of the seizure. Finally, Peters et al.[79] reported a case of SE induced in a human patient receiving subconvulsive doses of theophylline during electroconvulsive therapy (ECT).

These results suggest that loss of adenosine anticonvulsant systems can lead to the development of SE in both animals and humans; whether or not SE occurring spontaneously in humans is caused by such a loss of adenosine mechanisms is unknown. However, adenosine agonist drugs may prove useful in the treatment of SE and its associated nerve cell death.[84]

V. EFFECTS OF SEIZURES ON ADENOSINE LEVELS, ENZYMES, UPTAKE SITES, AND RECEPTORS

Seizures, induced by electroshock or bicuculline, produce large increases in brain adenosine levels that peak soon after seizure onset and return rapidly to baseline after seizure termination.[85-88] As already discussed, these elevations in adenosine levels are likely to be due to release of excitatory amino acids during seizures. Assuming that these elevated adenosine levels have access to extracellular neuronal adenosine receptors, then these results suggest that the anticonvulsant effects of adenosine are likely to be greatest after seizure onset.

One report has also shown that seizures, induced by leptazol, are associated with a large increase in the activity of the enzyme adenosine deaminase in the brain,[89] which might account for the rapid degradation of adenosine to inosine after seizures.[90]

Seizures also regulate adenosine receptor number and coupling to second messengers. Sattin[91] found that repeated electroconvulsive shock (ECS) led to increased adenosine-induced cAMP formation in the cortex of rats 2 d after the last fit, although Newman et al.[92] found that 2 d after a series of ECS, CADO-induced cAMP formation was reduced in the cortex of rats. The reason for these opposite results is unclear. The effect of seizures on adenosine receptor number depends upon the brain region studied. Wybenga et al.[93] found decreased numbers of A_1 receptors in the cerebellum 1 h after metrazol seizures, whereas Newman et al.[92] found an upregulation of cerebral cortical A_1 receptors 2 d after ECS. Daval and Sarfati[94] found a significant increase in A_1 receptor number in cerebral cortex 30 min after bicuculline-induced seizures, and Gleiter et al.[95] found an upregulation of A_1 receptors following chronic, but not acute, ECS, and these upregulated receptors appeared to be coupled to G proteins. A_2 receptor number in the striatum was not altered by ECS, although there was an upregulation of the adenosine uptake site in the striatum.[95]

Thus, seizures regulate various aspect of adenosine neurochemistry, including levels, degradation enzymes, uptake sites, receptor number, and coupling to second messengers.

VI. ADENOSINE RECEPTORS IN EPILEPTIC RAT AND HUMAN BRAIN

Adenosine is clearly a powerful anticonvulsant substance in the nervous system. Could loss of this adenosine system be involved in producing epilepsy? Ohisalo et al.[96] found normal levels of adenosine in the cerebrospinal fluid of patients with myoclonic epilepsy. Recently, it has been shown that endogenous adenosine can reduce epileptiform activity in the human epileptogenic cortex *in vitro*,[97] indicating that adenosine has anticonvulsant-like effects on human brain. Over the past year we have initiated a project funded by the New Zealand Neurological Foundation to look at A_1 adenosine receptors in the kindling model of complex partial seizures in rats and in biopsied specimens of human temporal lobe and hippocampus removed in the routine neurosurgical treatment of epilepsy.[98,99]

In our rat studies on amygdala-kindled seizures, our preliminary results suggest that 24 h or 1 month after kindling has been completed there are no major changes in A_1 receptor number in any brain region, although we have not yet analyzed these results with densitometry.[111] This preliminary result suggests that the increased seizure susceptibility in kindled rats is not due to loss of A_1 receptors. We are presently investigating other aspects of adenosine neurochemistry in kindled brains.

In our human studies, preliminary results suggest that in epileptic temporal cortex there are regions of loss of A_1 receptors.[99] Interestingly, benzodiazepine and excitatory amino acid receptors are not altered in regions of temporal cortex where there is A_1 receptor loss. This observation suggests that there is a selective loss of A_1 receptors in regions of epileptic temporal cortex.[99] Assuming that these preliminary results are confirmed in the larger study that we are presently undertaking, there are a number of possibilities that could account for this selective A_1 loss. Carbamazepine, which is the drug of choice for complex partial seizures, binds to A_1 receptors in rat and human brain.[100] Therefore, the loss of A_1 binding may be simply due to displacement by carbamazepine. However, this seems unlikely because the loss of A_1 binding does not occur throughout the entire temporal lobe specimen, and the tissue is preincubated so that the drug would be washed away. Another possibility is that adenosine released by seizure activity or during the disruption of the tissue at surgery displaces the radioactive ligand from the A_1 receptor. However, we preincubate and incubate our tissue sections with adenosine deaminase, which would be expected to degrade endogenous adenosine. However, recent evidence suggests that it is difficult to completely remove endogenous adenosine with adenosine deaminase,[101] so that this is still a possibility. Finally, this A_1 loss might represent a real loss of A_1 binding in epileptic temporal lobe and may be caused by a loss of cells upon which adenosine receptors are located or to a specific loss in A_1 binding without cell loss. We are presently investigating these possibilities. This A_1 loss is unlikely to be a consequence of the seizure itself because animal studies have shown that seizures upregulate, rather than downregulate, cortical A_1 receptors (see Section V).

VII. EFFECTS OF ANTICONVULSANT DRUGS ON ADENOSINE SYSTEMS

A number of anticonvulsant drugs are known to modulate aspects of adenosine neurochemistry. Carbamazepine binds to A_1 receptors and appears to have antagonistic properties;[14,45,100,102-105] benzodiazepines, meprobamate, and phenytoin are adenosine uptake blockers;[106,107] and aminophylline reverses the anticonvulsant effects of tizanidine.[108] However, although many anticonvulsant drugs affect adenosine neurochemistry, there is no conclusive evidence to suggest that their anticonvulsant effects are mediated by this action. Perhaps anticonvulsant drugs of the future will be based upon compounds that enhance brain adenosine systems.

VIII. CONCLUSIONS

Adenosine is a powerful inhibitory neuromodulator.[109] Adenosine is also a powerful endogenous anticonvulsant substance that restricts the spread and terminates seizures by activating extracellular receptors (probably A_1) located both pre- and postsynaptically on neurons. Adenosine is formed (from nucleotides) during seizures, perhaps in response to activation of excitatory amino acid receptors and/or uptake mechanisms,[50,110] and can inhibit release of excitatory amino acids and directly hyperpolarize neurons. Adenosine agonists may prove useful in treating epilepsy, especially SE, and adenosine may be involved in the etiology of status and other seizure disorders such as complex partial seizures.[97,99]

REFERENCES

1. **Maitre, M., Ciesielski, L., Lehman, A., Kempf, E., and Mandel, P.,** Protective effect of adenosine and nicotinamide against audiogenic seizures, *Biochem. Pharmacol.*, 23, 2807, 1974.
2. **Dunwiddie, T. V.,** Endogenously released adenosine regulates excitability in the *in vitro* hippocampus, *Epilepsia*, 21, 541, 1980.
3. **Haas, H. L., Jeffreys, J. G. R., Slater, N. T., and Carpenter, D. O.,** Modulation of low calcium induced field bursts in the hippocampus by monoamines and cholinomimetics, *Pflügers Arch. Physiol.*, 400, 28, 1984.
4. **Lee, K. S., Schubert, P., and Heinemann, U.,** The anticonvulsive action of adenosine: a postsynaptic, dendritic action by a possible endogenous anticonvulsant, *Brain Res.*, 321, 160, 1984.
5. **Ault, B. and Wang, C. M.,** Adenosine inhibits epileptiform activity arising in hippocampal area CA3, *Br. J. Pharmacol. Chemother.*, 87, 695, 1986.
6. **O'Shaughnessy, C. T., Aram, J. A., and Lodge, D.,** A_1 adenosine receptor-mediated block of epileptiform activity induced in zero magnesium in rat neocortex in vitro, *Epilepsy Res.*, 2, 294, 1988.
7. **Ashton, D., De Prins, E., Willems, R., Van Belle, H., and Wauquier, A.,** Anticonvulsant action of the nucleoside transport inhibitor, soluflazine, on synaptic and non-synaptic epileptogenesis in the guinea-pig hippocampus, *Epilepsy Res.*, 2, 65, 1988.
8. **Snyder, S. H., Katims, J. J., Annau, Z., Bruns, R. F., and Daly, W.,** Adenosine receptors and behavioral actions of methylxanthines, *Proc. Natl. Acad. Sci. U.S.A.*, 75, 3260, 1981.

9. **Dunwiddie, T. V. and Worth, T.,** Sedative and anticonvulsant effects of adenosine analogs in mouse and rat, *J. Pharmacol. Exp. Ther.,* 220, 70, 1982.

10. **Lapin, I. P.,** Structure-activity relationships in kynurenine, diazepam and some putative endogenous ligands of the benzodiazepine receptors, *Neurosci. Biobehav. Rev.,* 7, 107, 1983.

11. **Morrisett, R. A., Jope, R. S., and Carter-Snead, O., III,** Effects of drugs on the initiation and maintenance of status epilepticus induced by administration of pilocarpine to lithium-pretreated rats, *Exp. Neurol.,* 97, 192, 1987.

12. **Murray, T. F., Sylvester, D., Schultz, C. S., and Szot, P.,** Purinergic modulation of the seizure threshold for pentylenetetrazol in the rat, *Neuropharmacology,* 24, 761, 1985.

13. **Niglio, T., Popoli, P., Caporali, M. G., and Scotti de Carolis, A.,** Antiepileptic effects of N^6-L-phenylisopropyladenosine (L-PIA) on penicillin-induced epileptogenic focus in rabbits, *Pharmacol. Res. Commun.,* 20, 561, 1988.

14. **Popoli, P., Benedetti, M., and Scotti de Carolis, A.,** Anticonvulsant activity of carbamazepine and N^6-L-phenylisopropyladenosine in rabbits. Relationship to adenosine receptors in central nervous system, *Pharmacol. Biochem. Behav.,* 29, 533, 1988.

15. **Sellinger, O. Z., Schatz, R. A., Porta, R., and Wilens, T. E.,** Brain methylation and epileptogenesis: the case of methionine sulfoximine, *Ann. Neurol.,* 16, S115, 1984.

16. **Turski, W. A., Cavalheiro, E. A., Ikonomidou, C., Moraes Mello, L. E. A., Bortolotto, Z. A., and Turski, L.,** Effects of aminophylline and 2-chloroadenosine on seizures produced by pilocarpine in rats: morphological and electroencephalographic correlates, *Brain Res.,* 3611, 309, 1985.

17. **Franklin, P. H., Tripp, E. D., Zhang, G., Gale, K., and Murray, T. F.,** Adenosine receptor activation blocks seizures induced by bicuculline methiodide in the rat prepiriform cortex, *Eur. J. Pharmacol.,* 150, 207, 1988.

18. **Albertson, T. E., Stark, L. G., Joy, R. M., and Bowyer, J. F.,** Aminophylline and kindled seizures, *Exp. Neurol.,* 81, 703, 1983.

19. **Barraco, R. A., Swanson, T. H., Phillis, J. W., and Berman, R. F.,** Anticonvulsant effects of adenosine analogs on amygdaloid-kindled seizures in rats, *Neurosci. Lett.,* 46, 317, 1984.

20. **Bortolotto, Z. A., Mello, L. E. M., Turski, L., and Cavalheiro, E. A.,** Effects of 2-CLA on amygdaloid and hippocampal kindled seizures, *Arch. Int. Pharmacodyn. Ther.,* 277, 313, 1985.

21. **Dragunow, M. and Goddard, G. V.,** Adenosine modulation of amygdala kindling, *Exp. Neurol.,* 84, 654, 1984.

22. **Dragunow, M., Goddard, G. V., and Laverty, R.,** Is adenosine an endogenous anticonvulsant?, *Epilepsia,* 26(5), 480, 1985.

23. **Rosen, J. B. and Berman, R. F.,** Prolonged postictal depression in amygdala-kindled rats by the adenosine analog, L-phenylisopropyladenosine, *Exp. Neurol.,* 90, 549, 1985.

24. **Rosen, J. B. and Berman, R. F.,** Differential effects of adenosine analogs on amygdala, hippocampus, and caudate nucleus kindled seizures, *Epilepsia,* 28, 658, 1987.

25. **Roussinov, K. S., Lazarova, M., and Atanassova-Shopova, S.,** Experimental study of the effects of imidazol, papaverine and histamine on convulsive-seizure reactivity, *Acta Physiol. Pharmacol. Bulg.,* 2, 78, 1976.

26. **Dragunow, M.,** Seizure termination produced by papaverine, an inhibitor of adenosine uptake, *Proc. Univ. Otago Med. Sch.,* 62, 78, 1984.

27. **Dragunow, M.,** Time-dependent effects of papaverine on electrically-induced seizures in rats, *Neurosci. Lett.,* 74, 320, 1987.

28. **Ramzan, I.,** Proconvulsant effect of papaverine on theophylline-induced seizures in rats, *Clin. Exp. Pharmacol. Physiol.,* 16, 425, 1989.

29. **Londos, C., Cooper, D. M. F., and Wolff, J.,** Subclasses of external adenosine receptors, *Proc. Natl. Acad. Sci. U.S.A.,* 77, 2551, 1980.

30. **Van Calker, D., Muller, M., and Hamprecht, B.,** Adenosine regulates via two different receptors, the accumulation of cyclic AMP in cultured brain cells, *J. Neurochem.,* 33, 999, 1979.

31. **Ribeiro, J. A. and Sebastião, A. M.,** Adenosine receptors and calcium: basis for proposing a third (A_3) adenosine receptor, *Progr. Neurobiol.,* 26, 179, 1986.

32. **Chin, J. H. and DeLorenzo, R. J.,** Cobalt ion enhancement of 2-chloro[^{3}H]adenosine binding to a novel class of adenosine receptors in brain: antagonism by calcium, *Brain Res.,* 348, 381, 1985.

33. **Chin, J. H. and DeLorenzo, R. J.,** A3 adenosine binding sites: possible endogenous anticonvulsant receptor in brain?, *Ann. Neurol.,* 22, 209, 1987.

34. **Jarvis, M. F. and Williams, M.,** Direct autoradiographic localization of adenosine A_2 receptors in the rat brain using the A_2-selective agonist, [^{3}H]CGS 21680, *Eur. J. Pharmacol.,* 168, 243, 1989.

35. **Dragunow, M., Murphy, K., Leslie, R. A., and Robertson, H. A.,** Localization of adenosine A_1-receptors to the terminals of the perforant path, *Brain Res.,* 462, 252, 1988.

36. **Dunwiddie, T. V. and Fredholm, B. B.**, Adenosine receptors mediating inhibitory electrophysiological responses in rat hippocampus are different from receptors mediating cyclic AMP accumulation, *Arch. Pharmacol.*, 326, 294, 1984.

37. **Schubert, P.**, Modulation of synaptically evoked neuronal calcium fluxes by adenosine, in *Neurotransmitters and Cortical Function: from Molecules to Mind*, Avoli, M., Render, T. A., Dykes, R. W.,, and Gloor, P., Eds., Plenum Press, New York, 1988, 471.

38. **Dolphin, A. C. and Prestwich, S. A.**, Pertussis toxin reverses adenosine inhibition of neuronal glutamate release, *Nature*, 316, 148, 1985.

39. **Stratton, K. R., Cole, A. J., Pritchett, J., Eccles, C. U., Worley, P. F., and Baraban, J. M.**, Intrahippocampal injection of pertussis toxin blocks adenosine suppression of synaptic responses, *Brain Res.*, 494, 359, 1989.

40. **Sanders, R. C. and Murray, T. F.**, Temporal relationship in bicuculline seizure susceptibility in rats, *Neurosci. Lett.*, 101, 325, 1989.

41. **Szot, P., Sanders, R. C., and Murray, T. F.**, Theophylline-induced upregulation of A_1-adenosine receptors associated with reduced sensitivity to convulsants, *Neuropharmacology*, 26, 1173, 1987.

42. **Fastbom, J. and Fredholm, B. B.**, Effects of long-term theophylline treatment on adenosine A_1-receptors in rat brain: autoradiographic evidence for increased receptor number and altered coupling to G-proteins, *Brain Res.*, 507, 195, 1990.

43. **Green, R. M. and Stiles, G. L.**, Chronic caffeine ingestion sensitizes the A1 adenosine receptor-adenylate cyclase system in rat cerebral cortex, *J. Clin. Invest.*, 77, 222, 1986.

44. **Ramkumar, V., Bumgarner, J. R., Jacobson, K. A., and Stiles, G. L.**, Multiple components of the A_1 adenosine receptor-adenylate cyclase system are regulated in rat cerebral cortex by chronic caffeine ingestion, *J. Clin. Invest.*, 82, 242, 1988.

45. **Daval, J-L., Deckert, J., Weiss, S. R. B., Post, R. M., and Marangos, P. J.**, Upregulation of adenosine A1 receptors and forskolin binding sites following chronic treatment with caffeine or carbamazepine: a quantitative autoradiographic study, *Epilepsia*, 30(1), 26, 1989.

46. **Piredda, S. and Gale, K.**, A crucial epileptogenic site in the deep prepiriform cortex, *Nature*, 317, 623, 1985.

47. **Goodman, R. R., Kuhar, J. J., Hester, L., and Snyder, S. H.**, Adenosine receptors: autoradiographic evidence for their location on axon terminals of excitatory neurons, *Science*, 220, 967, 1983.

48. **Corradetti, R., Lo Conte, G., Moroni, F., Passani, M. B., and Pepeu, G.**, Adenosine decreases aspartate and glutamate release from rat hippocampal slices, *Eur. J. Pharmacol.*, 104, 19, 1984.

49. **Terrian, D. M., Hernandez, P. G., Rea, M. A., and Peters, R. I.**, ATP release, adenosine formation, and modulation of dynorphin and glutamic acid release by adenosine analogues in rat hippocampal mossy fiber synaptosomes, *J. Neurochem.*, 53, 1390, 1989.

50. **Bartrup, J. T. and Stone, T. W.**, Presynaptic actions of adenosine are magnesium-dependent, *Neuropharmacology*, 27, 761, 1988.

51. **Fredholm, B. B. and Lindgren, E.**, Effects of *N*-ethylmaleimide and forskolin on noradrenaline release from rat hippocampal slices. Evidence that prejunctional adenosine and α-receptors are linked to *N*-proteins but not to adenylate cyclase, *Acta Physiol. Scand.*, 130, 95, 1987.

52. **Duner-Engström, M. and Fredholm, B. B.**, Evidence that prejunctional adenosine receptors regulating acetylcholine release from rat hippocampal slices are linked to an N-ethylmaleimide-sensitive G-protein, but not to adenylate cyclase or dihydropyridine-sensitive Ca^{2+}-channels, *Acta Physiol. Scand.*, 134, 119, 1988.

53. **Hoehn, K. and White, T. D.**, Role of excitatory amino acid receptors in K^+- and glutamate-evoked release of endogenous adenosine from rat cortical slices, *J. Neurochem.*, 54, 256, 1990.

54. **Hagberg, H., Andersson, P., Butcher, S., Sandberg, M., Lehmann, A., and Hamberger, A.**, Blockade of N-methyl-D-aspartate-sensitive acidic amino acid receptors inhibits ischemia-induced accumulation of purine catabolites in the rat striatum, *Neurosci. Lett.*, 68, 311, 1986.

55. **Connick, J. H. and Stone, T. W.**, Quinolinic acid neurotoxicity: protection by intracerebral phenyliso-propyladenosine (PIA) and potentiation by hypotension, *Neurosci. Lett.*, 101, 191, 1989.

56. **Arvin, B., Neville, L. F., Pan, J., and Roberts, P. J.**, 2-Chloroadenosine attenuates kainic acid-induced toxicity within the rat striatum: relationship to release of glutamate and Ca^{2+} influx, *Br. J. Pharmacol.*, 98, 225, 1989.

57. **Simoes, A. P., Oliveira, P. C., Sebastião, A. M., and Ribeiro, J. A.**, N^6-Cyclohexyladenosine inhibits veratridine-stimulated ^{22}Na uptake by rat brain synaptosomes, *J. Neurochem.*, 50, 899, 1989.

58. **Proctor, W. R. and Dunwiddie, T. V.**, Pre- and postsynaptic actions of adenosine in the in vitro rat hippocampus, *Brain Res.*, 426, 187, 1987.

59. **Schubert, P. and Lee, K. S.**, Non-synaptic modulation of repetitive firing by adenosine is antagonized by 4-aminopyridine in a rat hippocampal slice, *Neurosci. Lett.*, 67, 334, 1986.

60. **Trussell, L. O. and Jackson, M. B.**, Adenosine-activated potassium conductance in cultured striatal neurons, *Proc. Natl. Acad. Sci. U.S.A.*, 82, 4857, 1985.

61. **El-Fakahany, E. E., Alger, B. E., Lai, W. S., Pitler, T. A., Worley, P. F., and Baraban, J. M.,** Neuronal muscarinic responses: role of protein kinase C, *FASEB J.,* 2, 2575, 1988.

62. **Stratton, K. R., Cole, A. J., Worley, P. F., and Baraban, J. M.,** Persistent block of adenosine action in the dentate gyrus following NMDA receptor activation, *Soc. Neurosci. Abstr.,* 14, 792, 1988.

63. **Worley, P. F., Heller, W. A., Snyder, S. H., and Baraban, J. M.,** Lithium blocks a phosphoinositide-mediated cholinergic response in hippocampal slices, *Science,* 239, 1428, 1988.

64. **Dunwiddie, T. V., Hoffer, B. J., and Fredholm, B. B.,** Alkylxanthines elevate hippocampal excitability: evidence for a role of endogenous adenosine, *Arch. Pharmacol.,* 316, 316, 1981.

65. **Alzheimer, C., Sutor, B., and ten Bruggencate, G.,** Transient and selective blockade of adenosine A$_1$ receptors by 8-cyclopentyl-1,3-dipropylxanthine (DPCPX) causes sustained epileptiform activity in hippocampal CA3 neurons of guinea pigs, *Neurosci. Lett.,* 99, 107, 1989.

66. **Ault, B., Olney, M. A., Joyner, J. L., Boyer, C. E., Notrica, M. A., Soroko, F. E., and Wang, C. M.,** Pro-convulsant actions of theophylline and caffeine in the hippocampus: implications for the management of temporal lobe epilepsy, *Brain Res.,* 426, 93, 1987.

67. **Frank, C., Sagratella, S., Benedetti, M., and Scotti de Carolis, A.,** Comparative influence of calcium blocker and purinergic drugs on epileptiform bursting in rat hippocampal slices, *Brain Res.,* 441, 393, 1988.

68. **Dragunow, M.,** Purinergic mechanisms in epilepsy, *Prog. Neurobiol.,* 31, 85, 1988.

69. **Dragunow, M.,** Adenosine: the brain's natural anticonvulsant?, *TIPS,* 18, 129, 1986.

70. **Dragunow, M.,** Endogenous anticonvulsant substances, *Neurosci. Biobehav. Rev.,* 10, 229, 1986.

71. **Dragunow, M. and Robertson, H. A.,** 8-Cyclopentyl 1,3-dimethylxanthine prolongs epileptic seizures in rats, *Brain Res.,* 417, 377, 1987.

72. **Burdette, L. J. and Dyer, R. S.,** Differential effects of caffeine, picrotoxin, and pentyleneterazol on hippocampal afterdischarge activity and wet dog shakes, *Exp. Neurol.,* 96, 381, 1987.

73. **Dragunow, M., Goddard, G. V., and Laverty, R.,** Proconvulsant effects of theophylline on hippocampal afterdischarges, *Exp. Neurol.,* 96, 732, 1987.

74. **Albertson, T. E., Joy, R. M., and Stark, L. G.,** Caffeine modification of kindled amygdaloid seizures, *Pharmacol. Biochem. Behav.,* 19, 339, 1983.

75. **Albertson, T. E. and Foulke, G. E.,** Interactions of aminophylline and three benzodiazepine compounds with amygdala-kindled seizures in rats, *Exp. Neurol.,* 97, 725, 1987.

76. **Albertson, T. E. and Joy, R. M.,** Modification of excitation and inhibition evoked in dentate gyrus by perforant path stimulation: effects of aminophylline and kindling, *Pharmacol. Biochem. Behav.,* 24, 85, 1986.

77. **Popoli, P., Sagratella, S., and Scotti de Carolis, A.,** An EEG and behavioural study on the excitatory properties of caffeine in rabbits, *Arch. Intl. Pharmacodyn. Ther.,* 290, 5, 1987.

78. **Trommer, B. L., Pasternak, J. F., and Suyeoka, G. M.,** Proconvulsant effects of aminophylline during amygdala kindling in developing rats, *Dev. Brain Res.,* 46, 169, 1989.

79. **Peters, S. G., Wochos, D. N., and Peterson, G. C.,** Status epilepticus as a complication of concurrent electroconvulsive and theophylline therapy, *Mayo Clin. Proc.,* 59, 568, 1984.

80. **Shapira, B., Lerer, B., Gilboa, D., Drexler, H., Kugelmass, S., and Calev, A.,** Facilitation of ECT by caffeine pretreatment, *Am. J. Psychiatry,* 144, 1199, 1987.

81. **Albertson, T. E.,** Effect of aminophylline on amygdaloid-kindled postictal inhibition, *Pharmacol. Biochem. Behav.,* 24, 1599, 1986.

82. **Treiman, D. M. and Handforth, A.,** Role of adenosine in limiting status epilepticus entry and severity in the electrogenic limbic status model, *Soc. Neurosci. Abstr.,* 15, 777, 1989.

83. **Eldridge, F. L., Paydarfar, D., Scott, S. C., and Dowell, R. T.,** Role of endogenous adenosine in recurrent generalized seizures, *Exp. Neurol.,* 103, 179, 1989.

84. **Dragunow, M. and Faull, R.,** Neuroprotective effects of adenosine, *TIPS,* 9, 193, 1987.

85. **Schultz, V. and Lowenstein, J. M.,** The purine nucleotide cycle: studies of ammonia production and interconversions of adenosine and hypoxanthine nucleotides and nucleosides by rat brain *in situ, J. Biol. Chem.,* 253, 1938, 1978.

86. **Schrader, J., Wahl, M., Kuschinsky, W., and Kreutzberg, G. W.,** Increase of adenosine content in cerebral cortex of the cat during bicuculline-induced seizures, *Pfluegers Arch. Gesamte Physiol. Menschen Tiere,* 387, 245, 1980.

87. **Winn, H. R., Welsh, J. E., Bryner, C., Rubio, R., and Berne, R. M.,** Brain adenosine production during the initial 60 seconds of bicuculline seizures in rats, *Acta Neurol. Scand.,* 60, 536, 1979.

88. **Chapman, A. G.,** Free fatty acid release and metabolism of adenosine and cyclic nucleotides during prolonged seizures, in *Neurotransmitters, Seizures and Epilepsy,* Morselli, P. L., Ed., Raven Press, New York, 1981, 165.

89. **Subrahmanyam, K., Murphy, B., Prasad, J., Shrivaastaw, K. P., and Sadasivudu, B.,** Adenosine deaminase in convulsions along with its regional, cellular and synaptosomal distribution in rat brain, *Neurosci. Lett.,* 48, 327, 1984.

90. **Lewin, E. and Bleck, V.,** Electroshock seizures in mice: effect on brain adenosine and its metabolites, *Epilepsia,* 50, 577, 1981.

91. **Sattin, A.,** Adenosine as a mediator of central nervous system efforts during psychiatric treatment, in *Neuropharmacology of CNS and Behavioural Disorders,* Palmer, G. C., Ed., Academic Press, New York, 1981, 645.

92. **Newman, M., Zohar, J., Kalian, M., and Belmaker, R. H.,** The effects of chronic lithium and ECT on A1 and A2 adenosine receptor systems in rat brain, *Brain Res.,* 291, 188, 1984.

93. **Wybenga, M. P., Murphy, M. G.,, and Robertson, H. A.,** Rapid changes in cerebellar adenosine receptors following experimental seizures, *Eur. J. Pharmacol.,* 75, 79, 1981.

94. **Daval, J-L. and Sarfati, A.,** Effects of bicuculline-induced seizures on benzodiazepine and adenosine receptors in developing rat brain, *Life Sci.,* 41, 1685, 1987.

95. **Gleiter, C. H., Deckert, J., Nutt, D. J., and Marangos, P. J.,** Electroconvulsive shock (ECS) and the adenosine neuromodulatory system: effect of single and repeated ECS on the adenosine A1 and A2 receptors, adenylate cyclase, and the adenosine uptake site, *J. Neurochem.,* 52, 641, 1989.

96. **Ohisalo, J. J., Murros, K., Fredholm, B. B., and Hare, T. A.,** Concentrations of GABA and adenosine in the CSF in progressive myoclonus epilepsy without Lafora's bodies, *Arch. Neurol.,* 40, 623, 1983.

97. **Kostopoulos, G., Drapeau, C., Avoli, M., Olivier, A., and Villemeure, J. G.,** Endogenous adenosine can reduce epileptiform activity in the human epileptogenic cortex maintained *in vitro, Neurosci. Lett.,* 106, 119, 1989.

98. **Mee, E. W., Dragunow, M., Jansen, K., and Faull, R. L. M.,** Excitatory amino acids and adenosine receptors and proto-oncogene c-fos in epileptic, human, temporal gyrus and hippocampus, *Int. J. Neurosci.,* 46(102), 28, 1989.

99. **Mee, E., Dragunow, M., Jansen, K., Faull, R., Frith, R., and Synek, B.,** Neurotransmitter receptors in spiking and non-spiking regions of human epileptic temporal cortex, *Int. J. Neurosci.,* in press.

100. **Dodd, P. R., Watson, W. E. J., and Johnston, G. A. R.,** Adenosine receptors in post-mortem human cerebral cortex and the effect of carbamazepine, *Clin. Exp. Pharmacol. Physiol.,* 13, 711, 1986.

101. **Linden, J.,** Adenosine deaminase for removing adenosine: how much is enough?, *TIPS,* 260, 1989.

102. **Skerritt, J. H., Davies, L. P., and Johnston, G. A. R.,** A purinergic component in the anticonvulsant action of carbamazepine?, *Eur. J. Pharmacol.,* 82, 195, 1982.

103. **Marangos, P. J., Post, R. M., Patel, J., Zander, K., Parma, A., and Weiss, S.,** Specific and potent interactions of carbamazepine with brain adenosine receptors, *Eur. J. Pharmacol.,* 93, 175, 1983.

104. **Marangos, P. J., Weiss, S. R. B., Montgomery, P., Patel, J., Narang, P. K., Cappabianca, A. M., and Post, R. M.,** Chronic carbamazepine treatment increases brain adenosine receptors, *Epilepsia,* 26, 493, 1985.

105. **Weir, R. L., Padgett, W., Daly, J. W., and Anderson, S. M.,** Interaction of anticonvulsant drugs with adenosine receptors in the central nervous system, *Epilepsia,* 25, 492, 1984.

106. **Wu, P. H., Phillis, J. W., and Bender, A. S.,** Do benzodiazepines bind at adenosine uptake sites in CNS?, *Life Sci.,* 28, 1023, 1981.

107. **Phillis, J. W. and Delong, R. E.,** A purinergic component in the central actions of meprobamate, *Eur. J. Pharmacol.,* 101, 295, 1984.

108. **De Sarro, G. B. and De Sarro, A.,** Antagonists of adenosine and alpha-2-adrenoceptors reverse the anticonvulsant effects of tizanidine in DBA/2 mice, *Neuropharmacology,* 28, 211, 1989.

109. **Phillis, J. W. and Wu, P. H.,** The role of adenosine and its nucleotides in central synaptic transmission, *Prog. Neurobiol.,* 16, 187, 1981.

110. **Hoehn, K. and White, T. D.,** Glutamate-evoked release of endogenous adenosine from rat cortical synaptosomes is mediated by glutamate uptake and not by receptors, *J. Neurochem.,* 54, 1716, 1990.

111. **Dragunow, M., Jansen, K., and Holmes, K.,** Unpublished observations.

Chapter 30

ADENOSINE AND SLEEP

Miodrag Radulovacki

TABLE OF CONTENTS

I. INTRODUCTION

Our interest in the possible hypnotic role of adenosine was stimulated by findings that behavioral stimulant effects of methylxanthines involve a blockade of central adenosine receptors.[1] In addition, experiments with iontophoretic application of adenosine showed that adenosine had depressant effects on the responses of neurons in several brain regions,[2] and general neurophysiologic effects of adenosine were shown to be inhibitory.[3,4] Thus, it was conceivable that stimulation of adenosine receptors may produce sedation or sleep. Moreover, preliminary data in dogs[5] indicated a possible hypnotic role for adenosine, whereas administration of adenosine into the brains of rats, cats, and fowls produced behavioral sleep.[6-8] Administration of relatively low doses of adenosine analogs to mice and rats produced marked sedation and hypothermia.[1,9] In accordance, administration of adenosine triphosphate to rabbits or mice caused sedation.[10,11]

These behavioral inhibitory actions of adenosine may be related to its inhibition of the release of excitatory amino acid glutamate[12] as well as acetylcholine,[13] dopamine,[14] norepinephrine,[15] and serotonin[16] into the synaptic cleft. Recent findings of Haas and Greene[17] showed that endogenous adenosine reduces neuronal excitability. There is no direct evidence whether or not endogenous adenosine is also a behavioral depressant, but our work with deoxycoformycin,[18] a potent inhibitor of adenosine deaminase, and caffeine,[19] an adenosine receptor antagonist,[1] leads us to believe that it may be.

II. ROLE OF ADENOSINE AND ADENOSINE ANALOGS IN SLEEP

A. A COMPARISON OF THE DOSE-RESPONSE EFFECTS OF PYRIMIDINE RIBONUCLEOSIDES AND ADENOSINE ON SLEEP IN RATS

We examined and compared the dose-response effects of intracerebroventricular (i.c.v.) infusion of the pyrimidine ribonucleosides cytidine and uridine and the purine ribonucleoside adenosine on sleep and waking in rats, since iontophoretic administration of cytidine and uridine was shown to be devoid of the depressant effects on brain neurons,[20] while iontophoretic application of adenosine depressed neuronal responses in several brain areas.[2] We administered all three drugs at doses of 1, 10, and 100 nmol and electroencephalographically monitored rats for 6 h following administration of the drugs. The results showed that i.c.v infusion of 1 and 10 nmol cytidine significantly suppressed sleep and that administration of uridine did not affect sleep. In contrast, adenosine exhibited significant hypnotic effects at all doses examined (see Table 1). All three doses of adenosine significantly reduced waking and increased total sleep (TS). Both the 1- and 100-nmol doses of adenosine also significantly reduced the latencies to the onset of rapid eye movement (REM) sleep.[21]

Our results showed that pyrimidine ribonucleosides did not produce hypnotic effects similar to those of adenosine. In addition, adenosine was shown to significantly increase TS, primarily through an enhancement of deep slow-wave sleep (SWS2), and to decrease waking at a dose as low as 1 nmol. The hypnotic effects of adenosine may be related to its depressant effects on neuronal activity, since the pyrimidine ribonucleosides lack both depressant and hypnotic effects.

B. EFFECTS ON SLEEP OF MICROINJECTIONS OF ADENOSINE TO PREOPTIC AREA IN RATS

For almost half of the century researchers have sought to find discrete groups of neurons which controlled and regulated the occurrence of mammalian sleep or its component stages. Although several brain regions which modulate sleep are now recognized, there is little consensus among current workers concerning the specific contribution of any neuronal group

TABLE 1
Dose-Response Effects of Intracerebroventricular Injection of Adenosine on Sleep and Wakefulness in Rats

Sleep state	Hours	Saline	Adenosine (nmol)		
			1	10	100
W	0—3	61.5 ± 10.6	28.7 ± 5.2[+]	35.7 ± 5.9[**]	25.2 ± 3.4[+]
	3—6	51.2 ± 17.0	21.8 ± 7.3	29.8 ± 8.2	9.5 ± 1.6[***]
	0—6	112.7 ± 26.6	50.5 ± 10.5[**]	65.5 ± 11.6[*]	34.7 ± 3.0[+]
SWS1	0—3	33.5 ± 1.1	33.3 ± 6.0	38.8 ± 4.9	34.8 ± 2.0
	3—6	42.7 ± 3.7	41.7 ± 5.6	44.7 ± 3.9	34.7 ± 5.6
	0—6	76.2 ± 3.7	75.0 ± 10.6	83.5 ± 7.6	69.5 ± 6.5
SWS2	0—3	75.2 ± 9.2	105.0 ± 7.0[**]	87.8 ± 9.8	105.0 ± 3.1
	3—6	71.3 ± 13.7	95.0 ± 7.5	89.5 ± 9.8	111.2 ± 4.6[***]
	0—6	146.2 ± 22.1	200.0 ± 13.6	177.3 ± 18.1	216.2 ± 7.0[***]
REM	0—3	9.0 ± 3.2	13.0 ± 2.2	17.7 ± 1.0	13.3 ± 3.3
	3—6	15.5 ± 5.0	21.5 ± 3.0	16.0 ± 2.0	18.7 ± 2.6
	0—6	24.5 ± 7.6	34.5 ± 4.3	33.7 ± 1.5	32.0 ± 5.2
TS	0—3	118.5 ± 10.6	151.3 ± 5.2[+]	144.3 ± 5.9[**]	154.8 ± 3.4[+]
	3—6	128.8 ± 17.0	158.2 ± 7.3	150.2 ± 8.2	170.5 ± 1.6[***]
	0—6	247.3 ± 26.6	309.5 ± 10.5[**]	294.5 ± 116.6[*]	325.3 ± 3.0[+]

Note: All values reported are means ± SEM in minimum of six rats per group. Significantly different from saline control group: [*]p <0.050; [**]p <0.010; [***]p <0.010; and [+]p <0.005.

(From Radulovacki, M., Virus, R. M., Rapoza, D., and Crane, R., *Psychopharmacology, 87,* 136, 1985. With permission.)

to the normal sleep cycle. One of the brain areas that was investigated as a potential sleep "center" was the preoptic area. This was because Sterman and Clemente[22] showed that both low- and high-frequency electrical stimulation of the preoptic area produced sleep. We have examined the effects on sleep of bilateral microinjections of adenosine in the preoptic area of the rat (see Table 2). The results showed that administration of 12.5 nmol adenosine increased deep SWS2, REM sleep, and TS during the first 3 h of polygraphic recording. The dose of 25 nmol adenosine did not affect sleep.[23,24] This finding is in accordance with our data with adenosine analogs where highest doses of L-phenylisopropyladenosine (L-PIA) and N^6-cyclohexyladenosine (CHA) were devoid of hypnotic effects, possibly due to their action on body temperature.

C. EFFECTS ON SLEEP OF ADENOSINE DEAMINASE INHIBITOR DEOXYCOFORMYCIN

We administered to rats deoxycoformycin, a potent inhibitor of adenosine deaminase,[25] which would be expected to elevate the levels of adenosine in the central nervous system (CNS), to determine if an increase in endogenous adenosine would promote sleep. Deoxycoformycin (0.5 or 2.0 mg/kg) was administered intraperitoneally (i.p.) and animals were polygraphically recorded for 6 h. The 0.5-mg/kg dose of the drug was shown to increase REM sleep and reduce REM sleep latency, while the dose of 2 mg/kg increased SWS2.[18] Those results were consistent with the results that we had previously reported for the adenosine analog L-PIA[19] and indicated a hypnotic role for endogenous adenosine.

D. EFFECTS ON SLEEP OF L-PIA, CHA, AND NECA

Following our studies with L-PIA, an adenosine A_1 receptor agonist, and deoxycoformycin, we examined the effects on sleep in rats of another adenosine A_1 receptor stimulant, CHA, and adenosine A_1 and A_2 receptor stimulant, 5'-N-ethylcarboxamidoadenosine (NECA),

TABLE 2
Effects on Sleep of Bilateral Microinjection of
Adenosine to Preoptic Area in the Rat

	Saline (n = 13)	Adenosine	
		12.5 nmol (n = 10)	25 nmol (n = 5)
0—3 h			
SWS1	40 ± 7	34 ± 7	34 ± 4
SWS2	63 ± 10	82 ± 7*	78 ± 4
REM	9 ± 2	15 ± 3*+	8 ± 2
TS	112 ± 7	131 ± 6*	120 ± 5
3—6 h			
SWS1	37 ± 7	31 ± 4	31 ± 5
SWS2	47 ± 7	59 ± 9	49 ± 7
REM	9 ± 2	13 ± 3	11 ± 4
TS	94 ± 6	104 ± 10	92 ± 9
0—6 h			
SWS1	77 ± 14	65 ± 12	65 ± 7
SWS2	111 ± 15	141 ± 15*	128 ± 7
REM	19 ± 3	28 ± 2	19 ± 4
TS	207 ± 11	235 ± 14	212 ± 6
1 h			
SWS2	10 ± 3	17 ± 3*	15 ± 2
TS	27 ± 4	28 ± 2	25 ± 4
Sleep latencies			
SWS1 onset	19 ± 3	14 ± 2	23 ± 6
SWS2 onset	25 ± 3	14 ± 7	25 ± 5
REM onset	71 ± 27	53 ± 13	39 ± 9

Note: * = Statistically ($p < 0.05$) different from saline control;
+ = statistically ($p < 0.05$) different from other drug concentrations. All analysis done by one-way ANOVA.

(From Ticho, S. and Radulovacki, M., unpublished observations.)

in a dose-related manner.[26] This was of interest since adenosine is rapidly metabolized by adenosine deaminase to inosine,[27] and the effects on sleep of metabolically stable adenosine analogs L-PIA, CHA, and NECA were expected to be of equal or of longer duration than those of adenosine. The effects of L-PIA, CHA, and NECA on sleep consisted of increased deep SWS2 from 6.6 to 45.7% in all doses used for CHA and NECA and for 0.1 and 0.3 μmol/kg of L-PIA, respectively. All three agents reduced REM sleep at 0.9 μmol/kg dose, whereas L-PIA at this dose increased waking as well. The results showed that the effect on sleep was obtained with nanomolar doses of the drugs and that it diminished or disappeared when the drug dose increased (0.9 μmol/kg). The only exception was administration of 0.9 μmol/kg NECA, which increased SWS2 and TS for the first 3 h of polygraphic recording. This suggests that stimulation of adenosine A_2 receptors by NECA is also relevant to behavioral inhibition and sleep.

III. ROLE OF ADENOSINE RECEPTORS IN SLEEP

A. CIRCADIAN VARIATION OF [³H]L-PIA BINDING IN RAT BRAINS
In view of the possible involvement of adenosine in the modulation of sleep and demonstrated circadian variations in the receptors for other putative neurotransmitter substances,[28] we examined possible circadian variations in adenosine receptors by examining [³H]L-PIA

binding to whole rat brain membranes. Our data indicated that the number (B_{max}) — but not the dissociation constant (K_d) — of adenosine receptors in rat brain exhibited a statistically significant circadian rhythm, with a major peak at 23.00 h and a minor peak at 11.00 h, 3 h after the onset of dark and light phases, respectively, of the diurnal cycle.[29] This biphasic circadian variation in the number of adenosine receptors, with the highest number during the rats' more active period and the next highest number when rats are less active, complicates the correlation of the B_{max} of adenosine receptors with the reported hypnogenic effects of adenosine and its congeners in rats.[26] This apparent paradox may, however, be explained in part by measurements of the circadian variation of adenosine concentrations in the rat brain.[30] During the light period (when rats spend most time asleep), rat brain adenosine concentrations were high, as were B_{max} values, suggesting significant stimulation of adenosine receptors, while brain adenosine concentrations were low during the dark phase of the diurnal cycle (during which rats exhibit most activity), suggesting very low levels of adenosine receptor stimulation despite the presence of the greatest number of receptors. Therefore, these results were not inconsistent with the hypothesis that adenosine acts as a positive modulator of sleep in rats.

B. [³H]L-PIA BINDING IN BRAINS FROM YOUNG AND OLD RATS

Aged rats are known to exhibit less TS and altered patterns of sleep as compared to young rats.[31,32] Since adenosine appeared to mediate sleep in rats, we investigated possible age-dependent changes in the binding of [³H]L-PIA to membrane preparations of whole brains from normal male Sprague-Dawley rats 12 and 84 weeks of age.[33] Two populations of binding sites, probably corresponding to adenosine A_1 and A_2 receptors, are reduced in old rats. Since treatment with both A_1 and A_2 agonists enhanced SWS2 or TS in young rats,[26] the reduced affinities of both A_1 and A_2 receptors observed in old rats may contribute to the reduced amounts and altered patterns of sleep reported in old rats by Rechtschaffen's group.[31,32]

C. [³H]L-PIA BINDING IN BRAIN REGIONS FROM REM SLEEP-DEPRIVED RATS

If adenosine or adenosine receptors have a role in sleep, then what would be the effect of REM sleep deprivation on both these parameters? We decided to investigate to what extent CNS adenosine concentration and adenosine A_1 receptors interact in the modulation of sleep and arousal using the "flowerpot" method of REM sleep deprivation.[34] That technique had been shown to selectively deprive rats of REM sleep and result in REM rebound following REM sleep deprivation.[35] If adenosine indeed has a role in sleep, then during the REM sleep rebound, when the amount of REM sleep almost doubles the normal values, either adenosine concentration or adenosine receptors, or both, could be expected to be elevated. We assayed endogenous adenosine in microwave-fixed brain tissue of rats deprived of REM sleep for 48 h and found no significant changes in adenosine concentration when compared to controls. However, those data did not reflect the role that intrasynaptic adenosine concentration might play in the modulation of adenosine receptors in the CNS. We determined adenosine receptor binding in the brains of rats deprived of REM sleep for 48 and 96 h using [³H]L-PIA. The data showed that adenosine A_1 receptors (B_{max}) were significantly increased in the cortex and corpus striatum and that the increase was sleep specific. We concluded that the effect of endogenous adenosine on sleep after REM sleep deprivation may not have been the result of its accumulation in specific brain areas, but could rather be a consequence of changes that occurred at the level of the adenosine A_1 receptor.

IV. RELATIONSHIP BETWEEN METHYLXANTHINES, ADENOSINE RECEPTORS, AND SLEEP

A. EFFECTS OF L-PIA AND CAFFEINE ON SLEEP IN RATS

If behavioral stimulant effects of methylxanthines involve a blockade of central adenosine receptors,[1] then the effects on sleep of adenosine or an adenosine analog would be abolished by caffeine. We tested this possibility and determined the effect of L-PIA on sleep in the presence of caffeine. The rats were implanted with electrodes for EEG recording and the effects of L-PIA (0.115 mg/kg), caffeine (15 mg/kg), and L-PIA + caffeine were recorded for 6 h. The results showed that administration of L-PIA increased SWS2 by 54 min, suggesting that stimulation of adenosine receptors promoted deep sleep. However, administration of L-PIA failed to produce the same effect in the presence of caffeine, a finding consistent with the hypothesis that the CNS stimulant effect of caffeine is due to its ability to antagonize depressant effects of endogenous adenosine.[19]

B. THE DOSE-RESPONSE EFFECTS OF CAFFEINE ON SLEEP IN RATS

In humans, caffeine produces insomnia, a reduction in TS time and an increase in wakefulness, and causes a suppression of REM sleep rebound in REM sleep-deprived rats.[36] Additionally, caffeine produced a biphasic effect on locomotor behavior in mice, with low doses resulting in profound depression.[37]

We investigated the possibility of whether or not low doses of caffeine in addition to depressing locomotor activity would also affect sleep,[38] as has been shown with low doses of apomorphine[39] or bromocriptine.[40] The rats were implanted with electrodes for polygraphic recording, and the effects of caffeine (0.125, 1.25, 12.5, and 25 mg/kg) were monitored by the EEG for 6 h. The results showed that the 12.5- and 25-mg/kg doses of caffeine increased wakefulness and decreased SWS1, SWS2, REM sleep, and TS time. The 0.125- and 1.25-mg/kg doses of caffeine increased light sleep (SWS1) at the expense of deep sleep (SWS2), and did not affect TS time. This finding was of interest since adenosine or adenosine agonists had been shown to increase SWS2 at the expense of waking or SWS1 without an increase in TS time. Thus, the obtained effects of low doses of caffeine on sleep suggested that caffeine administration antagonizes the effects of adenosine not only at the receptor level, but also at the behavioral level.

C. A COMPARISON OF THE EFFECTS OF CAFFEINE, CPT, AND ALLOXAZINE ON SLEEP IN RATS

Although the methylxanthine caffeine is a potent CNS stimulant, the molecular and cellular basis for this CNS stimulation is only incompletely understood. Considerable experimental evidence suggests that the most important of the proposed mechanisms of methylxanthine action for the production of CNS stimulation is adenosine receptor blockade.[1,41] Caffeine blocks both adenosine A_1 and A_2 receptors,[41] but *in vitro* radioligand binding studies have shown that caffeine has a slightly greater affinity at A_1 adenosine receptors than at A_2 receptors as indicated by the A_2K_i/A_1K_i ratio of 1.65.[42] Therefore, it was of interest for us to determine the role of A_1 and A_2 adenosine receptors in the stimulant effects of caffeine and to document it by the EEG.[43] We implanted rats for EEG recording and polygraphically monitored the effects of wakefulness of caffeine (12.5 mg/kg i.p.), 8-cyclopentyltheophylline (CPT), a selective A_1 receptor antagonist[44] (10, 20, and 40 mg/kg i.p.), alloxazine, an A_2 receptor antagonist[45] (12.5, 25, and 50 mg/kg i.p.), and CPT (20 mg/kg) + alloxazine (50 mg/kg) for 6 h. The results showed that both CPT and alloxazine injected individually produced sleep suppression qualitatively similar to that produced by caffeine, but of a lower magnitude. However, when 20 mg/kg CPT and 50 mg/kg alloxazine were injected together, their sleep-suppressant effect was of the same magnitude as that of 12.5 mg/kg caffeine.

The results supported the hypothesized involvement of adenosine receptor blockade in the effects of caffeine on sleep in rats. They further suggested that A_1 adenosine receptor blockade may be more important than A_2 receptor blockade, since behavioral effects of the selective *in vitro* antagonist CPT were generally similar to those of nonselective *in vitro* adenosine receptor antagonists caffeine and alloxazine.

V. RELATIONSHIP BETWEEN BENZODIAZEPINES AND CENTRAL ADENOSINE RECEPTORS

A. EFFECTS OF CHRONIC ADMINISTRATION OF DIAZEPAM ON ADENOSINE RECEPTORS IN THE RAT BRAIN

Benzodiazepines inhibit adenosine uptake[46] and thus potentiate actions of adenosine, i.e., depression of spontaneous firing of neurons,[47] elevation of cyclic adenosine 5'-monophosphate,[48] and inhibition of acetylcholine release.[49] Also, several effects of benzodiazepines are antagonized by caffeine and theophylline, adenosine receptor antagonists.[50] Since some of the actions of benzodiazepines may involve adenosine uptake inhibition, we tested the hypothesis that chronic administration of diazepam may alter adenosine receptor binding. Following chronic administration for 10 to 20 d of diazepam (5 mg/kg/d, subcutaneous pellets), adenosine receptors in different rat brain areas were assessed by radioligand binding studies using [^{3}H]R-PIA for A_1 receptors and the [^{3}H]NECA + [^{3}H]R-PIA assay for A_2 receptors. The results showed that chronic administration of diazepam for 10 d decreased A_1 receptors in the striatum by 46% and A_1 receptors in the hippocampus by 13% ($p < 0.05$).[51] The results were also in accordance with the postulate of Phillis and Wu,[2] who proposed that the sedative effect of benzodiazepines could be related to their inhibition of adenosine uptake. Thus, the data gave new evidence for a role of adenosine in the central actions of benzodiazepines by showing that diazepam modifies adenosine receptor binding.

B. EFFECTS OF PROLONGED ADMINISTRATION OF TRIAZOLAM ON CENTRAL ADENOSINE RECEPTORS IN RATS

Triazolam is a triazolobenzodiazepine derivative with a short plasma half-life which differs in several aspects from diazepam, a benzodiazepine with a long plasma half-life. O'Regan and Phillis[52] showed that triazolam, like diazepam,[47] potentiated adenosine-evoked depression of cerebral cortical neuronal firing in the rat and that those effects were blocked by the adenosine antagonist caffeine. Since benzodiazepines had been shown to inhibit the uptake of adenosine in cerebral cortical slices,[46] synaptosomes,[53] and primary cultures of neurons and astrocytes,[54] Phillis et al.[49] hypothesized that the effects of benzodiazepines on the adenosine system are through inhibition of the uptake of adenosine and/or an enhancement of adenosine release. Our previous work showed that chronic administration of diazepam decreased central adenosine receptors,[51] and we decided to examine whether or not central adenosine receptors would be affected by prolonged administration of triazolam as well. The result showed that following continuous subcutaneous administration of triazolam (0.5, 1, and 2 mg/d, pellets) for 10 d, radioligand binding to adenosine A_2 receptors in the rat striatum either decreased (31%, 2 mg/d) or increased (15%, 0.5 mg/d).[55] The data indicated that although triazolam has different pharmacokinetic properties than diazepam, its administration affected central adenosine receptors in a similar manner as diazepam.[51]

C. THE EFFECT OF ADENOSINE TRANSPORT INHIBITOR SOLUFLAZINE ON ADENOSINE RECEPTORS IN RAT BRAIN

If the downregulation of central adenosine receptors by prolonged administration of diazepam[51] and triazolam[55] was accomplished by the inhibition of adenosine uptake, then prolonged administration of adenosine transport inhibitors to rats should produce similar effects. We tested this postulate by administering to rats for 14 d a potent adenosine transport inhibitor, soluflazine[56] (1 μmol, 0.5 μl/h via ALZET® miniosmotic pumps), and examined its effects on central adenosine receptors in specific brain areas.[57] Soluflazine decreased adenosine A_1 radioligand binding in the hippocampus as measured by [^{3}H]R-PIA and lowered A_2 binding sites in the striatum, as estimated by the ''NECA minus R-PIA'' assay. The data showed that a specific adenosine transport inhibitor produced the same effect on adenosine receptors as benzodiazepines and further suggested a role for adenosine in CNS effects of benzodiazepines, as originally proposed by Phillis and Wu.[2]

VI. HYPOTHESIS FOR ADENOSINE HYPNOTIC ACTION

What do we propose as the mechanism for hypnotic action of adenosine? Since there are two types of adenosine receptors, i.e., A_1 and A_2, whose role in sleep is not yet clear, we start from evidence that stimulation of adenosine A_1 receptors by adenosine or adenosine-related compounds leads to a suppression of calcium influx into presynaptic nerve terminals,[58] possibly as a consequence of an inhibition of adenylate cyclase,[54,59] and this inhibits the release of brain neurotransmitters.[12-16] The end result of this process is the reduced amount of neurotransmitter at synapses in brain regions critical for sleep generation, which may lead to the induction of sleep. This should not be interpreted to mean that sleep is a passive process; sleep may be actively brought up by stimulation of adenosine A_1 receptors which would initiate the chain of events as described above.

VII. CONCLUSIONS

Adenosine and adenosine analogs were shown to produce hypnotic effects in several animal species. Although their effects on sleep architecture differ from the hypnotic effect of benzodiazepines, our studies showed that benzodiazepines interact with central adenosine receptors. We also reported that deprivation of REM sleep upregulates adenosine receptors in the brains of rats in a manner similar to that of long-term administration of caffeine.[60] This suggests the existence of an ''endocaffeine'', whose normal role would be to block adenosine receptors during prolonged sleep deprivation — a mechanism that could be responsible for the increased number of adenosine receptors.[61]

REFERENCES

1. **Synder, S. H., Katims, J. J., Annau, Z., Bruns, R. F., and Daly, J. W.,** Adenosine receptors and behavioral actions of methylxanthines, *Proc. Natl. Acad. Sci. U.S.A.,* 78, 3260, 1981.
2. **Phillis, J. W. and Wu, P. H.,** The role of adenosine and its nucleotides in central synaptic transmission, *Prog. Neurobiol.,* 16, 187, 1981.
3. **Phillis, J. W., Edstrom, J. P., Kostopoulos, G. K., and Kirkpatrick, J. R.,** Effects of adenosine and adenosine nucleotides on synaptic transmission in the cerebral cortex, *Can. J. Physiol. Pharmacol.,* 57, 1289, 1979.
4. **Stone, T. W.,** Physiological roles for adenosine and adenosine 5'-triphosphate in the nervous system, *Neuroscience,* 6, 523, 1981.
5. **Haulica, I., Ababei, L., Branisteanu, D., and Topoliceanu, F.,** Preliminary data on the possible hypnogenic role of adenosine, *J. Neurochem.,* 21, 1019, 1973.
6. **Buday, P. V., Carr, C. J., and Miya, T. S.,** A pharmacologic study of some nucleosides and nucleotides, *J. Pharm. Pharmacol.,* 13, 290, 1961.
7. **Feldberg, W. and Sherwood, S. L.,** Injections of drugs into the lateral ventricle of the cat, *J. Physiol. (London),* 123, 148, 1954.

8. **Marley. E. and Nistico, G.,** Effects of catecholamines and adenosine derivatives given into the brain of fowls, *Br. J. Pharmacol.,* 46, 619, 1972.

9. **Dunwiddie, T. V. and Worth, T.,** Sedative and anti-convulsant effects of adenosine analogs in mouse and rat, *J. Pharmacol. Exp. Ther.,* 220, 70, 1982.

10. **Bhattacharya, I. C., Goldstein, L., and Pfeiffer, C. C.,** Influence of acute and chronic nicotine administration on EEG reactivity to drugs in rabbits. I. Nucleosides and nucleotides, *Res. Commun. Chem. Pathol. Pharmacol.,* 1, 99, 1970.

11. **Mathieu-Levy, N.,** Contribution a l'etude du mechanism de la potentialisation du sommeil experimental par l'acide adenosine triphosphorique (ATP). Sur quelques actions d'ATP au niveau du systeme nerveux central, *Therapie,* 23, 1157, 1968.

12. **Dolphin, A. C., Prestwich, S. A., and Forda, S. R.,** Presynaptic modulation by adenosine analogues: relationship to adenylate cyclase, in *Adenosine Receptors and Modulation of Cell Function,* Stefanovich, V., Rudolphi, E., and Schubert, P., Eds., IRL Press, Oxford, 1985, 107.

13. **Sawynok, J. and Jhamandas, K. H.,** Inhibition of acetylcholine release from cholinergic nerves by adenosine, adenosine nucleotides, and morphine: antagonism by theophylline, *J. Pharmacol. Exp. Ther.,* 197, 379, 1976.

14. **Michaelis, M. L., Michaelis, E. K., and Myers, S. L.,** Adenosine modulation of synaptosomal dopamine release, *Life Sci.,* 24, 2083, 1979.

15. **Hedquist, P. and Fredholm, B. B.,** Effects of adenosine on adrenergic neurotransmission: prejunctional inhibition and postjunctional enhancement, *Naunyn-Schmiedeberg's Arch. Pharmacol.,* 293, 217, 1976.

16. **Harms, H. H., Warden, G., and Mulder, A. H.,** Effects of adenosine on depolarization-induced release of various neurotransmitters from slices of rat corpus striatum, *Neuropharmacology,* 18, 577, 1979.

17. **Haas, H. H. and Greene, R. W.,** Endogenous adenosine inhibits hippocampal CAI neurons: further evidence from extra- and intra-cellular recordings, *Naunyn-Schmiedeberg's Arch. Pharmacol.,* 337, 561, 1988.

18. **Radulovacki, M., Virus, R. M., Djuricic-Nedelson, M., and Green, R. D.,** Hypnotic effects of deoxycoformycin in rats, *Brain Res.,* 271, 392, 1983.

19. **Radulovacki, M., Miletich, R. S., and Green, R. D.,** N^6(L-Phenylisopropyl) adenosine (L-PIA) increases slow wave sleep (S_2) and decreases wakefulness in rats, *Brain Res.,* 246, 178, 1982.

20. **Phillis, J. W., Kostopoulos, G. K., and Limacher, J. J.,** Depression of corticospinal cells by various purines and pyrimidines, *Can. J. Physiol. Pharmacol.,* 52, 1226, 1974.

21. **Radulovacki, M., Virus, R. M., Rapoza, D., and Crane, R.,** A comparison of the dose response effects of pyrimidine ribonucleosides and adenosine on sleep in rats, *Psychopharmacology,* 87, 136, 1985.

22. **Sterman, M. B. and Clemente, C. D.,** Forebrain inhibitory mechanisms: critical synchronization induced by basal forebrain stimulation, *Exp. Neurol.,* 6, 91, 1962.

23. **Ticho, S. R., Virus, R. M., and Radulovacki, M.,** Intracerebral administration of adenosine to the medial preoptic area enhances sleep in rats, *Soc. Neurosci. Abstr.,* 14, 5241, 1988.

24. **Ticho, S. R. and Radulovacki, M.,** Effects on sleep of adenosine, adenosine analogs, triazolam, pentobarbital and GABA microinjected into the preoptic area of rats, *Soc. Neurosci. Abstr.,* 15, 242, 1989.

25. **Agarwal, R. P., Spector, T., and Parks, R. E.,** Tight-binding inhibitors. IV. Inhibition of adenosine deaminase by various inhibitors, *Biochem. Pharmacol.,* 26, 359, 1977.

26. **Radulovacki, M., Virus, R. M., Djuricic-Nedelson, M., and Green, R. D.,** Adenosine analogs and sleep in rats, *J. Pharm. Exp. Ther.,* 228, 268, 1984.

27. **Skolnick, P., Nimilkitpaisan, Y., Stalvey, I., and Daley, J. W.,** Inhibition of brain adenosine deaminase by 2'-deoxycoformycin and erythro-9-(2-hydroxy-3-nonyl)adenine, *J. Neurochem.,* 30, 1579, 1978.

28. **Kafka, M. S., Wirz-Justice, A., Naber, D., Moore, R. Y., and Benedito, M. A.,** Circadian rhythms in rat brain neurotransmitter receptors, *Fed. Proc., Fed. Am. Soc. Exp. Biol.,* 42, 2796, 1983.

29. **Virus, R. M., Baglajewski, T., and Radulovacki, M.,** Circadian variation of [^{3}H]N^6-(L-phenylisopropyl) adenosine binding in rat brain, *Neurosci. Lett.,* 46, 219, 1984.

30. **Chagoya DeSanchez, V., Hernandez-Munoz, R., Diaz-Munoz, M., Suarez, J., Vidrio, S., and Yanez, L.,** Circadian variations of adenosine and its physiological meaning in the energetic homeostasis of the cell and the sleep-wake cycle of the rat, in *Proc. 4th Int. Congr. Sleep Res.,* Society for Neuroscience, Washington D.C., 255, 1983.

31. **Rosenberg, R. S., Zepelin, H., and Rechtschaffen, A.,** Sleep in young and old rats, *J. Gerontol.,* 34, 525, 1979.

32. **Zepelin, H., Whitehead, W. E., and Rechtschaffen, A.,** Aging and sleep in the albino rat, *Behav. Biol.,* 7, 65, 1972.

33. **Virus, R. M., Baglajewski, T., and Radulovacki, M.,** [^{3}H]N^6-(L-phenylisopropyl) adenosine binding in brains from young and old rats, *Neurobiol. Aging,* 5, 61, 1984.

34. **Yanik, G. and Radulovacki, M.,** REM sleep deprivation upregulates adenosine A_1 receptors, *Brain Res.,* 402, 362, 1987.

35. **Mendelson, W. B., Guthrie, R. D., Frederick, G., and Wyatt, R. J.,** The flowerpot technique of rapid eye movement (REM) sleep deprivation, *Pharmacol. Biochem. Behav.,* 2, 543, 1974.

36. **Radulovacki, M., Walovitch, R., and Yanik, G.,** Caffeine produces REM sleep rebound in rats, *Brain Res.,* 201, 497, 1980.

37. **Snyder, S. and Sklar, P.,** Psychiatric progress, behavioral and molecular actions of caffeine: focus on adenosine, *J. Psychiatry Res.,* 18, 91, 1984.

38. **Yanik, G., Glaum, S., and Radulovacki, M.,** The dose-response effects of caffeine on sleep in rats, *Brain Res.,* 403, 177, 1987.

39. **Mereu, G. P., Scarnatti, E., Paglietti, E., Chessa, P., Chicara, G., and Gessa, G. I.,** Sleep induced by low doses of apomorphine in rats, *Electroencephalogr. Clin. Neurophysiol.,* 46, 214, 1979.

40. **Loew, D. M. and Spiegel, R.,** Polygraphic sleep studies in rats and humans. Their use in psychopharmacological research, *Arzneim. Forsch.,* 26, 1032, 1976.

41. **Daly, J. W., Butts-Lamb, P., and Padgett, W.,** Subclasses of adenosine receptors in the central nervous system: interactions with caffeine and related methylxanthines, *Cell. Mol. Pharmacol.,* 3, 69, 1983.

42. **Bruns, R. F., Lu, G. H., and Pugsley, T. A.,** Characterization of the A_2 adenosine receptor labelled by [^{3}H]NECA in rat striated membranes, *Mol. Pharmacol.,* 29, 331, 1986.

43. **Virus, R. M., Ticho, S., Pilditch, M., and Radulovacki, M.,** A comparison of the effects of caffeine, 8-cyclopentyl theophylline, and alloxazine on sleep in rats; possible roles of central nervous system adenosine receptors, *Neuropsychopharmacology,* 3, 243, 1990.

44. **Bruns, R. F., Fergus, J. H., Badger, E. W., Bristol, J. A., Santay, L. A., Hartman, J. D., Hays, Sj., and Huang, C. C.,** Binding of the A_1 selective adenosine antagonist 8-cyclopentyl-1-3-di-propylxanthine to rat brain membranes, *Naunyn-Schmiedeberg's Arch. Pharmacol.,* 335, 59, 1987.

45. **Bruns, R. F., Fergus, J. H., Badger, E. W., Bristol, J. A., Santay, L. A., and Hays, J. A.,** An antagonist ligand for adenosine A_2 receptors, *Naunyn-Schmiedeberg's Arch. Pharmacol.,* 335, 64, 1987.

46. **Phillis, J. W., Bender, A. S., and Wu, P. H.,** Benzodiazepines inhibit adenosine uptake into rat brain synaptosomes, *Brain Res.,* 195, 494, 1980.

47. **Phillis, J. W.,** Diazepam potentiation of purinergic depression of central neurons, *Can. J. Physiol. Pharmacol.,* 57, 432, 1979.

48. **Schultz, J.,** Adenosine 3', 5'-monophosphate in guinea cerebral cortical slices. Effect of benzodiazepines, *J. Neurochem.,* 22, 685, 1974.

49. **Phillis, J. W., Siemens, R. K., and Wu, P. H.,** Effects of diazepam on adenosine and acetylcholine release from rat cerebral cortex: further evidence for a purinergic mechanism in action of diazepam, *Br. J. Pharmacol.,* 70, 341, 1980.

50. **Polc, P., Bonetti, E. P., Pieri, L., Cumin, R., Angioi, R. M., Mohler, H., and Haefely, W. E.,** Caffeine antagonizes several central effects of diazepam, *Life Sci.,* 28, 2265, 1981.

51. **Hawkins, M., Pravica, M., and Radulovacki, M.,** Chronic administration of diazepam downregulates adenosine receptors in the rat brain, *Pharmacol. Biochem. Behav.,* 30, 303, 1988.

52. **O'Regan, M. H. and Phillis, J. W.,** Potentiation of adenosine-evoked depression on rat cerebral cortical neurons by triazolam, *Brain Res.,* 445, 376, 1988.

53. **Bender, A. S., Wu, P. H., and Phillis, J. W.,** The characterization of [^{3}H] adenosine uptake into rat cerebral cortical synaptosomes, *J. Neurochem.,* 35, 629, 1980.

54. **Van Calker, D., Muller, M., and Hambrecht, V.,** Adenosine regulates, via two different types of receptors, the accumulation of cyclic AMP in cultured brain cells, *J. Neurochem.,* 33, 999, 1979.

55. **Hawkins, M., Hajduk, P., O'Connor, S., Radulovacki, M., and Starz, K. E.,** Effects of prolonged administration of triazolam on adenosine A_1 and A_2 receptors in the brain of rats, *Brain Res.,* 445, 376, 1988.

56. **Van Belle, H.,** Myocardial purines during ischemia, reperfusion and pharmacological protection, *Mol. Physiol.,* 8, 615, 1985.

57. **O'Connor, S. D. and Radulovacki, M.,** The effect of soluflazine on adenosine in the rat, *Neuropsychopharmacology,* 30, 93, 1991.

58. **Ten Bruggencate, D., Steinberg, R., Stockle, H., and Nicholson, C.,** Modulation of extracellular Ca^{++}- and K^+-levels in the mammalian cerebellar cortex, in *Iontophoresis and Transmitter Mechanisms in the Mammalian Central Nervous System,* Ryall, R. W. and Kelly, J. S., Eds., Elsevier/North-Holland, Amsterdam, 1977, 442.

59. **Londos, C., Cooper, M. F., and Wolff, J.,** Subclasses of external adenosine receptors, *Proc. Natl. Acad. Sci. U.S.A.,* 77, 2551, 1980.

60. **Marangos, P. J., Boulenger, J.-P., and Patel, J.,** Effects of chronic caffeine on brain adenosine receptors: regional and ontogenic studies, *Life Sci.,* 34, 899, 1985.

61. **Radulovacki, M.,** Progress in sleep, *N. Engl. J. Med.,* 316, 1275, 1987.

Chapter 31

ADENOSINE AND PAIN

Jana Sawynok

TABLE OF CONTENTS

I. INTRODUCTION

A variety of neurotransmitter and neuromodulator substances have been implicated in the signaling and suppression of pain, both at peripheral nerve endings and centrally in the spinal cord and at supraspinal sites. There has been considerable recent interest in the role of adenosine in pain based on the following approaches: (1) systemic and intrathecal (i.t.) administration of adenosine analogs produces antinociception in a variety of tests commonly used to detect antinociceptive activity of centrally active drugs; (2) peripheral administration of adenosine can elicit pain in human subjects; (3) release of adenosine within the spinal cord may mediate a significant component of antinociception produced by spinally and supraspinally administered morphine; (4) adenosine potentiates spinal antinociception by α-adrenergic agonists; (5) release of adenosine 5′-triphosphate (ATP) from nonnociceptive sensory fibers may play an indirect role in inhibiting pain transmission in the spinal cord. The role of purines in nociception is the subject of a recent comprehensive review.[1] In this chapter, this role will be considered with an emphasis on recent developments in this field.

II. ADENOSINE ANALOGS AND ANTINOCICEPTION

A. ACTIVITY IN ANTINOCICEPTIVE TESTS

The systemic, i.t., and intracisternal or intracerebroventricular (i.c.v.) administration of analogs of adenosine produces antinociception in established tests for centrally active agents such as the tail-flick, hot-plate, and writhing tests, as well as more recently developed tests such as the substance P and *N*-methyl-D-aspartate (NMDA) scratching-response tests (Table 1). Antinociception is blocked by methylxanthines, indicating it is due to activation of cell-surface adenosine receptors. Following systemic adminstration, antinociception may be expressed predominantly at spinal sites of action.[6] A variety of analogs are active in nociceptive tests (Table 1), including those with adenosine A_1 receptor selectivity (e.g., R-PIA, CHA), as well as NECA which has been used as a prototype A_2 ligand in these studies but actually has comparable activity at both A_1 and A_2 receptors.[17] Activation of both A_1 and A_2 receptors has been implicated in spinal antinociception.[6,10,11,14] Recently, it was demonstrated that R-PIA and NECA penetrate the spinal cord to different degrees following i.t. application,[15] and this pharmacokinetic difference may account for some of the difference in activity observed in behavioral tests.

In addition to antinociception, most studies note that i.t. and i.c.v. administration of adenosine analogs produces impairment of motor activity in doses which are similar to those that produce antinociception. There is concern that antinociception in tests which rely on coordinated expression of motor activity (particularly the hot-plate and writhing tests) may be secondary to these motor effects (for example, see Herrick-Davis et al.).[7] However, these effects can be dissociated on the basis of dose and time course of action. Thus, with L-PIA and CHA, antinociception occurs in lower doses than does motor impairment.[11,14,18] In addition, motor impairment has a slower onset and longer duration of action than does antinociception, and both effects can be observed independently at different times following injection.[11,18] For NECA, motor and antinociceptive effects are more similar with respect to dose and time course of action.[11,18] Adenosine A_1 and A_2 receptors are distributed in both the dorsal and ventral spinal cord.[19,20] NECA penetrates more extensively into spinal cord tissue following i.t. injection,[15] perhaps accounting for its more prominent motor effects. Adenosine A_1-selective agonists may be more promising for the development of agents with therapeutic potential because of the separation of antinociceptive effects from motor effects.[18]

TABLE 1
Antinociceptive Activity of Analogs of Adenosine

Agent	Dose	Species; test	Ref.
Systemic administration			
R-PIA	0.2 mg/kg i.p.	Mouse; hot plate	2
CHA, CADO	0.5—5 mg/kg i.p.	Mouse; hot plate	3
R-PIA	0.3 mg/kg s.c.	Rat; tail flick	4
R-PIA	0.5—3 μmol/kg s.c.	Mouse; tail flick, acetic acid writhing	5
NECA, CHA, R-PIA	0.1—3 μmol/kg s.c.	Rat; tail flick	6
NECA, CHA, R-PIA, MECA, CPA, CADO	ED_{50} 0.5—8.2 mg/kg p.o.	Mouse; acetic acid writhing	7
Intrathecal administration			
NECA	0.5—5 nmol	Mouse; tail flick, hot plate	8
CADO	0.5—4 nmol	Mouse; tail flick, hot plate	9
NECA, CHA, R-PIA	0.5—10 nmol	Rat; tail flick, hot plate	10
NECA, CHA, R-PIA	3—100 nmol	Rat; tail flick	6
NECA, R-PIA, CADO	0.07—4.0 nmol	Mouse; tail flick, hot plate	11
AMP, ADP, ATP	ED_{50} 0.5—2.1 μg	Mouse; SP scratching	12
NECA, CPA, R-PIA, CHA, CADO	ED_{50} 3.56—13.2 pmol ED_{50} 5.59—36.2 pmol	Mouse; SP scratching NMDA, scratching	13
NECA, CHA, R-PIA	0.1—1 nmol	Rat; tail flick, hot plate, acetic acid, writhing	14
R-PIA, NECA	1 nmol	Rat; tail flick	15
Supraspinal administration			
ACC, NECA	0.01—0.1 i.cis.	Mouse; hot plate	16
CADO, ADO	1—100 i.cis.		
NECA, CHA, R-PIA, MECA, CPA, CADO	ED_{50} 0.8—37 μg/kg i.c.v.	Mouse; acetic acid writhing	7

Note: Abbreviations—Agents: ACC, 5'-N-cyclopropylcarboxamido adenosine; ADO, adenosine; ADP, adenosine 5'-diphosphate; AMP, adenosine 5'-monophosphate; ATP, adenosine 5'-triphosphate; CADO, 2-chloroadenosine; CHA, N^6-cyclohexyladenosine; CPA, N^6-cyclopentyladenosine; R-PIA, R-phenylisopropyladenosine; MECA, 5'-N-methylcarboxamidoadenosine; NECA, 5'-N-ethylcarboxamidoadenosine. Routes: i.cis., intracisternal; i.c.v., intracerebroventricular; i.p., intraperitoneal; p.o., oral; s.c., subcutaneous.

B. SITES AND MECHANISMS OF ACTION
1. Localization of Receptors

Adenosine A_1 and A_2 receptors have been localized in the spinal cord.[19,20] Binding studies indicate that levels of A_1 receptors in the dorsal spinal cord are greater than in the ventral spinal cord, but for A_2 receptors this difference is less prominent.[19,20] Autoradiography reveals the highest density of both A_1 and A_2 adenosine receptors in the substantia gelatinosa of the dorsal horn of the spinal cord.[21] Adenosine receptors appear to be located on neurons with their cell bodies in the spinal cord, but not on primary afferent nerve terminals or on terminals of descending pathways.[19,21]

2. Postsynaptic Actions

The i.t. administration of adenosine analogs inhibits the characteristic biting, licking, and scratching syndrome induced by i.t. injection of substance P and NMDA.[12,13] These responses have been interpreted as a direct postsynaptic activation of sensory pathways in the spinal cord, and this action may contribute to the spinal antinociception observed in other tests. However, as this effect occurs in doses much lower than does antinociception in the tail-flick and hot-plate tests (Table 1), this may represent only a partial mechanism of action. Inhibition of postsynaptic activation of sensory pathways by adenosine analogs is consistent with the predominant postsynaptic localization of adenosine receptors in the spinal cord (see above). Adenosine hyperpolarizes dorsal horn neurons in the spinal cord.[22] In other brain regions, hyperpolarization produced by adenosine is due to an increase in K^+ conductance, which is mediated by a G protein (for example, see Trussell and Jackson).[23] Adenosine can also reduce Ca^{2+} conductance in spinal cord neurons,[24] which potentially could cause inhibition of neurotransmitter release from interneurons.

3. Presynaptic Actions

Although adenosine receptors have not been identified on primary afferent nerve terminals in adults (see above), adenosine analogs inhibit Ca^{2+} currents in cultured dorsal root ganglion cells.[25,26] Although potentially this could cause inhibition of transmitter release from the nerve terminals of such neurons, adenosine analogs do not inhibit release of substance P from spinal cord slices.[27] Intrathecal injection of NECA to mice inhibits the algesic behavioral response elicited by i.t. injection of capsaicin, and it was suggested that this reflects a presynaptic action of NECA within the spinal cord.[28] However, this study did not find inhibition of the algesic action of substance P as noted in other studies.[12,13]

4. Second-Messenger Mediators

Adenosine A_1 and A_2 receptors have been reported linked to inhibition and stimulation, respectively, of adenylate cyclase activity in the spinal cord.[20] Inhibition of adenylate cyclase may be involved in spinal antinociception by CHA (but not NECA), because antinociception is reduced by pretreatment with the nonxanthine phosphodiesterase inhibitors RO 20 1724 and rolipram.[29] G proteins are implicated in spinal antinociception by both CHA and NECA, because i.t. pretreatment with pertussis toxin, which ADP-ribosylates and inactivates G_i and G_o, inhibits antinociception by both agents.[29]

III. NOCICEPTIVE ACTIVITY OF ADENOSINE

In contrast to the action of adenosine in the spinal cord where adenosine inhibits the transmission of nociceptive information, at sensory nerve endings adenosine stimulates nociception. Thus, adenosine and adenine nucleotides produce pain when applied to the human blister base preparation.[30] Administered intravenously, adenosine produces chest pain which resembles angina pectoris both in healthy volunteers[31] and in patients with coronary insufficiency.[32] This effect is blocked by aminophylline.[31] Intravenous administration of adenosine also produces epigastric pain[33] and pain in the head, jaw, neck, back, arms, chest, and abdomen.[34] Injection of adenosine into the brachial artery produces ischemic-like pain the forearm,[35] while theophylline decreases pain in the ischemic forearm test.[36] It has been proposed that adenosine, which is produced from the breakdown of ATP under ischemic conditions, activates sensory nerve endings as a means of protecting tissue in a precarious metabolic state.[35]

IV. INVOLVEMENT OF ADENOSINE IN SPINAL ANTINOCICEPTION BY OTHER AGENTS

A. MORPHINE

A significant component of the spinal antinociceptive action of morphine is mediated by release of adenosine. In behavioral studies, spinal antinociception by morphine is reduced by i.t. administration of methylxanthine adenosine receptor antagonists.[9,37,38] In biochemical studies, morphine releases adenosine from synaptosomes prepared from the dorsal half of the spinal cord[38] and the superfused spinal cord *in vivo*.[39] This adenosine appears to originate from capsaicin-sensitive, small-diameter primary afferent neurons as release in both paradigms is reduced by pretreatment with capsaicin.[39] Other investigators have demonstrated that morphine releases adenosine from the cortical surface *in vivo*[40] and enhances stimulus-evoked release of adenosine from cortical slices.[41,42]

When administered systemically, analogs of adenosine have been reported both to inhibit[43,44] and to potentiate the antinociceptive action of systemically administered morphine.[5] Antagonism of the action of morphine could be due to a peripheral action of the adenosine analogs on sensory nerve terminals whereby they stimulate activity (Section III), while the potentiation may represent a central antinociceptive action within the spinal cord which interacts with morphine (Section II). When adenosine analogs are administered i.c.v., there is no intrinsic activity in the tail-flick test, but an inhibition of the action of systemically administered morphine.[44] Adenosine inhibits neuronal activity in a variety of brain regions,[45,46] and it is conceivable that inhibitory actions of adenosine at supraspinal sites could counteract opioid stimulation of descending mechanisms involved in antinociception.[47,48]

B. 5-HYDROXYTRYPTAMINE

A component of the spinal antinociceptive action of 5-hydroxytryptamine (5-HT) also appears to be due to the spinal release of adenosine. Thus, theophylline blocks the spinal antinociceptive action of 5-HT,[49] and 5-HT releases adenosine from dorsal spinal cord synaptosomes[50] and from the spinal cord *in vivo*.[51] The adenosine released from synaptosomes originates as a nucleotide as the amount of adenosine detected is substantially reduced by 5'-nucleotidase inhibitors,[50] while *in vivo* the adenosine may originate as cyclic adenosine 5'-monophosphate (cAMP), as 5-HT also increases release of cAMP from the superfused spinal cord.[51] In both release paradigms, release is capsaicin sensitive, indicating it originates from small-diameter primary afferent neurons.[50,51]

C. NORADRENALINE

In contrast to morphine and 5-HT, release of adenosine does not directly mediate spinal antinociception by noradrenaline (NA), as spinal antinociception is not blocked by methylxanthines.[38,49] Although NA can release a nucleotide from spinal cord synaptosomes,[38] it is not clear where this originates from,[39] and release of adenosine is not observed from the superfused spinal cord *in vivo*.[51] Interestingly, adenosine analogs potentiate spinal antinociception by NA,[49,52] an effect which appears to be due to an interaction between adenosine A_2 receptors and α_2-adrenergic receptors.[49,53]

D. CALCIUM

At many sites within the central and peripheral nervous systems, calcium (Ca^{2+}) antagonizes opioid actions, including antinociception.[54] However, following i.t. administration, Ca^{2+} potentiates spinal antinociception by morphine[55,56] and NA,[56] and in higher dose actually induces antinociception.[55,56] Both the increase in the action of NA by Ca^{2+} and the intrinsic antinociception appear to be mediated by release of adenosine in the spinal cord as they are eliminated by i.t. pretreatment with 8-phenyltheophylline.[56] This adenosine originates from small-diameter primary afferent neurons as antinociception produced by Ca^{2+} is eliminated by i.t. pretreatment with capsaicin.[56] However, it is not known whether adenosine is released directly or indirectly as a nucleotide, or whether release is a direct effect of Ca^{2+} on afferent terminals or an indirect effect via interneurons. The antinociceptive action of Ca^{2+} may reflect an interaction between endogenously released adenosine and NA, as Ca^{2+} antinociception also is reduced by i.t. 6-hydroxydopamine (which depletes spinal cord NA levels) and phentolamine (an α-adrenergic antagonist).[56]

E. VIBRATION-INDUCED ANALGESIA

ATP may be released from low-threshold primary afferent neurons in the spinal cord and activate spinal neurons responding to nonnoxious stimulation.[57,58] Inhibitory effects of ATP and AMP also are observed on wide-dynamic range neurons, and these effects are blocked by theophylline, suggesting they result from conversion of the nucleotide to adenosine with a subsequent activation of adenosine receptors.[58] The inhibition of firing of spinal cord neurons evoked by noxious stimulation that is induced by mechanical vibration is blocked by caffeine,[59] and it has been proposed that this procedure activates large-diameter sensory afferents to release ATP, which is metabolized to adenosine, and this mediates the pain suppression produced by this physical manipulation.[59]

V. ROLE OF ADENOSINE IN SUPRASPINAL MECHANISMS OF ANTINOCICEPTION

In addition to mediating a component of antinociception following spinal administration of drugs, release of adenosine from the spinal cord also may play a significant role in antinociception following supraspinal administration of morphine. This was first suggested by DeLander and Hopkins,[9] who demonstrated that the antinociceptive effect of i.c.v. administration of morphine to mice was reduced by the i.t. administration of theophylline, caffeine, and 3-isobutyl-1-methylxanthine. This has been confirmed in rats using theophylline and the more potent and more selective adenosine receptor antagonist 8-phenyltheophylline.[60] However, a higher dose of theophylline also increases the antinociceptive action of i.c.v. morphine,[60] indicating the need for caution in interpreting behavioral data using agents with mixed pharmacological properties. Release of adenosine from the superfused spinal cord *in vivo* also has been demonstrated following the i.c.v. administration of antinociceptive doses of morphine.[60] When morphine is administered directly into the periaqueductal grey region, i.t. administration of 8-phenyltheophylline reverses antinociception in the hot-plate test.[61] This reduction is no longer seen following i.t. pretreatment with 5,7-dihydroxytryptamine (but not with 6-hydroxydopamine), which depletes spinal cord 5-HT levels,[61] suggesting the adenosine released from the spinal cord is dependent upon activation of descending serotonergic pathways. Release of adenosine from the spinal cord following i.c.v. administration of morphine is reduced by both 5,7-dihydroxytryptamine and methysergide, confirming this inference.[62] Other observations also provide evidence for this scheme. Thus, i.t. theophylline attenuates antinociception produced by stimulation of the ventromedial medulla,[63] an area in which descending serotonergic mechanisms are implicated in antinociception.[64]

VI. SUMMARY AND CONCLUSIONS

Adenosine appears to play a significant role both in the signaling of pain in the periphery by stimulating sensory nerve endings, and in pain suppression within the dorsal horn of the spinal cord (Figure 1). Following systemic administration, adenosine analogs produce antinociception in a number of tests, probably reflecting intrinsic activity within the spinal cord. Systemic administration of adenosine analogs also has been reported to inhibit the antinociceptive action of morphine, perhaps reflecting in this case a peripheral action of the adenosine analogs on sensory afferent nerve terminals. In the spinal cord, adenosine is released from capsaicin-sensitive, small-diameter primary afferent neurons by morphine (whereby it originates as adenosine per se), 5-HT (whereby it originates as nucleotide, probably cAMP), and Ca^{2+} (biochemical origin of adenosine not determined). Adenosine also is released from the spinal cord following supraspinal administration of morphine into the lateral cerebral ventricles and the periaqueductal grey, and this release is dependent on activation of descending serotonergic pathways. With respect to the potential for development

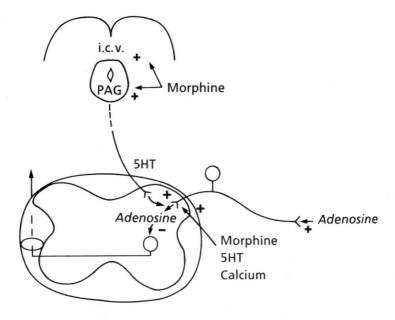

FIGURE 1.   Schematic diagram indicating the involvement of adenosine in nociception and antinociception. In the periphery, adenosine stimulates sensory nerve endings to promote the sensation of pain. In the dorsal horn of the spinal cord, adenosine suppresses pain transmission, primarily by activating receptors postsynaptic to primary afferent neurons. Adenosine can be released from capsaicin-sensitive nerve terminals by a local action of morphine (directly), 5-HT (indirectly), and Ca^{2+} (origin of adenosine not determined). Adenosine also can be released from the spinal cord by morphine administered supraspinally into the lateral cerebral ventricles or into the periaqueductal grey region. This release is dependent upon activation of descending serotonergic pathways. See text for references.

of adenosine analogs as analgesics, because of the duality of the action of adenosine following systemic administration reflecting both peripheral and central components of action, this potential may be limited for systemic routes of administration. However, following spinal administration (epidural, i.t.), there appears to be a greater potential for the development of A_1 agonists as antinociceptive agents than for A_2 agonists because of the clearer separation between motor effects and antinociception for these agents. There also may be some potential for spinal combinations of low doses of adenosine A_2-selective ligands with α_2-adrenergic agonists, as these agents produce synergistic interactions on nociceptive thresholds following spinal administration.

ACKNOWLEDGMENTS

I would particularly like to thank my colleagues Dr. M. I. Sweeney, Dr. T. D. White, and Ms. A. Reid for their valuable contributions to work described in this review. Work cited herein was supported by the Medical Research Council of Canada.

REFERENCES

1. **Sawynok, J. and Sweeney, M. I.,** The role of purines in nociception, *Neuroscience,* 32, 557, 1989.
2. **Vapaatalo, H., Onken, D., Neuvonen, P. J., and Westermann, E.,** Stereoselectivity in some central and circulatory effects of phenylisopropyl-adenosine (PIA), *Arzneim. Forsch.,* 25, 407, 1975.
3. **Crawley, J. N., Patel, J., and Marangos, P. J.,** Behavioral characterization of two long-lasting adenosine analogs: sedative properties and interaction with diazepam, *Life Sci.,* 29, 2623, 1981.
4. **Holmgren, M., Hedner, T., Nordberg, G., and Mellstrand, T.,** Antinociceptive effects in the rat of an adenosine analogue, N^6-phenylisopropyladenosine, *J. Pharm. Pharmacol.,* 35, 679, 1983.
5. **Ahlijanian, M. K. and Takemori, A. E.,** Effects of (-)-N^6-(R-phenylisopropyl)-adenosine (PIA) and caffeine on nociception and morphine-induced analgesia, tolerance and dependence in mice, *Eur. J. Pharmacol.,* 112, 171, 1985.
6. **Holmgren, M., Hedner, J., Mellstrand, T., Nordberg, G., and Hedner, T.,** Characterization of the antinociceptive effects of some adenosine analogues in the rat, *Naunyn-Schmiedeberg's Arch. Pharmacol.,* 334, 290, 1986.
7. **Herrick-Davis, K., Chippari, S., Luttinger, D., and Ward, S. J.,** Evaluation of adenosine agonists as potential analgesics, *Eur. J. Pharmacol.,* 162, 365, 1989.
8. **Post, C.,** Antinociceptive effects in mice after intrathecal injection of 5'-N-ethylcarboxamide adenosine, *Neurosci. Lett.,* 51, 325, 1984.
9. **DeLander, G. E. and Hopkins, G. J.,** Spinal adenosine modulates descending antinociceptive pathways stimulated by morphine, *J. Pharmacol. Exp. Ther.,* 239, 88, 1986.
10. **Sawynok, J., Sweeney, M. I., and White, T. D.,** Classification of adenosine receptors mediating antinociception in the rat spinal cord, *Br. J. Pharmacol.,* 88, 923, 1986.
11. **DeLander, G. E. and Hopkins, C. J.,** Involvement of A_2 adenosine receptors in spinal mechanisms of antinociception, *Eur. J. Pharmacol.,* 139, 215, 1987.
12. **Doi, T., Kuzuna, S., and Maki, Y.,** Spinal antinociceptive effects of adenosine compounds in mice, *Eur. J. Pharmacol.,* 137, 227, 1987.
13. **DeLander, G. E. and Wahl, J. J.,** Behaviour induced by putative nociceptive neurotransmitters is inhibited by adenosine or adenosine analogs coadministered intrathecally, *J. Pharmacol. Exp. Ther.,* 246, 565, 1988.
14. **Sosnowski, M., Stevens, C. W., and Yaksh, T. L.,** Assessment of the role of A_1/A_2 adenosine receptors mediating the purine antinociception, motor and autonomic function in the rat spinal cord, *J. Pharmacol. Exp. Ther.,* 250, 915, 1989.
15. **Fastbom, J., Post, C., and Fredholm, B. B.,** Antinociceptive effects and spinal distribution of two adenosine receptor agonists after intrathecal administration, *Pharmacol. Toxicol.,* 66, 69, 1990.
16. **Yarbrough, G. G. and McGuffin-Clineschmidt, J. C.,** In vivo behavioural assessment of central nervous system purinergic receptors, *Eur. J. Pharmacol.,* 76, 137, 1981.
17. **Bruns, R. F., Lu, G. H., and Pugsley, T. A.,** Characterization of the A_2 adenosine receptor labelled by [^{3}H]NECA in rat striatal membranes, *Mol. Pharmacol.,* 29, 331, 1986.
18. **Karlsten, R., Gordh, T., Hartvig, P., and Post, C.,** Effects of intrathecal injection of the adenosine receptor agonists R-phenylisopropyl-adenosine and N-ethylcarboxamide-adenosine on nociception and motor function in the rat, *Anesth. Analg.,* 71, 60, 1990.
19. **Geiger, J. D., LaBella, F. S., and Nagy, J. I.,** Characterization and localization of adenosine receptors in rat spinal cord, *J. Neurosci.,* 4, 2302, 1984.
20. **Choca, J. I., Proudfit, H. K., and Green, R. D.,** Identification of A_1 and A_2 receptors in the rat spinal cord, *J. Pharmacol. Exp. Ther.,* 242, 905, 1987.
21. **Choca, J. I., Green, R. D., and Proudfit, H. K.,** Adenosine A_1 and A_2 receptors of the substantia gelatinosa are located predominantly on intrinsic neurons: an autoradiographic study, *J. Pharmacol. Exp. Ther.,* 247, 757, 1988.
22. **Kangra, I., Randic, M., and Jeftinija, S.,** Adenosine and (-)-baclofen have a neuromodulatory role in the rat spinal dorsal horn, *Soc. Neurosci. Abstr.,* 13, 1134, 1987.
23. **Trussell, L. O. and Jackson, M. B.,** Dependent of an adenosine-activated potassium current in a GTP-binding protein in mammalian central neurons, *J. Neurosci.,* 7, 3306, 1987.
24. **Sah, D. W. Y.,** Neurotransmitter modulation of calcium current in rat spinal cord neurons, *J. Neurosci.,* 10, 136, 1990.
25. **Dolphin, A. C., Forda, S. R., and Scott, R. H.,** Calcium-dependent currents in cultured rat dorsal root ganglion neurons are inhibited by an adenosine analogue, *J. Physiol. (London),* 373, 47, 1986.
26. **MacDonald, R. L., Skerritt, J. H., and Werz, M. A.,** Adenosine agonists reduce voltage-dependent calcium conductance of mouse sensory neurons in cell culture, *J. Physiol. (London),* 370, 75, 1986.
27. **Vasko, M. R. and Ono, H.,** Adenosine analogs do not inhibit the potassium-stimulated release of substance P from rat spinal cord slices, *Naunyn-Schmiedeberg's Arch. Pharmacol.,* 342, 441, 1990.
28. **Hunskaar, S., Post, C., Gasmer, O. B., and Arwestrom, E.,** Intrathecal injection of capsaicin can be used as a behavioural nociceptive test in mice, *Neuropharmacology,* 25, 1149, 1986.

29. **Sawynok, J. and Reid, A.,** Role of G-proteins and adeylate cyclase in antinociception produced by intrathecal purines, *Eur. J. Pharmacol.,* 156, 25, 1988.

30. **Bleehen, T. and Keele, C. A.,** Observations on the algogenic actions of adenosine compounds on the human blister base preparation, *Pain,* 3, 367, 1977.

31. **Sylvén, C., Beermann, B., Jonzon, B., and Brandt, R.,** Angina pectoris-like pain provoked by intravenous adenosine in healthy volunteers, *Br. Med. J.,* 293, 227, 1986.

32. **Sylvén, C., Beermann, B., Edlund, A., Lewander, R., Jonzon, B., and Mogensen, L.,** Provocation of chest pain in patients with coronary insufficiency using the vasodilator adenosine, *Eur. Heart J.,* 9, 6, 1988.

33. **Watt, A. H., Lewis, D. J. M., Horne, J. J., and Smith, P. M.,** Reproduction of epigastric pain of duodenal ulceration by adenosine, *Br. Med. J.,* 294, 10, 1987.

34. **Conradsson, T. B., Dixon, C. M. S., Clarke, B., and Barnes, P. J.,** Cardiovascular effects of infused adenosine in man: potentiation by dipyridamole, *Acta Physiol. Scand.,* 129, 387, 1987.

35. **Sylvén, C., Jonzon, B., Fredholm, B. B., and Kaijser, L.,** Adenosine injection into the brachial artery produces ischaemia-like pain or discomfort in the forearm, *Cardiovasc. Res.,* 9, 674, 1988.

36. **Jonzon, B., Sylvén, C., and Kaijser, L.,** Theophylline decreases pain in the ischaemic forearm test, *Cardiovasc. Res.,* 9, 807, 1989.

37. **Jurna, I.,** Cyclic nucleotides and aminophylline produce different effects on nociceptive motor and sensory responses in the rat spinal cord, *Naunyn-Schmiedeberg's Arch. Pharmacol.,* 327, 23, 1984.

38. **Sweeney, M. I., White, T. D., and Sawynok, J.,** Involvement of adenosine in the spinal antinociceptive effects of morphine and noradrenaline, *J. Pharmacol. Exp. Ther.,* 243, 657, 1987.

39. **Sweeney, M. I., White, T. D., and Sawynok, J.,** Morphine, capsaicin and K^+ release purines from capsaicin-sensitive primary afferent nerve terminals in the spinal cord, *J. Pharmacol. Exp. Ther.,* 248, 497, 1989.

40. **Phillis, J. W., Jiang, Z. G., Cherlack, B. J., and Wu, P. H.,** The effect of morphine on purine and acetylcholine release from rat cerebral cortex: evidence for a purinergic component in morphine's action, *Pharmacol. Biochem. Behav.,* 13, 421, 1980.

41. **Fredholm, B. B. and Vernet, L.,** Morphine increases depolarization induced purine release from rat cortical slices, *Acta Physiol. Scand.,* 104, 502, 1978.

42. **Stone, T. W.,** The effects of morphine and methionine-enkephalin on the release of purines from cerebral cortex slices of rats and mice, *Br. J. Pharmacol.,* 74, 171, 1980.

43. **Gourley, D. R. H. and Beckner, S. K.,** Antagonism of morphine analgesia by adenosine, adenosine and adenine nucleotides, *Proc. Soc. Exp. Biol. Med.,* 144, 774, 1973.

44. **Mantegazza, P., Tammiso, R., Zambotti, F., Zecca, L., and Zonta, N.,** Purine involvement in morphine antinociception, *Br. J. Pharmacol.,* 83, 883, 1984.

45. **Phillis, J. W. and Wu, P. H.,** The role of adenosine and its nucleotides in central synaptic transmission, *Prog. Neurobiol.,* 16, 187, 1981.

46. **Dunwiddie, T. V.,** The physiological role of adenosine in the central nervous system, *Int. Rev. Neurobiol.,* 27, 63, 1985.

47. **Basbaum, A. I. and Fields, H. L.,** Endogenous pain control mechanisms: review and hypothesis, *Ann. Neurol.,* 4, 451, 1978.

48. **Yaksh, T. L., Hammond, D. L., and Tyce, G. M.,** Functional aspects of bulbospinal monoaminergic projections in modulating processing of somatosensory information, *Fed. Proc. Fed. Am. Soc. Exp. Biol.,* 40, 2786, 1981.

49. **DeLander, G. E. and Hopkins, C. J.,** Interdependence of spinal adenosinergic, serotonergic and noradrenergic systems mediating antinociception, *Neuropharmacology,* 26, 1791, 1987.

50. **Sweeney, M. I., White, T. D., and Sawynok, J.,** 5-Hydroxytryptamine releases adenosine from primary afferent nerve terminals in the spinal cord, *Brain Res.,* 462, 346, 1988.

51. **Sweeney, M. I., White, T. D., and Sawynok, J.,** 5-Hydroxytryptamine releases adenosine and cyclic AMP from primary afferent nerve terminals in the spinal cord *in vivo, Brain Res.,* 528, 55, 1990.

52. **Aran, S. and Proudfit, H. K.,** Antinociception produced by interactions between intrathecally-administered adenosine agonists and norepinephrine, *Brain Res.,* 513, 255, 1990.

53. **Aran, S. and Proudfit, H. K.,** Antinociception induced by intrathecal injection of sub-effective doses of clonidine and 5'-N-ethylcarboxamide adenosine, *Soc. Neurosci. Abstr.,* 12, 1017, 1986.

54. **Chapman, D. B. and Way, E. L.,** Metal ion interactions with opiates, *Annu. Rev. Pharmacol. Toxicol.,* 20, 553, 1980.

55. **Lux, F., Welch, S. P., Brase, D. A., and Dewey, W. L.,** Interaction of morphine with intrathecally administered calcium and calcium antagonists: evidence for supraspinal endogenous opioid mediation of intrathecal calcium-induced antinociception in mice, *J. Pharmacol. Exp. Ther.,* 246, 500, 1988.

56. **Sawynok, J., Reid, A., and Isbrucker, R.,** Adenosine mediates calcium-induced antinociception and potentiation of noradrenergic antinociception in the spinal cord, *Brain Res.,* 524, 187, 1990.

57. **Fyffe, R. E. and Perl, E. R.,** Is ATP a central synaptic mediator for certain primary afferent fibres from mammalian skin?, *Proc. Natl. Acad. Sci. U.S.A.,* 81, 6890, 1984.

58. **Salter, M. W. and Henry, J. L.,** Effects of adenosine 5′-monophosphate and adenosine 5′-triphosphate on functionally identified units in the cat spinal dorsal horn. Evidence for a differential effect of adenosine 5′-triphosphate on nociceptive vs. non-nociceptive units, *Neuroscience,* 15, 815, 1985.

59. **Salter, M. W. and Henry, J. L.,** Evidence that adenosine mediates the depression of spinal dorsal horn neurons induced by peripheral vibration in the cat, *Neuroscience,* 22, 631, 1987.

60. **Sweeney, M. I., White, T. D., and Sawynok, J.,** Adenosine release from the spinal cord may mediate antinociception by intracerebroventricular morphine, *Soc. Neurosci. Abstr.,* 15, 371, 1989.

61. **Sawynok, J., Espey, M. J., and Reid, A.,** 8-Phenyltheophylline reverses the antinociceptive action of morphine in the periaqueductal gray, *Neuropharmacology,* in press.

62. **Sweeney, M. I., White, T. D., and Sawynok, J.,** Intracerebroventricular morphine releases cyclic AMP from the spinal cord via a serotonergic mechanism, *Eur. J. Pharmacol.,* 183, 1450, 1990.

63. **Aran, S. and Proudfit, H. K.,** Intrathecal theophylline attenuates antinociception evoked by stimulation in the ventral medial medulla, *Soc. Neurosci. Abstr.,* 15, 152, 1989.

64. **Jensen, T. S. and Yaksh, T. L.,** Examination of spinal monoamine receptors through which brainstem opiate-sensitive systems act in the rat, *Brain Res.,* 363, 114, 1986.

VII. Potential Therapeutic Avenues for Purine Use

Chapter 32

ADENOSINE AND ASTHMA

W. H. Ng and M. K. Church

TABLE OF CONTENTS

I. INTRODUCTION

A possible role for adenosine in asthma has been the subject of debate for over a decade. The first indication that adenosine could play a part in inflammatory processes was in 1978 when Marquardt and colleagues[1] reported that micromolar concentrations of the purine nucleoside could potentiate immunologically and nonimmunologically activated inflammatory mediator release from rat peritoneal mast cells *in vitro*. Furthermore, they found that this augmentation could be blocked by theophylline, a methylxanthine used in the treatment of asthma. This led them and others[2] to suggest that the beneficial actions of methylxanthines in asthma were due to adenosine receptor antagonism rather than inhibition of cyclic adenosine 5′-monophosphate (cAMP) phosphodiesterase activity as was commonly supposed.[3] In support of this hypothesis was the fact that the therapeutic range of theophylline was well below the millimolar concentrations required for phosphodiesterase inhibition, but well within the micromolar concentrations required for extracellular adenosine P_1 receptor antagonism. Of even more significance, however, was the finding shortly thereafter that aerosol challenge with adenosine provoked bronchoconstriction in atopic and nonatopic asthmatics, but not in normal, nonasthmatic subjects.[4] Subsequently, much work has been performed to elucidate the mechanism underlying the pulmonary effects of adenosine and related adenine nucleotides in asthma. In this chapter, we describe the actions of adenosine inhalation in asthmatics and review the possible mechanisms by which it causes bronchoconstriction.

II. THE GENERATION AND RELEASE OF ADENOSINE

Most animal tissues possess the enzymes to produce and metabolize adenosine, although different cell types have varying capacities to produce and release this purine nucleoside.[5] In cells which have a high capacity to produce adenosine, the major source is from the dephosphorylation of cytosolic AMP by a cytosolic 5′ nucleotidase,[6] whereas in cells with a low capacity, such as erythrocytes and lymphocytes, the importance of a membrane-bound 5′ nucleotidase with access to extracellular AMP is apparent.[7] Exchange of adenosine between extracellular and intracellular compartments is achieved by a facilitated diffusion mechanism which is sensitive to blockade by compounds such as dipyridamole and nitrobenzylthioinosine.[5] Under normoxic conditions, adenosine is normally rephosphorylated to form AMP and other high-energy phosphates. However, when energy requirements are in excess of their supply, then adenosine and related purines are released into the extracellular local environment. Evidence for this occurring under conditions which prevail in areas of inflammation and in the asthmatic lung is provided by the demonstration that hypoxia or challenge with immunological and nonimmunological stimuli cause adenosine release *in vitro* from cells and tissues such as human leukocytes, mouse mast cells, and rat and dog lungs.[8-11] Moreover, Mann et al.[12] reported that plasma adenosine concentrations were markedly increased in atopic asthmatic subjects following bronchoprovocation with allergen, rising to 50% above the baseline of 5.4 ± 0.9 ng/ml 2 min after challenge, and reaching 290% above baseline at 45 min after challenge.

III. EFFECT OF INHALED ADENOSINE ON THE AIRWAYS

The effects of bronchoprovocation with adenosine in man were first reported by Cushley et al. in 1983.[4] They found that inhalation of adenosine provoked bronchoconstriction in atopic and nonatopic asthmatics, but not in normal subjects. The time course of bronchoconstriction was similar to that following histamine and methacholine challenge, a drop in specific airways conductance or forced expiratory volume in 1 s being evident within the first minute after challenge and a maximum effect being reached at 2 to 5 min postchallenge.

The fact that bronchoprovocation with the related nucleoside guanosine did not affect lung function led them to suggest that adenosine-induced bronchoconstriction in asthmatic subjects was a specific pharmacological effect. That adenosine was the active mediator was demonstrated by the failure of its major metabolite inosine to provoke bronchoconstriction.[12,13] The precursors, adenosine monophosphate (AMP) and adenosine diphosphate, (ADP) however, exhibited similar dose-dependent brochoconstriction,[12] and it is likely that their effects are mediated by adenosine following their degradation by extracellular nucleotidases.[5]

Further evidence that adenosine-induced bronchoconstriction is a specific effect and not due to nonspecific irritation of the airways was demonstrated by pharmacological studies. If the actions of adenosine were receptor mediated, then drugs which interfere with receptor binding and adenosine metabolism should modulate the response to adenosine *in vivo*. Thus the methylxanthine theophylline, which as mentioned before is an inhibitor of extracellular adenosine P_1 receptors, has been shown to inhibit the effects of adenosine and histamine bronchoconstriction in asthmatic subjects, being roughly two to three times more effective against the former provicant than against the latter.[14] However, although theophylline appears to exhibit a degree of selectivity in blocking adenosine-induced bronchoconstriction, it is not clear whether or not its therapeutic value in asthma is due to adenosine receptor antagonism. Recent studies with another xanthine derivative, enprofylline, argue against this. Enprofylline is several times more potent than theophylline as a bronchodilator *in vivo*[15] and as a relaxant of airways smooth muscle *in vitro*,[16] although it is a very weak antagonist at cell-surface purinoceptors[17] and without effect against adenosine-induced bronchoconstriction in guinea pig isolated lung *in vitro*[18] or in man *in vivo*.[19]

Similarly, if adenosine is a mediator in asthma, then its effects *in vivo* should be enhanced by blocking its uptake into cells using dipyridamole. When 50 mg dipyridamole was administered by intravenous infusion, only a small enhancement of adenosine-induced bronchoconstriction occurred in asthmatic subjects, which was not statistically significant.[20] Administration of 2.5 mg dipyridamole by inhalation, however, significantly enhanced the effects of adenosine challenge in asthmatics, suggesting that adenosine may exert its effects by a cell-surface receptor-mediated mechanism.[21] Indeed, it has recently been reported by Eagle and Boucher[22] that infusion of dipyridamole can precipitate severe asthma in susceptible individuals, supporting the idea that elevation of endogenous levels of adenosine is sufficient to trigger bronchoconstriction in some asthmatics. In contrast, Larsson and Sollevi[23] reported that intravenous infusion of adenosine failed to provoke any changes in lung function in asthmatic subjects, although it was acknowledged by these authors that higher concentrations of adenosine may have been required to provoke a response in the lung.

IV. THE MECHANISM OF ACTION OF ADENOSINE-INDUCED BRONCHOCONSTRICTION

A. EVIDENCE FOR A DIRECT CONTRACTILE EFFECT ON BRONCHIAL SMOOTH MUSCLE

But how does adenosine mediate bronchoconstriction in asthmatic subjects? One possibility is that adenosine, like histamine and methacholine, produces a direct contractile effect on airways smooth muscle. However, *in vitro* experiments using guinea pig airways smooth-muscle preparations have provided conflicting results as to the effect of adenosine on airways smooth muscle. Some groups have reported that adenosine relaxed tracheal preparations whether under resting tone[24] or those which have been precontracted with histamine or carbachol.[25,26] In contrast, other studies have reported that the response elicited by adenosine was dependent on whether or not the tracheal tissue had been precontracted[27] and how it was cut, spirals responding with contraction and transverse strips with relaxation.[28] Furthermore, several groups also observed biphasic, concentration-dependent effects of adenosine and its analogs; contracting at concentrations lower than 10^{-6} M and relaxing at higher concentrations.[27,29] Two preparations of guinea pig whole isolated perfused lungs, on the other hand, have been reported to consistently contract to adenosine,[30,31] the contraction being enhanced by prior sensitization of the animal to ovalbumin.[31]

With human airways tissue, adenosine produced a weak but consistent contraction in bronchi[16,32] and lung parenchymal strips[33] from nonasthmatic subjects. The relative potency of adenosine analogs in producing contraction was NECA > adenosine > L-PIA, indicative of an A_2 adenosine receptor-mediated action.[32] Dose-dependent contractions to adenosine have also been reported in bronchi from two birch pollen-sensitive asthmatics,[34] and in a more recent study using bronchi from both asthmatic and nonasthmatic subjects, Bjorck et al.[35] found that this contractile action was sensitive to the A_1 antagonist, 8-cyclopentyl-theophylline, and could be abolished by a combination of the histamine antagonists, me-pyramine or metiamide, and the lipoxygenase inhibitor, ICI-198 615. This suggests that adenosine-induced contraction of human bronchi is dependent on secondary mediator release from inflammatory cells in the airways.

B. EVIDENCE FOR THE INVOLVEMENT OF MAST CELL MEDIATORS

An indirect effect of adenosine involving proinflammatory mediators such as histamine and prostaglandins would be consistent with the results from clinical studies. A central role for histamine, for example, has been established by the demonstration by Holgate and co-workers that premedication with terfenadine, a histamine H_1 receptor antagonist, taken as a 180-mg oral dose 3 h before challenge, inhibited over 80% of the response to AMP bronchoprovocation in atopic and nonatopic asthmatic subjects.[36,37] Identical results were found after premedication with another selective, but chemically distinct histamine receptor antagonist, astemizole.[38] More direct evidence has been obtained by the measurement of plasma histamine levels following AMP challenge. In an earlier study using atopic asthmatics, Cushley and Holgate[39] reported that AMP-induced bronchoconstriction was not accompanied by an increase in circulating levels of histamine and neutrophil chemotactic activity, two markers of mast cell activation. However, in this study, the subjects had markedly hyper-responsive airways when assessed by methacholine provocation, so only small increases in endogenous mediator release would have been required to provoke bronchoconstriction in these subjects. In contrast, in a study with atopic nonasthmatic subjects with less hyperres-ponsive airways, we found a small, but significant, increase in plasma histamine concomitant with the onset of AMP-induced bronchoconstriction.[40] More direct measurement of histamine levels in bronchoalveolar lavage are, however, required to obtain definitive proof of ade-nosine-induced histamine release in the airways.

There is also indirect evidence to support a role for histamine in the pulmonary response to adenosine and AMP challenge in asthmatic subjects. For example, Daxun et al.[41] showed that repeated challenge with AMP resulted in a tachyphylaxis to the stimulus which could not be accounted for by a reduced sensitivity to histamine. Furthermore, Finnerty and Holgate[42] reported that exercise was cross tachyphylactic with AMP challenge. Since ex-ercise-induced bronchoconstriction is also thought to be mediated partly by histamine,[43] these results are compatible with the hypothesis that preformed mast cell mediators such as histamine are being depleted. However, such an interpretation would not be consistent with the recent finding that in patients rendered tachyphylactic to adenosine, the response to subsequent allergen challenge was enhanced rather than suppressed.[44]

In addition to a histamine component in the response to adenosine or AMP broncho-provocation, it is apparent that newly generated mediators such as PGD_2 may also be involved. Thus it has been found that pretreatment with cyclooxygenase inhibitors such as indomethacin and flurbiprofen also protected against the effect of AMP challenge.[45,46] The effects of the histamine antagonist terfenadine and flurbiprofen on AMP-induced bronchoprovocation were reported not to be additive, however, suggesting that prostanoids modulate the release or action of histamine in some way, or *vice versa*.[46]

The cellular basis of adenosine-induced bronchoconstriction has nevertheless not been fully resolved. It has long been known that adenosine by itself has no effect on mediator release from mast cells or basophils, the two most likely sources of histamine, although there is much evidence to show that the purine nucleoside is capable of enhancing release activated by other stimuli. Marquardt et al.[1] first reported this activity of adenosine in immunologically stimulated rat serosal mast cells, and subsequent studies have confirmed that preincubation with adenosine augmented preformed mediator release from rat peritoneal mast cells,[2,47,48] as well as from mouse cultured bone marrow-derived mast cells,[49] human lung mast cells,[50,51] and guinea pig lung fragments.[52] As with the response of airways smooth muscle, however, it is apparent that adenosine may have a dual effect depending on the experimental conditions. For example, Nishibori et al.[47] reported that with rat peritoneal mast cells, incubation with submicromolar concentrations of adenosine led to inhibition of histamine release, whereas incubation with higher concentrations of adenosine produced enhancement. Conversely, we have shown a dual effect in human basophils and mast cells which was dependent on the time of addition of adenosine relative to the addition of the secretagogue. Previously, studies on basophils had demonstrated only an inhibitory action of adenosine and its analogs on histamine release.[53] We found, however, that addition of adenosine at or prior to immunological challenge led to inhibition, whereas addition post-challenge resulted in a small, but significant, enhancement. Maximal inhibition and enhancement were observed at 15 min pre- and postchallenge, respectively. Similar results were obtained with human lung mast cells, except maximal enhancement was observed at 5 min postchallenge.[50]

In addition to enhancing the release of preformed mediators, it has also been reported recently that adenosine potentiates the release of newly generated lipid mediators such as PGD_2 and leukotriene (LT) C_4 from mast cells. Marquardt et al.[49] originally reported that adenosine had no effect on newly generated mediator release from mouse bone marrow-derived mast cells, and Vigano et al.[54] reported only the inhibition of immunologically activated LT and PGD_2 release from human lung fragments. However, recent studies have shown an enhancement of immunologically stimulated LTC_4 release from human lung mast cells.[51,55]

C. EVIDENCE FOR THE INVOLVEMENT OF NEURONAL REFLEXES

Although the results from clinical studies clearly implicate a role for secondary mast cell mediator release in the mechanism of action of adenosine in asthma, it is clear that this alone is not sufficient to account for all the evidence accumulated thus far. For example, as mentioned above, we found that AMP bronchoprovocation in atopic nonasthmatic subjects led to a significant increase in plasma histamine levels, but this was much smaller than the one observed following an equipotent dose of allergen in the same subjects.[40] Furthermore, histamine H_1 receptor antagonists and cyclooxygenase inhibitors do not provide complete protection against the effects of AMP inhalation either alone or in combination,[46] suggesting that there is a mast cell-independent component in the actions of adenosine.

Another possible mechanism by which adenosine may provoke bronchoconstriction is by modulation of neuronal reflexes in the asthmatic lung. Airway caliber is influenced by a bronchoconstrictory cholinergic pathway and a bronchodilatory adrenergic component which may be mediated by circulating catecholamines, since adrenergic innervation to the airways in primates is sparse.[56] In addition, there is a third nonadrenergic, noncholinergic pathway which may be both excitatory and inhibitory. The neurotransmitter in this pathway was once thought to be purines, although it is now considered that neuropeptides such as vasoactive intestinal peptide, substance P, and neurokinins A and B mediate these actions.[57]

However, although there is little evidence for a purinergic neural pathway in the lung, adenosine may still influence airway caliber by interaction with other neural pathways. The modulatory effects of adenine nucleosides and nucleotides on neural transmission in the peripheral and central nervous systems have been well documented since Ginsborg and Hirst[58] first demonstrated their depressant action on synaptic transmission in the rat neuromuscular junction using electrophysiological techniques. Subsequently, it has been established that adenosine may inhibit the release of many different neurotransmitters, including noradrenaline, acetylcholine, glutamate, dopamine, serotonin, and γ-aminobutyric acid from various neuronal pathways by stimulation of presynaptic A_1 adenosine receptors.[59] Furthermore, there is increasing evidence that purines are released from nerve endings following nerve stimulation,[60] raising the possibility that adenine nucleotides and nucleosides may exert postsynaptic as well as presynaptic actions in nonpurinergic neural pathways.

A modulatory role for purines in neural transmission therefore appears to be the rule in both the peripheral and the central nervous systems rather than the exception. Evidence for this occurring in airways tissue is conflicting, however. *In vitro*, adenosine was found to have no effect on cholinergic contractile responses elicited by electrical field stimulation in fresh post-mortem specimens of human tracheal smooth muscle[61] or guinea pig trachea[62] *in vitro*. In contrast, Gustafsson et al.[63] reported that adenosine and its analogs enhanced the constrictor response to transmural nerve stimulation in rabbit bronchial smooth muscle by an A_2 purinoceptor-mediated action. Similarly, Sakai and colleagues[64] recently reported that adenosine and its analogs produced a dose-dependent enhancement of contractile responses elicited by electrical field stimulation in canine isolated bronchial smooth-muscle segments. The rank order potency of adenosine analogs in this tissue was NECA > adenosine > N-cyclohexyladenosine, suggesting that, as in rabbit bronchial smooth muscle, this effect in canine tracheal smooth muscle was also mediated by A_2 purinoceptors. The mechanism of action of adenosine in producing this enhancement is thought to be through activation of prejunctional purinoceptors to augment or accelerate the release of acetylcholine from cholinergic nerve terminals, since adenosine had no effect on contraction stimulated by exogenously added acetylcholine. Suppression of β-adrenergic relaxation in these experiments is unlikely, since the β-antagonist propranolol was present throughout. Moreover, they found that adenosine potentiated histamine-induced contraction by a mechanism which was sensitive to blockade by the muscarinic antagonist atropine. Thus in isolated canine bronchial smooth muscle *in vitro*, adenosine augments contraction by modulating the cholinergic component of histamine-induced contraction as well as enhancing neurotransmitter release initiated by electrical field stimulation.

An alternative mechanism by which adenosine could influence pulmonary function is by modulating β-adrenergic control of the airways, although evidence for this is largely circumstantial. Apart from the direct bronchodilator effect on airways smooth muscle,[16] β-adrenergic stimulation may also influence the viscosity of tracheobronchial secretions by increasing water and chloride fluxes,[65] and modulate the release of inflammatory mediators which contract smooth muscle.[66] Therefore, a defect in the β-adrenergic modulation of bronchial tone, or an attenuation of the responses to β-adrenergic stimulation, could have profound effects on the airways.

Since adenosine has been reported to have a modulatory effect on many adrenergically mediated actions, it is possible that the bronchial response to adenosine inhalation may partly be due to modulation of this homeostatic mechanism. Apart from the well-documented modulation of the adrenergic system at the level of presynaptic neurotransmitter release,[60] adenosine has also been found to modulate adrenergic responses at the postsynaptic or effector cell level in many tissues *in vitro*. For example, adenosine and its analogs have been found to inhibit β-adrenoceptor stimulation of cAMP accumulation, cAMP-dependent protein kinase activity, glycogenolysis and contractility in rat isolated heart preparations,[67] and lipolysis and respiration in rat and hamster adipocytes.[68,69] Moreover, Kioumis et al.[70] reported that adenosine inhibited β-adrenoceptor-mediated accumulation of cAMP in guinea pig trachea, and we found that adenosine and its analogs reversed the inhibition of immunologically activated histamine release from human tonsillar mast cells by adrenaline.[71] Thus, inhibition of β-adrenergic relaxation of airways smooth muscle and reversal of sympathetic inhibition of histamine release from mast cells may represent further mechanisms of action of adenosine.

However, whether or not adenosine modulates neuronal reflexes in the lung *in vivo* and whether or not this underlies the response to adenosine inhalation in asthma is unclear. Evidence for an involvement of cholinergic pathways has been found in the rat by Pauwels and Van der Straeten,[72] who showed that adenosine-induced bronchoconstriction in this species could be inhibited by atropine. In contrast, the existence of a cholinergic component in man is more controversial. Mann et al.[73] and Kung and Diamond[74] reported no inhibition of adenosine-induced bronchoconstriction following inhalation of ipratropium bromide or atropine, respectively, whereas Okayama et al.[75] reported that premedication with atropine did provide some degree of protection. Likewise, direct evidence for interaction with peptidergic pathways or adrenergic modulation of bronchial tone is similarly lacking.

V. CONCLUSIONS

While it is clear that exogenous adenosine may induce bronchoconstriction when inhaled by asthmatic subjects, the mechanism by which it does so is far from clear. Pharmacological studies have suggested enhanced mediator release from pulmonary mast cells as a possible mechanism, but detailed *in vitro* studies with isolated cells have failed to support this. A further possible mechanism involving modulation of neuronal pathways in the lung is also without definitive proof. Of more interest, however, is the question of the importance of adenosine as a mediator in asthma. Although studies with enprofylline have been interpreted as evidence against a role of adenosine, until we understand the mechanism of action of the purine nucleoside and develop appropriate potent antagonists of its action, this question cannot be answered satisfactorily.

REFERENCES

1. **Marquardt, D. L., Parker, C. W., and Sullivan, T. J.,** Potentiation of mast cell mediator release by adenosine, *J. Immunol.,* 120, 871, 1978.
2. **Fredholm, B. B. and Sydbom, A.,** Are the anti-allergic actions of theophylline due to antagonism at the adenosine receptor, *Agents Actions,* 10, 145, 1980.
3. **Butcher, R. W. and Sutherland, E. W.,** Adenosine 3′,5′ monophosphate in biologic materials, *J. Biol. Chem.,* 237, 1244, 1962.
4. **Cushley, M. J., Tattersfield, A. E., and Holgate, S. T.,** Inhaled adenosine and guanosine on airway resistance in normal and asthmatic subjects, *Br. J. Clin. Pharmacol.,* 15, 161, 1983.
5. **Arch, J. R. S. and Newsholme, E. A.,** The control of the metabolism and the hormonal role of adenosine, in *Essays in Biochemistry,* Campbell, P. W. and Aldridge, W. N., Eds., Vol. 14, Academic Press, New York, 1978, 82.
6. **Collinson, A. R., Peuhkurinen, K. J., and Lowenstein, J. M.,** Regulation and function of 5′-nucleotidases, in *Topics and Perspectives in Adenosine Research,* Gerlach, E. and Becker, B. F., Eds., Springer-Verlag, Berlin, 1987, 133.
7. **Barankiewicz, J., Dosch, H. M., and Cohen, A.,** Extracellular nucleotide catabolism in human B and T lymphocytes. The source of adenosine production, *J. Biol. Chem.,* 263, 7094, 1988.
8. **Fredholm, B. B.,** The release of adenosine from rat lung by antigen and compound 48/80, *Acta Physiol. Scand.,* 111, 507, 1981.
9. **Mann, J. S., Renwick, A. G., and Holgate, S. T.,** Release of adenosine and its metabolites from activated human leucocytes, *Clin. Sci.,* 70, 461, 1986.

10. **Marquardt, D. L., Gruber, H. E., and Wasserman, S. I.,** Adenosine release from stimulated mast cells, *Proc. Natl. Acad. Sci. U.S.A.,* 81, 6192, 1984.

11. **Mentzer, R. M., Rubio, R., and Berne, R. M.,** Release of adenosine from hypoxic canine lung tissue and its possible role in the pulmonary circulation, *Am. J. Physiol.,* 229, 1625, 1975.

12. **Mann, J. S., Holgate, S. T., Renwick, A. G., and Cushley, M. J.,** Airway effects of purine nucleosides and nucleotides and release with bronchial provocation in asthma, *J. Appl. Physiol.,* 61, 1667, 1986.

13. **Cushley, M. J. and Holgate, S. T.,** Adenosine induced bronchoconstriction in asthma: specificity and relationship to airway reactivity, *Thorax,* 38, 705, 1983.

14. **Cushley, M. J., Tattersfield, A. E., and Holgate, S. T.,** Adenosine induced bronchoconstriction in asthma: antagonism by inhaled theophylline, *Am. Rev. Respir. Dis.,* 129, 380, 1984.

15. **Persson, C. G. A. and Kjellin, G.,** Enprofylline, a principally new anti-asthmatic xanthine, *Acta Pharmacol. Toxicol.,* 49, 313, 1981.

16. **Finney, M. J. B., Karlsson, J. A., and Persson, C. G. A.,** Effect of bronchoconstrictors and bronchodilators on a novel human small airway preparation, *Br. J. Pharmacol.,* 85, 29, 1985.

17. **Persson, C. G. A., Karlsson, J. A., and Erjefalt, J.,** Differentiation between bronchodilation and universal adenosine antagonism amongst xanthine derivatives, *Life Sci.,* 30, 2181, 1982.

18. **Pauwels, R. and Van der Straeten, M.,** Experimental asthma and xanthines, in *Anti-Asthma Xanthines and Adenosine,* Andersson, K. E. and Persson, C. G. A., Eds., Excerpta Medica, Amsterdam, 1985, 97.

19. **Kroll, F., Karlsson, J. A., Ryrfeldt, A., and Persson, C. G. A.,** Adenosine-induced bronchoconstriction in the guinea pig isolated lung: interaction with theophylline and enprofylline, *Pulmonary Pharmacol.,* 1, 85, 1988.

20. **Cushley, M. J., Tallant, N., and Holgate, S. T.,** The effect of dipyridamole on histamine- and adenosine-induced bronchoconstriction in normal and asthmatic subjects, *Eur. J. Respir. Dis.,* 67, 185, 1985.

21. **Crimi, N., Palermo, F., Oliveri, R., Maccarrone, C., Palermo, B., Vancheri, C., Polosa, R., and Mistretta, A.,** Enhancing effect of dipyridamole inhalation on adenosine-induced bronchospasm in asthmatic patients, *Allergy,* 43, 179, 1988.

22. **Eagle, K. A. and Boucher, C. A.,** Intravenous dipyridamole infusion causes severe bronchospasm in asthmatic patients, *Chest,* 95, 258, 1989.

23. **Larsson, K. and Sollevi, A.,** Influence of infused adenosine on bronchial tone and bronchial reactivity in asthma, *Chest,* 93, 280, 1988.

24. **Coleman, R. A.,** Effects of some purine derivatives on the guinea pig trachea and their interaction with drugs that block adenosine uptake, *Br. J. Pharmacol.,* 57, 51, 1976.

25. **Brown, C. M. and Collis, M. G.,** Evidence for an A2/Ra adenosine receptor in guinea pig trachea, *Br. J. Pharmacol.,* 76, 381, 1982.

26. **Abe, M., Katzuragi, T., and Furukawa, T.,** Discrimination by dipyridamole of two types of response to adenosine and ATP of the carbachol induced contraction of guinea pig trachea, *Eur. J. Pharmacol.,* 90, 29, 1983.

27. **Caparrotta, L., Cillo, F., Fassina, G., and Gaion, R. M.,** Dual effect of (-)-N^6-phenylisopropyladenosine on guinea-pig trachea, *Br. J. Pharmacol.,* 83, 23, 1984.

28. **Satchell, D. and Smith, R.,** Adenosine causes contractions in spiral strips and relaxations in transverse strips of guinea pig trachea: studies on mechanism of action, *Eur. J. Pharmacol.,* 101, 243, 1984.

29. **Farmer, S. G., Canning, B. J., and Wilkins, D. E.,** Adenosine-mediated contraction and relaxation of guinea pig isolated tracheal smooth muscle: effects of adenosine antagonists, *Br. J. Pharmacol.,* 95, 371, 1988.

30. **Kroll, F., Karlsson, J. A., Persson, C. G. A., and Ryrfeldt, A.,** Interactions between xanthines, mepyramine and adenosine in the guinea-pig lung, in *Anti-Asthma Xanthines and Adenosine,* Andersson, K. E. and Persson, C. G. A., Eds., Current Clinical Practice Series No. 19, Excerpta Medica, Amsterdam, 1985, 193.

31. **Thorne, J. and Broadley, K. J.,** A bronchoconstrictor response to adenosine of guinea pig perfused lungs, *Br. J. Pharmacol.,* 93 (Suppl.), 278P, 1988.

32. **Holgate, S. T., Cushley, M. J., Mann, J. S., Hughes, P. J., and Church, M. K.,** The action of purines on human airways, *Arch. Int. Pharmacodyn. Ther.,* 280 (Suppl.), 240, 1986.

33. **Napier, F. E. and Temple, D. M.,** Theophylline's inhibition of antigen-induced contraction of human parenchymal strips is independent of adenosine antagonism, *Eur. J. Pharmacol.,* 142, 253, 1987.

34. **Dahlen, S. E., Hansson, G., Hedqvist, P., Bjorck, T., Granstrom, E., and Dahlen, B.,** Allergen challenge of lung tissue from asthmatics elicits bronchial contraction that correlates with the release of leukotrienes C4, D4, and E4, *Proc. Natl. Acad. Sci. U.S.A.,* 80, 1712, 1983.

35. **Bjorck, T., Cederqvist, B., Gustafsson, L. E., and Dahlen, S. E.,** Adenosine contracts human bronchi by an indirect mechanism involving histamine and leukotrienes, *Am. Rev. Respir. Dis.,* 141, A470, 1990.

36. **Rafferty, P., Beasley, R., Southgate, P., and Holgate, S.,** The role of histamine in allergen and adenosine-induced bronchoconstriction, *Int. Arch. Allergy Appl. Immunol.,* 82, 292, 1987.

37. **Phillips, G. D., Rafferty, P., Beasley, C. R. W., and Holgate, S. T.,** The effect of oral terfenadine on the bronchoconstrictor response to inhaled histamine and adenosine 5′-monophosphate in non-atopic asthma, *Thorax,* 42, 939, 1987.

38. **Rafferty, P., Beasley, C. R., and Holgate, S. T.,** The contribution of histamine to bronchoconstriction produced by inhaled allergen and adenosine 5′-monophosphate in asthma, *Am. Rev. Respir. Dis.,* 136, 369, 1987.

39. **Cushley, M. J. and Holgate, S. T.,** Adenosine induced bronchoconstriction in asthma: role of mast cell mediator release, *J. Allergy Clin. Immunol.,* 75, 272, 1985.

40. **Phillips, G. D., Ng, W. H., Church, M. K., and Holgate, S. T.,** The response of plasma histamine to bronchocoprovocation with methacholine, adenosine 5′-monophosphate, and allergen in atopic non-asthmatic subjects, *Am. Rev. Respir. Dis.,* 141, 9, 1990.

41. **Daxun, Z., Rafferty, P., Richards, R., and Holgate, S. T.,** Airway refractoriness to the bronchoconstrictor effect of inhaled adenosine 5′ monophosphate, *Clin. Sci.,* 73 (Suppl. 17), 4P, 1987.

42. **Finnerty, J. P. and Holgate, S. T.,** Effect of repeated bronchial challenges with adenosine mono-phosphate (AMP) on exercise induced bronchoconstriction in asthma, *Thorax,* 44 (Abstr.), 865P, 1989.

43. **Barnes, P. J. and Brown, M. J.,** Venous plasma histamine in exercise- and hyperventilation-induced asthma in man, *Clin. Sci.,* 61, 159, 1981.

44. **Phillips, G. D., Bagga, P. K., Djukanovic, R., and Holgate, S. T.,** The influence of refractoriness to adenosine 5′-monophosphate on allergen provoked bronchoconstriction in asthma, *Am. Rev. Respir. Dis.,* 140, 321, 1989.

45. **Crimi, N., Palermo, F., Polosa, R., Oliveri, R., Maccarrone, C., Palermo, B., and Mistretta, A.,** Effect of indomethacin on adenosine-induced bronchoconstriction, *J. Allergy Clin. Immunol.,* 83, 921, 1989.

46. **Phillips, G. D. and Holgate, S. T.,** The effect of oral terfenadine alone and in combination with flurbiprofen on the bronchoconstrictor response induced by inhaled adenosine 5′-monophosphate in non-atopic asthma, *Am. Rev. Respir. Dis.,* 139, 463, 1989.

47. **Nishibori, M., Shimamura, K., Yokoyama, H., Tsutsumi, K., and Saeki, K.,** Differential effects of adenosine on histamine secretion induced by antigen and chemical stimuli, *Arch. Int. Pharmacodyn. Ther.,* 265, 17, 1983.

48. **Church, M. K. and Hughes, P. J.,** Adenosine potentiates immunological histamine release from rat mast cells by a novel cyclic-AMP independent cell surface action, *Br. J. Pharmacol.,* 85, 3, 1985.

49. **Marquardt, D. L., Walker, L. L., and Wasserman, S. I.,** Adenosine receptors on mouse bone marrow-derived mast cells: functional significance and regulation by aminophylline, *J. Immunol.,* 133, 932, 1984.

50. **Hughes, P. J., Holgate, S. T., and Church, M. K.,** Adenosine inhibits and potentiates IgE-dependent histamine release from human lung mast cells by an A2-purinoceptor mediated mechanism, *Biochem. Pharmacol.,* 33, 3847, 1984.

51. **Marone, G., Cirillo, R., Genovese, A., Marino, O., and Quattrin, S.,** Human basophil/mast cell releasability. VII. Heterogeneity of the effect of adenosine on mediator secretion, *Life Sci.,* 45, 1745, 1989.

52. **Welton, A. F. and Simko, B. A.,** Regulatory role of adenosine in antigen-induced histamine release from the lung tissue of actively sensitized guinea pigs, *Biochem. Pharmacol.,* 29, 1085, 1980.

53. **Marone, G., Findlay, S. R., and Lichtenstein, L. M.,** Adenosine receptors on human basophils: modulation of histamine release, *J. Immunol.,* 123, 1473, 1979.

54. **Vigano, T., Toia, A., Galli, G., Berti, F., Crivellari, M. T., Mezzetti, M., and Folco, G. C.,** Adenosine and eicosanoid release from immunologically challenged human lung fragments, *Adv. Prostaglandin, Thromboxane, Leukotriene Res.,* 17, 992, 1987.

55. **Peachell, P. T., Lichtenstein, L. M., and Schleimer, R. P.,** Personal communication.

56. **Ind, P. W. and Barnes, P. J.,** Adrenergic control of airways in asthma, in *Asthma: Basic Mechanisms and Clinical Management,* Barnes, P. J., Rodger, I. W., and Thomson, N. C., Eds., Academic Press, London, 1988, 357.

57. **Barnes, P. J.,** Airways neuropeptides, in *Asthma: Basic Mechanisms and Clinical Management,* Barnes, P. J., Rodger, I. W., and Thomson, N. C., Eds., Academic Press, London, 1988, 395.

58. **Ginsborg, B. L. and Hirst, D. S.,** The effect of adenosine on the release of transmitters from the phrenic nerve of the rat, *J. Physiol. (London),* 224, 629, 1972.

59. **Fredholm, B. B. and Dunwiddie, T. V.,** How does adenosine inhibit transmitter release?, *Trends Pharmacol. Sci.,* 9, 130, 1988.

60. **Fredholm, B. B. and Hedqvist, P.,** Modulation of neurotransmission by purine nucleotides and nucleosides, *Biochem. Pharmacol.,* 29, 1635, 1980.

61. **Bai, T. R., Lam, R., and Prasad, F. Y. F.,** Effects of adrenergic agonists and adenosine on cholinergic neurotransmission in human tracheal smooth muscle, *Pulmonary Pharmacol.,* 1, 193, 1989.

62. **Grundstrom, N., Andersson, R. C. G., and Wikbirg, J. E. S.,** Investigation of possible presynaptic effects of adenosine and noradrenaline on cholinergic neuro-transmission in guinea pig trachea, *Acta Pharmacol. Toxicol.,* 49, 158, 1981.

63. **Gustafsson, L. E., Wiklund, N. P., and Cederqvist, B.,** Apparent enhancement of cholinergic transmission in rabbit bronchi via adenosine A2 receptors, *Eur. J. Pharmacol.,* 120, 179, 1986.

64. **Sakai, N., Tamaoki, J., Kobayashi, K., Katayama, M., and Takizawa, T.,** Adenosine potentiates neurally- and histamine-induced contraction of canine airway smooth muscle, *Int. Arch. Allergy Appl. Immunol.,* 90, 280, 1989.

65. **Nadel, J. A.,** Autonomic control of airway smooth muscle and airway secretion, *Am. Rev. Respir. Dis.,* 115, 117, 1977.

66. **Subramanian, N.,** Inhibition of immunological and non-immunological histamine release from human basophils and lung mast cells by formoterol, *Arzneim. Forsch.,* 36, 502, 1986.

67. **Dobson, J. G.,** Adenosine inhibition of catecholamine elicited contractile and glycogenolytic responses in the rat heart, *Fed. Proc., Fed. Am. Soc. Exp. Biol.,* 39, 1173, 1980.

68. **Szillat, D. and Bukowiecki, L. J.,** Control of brown adipose tissue lipolysis and respiration by adenosine, *Am. J. Physiol.,* 245, E555, 1983.

69. **Schimmel, R. J. and McCarthy, L.,** Role of adenosine as an endogenous regulator of respiration in hamster brown adipocytes, *Am. J. Physiol.,* 246, C301, 1984.

70. **Kioumis, I., Grandordy, B., and Barnes, P. J.,** Interaction between adenosine receptors and β-adrenoceptors in airways, *Br. J. Pharmacol.,* 92 (Suppl.), 592P, 1987.

71. **Ng, W. H., Polosa, R., and Church, M. K.,** Adenosine bronchoconstriction in asthma: investigations into its possible mechanism of action, *Br. J. Clin. Pharmacol.,* 30, 895, 1990.

72. **Pauwels, R. and Van der Straeten, M.,** An animal model for adenosine-induced bronchoconstriction, *Am. Rev. Respir. Dis.,* 136, 374, 1987.

73. **Mann, J. S., Cushley, M. J., and Holgate, S. T.,** Adenosine-induced bronchoconstriction in asthma: role of parasympathetic stimulation and adrenergic inhibition, *Am. Rev. Respir. Dirs.,* 132, 1, 1985.

74. **Kung, M. and Diamond, L.,** Adenosine-induced bronchoconstriction in human asthmatics: effects of pretreatment with indomethacin or atropine sulfate, *Pulmonary Pharmacol.,* 3, 17, 1990.

75. **Okayama, M., Ma, J. Y., Hataoka, I., Kimura, K., Iijima, H., Inoue, H., and Takishima, T.,** Role of vagal nerve activity on adenosine-induced bronchoconstriction in asthma, *Am. Rev. Respir. Dis.,* 133 (Suppl.), A93, 1986.

Chapter 33

ADENOSINE-BASED THERAPEUTICS IN NEUROLOGIC DISEASE

P. J. Marangos and L. Miller

TABLE OF CONTENTS

I. INTRODUCTION

The contents of this book provide ample proof for the importance of the ubiquitous purine adenosine in the control of cellular function. Very many, if not all of the effects of adenosine can be reduced to a relatively simple endpoint involving the autoregulation of cellular metabolism to a level consistent with energy supply. In this regard, conditions which result in increased adenosine levels are characerized by decreases in the universal metabolic fuel ATP. The generalized reduction of cell or organ function caused by this increase in adenosine results in a sparing of cells during traumatic situations (i.e., ischemia, seizures) where ATP breakdown exceeds its reformation. This conceptualization of the role adenosine plays in traumatized tissue has been formalized by Newby,[1] who has coined the term "retaliatory metabolite" to describe adenosine. The intimate and reciprocal relationship between ATP and adenosine levels imparts a logic to the depressant effects of adenosine on the level of cellular function.

Since the mid-1970s, substantial progress has been made regarding the mechanisms involved in the modulation of neural function by adenosine. The pathfinding studies of Burnstock[2] set forth the concept of purinergic neurons which could be classified as either binding adenosine (via P_1 receptors) or ATP (via P_2 receptors) as the neuroactive agent. As with many other neural receptor systems, purinergic receptors have now been shown to display considerable diversity, with P_1 systems now termed adenosine A_1 or A_2 receptors and the P_2 system constituting at least four receptor subtypes.[3] The current view is that adenosine serves as a neuromodulator (rather than a neurotransmitter) via its actions on A_1 or A_2 receptors and that ATP acts as a cotransmitter released with other neurotransmitters.[4,6]

In brain it is now apparent that adenosine is a major nonpeptide neuromodulator which acts via A_1 and possible A_2 receptors.[4,5] Although receptor classification was initially defined by their effects on cyclic adenosine 5'-monophosphate (cAMP) levels (A_1 receptors mediate decreases and A_2 receptors increases in cAMP levels),[4,5] the relatively recent development of specific radioreceptor assays and high-affinity ligands for each receptor subtype has now precluded this criteria as a discriminator of adenosine receptor subtypes.[6] A_1 and A_2 receptors are now defined on the basis of purely pharmacologic criteria[6] rather than effector responses. This is in some respects fortunate, since the relevance of cAMP as a second messenger or effector for adenosine-mediated processes in neural tissue has recently been called into question.[7,8]

II. ADENOSINE AND NEURAL FUNCTION

The sedative, anticonvulsant,[5,6] and more recently discovered neuroprotective effects[9,10] of adenosine and adenosine receptor agonists are now quite well documented and have generated considerable interest regarding the potential therapeutic applications of such agents.[10] Mechanistically, it appears that adenosine functions as a neuromodulator and not as a neurotransmitter, since it regulates the activity of numerous other neurotransmitter systems. It is likely that conditions which result in increased local extracellular levels of adenosine from either neuronal or glial origin exert tonic effects on surrounding adenosine receptor-containing neurons, which result in a decrease in the calcium-dependent release of a variety of neurotransmitters.[11,12] Conditions that predispose towards elevated extracellular levels of adenosine would include seizures, ischemia, and generalized metabolic trauma. Such situations result in net ATP depletion with resultant accumulation of adenosine, inosine, and hypoxanthine.[13]

It is important to stress that although the formation of adenosine in traumatized tissue may serve a homeostatic protective function by adjusting cellular metabolism to a level consistent with the supply of metabolic fuel, the capacity of this system to raise adenosine levels is limited. The major limitation results from the short tissue half-life of extracellular adenosine. Adenosine is rapidly taken up by neural and glial cells and it is metabolized by several enzymes, predominantly via phosphorylation and deamination.[14] The effects of endogenously generated adenosine can therefore be expected to be of rather short duration and highly localized to the trauma site. Therapeutic strategies involving more prolonged stimulation of the adenosine system in brain via an increase in its half-life might therefore be viewed as a potentiation or amplification of beneficial homeostatic mechanisms that will be useful in the treatment of seizures, ischemia, and generalized head trauma.

Although initially it was hypothesized that the mechansim of adenosine action as a neuromodulator involved cAMP,[7] it is now believed that the purine riboside probably exerts its effects via either neuronal calcium or potassium channels. The rather universal inhibitory effect of adenosine on calcium-dependent neurotransmitter release[11,12] was the initial suggestion for such a mode of action, and this has recently been supported by a number of studies where the effects of adenosine on calcium- and potassium-mediated membrane events have been directly studied.[15] Although still somewhat controversial, the emerging view is that adenosine apparently functions as an endogenous presynaptic calcium channel blocker, either via a direct action on presynaptic calcium channels or indirectly as a modulator of neuronal potassium channels.[15] In the latter scenario, adenosine is viewed as a potassium channel opener which would have the effect of hyperpolarizing the neuron (potassium would follow its concentration gradient and exit the cell) and closing voltage-dependent calcium channels. The net result would be a reduction in calcium influx and a decrease in calcium-dependent neurotransmitter release.

Although it is still somewhat controversial whether the predominant effect of adenosine is on potassium or calcium channels, it is clear that adenosine and adenosine A_1 receptor agonists are potent inhibitors of neurotransmitter release.[11,12] This property of adenosine has now been reported in a number of different systems, with many different neurotransmitter candidates ranging from the catecholamines to the amino acids. Burke and Nadler[12] recently reported on the effects of a number of different adenosine agonists on glutamate and aspartate release from potassium-stimulated hippocampal slices. In their study, L-N^6-phenylisopropyladenosine (L-PIA) and N^6-cyclohexyladenosine (CHA) were the most potent inhibitors of excitatory amino acid (EAA) release with potencies similar to those observed with these agents in receptor binding studies involving adenosine A_1 receptors. The effects of adenosine and adenosine agonists were blocked by adenosine receptor antagonists, clearly showing mediation of the effect by adenosine receptors. Interestingly, in the same report the adenosine transport inhibitor dipyridamole, which increases brain extracellular adenosine levels,[13] also inhibited glutamate and aspartate release. The dipyridamole-induced inhibition of EAA release is indirect, since it probably results from increased extracellular adenosine. This effect of adenosine uptake blockers will be further discussed below as it relates to the potential of these agents as neuroprotectives.

III. MECHANISMS OF NEURODEGENERATION

During the past 6 years, enormous progress has been made regarding the mechanisms involved in ischemia-induced neural tissue damage. It is now clear that a significant component of the neurodegeneration resulting from ischemia (stroke, head trauma) actually occurs postreperfusion rather than during the ischemic event. Such a situation also appears to be true in other tissues, such as the cardiovascular system.[16] In brain, the major agents responsible for postreperfusion neurodegeneration appear to be the EAA transmitters, which have been shown to be released at accelerated rates during ischemic trauma.[17,18] This process has been termed excitotoxicity, since the increased EAA release actually results in a situation where neurons using glutamate and aspartate as neurotransmitters are believed to be stim-

TABLE 1
Excitatory Amino Acid Receptor Subtypes

Subtype	Agonists	Antagonists	Modulators	Current
N-Methyl-D-aspartate (NMDA)	NMDA, glutamate, aspartate, quinolinate, ibotenate	APV APH CPP MK-801 Dextrorphan Mg^{++} Zn^{++} PCP Ketamine Kynurenate HA-966 7-CL-KYN	Glycine D-Serine Spermine	$Ca^{+2}/Na^+/K^+$
Kainic acid (KA)	Kainic acid, domoate	DNQX	—	Na^{++}
AMPA	Quisqualate, glutamate, aspartate, AMPA, kainate	CNQX	—	Na^+/K^+
ACPD	ACPD, ibotenate, quisqualate	—		IP3 formation

ulated to the point of metabolic insufficiency and death. It is important to stress several aspects of excitotoxicity that have recently come to light. First, the process is mediated by several subtypes of the EAA receptor.[19] There are at least four EAA receptor subtypes termed the N-methyl-D-aspartate (NMDA), the kainate, AMPA and ACPD receptors. The pharmacology of EAA receptor subtypes is summarized in Table 1. Recent data indicate that it is not sufficient to block just one of the receptor subtypes in order to effect a complete blockade of ischemia-induced neurodegeneration.[19] It had been thought until recently that blockade of the NMDA receptor subtype was sufficient to inhibit the excitotoxic process, and indeed the pharmaceutical industry responded rapidly with agents such as the Merck® compound MK-801, which has been shown to be neuroprotective.[20] These recent findings,[19] coupled with the central toxicity of EAA receptor blockers,[21] have made subtype-specific EAA receptor blockers a less desirable goal for drug development strategies in stroke and head trauma.

A second aspect of excitotoxicity that is important to consider from the standpoint of drug discovery is that the process appears to be delayed in character. It has been shown that postreperfusion neural tissue damage takes several days to manifest itself.[22] Essentially, this suggests that pharmacologic interaction postischemia is feasible and that therapeutic strategies are plausible.

IV. ADENOSINE AS A NEUROPROTECTIVE AGENT

The rationale for the neuroprotective properties of adenosine is now quite pervasive. Adenosine levels increase during and remain elevated for a short time period after neural trauma in a manner proportional and reciprocal to that of ATP depletion. Such a situation would seem to ideally suit adenosine to its role as a homeostatic neuroprotective metabolite. Some of the known effects of adenosine in brain are summarized in Table 2.

These documented effects of adenosine (Table 2) on brain function would seem to argue for the purine playing a key role in the endogenous defenses of the brain against the damaging effects of ischemia. However, as with all homeostatic mechanisms, it is important that they not be so potent as to compromise the organism. In this regard the adenosine-generating capacity of the brain is limited by the ATP level, and perhaps more importantly by the rapid

TABLE 2
Effects of Adenosine on Neural Tissue

Mediates cerebral hyperemia postischemia
Reduces platelet and neutrophil adhesion
Inhibits glutamate and aspartate release
Inhibits calcium influx into neurons

TABLE 3
Adenosine-Based Therapeutic Strategies

I. Metabolic agents (inhibitors of adenosine metabolism)
    Endogenous adenosine is the active agent (not receptor subtype selective)
    Self limiting
    Coupled to ATP degradation
II. Transport blockers
    Adenosine is the active agent, self limiting
    Somewhat coupled to ATP degradation
III. Adenosine receptor agonists
    Direct acting, subtype specific
    Not coupled to ATP catabolism
    Substantial peripheral side effects

metabolism of the purine to agents such as inosine (via adenosine deaminase), which are pharmacologically inactive at the adenosine receptor. The other process that seriously compromises the potential protective effect of adenosine postischemia is the rapid uptake of the purine by both neurons and glia (for review see Deckert et al.).[14] Adenosine transporters are present on most cells and are especially numerous on human red cells. Metabolism and uptake can therefore be viewed as two substantial governors of adenosinergic neuroprotection. Such a situation does, however, suggest a therapeutic strategy for the treatment of neurologic disorders characterized by ischemia that would involve the partial or complete blockade of these processes (i.e., inhibitors of adenosine-metabolizing enzymes or the blocking of adenosine uptake). Such agents can be viewed as adenosine-regulating agents or ARAs, and would have a site- and event-specific action. Studies by Phillis and O'Regan[23] have in fact already documented the beneficial effects of the adenosine deaminase inhibitor deoxycoformycin (DCF) in a gerbil cerebral ischemia model. Although adenosine deaminase inhibitors are not appropriate for chronic treatment, the ARA approach to the treatment of ischemic neurological disorders does appear to be viable.

An alternative approach to inhibitors of adenosine metabolism or uptake involves the use of adenosine receptor agonists. Such an approach has advantages and disadvantages associated with it. The agonist approach can be viewed as essentially nonlimiting in that the receptor can be totally occupied for as long a period as desired. The metabolic and uptake blocker strategy is much more restricted in character due to the limited ability of tissue to generate the purine. In addition, the agonist approach is not dependent on ATP degradation, and for that reason may be more useful in acute settings when administered postreperfusion. Certainly, this has in fact been shown to be the case in animal studies, as will be discussed below. The disadvantages associated with the adenosine receptor agonist strategy are the substantial peripheral side effects associated with these agents (i.e., hypotension, bradycardia, inhibition of renal function). The different adenosinergic therapeutic strategies for neuroprotective therapy are summarized in Table 3.

Although the theoretical and mechanistic rationale for a therapeutic strategy is important, it is of course imperative to demonstrate efficacy in animal model systems and ultimately in humans. In this regard, adenosine-related strategies have over the past 3 years been shown to be highly efficacious in a number of diverse animal model systems, which include gerbil global ischemia,[24-26] rat global ischemia,[27] and a variety of excitotoxin-induced scenarios.[28]

The first report suggesting that adenosine had neuroprotective properties came from the laboratoroy of Rudolphi et al.,[29] where it was shown that pretreatment of gerbils with theophylline (30 mg/kg) enhanced the hippocampal neural damage resulting from transient forebrain ischemia. The most dramatic potentiation of cell damage was observed with 2 min of global ischemia, where the nontreated animals had essentially no CA1 cell necrosis and the theophylline-pretreated animals sustained very substantial damage. The clear implications of this work which has been recently replicated and extended in a different laboratory[30] is that endogenous adenosine exerts neuroprotective effects, and that this purine probably does function homeostatically as a defense mechanism against ischemically induced cell death.

Intracerebral injection of the adenosine analog 2-chloroadenosine (CADO) and intracerebroventricular injection of CHA[26,27] have also been shown to provide marked neuroprotective effects. The CADO study utilized a 10-min ischemic period in rats with multiple injections administered 1 min, 4 h, and 10 h postreperfusion. The CHA study employed a 30-min period of global ischemia with the drug administered 15 min after ischemia. In both studies, core temperature was maintained for 2 h postischemia, and marked protective effects were observed as reflected by decreased CA1 cell necrosis. More recent studies with CHA administered intraperitoneally 5 min postischemia[25] have also shown substantial neuroprotective effects, as well as an increase in survival.[31] These studies employed a 30-min period of ischemia in the gerbil, followed by a 7- to 10-d evaluation period.

Adenosine agonists have also been shown effective in rat global and focal ischemia[27,32] and in a rat quinolinic acid-induced neurotoxicity model.[28] In the latter case, only intrahippocampal adminstration of R-PIA was effective. It therefore appears that adenosinergic neuroprotective strategies are effective to some extent across species.

In addition to the growing literature on adenosine agonists being protective when administered directly to gerbils and rats, there are now reports of caffeine-induced adenosine receptor upregulation also affording neuroprotection in gerbil global ischemia systems.[33] These studies provide persuasive evidence that the adenosine system in brain does appear to confer neuroprotective properties to brain areas sensitive to excitotoxic damage. They also support the direct adenosine agonist studies and tend to diffuse arguments regarding the nonspecific aspects of adenosinergic neuroprotection, i.e., hypotension, hypothermia, etc.

As already mentioned, Phillis and O'Regan[23] have recently shown that the adenosine deaminase inhibitor DCF is neuroprotective in the gerbil ischemia model.[23] Such protection was only observed with preischemic applications of the inhibitor, but nevertheless further supports the concept of adenosinergic neuroprotection and suggests that inhibitors of adenosine degradation might be useful in various CNS trauma settings. The recent observation by Park and Gidday[13] that dipyridamole increases postischemic adenosine levels in piglet brain suggests that adenosine transport blockers which permeate the brain may also be useful neuroprotectives, and is consistent with Burke and Nadler's[12] data documenting the inhibition of EAA release by dipyridamole in hippocampal slices. It therefore appears that a variety of adenosinergic neuroprotective strategies are possible, some of which may be more suited to certain indications than others.

In addition to the whole-animal data reviewed above, there has also been a report of adenosine and CHA being neuroprotective in a primary neural culture system.[34] In this study, adenosine and CHA markedly blocked postischemic cell lysis, and this effect was reversed by adenosine receptor antagonists. Adenosine displayed an EC_{50} of 100 μM, and CHA a value of about 100 nm.

It remains to be seen whether one or more of the three potential adenosinergic therapeutic strategies discussed will prove useful for the treatment of stroke and/or head trauma. There are potential problems associated with each approach, especially as regards adenosine agonists, which have profound side effects. In this regard, these side effects can be blocked by the use of adenosine antagonists that do not permeate the brain and therefore do not block the neuroprotective effect of the agonist.[31] Whether or not the use of a drug combination approach employing a centrally acting agonist and a brain-impermeable antagonist is feasible is not presently clear. Also, the hypothermic properties of systemically administered adenosine agonist do confer an additional benefit, since hypothermia (which is partly peripherally mediated) has been shown to be cerebroprotective in a number of studies.[35,36]

This raises the issue of the contribution of hypothermia to the neuroprotective effects of therapeutic strategies in general. Hypothermia has also recently been shown to account for a major component of the neuroprotection observed with some of the EAA receptor blockers.[36] It is important in these studies to consider the animal model being employed, since species differences, ischemia duration, and the means of temperature regulation are all important variables to be assessed when multiple protective mechanisms are being quantified.

The actual contribution of adenosine-induced hypothermia to the observed neuroprotection remains to be accurately assessed. Certainly, many of the studies documenting the neuroprotective properties of adenosine are temperature independent. These include the cell culture data,[34] the studies where low doses of agonists were injected directly into the hippocampus,[27] and the caffeine-induced adenosine receptor upregulation studies.[33] This, coupled with the excellent data establishing the adenosine-mediated inhibition of EAA release,[11,12] supports the concept that both hypothermia and EAA-related mechanisms are probably components of the neuroprotective effects of adenosine and its derivatives.

V. OTHER CENTRAL NERVOUS SYSTEM THERAPEUTIC USES OF ADENOSINE

In general, adenosine receptor agonists will probably have very limited utility in indications other than acute stroke or central nervous system (CNS) trauma. The multitude of systemic effects observed with these agents has consistently complicated their therapeutic application. A more subtle approach to adenosinergic therapeutic development involving the ARA approach would, however, be useful in the development of new classes of anticonvulsant and sedative agents. Over the past decade, adenosine has been shown to have substantial sedating and antiseizure properties (for review see Marangos and Boulenger[5] and Snyder[6]). Available information strongly suggests that the adenosine system in brain serves to regulate at least partially both sleep-wake cycles and seizure-like activity. Increased adenosine levels observed at seizure foci are probably the result of ATP catabolism and have been implicated in the attenuation of seizure spread.[5] This is consistent with the retaliatory metabolite[1] role of adenosine, since the end result is a downregulation of cellular metabolism to levels consistent with ATP levels. The therapeutic targeting of this homeostatic seizure-mediating system is therefore reasonable and might involve intervention with agents that would increase both the degree and the half-life of endogenously generated adenosine. This would involve either the use of adenosine uptake blockers or inhibitors of adenosine-utilizing enzymes, i.e., the ARA strategy. It is likely that the ARA approach to adenosinergic therapeutics will prove to be much more elegant and practical than previously employed strategies. There is certainly a medical need for more effective antiepileptic drugs with greater efficacy and less tolerance potential than currently used agents. Seizure medications are administered chronically and therefore must be orally bioavailable and have minor side effects. Whether or not the ARA agents discussed will meet these criteria remains to be seen. What does, however, appear to be almost certain is that currently available adenosine agonists will probably not be suitable in this situation.

Regarding the development of adenosinergic sedative agents, it would also appear that a significant opportunity exists here, as well. Currently available sedatives leave much to be desired in that they do not induce natural sleep and tend to have addiction potential associated with them. Several studies have now documented the sedating effects of adenosine and adenosine agonists,[37,38] and more importantly it appears that the purine does have a homeostatic role in the regulation of sleep cycles, since it results in a natural form of sleep as judged by electroencephalographic studies.[37,38] As with the seizure indication, it is difficult to envision using agonists as sedatives, but the ARA strategy should be much more suitable. Recent studies with adenosine uptake blockers have in fact shown some promise in this regard,[38] but more work is required to prove the value of this approach.

As regards the chronic neurodegenerative diseases, there is emerging evidence supporting a role for excitotoxicity in Parkinson's,[39] Alzheimer's, and Huntington's diseases.[40] It is therefore possible that employing an adenosinergic strategy in these disorders might delay or stop disease progression. Preliminary data obtained in our laboratory support such a contention in a mouse toxin model of Parkinson's disease where we have shown a protective effect of CHA in both MPTP- and methamphetamine-induced striatal degeneration.[41] Further studies will be required in order to demonstrate the feasibility of such an approach for the treatment of neurodegenerative diseases, and again will probably involve ARA agents which increase adenosine levels either metabolically or via uptake blockade, in a site- and event-specific manner.

Adenosinergic therapeutic strategies have been difficult to exploit successfully in cardiovascular medicine, largely due to the poor side-effect profile of agonists and dipyridamole. In CNS therapeutics the situation may be somewhat different, since some of the indications to be targeted (acute stroke and head trauma) require acute therapeutics and will tolerate somewhat more of a side-effect profile. The utilization of ARA strategies for some of the other CNS indications such as seizures, sedation, and neurodegenerative disease may also confer a greater utility to the adenosinergic approach, in that these agents can be expected to have more subtle site- and event-specific actions, with fewer side effects, that would better suit them to chronic treatment paradigms. The pursuit of ARA-related approaches to adenosinergic CNS therapeutics therefore holds potential promise, since it may provide a means of potentiating the homeostatic function of adenosine to a level capable of affording substantial therapeutic benefit in a variety of disease states.

VI. ABBREVIATIONS

ACPD	Trans-1-amino-1,3-cyclopentanedicarboxylic acid
AMPA	α-Amino-3-hydroxy-5-methyl-4-isoxazole propionic acid
APH	2-Amino-7-phosphonoheptanoate
APV	2-Amino-5-phosphonovalerate
CNQX	6-Cyano-7-nitroquinoxaline-2,3-dione
CPP	3-((-)-2-Carboxypiperzin-4-yl)propyl-1-phosphonate
DNQX	6,7-Dinitroquinoxaline-2,3-dione
KA	Kainic acid
KYN	Kynurenic acid
NMDA	*N*-Methyl-D-aspartate
PCP	Phencyclidine

REFERENCES

1. **Newby, A. C.**, Adenosine and the concept of retaliatory metabolites, *Trends Biochem. Sci.*, 9, 42, 1984.
2. **Burnstock, G.**, Purinergic nerves, *Pharmacol. Rev.*, 24, 509, 1972.
3. **Williams, M. and Cusack, N. J.**, Neuromodulatory roles of purine nucleosides and nucleotides: their receptors and ligands, *Neurotransmissions,* 6(1), 1, 1990.
4. **Dunwiddie, T. V.**, The physiological role of adenosine in the central nervous system, *Int. Rev. Neurobiol.*, 27, 63, 1985.
5. **Marangos, P. J. and Boulenger, J. P.**, Basic and clinical aspects of adenosinergic neuromodulation, *Neurosci. Biobehav. Rev.*, 9, 421, 1985.
6. **Snyder, S. H.**, Adenosine as a neuromodulator, *Annu. Rev. Neurosci.*, 8, 103, 1985.
7. **Fredholm, B. B. and Dunwiddie, T. V.**, How does adenosine inhibit neurotransmitter release?, *Trends Pharmacol. Sci.*, 9, 130, 1988.
8. **Fredholm, B. B., Proctor, W., Van der Ploeg, I., and Dunwiddie, T. V.**, In vivo pertussis toxin treatment attenuates some, but not all, adenosine A$_1$ effects in slices of rat hippocampus, *Eur. J. Pharmacol.*, 172, 249, 1989.
9. **Dragunow, M. and Faull, R. L. M.**, The neuroprotective effects of adenosine, *Trends Pharmacol. Sci.*, 7, 194, 1988.
10. **Marangos, P. J.**, Adenosinergic approaches to stroke therapeutics, *Med. Hypothesis,* 32, 45, 1990.
11. **Fastbom, J. and Fredholm, B. B.**, Inhibition of ^{3}H glutamate release from rat hippocampal slices by L-PIA, *Acta Physiol. Scand.*, 125, 121, 1985.
12. **Burke, S. P. and Nadler, J. V.**, Regulation of glutamate and aspartate release from slices of the hippocampal CA-1 area: effects of adenosine and baclofen, *J. Neurochem.*, 51, 1541, 1988.
13. **Park, T. S. and Gidday, J. M.**, Effect of dipyridamole on cerebral extracellular adenosine level in vivo, *J. Cereb. Blood Flow Metab.*, 10, 424, 1990.
14. **Deckert, J., Morgan, P. F., and Marangos, P. J.**, Adenosine uptake site heterogeneity in the mammalian CNS? Uptake inhibitors as probes and potential neuropharmaceuticals, *Life Sci.*, 42, 1331, 1988.
15. **Fredholm, B. B., Fastbom, J., Duner-Engstrom, M., Hy, P. S., Van der Ploeg, I., Altiok, N., Gerwins, P., Kvanta, A., and Nordstedt, C.**, Mechanisms of adenosine action, in *Purines in Cellular Signaling, Targets for New Drugs,* Jacobsen, K. A., Daly, J. W., and Manganiello, V., Eds., Springer-Verlag, New York, 1990, 184.
16. **Engler, R. L.**, Free radical and granulocyte-mediated injury during myocardial ischemia and reperfusion, *Am. J. Cardiol.*, 63, 19E, 1989.
17. **Rothman, S. M. and Olney, J. W.**, Excitotoxicity and the NMDA receptor, *Trends Neurosci.*, 10, 299, 1987.
18. **Cotman, C. W. and Iversen, L. L.**, Excitatory amino acids in the brain, focus on NMDA receptors, *Trends Neurosci.*, 10, 263, 1987.
19. **Sheardown, M. J., Nielsen, E. D., Hansen, A. J., Jacobsen, P., and Honore, T.**, 2,3-Dihydroxy-6-nitro-7-sulfamoyl-benzo(F)quinoxaline: a neurotoxicant for cerebral ischemia, *Science,* 247, 571, 1990.
20. **Simon, R. P., Swan, J. H., Griffiths, T., and Meldrum, B. S.**, Blockade of NMDA receptors may protect against ischemic damage in the brain, *Science,* 226, 850, 1984.
21. **Crain, B. J., Westerkam, W. D., Harrison, A. H., and Nadler, J. V.**, Selective neuronal death after transient forebrain ischemia in the mongolian gerbil: a silver impregnation study, *Neuroscience,* 27, 387, 1988.
22. **Swan, J. H., Evans, M. C., and Meldrun, B. S.**, Long term development of selective neuronal loss and the mechanism of protection by 2-amino-7-phosphonoheptanoate in the rat model of incomplete forebrain ischemia, *J. Cereb. Blood Flow Metab.*, 8, 64, 1988.
23. **Phillis, J. W. and O'Regan, M. H.**, Deoxycoformycin antagonizes ischemia-induced neuronal degeneration, *Brain Res. Bull.*, 22, 537, 1989.
24. **Von Lubitz, D. K. J. E., Dambrosia, J. M., and Redmond, D. J.**, Protective effect of cyclohexyladenosine in treatment of cerebral ischemia in gerbils, *Neuroscience,* 30, 451, 1989.
25. **Daval, J. L., Von Lubitz, D. K. J. E., Deckert, J., Redmond, D. J., and Marangos, P. J.**, Protective effect of cyclohexyladenosine on adenosine A$_1$ receptor, guanine nucleotide and forskolin binding sites following transient brain ischemia: a quantitative autoradiographic study, *Brain Res.*, 491, 212, 1989.
26. **Von Lubitz, D. K. J. E., Dambrosia, J. M., Kempski, O., and Redmond, D. J.**, Post-ischemically applied cyclohexyladenosine protects against neuronal deaths in the CA-1 region of the hippocampus in gerbil, *Stroke,* 19, 1133, 1988.
27. **Evans, M. C., Swan, J. H., and Meldrum, B. S.**, An adenosine analogue, 2-chloroadenosine, protects against long term development of ischaemic cell loss in the rat hippocampus, *Neurosci. Lett.*, 83, 287, 1987.
28. **Connick, J. H. and Stone, T. W.**, Quinolinic acid neurotoxicity: protection by intracerebral phenyliospropyladenosine (PIA) and potentiation by hypotension, *Neurosci. Lett.*, 101, 191, 1989.

29. **Rudolphi, K. A., Keil, M., and Hinze, H. J.,** Effect of theophylline on ischemically-induced hippocampal damage in mongolian gerbils: a behavioral and histopathological study, *J. Cereb. Blood Flow Metab.,* 7, 74, 1987.

30. **Pinard, E., Riche, D., Puiroud, S., and Seylaz, J.,** Theophylline reduces cerebral hyperaemia and enhances brain damage induced by seizures, *Brain Res.,* 511, 303, 1990.

31. **Von Lubitz, D. K. J. E. and Marangos, P. J.,** Cerebral ischemia in gerbils: post-ischemic administration of cyclohexyladenosine and 8-sulphophenyl-theophylline, *J. Mol. Neurosci.,* 2, 53, 1990.

32. **Marangos, P. J., Von Lubitz, D. K. J. E., Daval, J. L., and Deckert, J.,** Adenosine: its relevance to the treatment of brain ischemia and trauma, in *Current and Future Trends in Anticonvulsant, Stroke and Anxiety Therapy,* Meldrum, B. S. and Williams, M., Eds., Alan R. Liss, New York, 1990, 331.

33. **Rudolphi, K. A., Keil, M., Fastbom, J., and Fredholm, B. B.,** Ischemic damage in gerbil hippocampus is reduced following upregulation of adenosine A_1 receptors by caffeine treatment, *Neurosci. Lett.,* 103, 275, 1989.

34. **Goldberg, M. P., Monyer, H., Weiss, J. H., and Choi, D. W.,** Adenosine reduces cortical neuronal injury induced by oxygen or glucose deprivation in vitro, *Neurosci. Lett.,* 89, 323, 1988.

35. **Busto, R., Dietrich, D., Globus, M., and Ginsberg, M. D.,** The importance of brain temperature in cerebral ischemic injury, *Stroke,* 20, 1113, 1989.

36. **Ikonomidou, C., Mosinger, J. L., and Olney, J. W.,** Hypothermia enhances protective effects of MK-801 against hypoxic/ischemic brain damage in infant rats, *Brain Res.,* 487, 184, 1989.

37. **Radulovacki, M., Miletich, R. S., and Green, R. O.,** N^6-L-Phenyl-isopropyladenosine (L-PIA) increases slow wave sleep (S_2) and decreases wakefulness in rats, *Brain Res.,* 246, 178, 1982.

38. **Wauquier, A., Van Belle, H., Van den Broeck, W. A. E., and Janssen, P. A. J.,** Sleep improvement in dogs after oral administration of mioflazine, a nucleoside transport inhibitor, *Psychopharmacology,* 91, 434, 1987.

39. **Sonsella, P. K., Nicklas, W. J., and Heikkila, R. E.,** Role for excitatory amino acids in methamphetamine-induced nigrostriatal dopaminergic toxicity, *Science,* 243, 398, 1989.

40. **Maragos, W. F., Greenamyre, J. T., Penney, J. B., and Young, A. B.,** Glutamate dysfunction in Alzheimer's disease: an hypothesis, *Trends Neurosci.,* 10, 65, 1987.

41. **Loftus, T. M. and Marangos, P. J.,** Protective effects of an adenosine receptor agonist in animal models of Parkinsons disease, in *Purines in Cellular Signaling, Targets for New Drugs,* Jacobsen, K. A., Daly, J. W., and Manganiello, V., Eds., Springer-Verlag, New York, 1990, 410.

Chapter 34

MANIPULATION OF PURINERGIC TONE AS A MECHANISM FOR CONTROLLING ISCHEMIC BRAIN DAMAGE

Karl Albert Rudolphi

TABLE OF CONTENTS

I. INTRODUCTION

Ischemic brain damage, in particular stroke, is one of the biggest unsolved medical and socioeconomic problems in the industrialized countries. Stroke is the second or third major cause of death and a large group of the surviving stroke victims remains permanently disabled.[1,2]

Despite intensive efforts in basic science and pharmaceutical research, until now, there is no convincing successful medical treatment for prevention or posttreatment of ischemic brain damage.

II. THE PROTECTIVE ROLE OF ADENOSINE AGAINST ISCHEMIC BRAIN DAMAGE — A HYPOTHESIS

Amongst the different pharmacological strategies which are currently being investigated, such as Ca^{++} entry blockers or glutamate receptor antagonist,[5] the potential neuroprotective effects of the endogenous neuromodulator adenosine have recently gained increasing interest. During the past 5 years, strong evidence has accumulated for a beneficial role of endogenous adenosine in brain ischemia (for review, see Dragunow and Faull[3] and Phillis[4]). This hypothesis is based on the concept of adenosine acting as a retaliatory metabolite or homeostatic neuromodulator,[5] which is formed and released when there is a decreased energy supply in relation to energy demand. This substance then modulates the biosystem to restore this imbalance by decreasing energy consumption or demand and simultaneously increasing its availability.

The pathophysiology of brain ischemia is characterized by a complex cascade of hemodynamic, electrophysiological, and biochemical processes which include many interwoven vicious circles (for review, see Siesjö[6] and Raichle[7]). The decrease of cerebral blood flow below a critical threshold results in energy failure, tissue acidosis, disturbed ion homeostasis characterized by enhanced cellular K^+-efflux and a Na^+ and Ca^{++}-influx, membrane depolarization, and cytotoxic edema. A massive release of the excitatory amino acid neurotransmitters, glutamate and aspartate, and the activation of glutamate receptors (including NMDA receptors) has recently been ascribed a major pathophysiological significance.[8] This triggers further membrane depolarization and additional accumulation of free cytosolic Ca^{++} by cellular influx, release from intracellular compartments, and disturbed extrusion mechanisms. The Ca^{++} accumulation then seems to play a key role in propagation of the process towards irreversible neuronal damage, in that it leads to an activation of a series of neurotoxic processes, such as lipid peroxidation, free-radical generation, activation of proteolytic enzymes, and pathological gene activation.[9] Brain ischemia often also includes the formation of vasogenic tissue edema,[10] which in stroke patients is one of the most serious complications and often accelerates the fatal outcome.

It must be stressed that the relative importance of these factors depends highly on the type and the duration of the ischemic insult — whether it is regional permanent ischemia as in the situation of stroke, global transient as in cardiac arrest, or focal and temporary as in transient ischemic attacks.[11]

Figure 1 shows a schematic picture of the presumed neuroprotective mechanisms of adenosine.

Both adenosine A_1 and A_2 receptors are present in the brain.[12] Autoradiographic studies with the metabolically stable adenosine analog cyclohexyladenosine (CHA) or the highly A_1-selective receptor antagonist dipropylcyclopentylxanthine (DPCPX) repeatedly demonstrated A_1 receptors in several species, including man.[13-15] Interestingly, a very high density of A_1 receptors is found in the dendritic regions of the hippocampus[13] — a brain structure particularly vulnerable to ischemia[16] and enriched with glutamate receptors of the NMDA

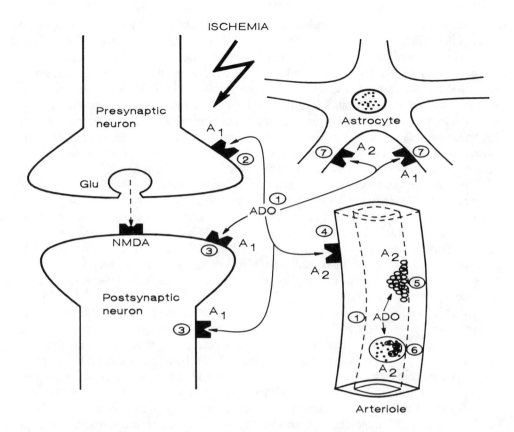

FIGURE 1. The protective role of adenosine during brain ischemia. Ischemia leads to a rapid cell membrane depolarization and a massive release of excitatory amino acids (EAA), including glutamate, which causes excessive neuronal firing, sustained membrane depolarization, intracellular Ca^{++}-accumulation, and neurotoxic effects via NMDA- and other EAA-receptors. Ischemia and increased nerve cell activity lead to an increase of extracellular and intravascular adenosine (1).

Adenosine acts to restore the metabolic energy balance and limit the neuronal damage . . . **reducing O_2 and glucose demand via adenosine A_1-receptors:** by presynaptic reduction of EAA release through inhibition of Ca^{++}-influx and opening of K^+-ion channels (2); and by postsynaptic inhibition of Ca^{++}-influx, inhibition of sustained membrane depolarization by opening of K^+-ion channels and counteracting excessive neuronal firing (3) . . . **enhancing O_2 and glucose supply via adenosine A_2-receptors:** by vasodilation, increasing cerebral blood flow (4); by inhibition of platelet aggregation (5); and by inhibiting the activation, endothelial adhesion, and capillary plugging of neutrophils (6). The role of A_1- and A_2-receptor activation on astrocytes during ischemia is still unknown (7).

type.[17] A_1 receptors are located pre- and postsynaptically, but also extrasynaptically along the neuronal dendrites.[18,19] Adenosine A_2 receptors are mainly located on smooth-muscle cells and probably also on the endothelium of brain blood vessels mediating the vascular effects of adenosine.[20,21] A very selective distribution of high-affinity A_2 receptors has recently been demonstrated in rat brain in the striatum and the olfactory tubercle.[22] The functional role of these receptors remains unclear. This is also true for the A_1 and A_2 receptors which have been found in rat astrocytes.[23]

After short-term global forebrain ischemia, autoradiographic studies in Mongolian gerbils showed a significant decrease in the number of hippocampal A_1 receptors, guanine nucleotide, and forskolin binding sites in the CA1 sector following the same time course as the development of the selective delayed pyramidal cell necrosis.[24,25] However, there seems also to be a very early downregulation of A_1 receptors occurring during the first hours before neuronal necrosis can be observed light microscopically.[26]

During brain ischemia, adenosine is preferentially formed by a rapid breakdown of intracellular ATP in the brain cells and released into the extracellular space.[27] Using microdialysis or cerebrospinal fluid-sampling methods, several authors have reported that immediately after the start of an ischemic or hypoxic episode the extracellular adenosine concentration rises up to 200 times from resting physiological levels of 0.1 to 1 μM.[28-32] Part of this increase may be attributed to ATP breakdown following concomitant glutamatergic neuroexcitation, because the amount of adenosine released in rat brain during complete forebrain ischemia could be significantly reduced by pretreating with the competitive glutamate receptor antagonist 2-amino-5-phosphonovalerate.[33] There is also evidence for an increase of intravascular adenosine concentration in ischemic tissues.[34]

What may be the neuroprotective mechanisms of this strongly enhanced extracellular and intravascular adenosine?

There is strong evidence that adenosine presynaptically inhibits the release of many neurotransmitters, including the potentially excitotoxic amino acids glutamate and aspartate.[35,36] This effect of adenosine is mediated by presynaptic A_1 receptors linked via G proteins to both Ca^{++} and K^+ ion channels.[37-40] Adenosine in concentrations reached during brain ischemia effectively blocks voltage-dependent Ca^{++} channels and counteracts membrane depolarization by opening K^+ ion channels.[41]

Postsynaptically adenosine, in addition to its Ca^{++}-channel blockade, stabilizes the neuronal membrane potential by opening K^+ channels.[42] Adenosine may thus pre- and postsynaptically diminish the ischemia-triggered early glutamatergic neuroexcitation leading to a decrease in energy demand of the compromized brain tissue. Hence, more importantly, adenosine may counteract the deleterious neurotoxic accumulation of free cytosolic calcium by its direct inhibitor effect on N-type Ca^{++} ion channels and by interrupting the sustained vicious circle of membrane depolarization and glutamate receptor activation.

In addition to this important neuromodulatory effect, adenosine acting via A_2 receptors in the cerebrovasculature and on the platelets may increase oxygen and glucose supply by its vasodilatory[43] and antithrombotic effects.[44] Indeed, it has been shown that adenosine seems to be responsible for the postischemic hyperemia.[45] In addition, it may be possible that intravascular adenosine released from endothelial cells or blood cells inhibits the activation of neutrophilic granulocytes via A_2 receptors, thereby reducing the release of free O_2 radicals and other cytotoxic substances and preventing the plugging of capillaries.[46] This seems to be an important pathophysiological factor in myocardial ischemia,[47] but has still to be proven experimentally for the brain.

Until now, nothing is known about the physiological or pathophysiological relevance of A_1 and A_2 receptors on astrocytes, although it has been shown that adenosine via A_2 receptors may induce astrocytic edema.[48] Hösli and co-workers[49] have recently demonstrated that the stimulation of A_1 receptors hyperpolarizes the astrocytic cell membrane, and it is tempting to speculate that this improves the known property of astrocytes to take up excessive extracellular potassium and glutamate.[50,51]

III. PHARMACOLOGICAL MANIPULATION OF THE ADENOSINE SYSTEM DURING BRAIN ISCHEMIA

In order to investigate the role of adenosine in brain ischemia and its therapeutic potential, a series of pharmacological studies have been published describing the effects of adenosine or metabolically stable adenosine analogs, adenosine receptor antagonists, or compounds which inhibit the physiological mechanisms of adenosine inactivation on the degree of experimentally induced ischemic brain damage.

TABLE 1
Effects of Adenosine Agonists on Ischemic Brain Damage

Substance	Dose	Type of ischemia/hypoxia	Species	Effects	Ref.
Adenosine	30—1000 μM	N$_2$-hypoxia/ hypoglycemia	Mouse neurons	Reduced neuronal damage and LDH[a] release	54
Cyclohexyl- adenosine (CHA)	125 pM i.c.v. post	30-min bilateral carotid occlusion	Gerbil	Reduced neuronal damage in hippocampus and caudate nucleus	53
	10 μM in medium	N$_2$-hypoxia/ hypoglycemia	Mouse neurons	Reduced neuronal damage and LDH release	54
	2 mg/kg i.p. post	20-min bilateral carotid occlusion	Gerbil	Reduced hippocampal damage; protection of A$_1$ receptors, guanine, nucleotide and forskolin binding sites	25
R-Phenyliso- propyladenosine (R-PIA)	10 μM/kg i.p. post	30-min four- vessel occlusion	Rat	Reduced hippocampal damage	55
	0.1 and 0.3 mg/kg i.p. pre	Permanent MCAO[b]	Rat	Reduced infarct volume	56
2-Chloroadenosine	15 μM focal pre + post	10-min bilateral carotid occlusion	Rat	Reduced CA1 damage in hippocampus	52

a   LDH = lactate dehydrogenase.
b   MCAO = middle cerebral artery occlusion.

A. ADENOSINE AND ADENOSINE AGONISTS

In most experiments, metabolically stable adenosine analogs have been used because adenosine itself is very rapidly taken up by nucleoside transport mechanisms and degraded by adenosine deaminase. Its plasma half-life after intravenous bolus administration is in the range of 3 to 6 s.[34] Table 1 summarizes the studies conducted with adenosine agonists.

In 1987, Evans et al.[52] reported that focal injections of $1.5 \times 10^{-5} M$ of the metabolically stable adenosine analog 2-chloroadenosine into the hippocampus immediately before and 4 and 10 h after 10 min of incomplete forebrain ischemia induced by bilateral carotid occlusion and hypovolemic hypotension protected hippocampal CA1 neurons. However, delayed focal injections of 2-chloroadenosine 10 or 24 h after ischemia failed to protect.

CHA, a more A$_1$ receptor-selective adenosine derivative, increased survival rate and protected the hippocampus and caudate nucleus from ischemic damage when injected intra-cerebroventricularly 15 min after ischemia in a dose of 125 pM.[53]

Intraperitoneal injection of 2 mg/kg CHA 5 min after 20-min forebrain ischemia in gerbils significantly decreased the necrosis of pyramidal neurons in the CA1 sector of the hip-pocampus. In addition, the autoradiographic determination of [^{3}H]-CHA-, [^{3}H]-Gpp(NH)p-, and [^{3}H]-forskolin binding showed that CHA prevented the ischemia-induced reduction of A$_1$ receptors and of signal transduction elements (G proteins and adenylate cyclase).[25]

Protective effects of CHA have also been observed in an *in vitro* experiment when mouse dissociated cortical cell cultures were subjected to hypoxia or hypoglycemia. CHA at a concentration of 10 μM in the culture medium significantly ameliorated the morphological appearance of the neurons and attenuated the level of lactate dehydrogenase released by damaged or destroyed neurons. In this study the protective effect of CHA was about ten times stronger than that of adenosine.[54] This *in vitro* experiment, lacking the vascular component, indicates that adenosine may exert its protective effects to a large extent via direct actions on the neurons or glial cells.

R-Phenylisopropyladenosine (R-PIA), another relatively A_1-specific adenosine agonist, also reduced ischemic hippocampal damage when administered in a dose of 10^{-5} mol/kg i.p. after 30 min of forebrain ischemia induced by simultaneous bilateral occlusion of the vertebral and carotid arteries in rats (four-vessel occlusion).[55]

A preliminary report has described that 30-min pretreatment with 0.3 mg/kg of R-PIA led to a significant reduction of the infarct volume following permanent middle cerebral artery occlusion in rats.[56] So far, this is the first and only report of the effects of an adenosine analog in an experimental model of focal brain ischemia, which very closely resembles the situation of a stroke in humans. In the future, more studies of this kind are necessary to confirm the potential protective effect of adenosine in the real clinical situation.

In summary, all experiments published to date showed a protective potential, both in global and regional models, when employing adenosine agonists. Unfortunately, when it comes to therapeutic applicability, it is difficult to administer adenosine or its metabolically stable analogs, even the relatively selective A_1 agonists, because all suffer from strong cardiovascular side effects, such as hypotension and cardiodepression.[34]

Finally, another word of caution has to be added when discussing the anti-ischemic effects of these compounds. All of them can have dose-dependent hypothermic effects or may reinforce the hypothermic effects of general anesthetics.[57,58] Given the recent evidence of strong protective effects of even low levels of hypothermia against ischemic brain damage,[59] all future studies should include a careful adjustment of normothermia throughout the experimental period.

B. ADENOSINE ANTAGONISTS

How do adenosine receptor antagonists influence ischemic brain damage? The published data are summarized in Table 2.

The first experiments with conflicting results have been conducted with the classical natural adenosine receptor antagonist theophylline. Originally, theophylline was thought to be a cerebral vasodilator,[60-63] and it has been shown to inhibit acute cerebral vasospasm following subarachnoid hemorrhage in monkeys and cats.[64] Contrarily, cerebral vasoconstriction has been observed during theophylline treatment in humans.[65] Since such vasoconstrictive effects were only observed in healthy brain areas, some authors concluded that the drug may increase the CBF in ischemic cortical areas indirectly by an "inverse steal effect".[65,66] However, until now, clinical benefit of theophylline in stroke patients has not been proven.[67-70]

Concerning studies in experimental animals, there are also somewhat conflicting results with theophylline. Kogure et al.[71] showed that pretreatment with theophylline (100 mg/kg i.p.) in embolic brain ischemia in rats reduced the edema formation and the mortality rate. Seida et al.[72] reported similar protective effects in a model of 1 h transient focal cortical brain ischemia by middle cerebral artery occlusion in rats with reperfusion. They found that pretreatment with theophylline reduced the brain edema and the neuronal damage. Such protective effects could not be observed by McGraw et al.,[73] who found a significantly higher mortality in gerbils with unilateral permanent carotid occlusion when 100 mg/kg theophylline was given intraperitoneally 1 h after the beginning of ischemia. An increased mortality was also found following intraperitoneal administration of 50 mg/kg theophylline 3 min after a period of 50 min bilateral carotid occlusion in gerbils.[74] The observation that weanling mice subjected to anoxia had a higher mortality following the intraperitoneal administration of 100 mg/kg theophylline and that this effect may be due to an increased cerebral metabolic rate further supports the assumption that theophylline might be deleterious in cases of cerebral ischemia or hypoxia.[75]

Rudolphi et al.[76,77] found that 10 min pretreatment with theophylline in a dose of 30 mg/kg p.o., at which the drug acts predominantly as an adenosine receptor antagonist,

TABLE 2
Effects of Adenosine Antagonists on Ischemic Brain Damage

Substance	Dose (mg/kg)	Type of ischemia	Species	Effects	Ref.
Theophylline	100 i.p.	Embolic	Rat	Reduced Edema and mortality	71
	100 i.p. 1 h post	Permanent unilateral carotid occlusion	Gerbil	Increased mortality	73
	50 i.p. 3 min post	50-min bilateral carotid occlusion	Gerbil	Increased mortality	74
	30 p.o. pre	2—5-min bilateral carotid occlusion	Gerbil	Increased CA1 necrosis	76,77
	24 i.p. pre + post	15-min bilateral carotid occlusion	Rat	Increased hippocampal and neocortical damage	80
	32 i.p. pre	2-min bilateral carotid occlusion	Gerbil	Increased CA1 damage	78
	0.3/min i.v. inf. intra + post	1-h MCAO	Cat	Reduced reactive hyperemia, BBB[a] disruption, edema, and neuronal damage	72
	30 p.o. 30 min post	Focal cortical photochemical	Rat	Increased infarct volume	81
Caffeine	30 p.o. pre	3-min bilateral carotid occlusion	Gerbil	Increased CA1 necrosis	82
	0.2% in drinking water 4 weeks pre	5-min bilateral carotid occlusion	Gerbil	Increased number of A_1 receptors, reduced CA1 necrosis	83
DPCPX-selective A_1 antagonist	10 p.o. pre	3-min bilateral carotid occlusion	Gerbil	Increased CA1 necrosis	82

[a] BBB = blood-brain barrier.

significantly aggravated the damage of pyramidal neurons in the CA1 subfield of the gerbil hippocampus following 2 to 3 min of bilateral carotid occlusion. This was confirmed by DeLeo et al.[78] and Dux et al.,[79] who also found a significant increase of ischemic hippocampal damage in gerbils when pretreating with 32 or 20 mg/kg i.p. following 2 or 5 min of forebrain ischemia, respectively. Similar findings were reported by Wieloch et al.[80] using theophylline in a dose of 24 mg/kg i.p. immediately before and 9 and 20 h following 15 min of bilateral carotid occlusion plus hypotension in rats.

Deleterious effects of theophylline were also recently found in a model of photochemically induced focal cortical brain ischemia in rats.[81] Theophylline (30 mg/kg p.o.) administered 30 min after the induction of ischemia significantly increased the brain infarct volume.

Acute pretreatment with caffeine (30 mg/kg p.o.), a nonspecific adenosine receptor antagonist like theophylline, also increased ischemic hippocampal damage in gerbils, and the highly selective A_1 receptor antagonist DPCPX was about three times more potent (Rudolphi et al.[82]).

In summary, most experiments with acute treatment with adenosine receptor antagonists increased ischemic brain damage. It is interesting that in both studies where theophylline showed protection, experimental models of brain ischemia were used in which edema formation plays an important role. This edema formation is thought to be partially caused by the strong reactive postischemic hyperemia which is attributed to the vasodilatory effect of

TABLE 3
Effects of Adenosine Deaminase Inhibitor and Adenosine Uptake Inhibitor on
Ischemic Brain Damage

Substance	Dose (mg/kg)	Type of ischemia	Species	Effects	Ref.
Adenosine deaminase inhibitor					
Deoxycoformycin	0.5 i.p. pre	Transient bilateral carotid occlusion	Gerbil	Reduced hippocampal damage, reduced postischemic hypermotility	95
	0.5 i.p. pre + post	20-min bilateral carotid occlusion + hypotension	Rat	Increased EC[a] adenosine, no effect on hippocampal damage	94
Adenosine uptake inhibitors					
Nitrobenzylthioinosine (NBTI)	10 i.p. pre	10-min bilateral carotid occlusion	Gerbil	No effect	78
Hydroxy-nitro-benzylthio-guanosine (NBTG)	10 i.p. pre	10-min bilateral carotid occlusion	Gerbil	No effect	78

[a]   EC = Extracellular.

the increased adenosine concentration. Adenosine antagonists like theophylline have been shown to reduce the postischemic hyperemia and may thereby inhibit edema formation and the successive neuronal damage.

As already mentioned above, Lee et al.[26] proposed that an early downregulation of adenosine receptors and thereby a weakening of the protective action of endogenous adenosine may be responsible for the selective delayed ischemic neuronal death in the hippocampus. Rudolphi et al.[83] upregulated the number of adenosine receptors in the brain by 4 weeks of chronic caffeine treatment (0.2% in drinking water *ad libitum*) and studied the effects of short-term ischemia under this situation. The chronic caffeine treatment caused a significant increase in the binding of the adenosine A_1 receptor ligand [^{3}H]CHA to several brain regions, including the hippocampal CA1 area, in Mongolian gerbils. Animals subjected to such treatment exhibited significantly less neuronal damage in the CA1 region following 5 min bilateral carotid occlusion than did control animals.

C. INHIBITORS OF ADENOSINE INACTIVATION

As already mentioned above, it is difficult to use adenosine agonists to treat brain ischemia because of their strong side effects.

Instead, it could be possible to augment the purported protective effects of adenosine by using agents which decrease adenosine inactivation. Such substances would increase the chance of endogenously formed adenosine, either by inhibiting adenosine degradation or blocking its cellular reuptake, to interact with its A_1 and/or A_2 receptors at the exterior surface of the cell membrane. However, only a few experimental studies have been published, and again the results are somewhat contradictory. They are summarized in Table 3.

1. Nucleoside Transport Inhibitors

The classical, potent adenosine uptake inhibitor dipyridamole may not be useful in the case of brain ischemia because of its poor permeability through the blood-brain barrier,[34] although it has been shown to produce a small but significant increase in hypoxia-induced adenosine release into brain interstitial fluid.[84] Two other potent adenosine transport inhibitors, nitrobenzylthioinosine (NBTI) or hydroxy-nitrobenzylthioguanosine (NBTG), failed to protect against hippocampal damage when administered 15 min prior to a 10-min bilateral carotid occlusion in gerbils.[78] However, there is a wide variety of chemically diverse compounds, although of lower potency than dipyridamole and NBTI or NBTG, which have also been reported to inhibit adenosine uptake.[85,86] Amongst them a novel xanthine derivative, propentofylline, has recently been shown by DeLeo et al.[78,87,88] and Dux et al.[79] to prevent ischemia-induced calcium loading and subsequent necrosis of hippocampal neurons on Mongolian gerbils. The compound was administered in a dose of 10 mg/kg i.p. 15 min before or 1 h after 5- to 12-min bilateral carotid occlusion.

In vitro, propentofylline is only a moderately strong inhibitor of adenosine uptake when compared with dipyridamole.[86,89,90] However, Andiné et al.[91] examined with the microdialysis technique the influence of this drug on the extracellular concentration of adenosine, inosine, hypoxanthine, and xanthine, as well as glutamate and aspartate in the hippocampus before, during, and after transient forebrain ischemia in rats induced by four-vessel occlusion. Pretreatment with propentofylline in a dose of 10 mg/kg i.p., which has previously been shown to reduce ischemic hippocampal damage, augmented the extracellular increase of adenosine in the CA1 sector of the hippocampus during ischemia, whereas the increase of the other purines was attenuated. Simultaneously, the increases of extracellular glutamate and to some extent aspartate were diminished. However, the fact that the protective effect of propentofylline was only slightly reduced in the presence of the adenosine receptor antagonist theophylline suggests that it may partially be due to other, still-unknown mechanisms.[78]

2. Adenosine Deaminase Inhibitors

Another way to elevate the levels of endogenous adenosine is by inhibiting the enzyme adenosine deaminase (ADA), which catalyzes the degradation of adenosine.[92,93] Busto et al.[94] reported that the ADA inhibitor, deoxycoformycin, given in a dose of 0.5 mg/kg i.p. 30 min before bilateral carotid occlusion plus hypotension, increased the extracellular adenosine concentration in the ischemic brain of rats, but failed to protect against hippocampal neuronal necrosis. On the other hand, Phillis and O'Regan[95] showed a significant neuroprotective effect of deoxycoformycin by pretreatment with 0.5 mg/kg i.p. in gerbils. The reason for these contradictory results is not clear.

It has also been discussed whether or not the protective effects of the xanthine oxidase inhibitors allopurinol and oxypurinol against ischemia reperfusion brain damage may not only be due to the reduction of free-radical formation, but also because the drug increased adenosine release during hypoxia/ischemia.[96,97]

Finally, two alternative approaches to reinforce the possible protective effects of endogenous adenosine have very recently been discussed. The use of the adenosine precursor, 5-amino-imidazole-4-carboxamide riboside (AICAR), showed protective effects in experimental myocardial ischemia,[98] and so-called "allosteric enhancers", 2-amino-3-benzoyl-thiophenes, increased the binding of CHA specifically to A_1 adenosine receptors and potentiated the inhibitory effect of cyclopentyladenosine on adenylate cyclase in a FRTC-5 thyroid cell line.[99] However, functional studies testing these drugs in models of experimental brain ischemia are not yet available.

IV. CONCLUSION

Despite some conflicting results, the majority of experimental studies, which have been published until now, strengthens the hypothesis that the endogenous neuromodulator adenosine has protective effects in brain ischemia. Nevertheless, most experiments with adenosine agonists, antagonists, or inhibitors of adenosine inactivation have been conducted in models of global forebrain ischemia with reperfusion. It is absolutely necessary, however, that the results from these studies be confirmed by using clinically more relevant models of permanent focal cortical brain ischemia.

In addition, a number of important questions have still to be answered:

1. In temporary as well as in permanent brain ischemia, the increase of extracellular adenosine is transient. What is the therapeutic time window for the use of drugs which reinforce the neuroprotective effect of endogenous adenosine?
2. What are the dynamic changes of the extracellular adenosine concentrations in the clinically most relevant form of brain ischemia in stroke, especially in the penumbra? The protective net effect of endogenous adenosine in this still viable area surrounding the irreversibly damaged tissue needs to be determined.
3. What are the effects of adenosine on astrocytes, and does adenosine promote astrocytic swelling and ischemic brain edema?
4. To which extent do the effects of adenosine on the cerebrovasculature, platelets, and white blood cells contribute to its protective potential?

The pharmacological reinforcement of the role of adenosine as an endogenous neuroprotective modulator could lead to progress in the drug therapy of ischemic brain disorders. Especially blood-brain barrier-permeable drugs, which inhibit the inactivation or enhance the functions of adenosine, may be of value because their actions would be predominant in endangered areas where there is an elevated adenosine concentration. This would then minimize the interference with physiologically relevant processes and limit the risks for untoward effects.

REFERENCES

1. World Health Statistics Annual 1989, World Health Organization, Geneva, 1989.
2. **Broderick, J. P., Phillis, St.J., Whisnant, J. P., O'Fallon, W. M., and Bergstralh, E. J.,** Incidence rates of stroke in the eighties: the end of the decline in stroke?, *Stroke,* 20, 577, 1989.
3. **Dragunow, M. and Faull, R. L. M.,** Neuroprotective effects of adenosine, *Trends Pharmacol. Sci.,* 9, 193, 1988.
4. **Phillis, J. W.,** Adenosine in the control of the cerebral circulation, *Cerebrovasc. Brain Metab. Rev.,* 1, 26, 1989.
5. **Newby, A. C.,** Adenosine and the concept of "retaliatory metabolites", *Trends Biochem. Sci.,* 2, 42, 1984.
6. **Siesjö, B. K.,** Cell damage in the brain: a speculative synthesis, *J. Cereb. Blood Flow Metab.,* 1, 155, 1981.
7. **Raichle, M. E.,** The pathophysiology of brain ischemia, *Ann. Neurol.,* 13, 2, 1983.
8. **Simon, R. P., Swan, J. H., Griffiths, T., and Meldrum, B. S.,** Blockade of N-methyl-D-aspartate receptors may protect against ischemic damage in the brain, *Science,* 226, 850, 1984.
9. **Siesjö, B. K. and Bengtsson, F.,** Calcium fluxes, calcium antagonists, and calcium related pathology in brain ischemia, hypoglycemia, and spreading depression: a unifying hypothesis, *J. Cereb. Blood Flow Metab.,* 9, 127, 1989.
10. **Hossmann, K. A.,** The pathophysiology of ischemic brain swelling, in *Brain Edema,* Inaba, Y., Klatzo, I., and Spatz, M., Eds., Springer-Verlag, Berlin, 1985, 367.
11. **Hossmann, K. A.,** Treatment of experimental cerebral ischemia, *J. Cereb. Blood Flow Metab.,* 2, 275, 1982.
12. **Daly, J. W.,** Adenosine receptors, *Adv. Cyclic Nucleotide Protein Phosphorylation Res.,* 19, 29, 1985.
13. **Fastbom, J., Pazos, A., Probst, A., and Palacios, J. M.,** Adenosine A_1-receptors in human brain: characterization and autoradiographic visualization, *Neurosci. Lett.,* 65, 127, 1986.
14. **Fastbom, J., Pazos, A., and Palacios, J. M.,** The distribution of adenosine A1 receptors and 5'-nucleotidase in the brain of some commonly used experimental animals, *Neuroscience,* 22, 813, 1987.

15. **Fastbom, J. and Fredholm, B. B.,** Effects of long-term theophylline treatment on adenosine A1-receptors in rat brain: autoradiographic evidence for increased receptor number and altered coupling to G-proteins, *Brain Res.,* 507, 195, 1990.

16. **Kirino, T.,** Delayed neuronal death in the gerbil hippocampus following ischemia, *Brain Res.,* 239, 257, 1982.

17. **Bowery, N. G., Wong, E. H. F., and Hudson, A. L.,** Quantitative autoradiography of [³H]-MK-801 binding sites in mammalian brain, *Br. J. Pharmacol.,* 93, 944, 1988.

18. **Schubert, P., Lee, K. S., Tetzlaff, W., and Kreutzberg, G. W.,** Post-synaptic modulation of neuronal firing pattern by adenosine, in *Molecular Basis of Nerve Activity,* Changeux, P. and Hucho, E., Eds., Walter de Gruyter, Berlin, 1985, 283.

19. **Tetzlaff, W., Schubert, P., and Kreutzberg, G. W.,** Synaptic and extra-synaptic localization of adenosine binding sites in the rat hippocampus, *Neuroscience,* 21, 869, 1987.

20. **Edvinsson, L. and Fredholm, B. B.,** Characterization of adenosine receptors in isolated cerebral arteries of cat, *Br. J. Pharmacol.,* 80, 631, 1989.

21. **McBean, D. E., Harper, A. M., and Rudolphi, K. A.,** Effects of adenosine and its analogues on porcine basilar arteries: are only A₂ receptors involved?, *J. Cereb. Blood Flow Metab.,* 8, 40, 1988.

22. **Jarvis, M. F., Jackson, R. H., and Williams, M.,** Autoradiographic characterization of high-affinity adenosine A₂ receptors in the rat brain, *Brain Res.,* 484, 111, 1989.

23. **Hösli, E. and Hösli, L.,** Autoradiographic studies on the uptake of adenosine and on binding of adenosine analogues in neurons and astrocytes of cultured rat cerebellum and spinal cord, *Neuroscience,* 24, 621, 1988.

24. **Onodera, H., Sato, G., and Kogure, K.,** Quantitative autoradiographic analysis of muscarinic cholinergic and adenosine A₁ binding sites after transient forebrain ischemia in the gerbil, *Brain Res.,* 415, 309, 1987.

25. **Daval, J. L., Von Lubitz, D. K. J. E., Deckert, J., Redmond, D. J., and Marangos, P. J.,** Protective effect of cyclohexyladenosine on adenosine A₁ receptors, guanine nucleotide and forskolin binding sites following transient brain ischemia: a quantitative autoradiographic study, *Brain Res.,* 491, 212, 1989.

26. **Lee, K. S., Tetzlaff, W., and Kreutzberg, G. W.,** Rapid down regulation of hippocampal adenosine receptors following brief anoxia, *Brain Res.,* 380, 155, 1986.

27. **Berne, R. M., Rubio, R., and Curnish, R. R.,** Release of adenosine from ischemic brain. Effect on cerebral vascular resistance and incorporation into cerebral adenine nucleotides, *Circ. Res.,* 35, 262, 1974.

28. **Hagberg, H., Andersson, P., Lacarewicz, J., Jacobson, I., Butcher, S., and Sandberg, M.,** Extracellular adenosine, inosine, hypoxanthine, and xanthine in relation to tissue nucleotides and purines in rat striatum during transient ischemia, *J. Neurochem.,* 49, 227, 1987.

29. **Zetterström, T., Vernet, L., Ungerstedt, U., Tossman, U., Jonzon, B., and Fredholm, B. B.,** Purine levels in the intact brain, studies with an implanted perfused hollow fibre, *Neurosci. Lett.,* 29, 111, 1982.

30. **Van Wylen, D. G. L., Park, T. S., Rubio, R., and Berne, R. M.,** Increases in cerebral interstitial fluid adenosine concentration during hypoxia, local potassium infusion, and ischemia, *J. Cereb. Blood Flow Metab.,* 6, 522, 1986.

31. **Phillis, J. W., Walter, G. A., O'Regan, M. H., and Stair, R. E.,** Increases in cerebral cortical perfusate adenosine and inosine concentrations during hypoxia and ischemia, *J. Cereb. Blood Flow Metab.,* 7, 679, 1987.

32. **Hillered, L., Hallström, A., Segersvärd, S., Persson, L., and Ungerstedt, U.,** Dynamics of extracellular metabolites in the striatum after middle cerebral artery occlusion in the rat monitored by intracerebral microdialysis, *J. Cereb. Blood Flow Metab.,* 9, 607, 1989.

33. **Hagberg, H., Andersson, P., Ostwald, Ch., Butcher, S., Sandberg, M., Lehmann, A., and Hamberger, A.,** Ischemia-evoked release of neuroactive compounds and acute effect of N-methyl-D-aspartate receptor blockade, in *Pharmacology of Cerebral Ischemia,* Krieglstein, J., Ed., Elsevier, Amsterdam, 1986, 268.

34. **Sollevi, A.,** Cardiovascular effects of adenosine in man: possible clinical implications, *Progr. Neurobiol.,* 27, 319, 1986.

35. **Corradetti, R., Lo Conte, G., Moroni, F., Passani, M. B., and Pepeu, G.,** Adenosine decreases aspartate and glutamate release from rat hippocampal slices, *Eur. J. Pharmacol.,* 104, 19, 1984.

36. **Dolphin, A. C. and Archer, E. R.,** An adenosine agonist inhibits and a cyclic AMP analogue enhances the release of glutamate but not GABA from slices of rat dentate gyrus, *Neurosci. Lett.,* 43, 49, 1983.

37. **Phillis, J. W. and Wu, P. H.,** The role of adenosine and its nucleotides in central synaptic transmission, *Progr. Neurobiol.,* 16, 187, 1981.

38. **Fredholm, B. B. and Hedqvist, P.,** Modulation of neurotransmission by purine nucleotides and nucleosides, *Biochem. Pharmacol.,* 29, 1635, 1980.

39. **Dunwiddie, T.,** The physiological role of adenosine in the central nervous system, *Int. Rev. Neurobiol.,* 27, 63, 1985.

40. **Fredholm, B. B. and Dunwiddie, T.,** How does adenosine inhibit transmitter release?, *Trends Pharmacol. Sci.,* 9, 130, 1988.

41. **Schubert, P.**, Synaptic and non-synaptic modulation by adenosine: a differential action on K- and Ca-fluxes, in *Adenosine: Receptors and Modulation of Cell Function*, Stefanovich, V., Rudolphi, K., and Schubert, P., Eds., IRL Press, Oxford, 1985, 117.

42. **Schubert, P., Heinemann, U., and Kolb, R.**, Differential effects of adenosine on pre- and postsynaptic calcium fluxes, *Brain Res.*, 376, 382, 1986.

43. **Collis, M. G.**, The vasodilator role of adenosine, *Pharmacol. Ther.*, 41, 143, 1989.

44. **Born, G. V. R. and Cross, M. J.**, The aggregation of blood platelets, *J. Physiol.*, 168, 178, 1963.

45. **Morii, S., Ngai, A. C., Ko, K. R., and Winn, H. R.**, Role of adenosine in regulation of cerebral blood flow: effects of theophylline during normoxia and hypoxia, *Am. J. Physiol.*, 253, H165, 1987.

46. **Cronstein, B. N., Levin, R. I., Belanoff, J., Weissmann, G., and Hirschhorn, R.**, Adenosine: an endogenous inhibitor of neutrophil-mediated injury to endothelial cells, *J. Clin. Invest.*, 78, 769, 1986.

47. **Engler, R.**, Consequences of activation and adenosine-mediated inhibition of granulocytes during myocardial ischemia, *Federation Proc.*, 46, 2407, 1987.

48. **Bourke, R. S., Kimbelberg, H. K., and Daze, M. A.**, Effects of inhibitors and adenosine on (HCO_3^-/CO_2) stimulated swelling and Cl^- in brain slices and cultured astrocytes, *Brain Res.*, 154, 196, 1978.

49. **Hösli, L., Hösli, E., and Della Briotta, G.**, Electrophysiological evidence for adenosine receptors on astrocytes of cultured rat central nervous system, *Neurosci. Lett.*, 79, 108, 1987.

50. **Drejer, J., Benveniste, H., Diemer, N. H., and Schousboe, A.**, Cellular origin of ischemia-induced glutamate release from brain tissue in vivo and in vitro, *J. Neurochem.*, 45, 145, 1985.

51. **Kauppinen, R. A., Enkvist, K., Holopainen, I., and Akerman, K. E. O.**, Glucose deprivation depolarizes plasma membrane of cultured astrocytes and collapses transmembrane potassium and glutamate gradients, *Neuroscience*, 26, 280, 1988.

52. **Evans, M. C., Swan, J. H., and Meldrum, B. S.**, An adenosine analogue, 2-chloroadenosine, protects against long term development of ischaemic cell loss in the rat hippocampus, *Neurosci. Lett.*, 83, 287, 1987.

53. **von Lubitz, D. K. J. E., Dambrosia, J. M., Kempski, O., and Redmond, D. J.**, Cyclohexyl adenosine protects against neuronal death following ischemia in the CA1 region of gerbil hippocampus, *Stroke*, 19, 1133, 1988.

54. **Goldberg, M. P., Monyer, H., Weiss, J. H., and Choi, D. W.**, Adenosine reduces cortical neuronal injury induced by oxygen or glucose deprivation in vitro, *Neurosi. Lett.*, 89, 323, 1988.

55. **Block, G. A. and Pulsinelli, W. A.**, The adenosine agonist R-phenyl-isopropyladenosine attenuates ischemic neuronal damage, *J. Cereb. Blood Flow Metab.*, 7 (Suppl. 1), S258, 1987.

56. **Bielenberg, G. W.**, R-PIA protects cortical tissue in permanent MCA occlusion in the rat, *J. Cereb. Blood Flow Metab.*, 9 (Suppl. 1), S645, 1989.

57. **Jonzon, B., Bergquist, A., Li, Y. O., and Fredholm, B. B.**, Effects of adenosine and two stable adenosine analogues on blood pressure, heart rate and colonic temperature in the rat, *Acta Physiol. Scand.*, 126, 491, 1986.

58. **Levine, A. S., Grace, M., Krahn, D. D., and Billington, C. J.**, The adenosine agonist N^6-R-phenyli-sopropyladenosine (R-PIA) stimulates feeding in rats, *Brain Res.*, 477, 280, 1989.

59. **Busto, R., Dietrich, W. D., Globus, M. Y. T., and Ginsberg, M. D.**, The importance of brain temperature in cerebral ischemic injury, *Stroke*, 20, 1113, 1989.

60. **Wolff, H. G.**, The cerebral circulation, *Physiol. Rev.*, 16, 545, 1936.

61. **Noell, W.**, Über die Wirkung des Theophyllins und der verschiedenen Lösungsvermittler auf die Gehirndurchblutung, *Z. Gesamte Exp. Med.*, 110, 589, 1942.

62. **Dumke, P. R. and Schmidt, C. F.**, Quantitative measurement of cerebral blood flow in the Macacque monkey, *Am. J. Physiol.*, 138, 421, 1943.

63. **Gottstein, U., Bernsmeier, A., Gebeling, H., and Steiner, K.**, Zur Behandlung zerebraler Durchblutungsstörungen mit Euphyllin (Theophyllin-Äthylendiamin), *Med. Klin.*, 56, 1598, 1961.

64. **Flamm, E. S., Kim, J., Lin, J., and Ransohoff, J.**, Phosphodiesterase inhibitors and cerebral vasospasm, *Arch. Neurol.*, 32, 569, 1975.

65. **Gottstein, U. and Paulson, D. B.**, The effect of intracarotid aminophylline infusion on the cerebral circulation, *Stroke*, 3, 560, 1972.

66. **Skinhøj, E. and Paulson, O. B.**, The mechanism of action of aminophylline upon cerebrovascular disorders, *Acta Neurol. Scand.*, 46, 129, 1970.

67. **Hadorn, W.**, Berichterstattung über die Euphyllin-Behandlung des Hirnschlags aufgrund der statistischen Verarbeitung von 705 Fällen, *Schweiz. Med. Wochenschr.*, 90, 1301, 1960.

68. **Gottstein, U.**, Behandlung der cerebralen Mangeldurchblutung. Eine kritische Übersicht, *Internist*, 15, 575, 1974.

69. **Herrschaft, H.**, Möglichkeiten der medikamentösen Behandlung der zerebralen Mangeldurchblutung, *Therapiewoche*, 27, 4525, 1979.

70. **Olsson, J. E.**, Recent advances in the treatment of cerebrovascular diseases, *Acta Neurol. Scand.*, 62 (Suppl. 78), 77, 1980.

71. **Kogure, K., Scheinberg, P., Busto, R., and Reinmuth, O. M.,** An effect of aminophylline in experimental cerebral ischemia, *Arch. Neurol.,* 32, 352, 1975.

72. **Seida, M., Wagner, H. G., Vass, K., and Klatzo, I.,** Effect of aminophylline on postischemic edema and brain damage in cats, *Stroke,* 19, 1275, 1988.

73. **McGraw, C. P., Crowell, G. F., and Howard, G.,** Effect of aminophylline on cerebral infarction in the Mongolian gerbil, *Stroke,* 9, 477, 1978.

74. **Jarrott, D. M. and Domer, F. R.,** A gerbil model of cerebral ischemia suitable for drug evaluation, *Stroke,* 11, 203, 1980.

75. **Thurston, J. H., Hauhart, H. E., and Dirgo, J. A.,** Aminophylline increases cerebral metabolic rate and decreases anoxic survival in young mice, *Science,* 201, 649, 1978.

76. **Rudolphi, K. A., Keil, M., Westhöfer, U., and Hinze, H. J.,** Effect of theophylline on ischemically induced hippocampal damage in Mongolian gerbils, in *Pharmacology of Cerebral Ischemia,* Krieglstein, J., Ed., Elsevier, Amsterdam, 1986, 358.

77. **Rudolphi, K. A., Keil, M., and Hinze, H. J.,** Effect of theophylline on ischaemically induced hippocampal damage in Mongolian gerbils: a behavioural and histopathological study, *J. Cereb. Blood Flow Metab.,* 7, 74, 1987.

78. **DeLeo, J., Schubert, P., and Kreutzberg, G. W.,** Propentofylline (HWA 285) protects hippocampal neurons of Mongolian gerbils against ischemic damage in the presence of an adenosine antagonist, *Neurosci. Lett.,* 84, 307, 1988.

79. **Dux, E., Fastbom, J., Ungerstedt, U., Rudolphi, K., and Fredholm, B. B.,** Protective effect of adenosine and a novel xanthine derivative propentofylline on the cell damage after bilateral carotid occlusion in the gerbil hippocampus, *Brain Res.,* 516, 248, 1990.

80. **Wieloch, T., Koide, T., and Westerberg, E.,** Inhibitory neurotransmitters and neuromodulators as protective agents against ischemic brain damage, in *Pharmacology of Cerebral Ischemia,* Krieglstein, J., Ed., Elsevier, Amsterdam, 1986, 191.

81. **Grome, J. J., Gojowczyk, G., and Hofmann, W.,** Theophylline administration exacerbates the volume of infarction following focal cerebral ischemia in the rat, *Neurosci. Lett.,* 36 (Suppl. 2), S94, 1989.

82. **Rudolphi, K. A., Keil, M., Fastbom, J., Grome, J. J., and Fredholm, B. B.,** Contrary effects of acute versus chronic caffeine administration on ischemically induced hippocampal damage in the gerbil. Involvement of adenosine receptors?, in *Purines in Cellular Signaling,* Jacobson, K. A., Daly, J. W., and Manganiello, V., Eds., Springer-Verlag, Berlin, 1990, 411.

83. **Rudolphi, K. A., Keil, M., Fastbom, J., and Fredholm, B. B.,** Ischaemic damage in gerbil hippocampus is reduced following upregulation of adenosine (A_1) receptors by caffeine treatment, *Neurosci. Lett.,* 103, 275, 1989.

84. **Phillis, J. W., O'Regan, M. H., and Walter, A.,** Effects of two nucleoside transport inhibitors, dipyridamole and soluflazine, on purine release from the rat cerebral cortex, *Brain Res.,* 481, 309, 1989.

85. **Phillis, J. W. and Wu, P. H.,** The effect of various centrally active drugs on adenosine uptake by the central nervous system, *Comp. Biochem. Physiol.,* 72C, 179, 1982.

86. **Fredholm, B. B. and Lindström, K.,** The xanthine derivative 1-(5'-oxo-hexyl)-3-methyl-7-propylxanthine (HWA 285) enhances the action of adenosine, *Acta Pharmacol. Toxicol.,* 58, 187, 1986.

87. **DeLeo, J., Toth, L., Schubert, P., Rudolphi, K., and Kreutzberg, G. W.,** Ischemia-induced neuronal cell death, calcium accumulation, and glial response in the hippocampus of the Mongolian gerbil and protection by propentofylline (HWA 285), *J. Cereb. Blood Flow Metab.,* 7, 745, 1987.

88. **DeLeo, J., Schubert, P., and Kreutzberg, G. W.,** Protection from ischemic brain damage by posttreatment with propentofylline: a histopathological and behavioral study in Mongolian gerbils, *Stroke,* 19, 1535, 1988.

89. **Porsche, E.,** Effects of methylxanthine derivatives on the adenosine uptake in human erythrocytes, *IRCS Med. Sci.,* 10, 389, 1982.

90. **Stefanovich, V.,** Uptake of adenosine by isolated bovine cortex microvessels, *Neurochem. Res.,* 8, 1459, 1983.

91. **Andiné, P., Rudolphi, K. A., Fredholm, B. B., and Hagberg, H.,** Effect of propentofylline (HWA 285) on extracellular purines and excitatory amino acids in CA1 of rat hippocampus during transient ischemia, *Br. J. Pharmacol.,* 100, 814, 1990.

92. **Phillis, J. W., O'Regan, M. H., and Walter, G. A.,** Effects of deoxycoformycin on adenosine, inosine, hypoxanthine, xanthine and uric acid release from the hypoxaemic rat cerebral cortex, *J. Cereb. Blood Flow Metab.,* 8, 733, 1988.

93. **Padua, R., Geiger, J. D., Dambock, S., and Nagy, J. I.,** 2'-Deoxycoformycin inhibition of adenosine deaminase in rat brain: in vivo and in vitro analysis of specificity, potency, and enzyme recovery, *J. Neurochem.,* 54, 1169, 1990.

94. **Busto, R., Globus, M. Y. T., Dietrich, W. D., Valdes, I., Santiso, M., and Ginsberg, M. D.,** The increase in endogenous adenosine by adenosine deaminase inhibitor fails to protect against ischemia-induced neuronal damage, *J. Cereb. Blood Flow Metab.,* 9 (Suppl. 1), S267, 1989.

95. **Phillis, J. W. and O'Regan, M. H.,** Deoxycoformycin antagonizes ischemia-induced neuronal degeneration, *Brain Res. Bull.,* 22, 537, 1989.

96. **Phillis, J. W.,** Oxypurinol attenuates ischemia-induced hippocampal damage in the gerbil, *Brain Res. Bull.,* 23, 467, 1989.

97. **Helfman, C. and Phillis, J. W.,** Oxypurinol administered post-ischaemia prevents brain injury in the gerbil, *Med. Sci. Res.,* 17, 969, 1989.

98. **Mitsos, S. E., Jolly, S. R., and Lucchesi, B. R.,** Protective effects of AICA-riboside in the globally ischemic isolated cat heart, *Pharmacology,* 31, 121, 1985.

99. **Bruns, R. F. and Fergus, J. H.,** Allosteric enhancers of adenosine A_1 receptor binding and function, in *Adenosine Receptors in the Nervous System,* Ribeiro, J. A., Ed., Taylor & Francis, London, 1989, 53.

Index

INDEX